Volcanotectonics

A volcanic eruption occurs when a magma-filled fracture propagates from its source to the surface. Analysing and understanding the conditions that allow this to happen constitute a major part of the scientific field of volcanotectonics. This new volume introduces this cutting-edge and interdisciplinary topic in volcanological research, which incoporates principles and methods from structural geology, tectonics, volcano-deformation studies, physical volcanology, seismology, and physics. It explains and illustrates the physical processes that operate inside volcanoes and which control the frequencies, locations, durations, and sizes of volcanic eruptions. Featuring a clear theoretical framework and helpful summary descriptions of various volcanic structures and products, as well as many worked examples and exercises, this book is an ideal resource for students, researchers, and practitioners seeking an understanding of the processes that give rise to volcanic deformation, earthquakes, and eruptions.

Agust Gudmundsson is Chair of Structural Geology at the Department of Earth Sciences, Royal Holloway, University of London. He has been instrumental in founding the field of volcanotectonics, which combines principles from structural geology, field geology, and physics. He is a member of Academia Europaea, and has written some 200 scientific papers and two previous books: *Rock Fractures in Geological Processes* (Cambridge University Press, 2011) and *The Glorious Geology of Iceland's Golden Circle* (Springer, 2017).

Volcanotectonics

Understanding the Structure, Deformation, and Dynamics of Volcanoes

AGUST GUDMUNDSSON

Royal Holloway, University of London

CAMBRIDGE
UNIVERSITY PRESS

CAMBRIDGE
UNIVERSITY PRESS

University Printing House, Cambridge CB2 8BS, United Kingdom

One Liberty Plaza, 20th Floor, New York, NY 10006, USA

477 Williamstown Road, Port Melbourne, VIC 3207, Australia

314–321, 3rd Floor, Plot 3, Splendor Forum, Jasola District Centre,
New Delhi – 110025, India

79 Anson Road, #06–04/06, Singapore 079906

Cambridge University Press is part of the University of Cambridge.

It furthers the University's mission by disseminating knowledge in the pursuit of
education, learning, and research at the highest international levels of excellence.

www.cambridge.org
Information on this title: www.cambridge.org/9781107024953
DOI: 10.1017/9781139176217

First published 2020

Printed in the United Kingdom by TJ International Ltd, Padstow Cornwall

A catalogue record for this publication is available from the British Library.

Library of Congress Cataloging-in-Publication Data
Names: Gudmundsson, Agust, author.
Title: Volcanotectonics : understanding the structure, deformation, and dynamics of volcanoes /
Agust Gudmundsson (Royal Holloway, University of London).
Other titles: Volcanotectonics
Description: Cambridge : Cambridge University Press, 2020. | Includes bibliographical
references and index.
Identifiers: LCCN 2019037755 (print) | LCCN 2019037756 (ebook) | ISBN 9781107024953 (hardback) |
ISBN 9781139176217 (ebook)
Subjects: LCSH: Volcanic activity prediction. | Volcanic hazard analysis. | Volcanic eruptions.
| Volcanoes. | Geology, Structural. | Morphotectonics.
Classification: LCC QE527.5 .G83 2020 (print) | LCC QE527.5 (ebook) | DDC 551.21–dc23
LC record available at https://lccn.loc.gov/2019037755
LC ebook record available at https://lccn.loc.gov/2019037756

ISBN 978-1-107-02495-3 Hardback

Additional resources for this publication at www.cambridge.org/volcanotectonics

Contents

Colour plates are to be found between pp. 322 and 323.

Preface

Volcanic eruptions occur when magma is able to form a path from a source to the surface. The source is normally a magma chamber while the path is a fracture generated by the pressure of the magma, a magma-driven fracture. It is thus the fluid pressure of the magma that ruptures the rock and makes it possible for the fracture to propagate from the source chamber to the surface. Most magma-driven fractures (such as dikes), however, do not reach the surface to erupt but become arrested (stop their propagation) at various depths in the crust. The arrest is commonly at a contact between layers whose mechanical properties and local stresses are unfavourable to the fracture propagation. Successful forecasting of volcanic eruptions depends thus not only on a deep understanding of the geology but also of the physics of volcanoes.

Volcanotectonics is a comparatively new scientific field that combines various methods and techniques of geology and physics so as to understand the structure and behaviour of polygenetic (central) volcanoes and the conditions for their eruptions. More specifically, volcanotectonics uses the techniques and methods of tectonics, structural geology, geophysics, and physics to collect data on volcanoes, as well as to analyse and interpret the physical processes that generate those data. The focus is on processes responsible for periods of volcanic unrest, caldera collapses, and eruptions.

For basic science, one principal aim of volcanotectonics is to develop methods for reliable forecasting of eruptions. Accurate forecasting as regards the location, time, and magnitude of eruptions has long been a major goal in volcanology. Volcanotectonics provides a theoretical framework and understanding of the physical processes that take place inside volcanoes prior to eruptions, thereby offering methods and techniques that allow us to use data obtained during unrest periods to forecast eruptions. For applied science and human society, another principal aim of volcanotectonics is to develop methods for preventing very large eruptions. This second aim may come as a surprise to some, but is of fundamental importance for the future of human civilisation. Very large eruptions, whose eruptive volumes may be of the order of hundreds or thousands of cubic kilometres, provide an existential threat to human civilisation.

Because of its focus on physical processes that occur inside volcanoes prior to eruptions, the emphasis in volcanotectonics is in combining data from active and inactive (fossil) volcanoes. The latter are commonly partly deeply eroded and allow us to observe in detail the propagation paths of dikes and inclined sheets that supply magma to most volcanic eruptions. Furthermore, some eroded fossil volcanoes make it possible to obtain a three-dimensional view of the ring-faults of collapse calderas, as well as of ring-dikes, and feeder-dikes. By combining data from structures being formed during unrest periods with well-exposed structures of the same type in deeply eroded volcanoes, we obtain a four-dimensional view of these structures – namely, their three spatial dimensions as well as their development through time.

The concept of a magma chamber is a central topic in the book. A magma chamber is the 'heart of the volcano'; without a magma chamber to channel magma to a limited area on the surface no polygenetic volcanic edifices would form. Volcanic unrest with the potential of an eruption normally begins with the rupture of the source chamber and the injection of a dike/sheet. Similarly, for a ring-fault to form and the caldera floor to subside, a magma chamber is a necessary – but not a sufficient – condition. Parts of many fossil magma chambers and associated dikes/sheets and ring-faults are well exposed. They offer excellent opportunities to test models on magma-chamber shapes, ruptures, dike injection, ring-fault formation, and associated inflation and deflation during unrest periods.

The purpose of this book is to provide an overview of the scientific field of volcanotectonics. The book is primarily aimed at, first, undergraduate and graduate students in geology, geophysics, and geochemistry and, second, civil authorities, scientists, engineers, and other professionals who deal with volcanoes and the associated hazards in their work. The book has been designed so that it can be used (1) for an independent study, (2) as a textbook for a course on volcanotectonics, and (3) as a supplementary text for general courses on volcanology, structural geology, geology, geophysics, geothermics, and natural hazards.

Each chapter begins with an overview of the aims and ends with a summary of the main topics discussed. In addition there is a list of symbols used in the chapter. Important concepts and conclusions are in bold face. In volcanotectonics the focus is on quantitative results. This is reflected in the 68 worked examples (solved probems), most of which include calculations. In addition there are 253 exercises (supplementary problems), many of which also require calculations. The examples and exercises are meant to provide a deeper knowledge of the basic principles of volcanotectonics and their use for understanding the formation of volcanoes, the physical processes that maintain their activities, and providing reliable eruption forecasts. While volcanic activity cannot be understood or forecasted without basic knowledge of the relevant physics, the physics presented in the book is mostly elementary and explained in detail. The only exception is part of Chapter 10, where more advanced physics is introduced to explain the propagation paths of magma-driven fractures.

I have taught much of the material in the book at various universities over the past 20 years to earth-science students in Norway, Germany, and England. In particular, many of the chapters form the basis of an undergraduate course on volcanology which I have taught in the past six years in England. Based on this experience, most of the material in the book should be suitable for earth-science students with a very modest knowledge of mathematics and physics.

Acknowledgements

I would like to thank Shigekazu Kusumoto and Jason Morgan for very helpful comments on Chapter 10. This chapter contains the most advanced physics and mathematics in the book, as well as many new ideas. Earlier versions of many chapters were also read by Valerio Acocella, John Browning, and Abdelsalam Elshaafi, who provided many helpful suggestions.

Many illustrations in the book have been remade from various sources, all of which are cited in the figure captions and in the reference lists. I thank the publishers for permission to use the illustrations. Many public-domain images are from NASA and the US Geological Survey while others are from the Flickr image hosting service. The authors of all images are indicated, except for the photographs taken by me.

Most of the illustrations have been modified many times, but were originally made by the technical staff and students and colleagues in Iceland, Norway, France, Germany, Italy, Spain, and the UK. Similarly, many of the numerical models in the book and in related papers were made in collaboration with colleagues and students. Any list of names of people who have contributed to illustrations or numerical models would necessarily be incomplete. I therefore prefer to offer a warm thank you to all those who have contributed to the illustrations and/or numerical models in the book.

I thank my wife, Nahid Mohajeri, for making the many original drawings for the book. She has, in addition, redrawn many of the earlier illustrations.

Although this book project has not received direct funding as such, many of the ideas and theoretical and applied results presented in the book were obtained during funded projects. In particular, some of the results presented here derive from various projects funded by the Icelandic Science Foundation, the Research Council of Norway, Tectonor (Norway), the Volkswagen Foundation (Germany), the Natural Environment Research Council (UK), and the European Research Council.

At Cambridge University Press, Zoë Pruce, Zoë Lewin, Esther Miguéliz Obanos, and Susan Francis have been very helpful, positive, and patient during the time it took to complete the book. I take this opportunity to thank the Cambridge University Press for a splendid collaboration.

Introduction

1.1 Aims

Volcanoes are of many types and behave in different ways. Different behaviour is partly because volcanoes are located in different tectonic environments. Many are associated with divergent plate boundaries, others with convergent plate boundaries, and some with transform-fault plate boundaries. In addition, there are volcanoes located within plate interiors, far from plate boundaries. To understand volcano behaviour with a view of being able to forecast volcanic eruptions we must use a variety of scientific techniques and approaches, primarily those of volcanotectonics. The main techniques and approaches for data collection, analysis, and interpretation are discussed in detail in later chapters, but they are briefly summarised here. The primary aims of this chapter are to:

- Provide a definition of a volcano.
- Define and explain the scope of the scientific field of volcanotectonics.
- Explain briefly how volcanotectonic data are collected and analysed.
- Discuss structural geological/tectonic techniques used for understanding volcanoes.
- Discuss some main geophysical techniques used in volcanotectonics.
- Explain accuracy, significant figures, and rounding of numbers in volcanotectonics.
- Summarise the basic units and prefixes used in volcanotectonics.

1.2 Definition of a Volcano

An **active volcano** is a vent that transports magma and volatiles to the Earth's surface (Fig. 1.1). The accumulation of volcanic materials around the vent gives rise to a small volcano, commonly a crater cone, either a cinder/scoria cone or a spatter cone, and similar structures. Volcanoes formed in single eruptions are named **monogenetic.** Many of these are volcanic fissures, composed of many spatter and cinder/scoria cones, while others are single craters (Fig. 1.1a) or cinder cones (Fig. 1.1b). Repeated eruptions through a single or, more commonly, many nearby vents result in the accumulation of volcanic materials into a volcanic edifice (Figs. 1.2, 1.3). Volcanoes formed in many eruptions are named **polygenetic**. Polygenetic volcanoes forming edifices may be divided into two main types: **basaltic edifices** and **stratovolcanoes** (also named composite volcanoes). Well-known

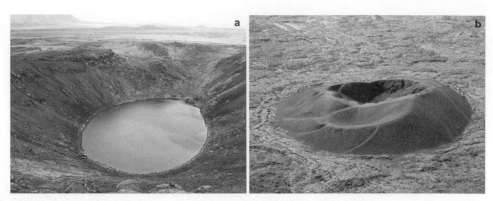

Fig. 1.1   A volcano is vent that transports magma and volatiles to the surface. The volcanic landform may be either negative (a depression) or positive (a hill or a mountain). (a) Kerid, a collapsed scoria-and-spatter cone, formed some 9000 years ago in Iceland, may be regarded as a pit crater. It is elliptical in plan view. It has a maximum diameter of about 300 m and a minimum diameter of about 170 m. It forms a depression, 50 m deep, that is partly filled with groundwater. (b) Cinder (scoria) cone from the Enclose Fouque collapse in Piton de la Fournaise, Reunion. The cone is approximately 150 m wide and 30 m high. Photo: Valerio Acocella.

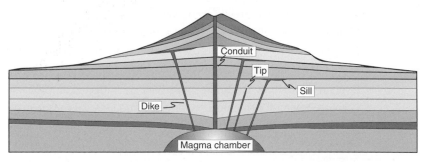

Fig. 1.2   Schematic cross-section through a stratovolcano. Every stratovolcano is supplied with magma from one or more shallow magma chambers. During magma-chamber rupture, a dike (or an inclined sheet) becomes injected into the roof of the chamber. Most dikes/sheets become arrested at contacts between layers, some deflecting into sills, while others thin out (taper away) in vertical sections. A minority of injected dikes reaches the surface to supply magma to eruptions. Some major conduits in stratovolcanoes are partly composed of many dikes (and partly of volcanic breccia).

examples of basaltic edifices are the volcanoes Mauna Loa and Kilauea, forming part of the Big Island of Hawaii, and Fernandina, forming part of the Galapagos Islands. Well-known stratovolcanoes include Fuji in Japan, Teide on Tenerife (Canary Islands), Semeru in Indonesia, and Augustine in Alaska (Fig. 1.3).

Volcanoes are regarded as active if they have erupted within a certain comparatively recent time period. The length of that period has been variously determined and there is, as yet, no agreed definition. Earlier, a volcano was regarded as active only if it had erupted in **historical time**. This definition is unsatisfactory for several reasons. First, historical time – which normally refers to the time covered by written documents – varies greatly between

Fig. 1.3 Stratovolcanoes are formed by repeated eruptions within a limited surface area. They commonly rise high above their surroundings and have a cone shape, as exemplified here by the Augustine Volcano, forming one of the islands offshore Alaska (United States), with an elevation (height above sea level) of 1260 m and a maximum diameter at sea level of about 12 km. A highly active volcano with frequent eruptions, this photograph shows gas rising during the 2005–2006 eruption. Photo: USGS/Cyrus Read. A black and white version of this figure will appear in some formats. For the colour version, please refer to the plate section.

different volcanic areas. For instance, historical time spans several thousand years in many volcanically active Mediterranean countries, such as Italy and Greece, but is very short in other volcanically active areas such as western continental United States and Hawaii and the continent of Antarctica. Second, it has happened many times that a volcano which, in the above definition, was regarded as inactive, suddenly erupted. For example, the ice-covered Fourpeaked Mountain in Alaska, located some 100 km south of the highly active Augustine Volcano, had not erupted for some 10 000 years prior to its September 2006 eruption.

In recent years the reference period used is not historical time but rather the **Holocene**. In this definition, a **volcano is active** if it has erupted during the Holocene, that is, sometime during the past **11 700 years**. This definition can generally be used, but with caution. For example, some very large and powerful volcanoes such as the Yellowstone caldera in the western United States erupt less frequently than once every 11 700 years. The last significant lava-flow eruption in the caldera was about 70 000 years ago, and the most recent (mostly steam) explosions occurred about 14 000 years ago. Yet, the Yellowstone caldera should be regarded as an active volcano.

Volcanoes that are no longer active are referred to as **extinct.** Another term that is used for an extinct volcano is a **fossil volcano.** A fossil volcano is normally much eroded so that parts of its internal structure can be observed (Fig. 1.2). When the magma source chamber of a fossil volcano is partly exposed as a pluton, that pluton is referred to as a **fossil magma chamber**. A volcano known to have erupted in Holocene but which has not erupted for a considerable time and shows little or no unrest is a **dormant volcano.**

As indicated, there are generally no absolute answers to whether a volcano should be regarded as active or extinct. A volcano may be regarded as **potentially active** so long as it has a chance of **receiving magma** from a source. The magma does not have to reach the surface for the volcano to be considered active: magma intrusions, such as dikes and sills (Fig. 1.2), are a clear indication that the volcano is active. To decide if a volcano is active, it is thus not enough to know its eruption frequency and history; we also need to know its magma intrusion frequency. Information on the latter can normally only be obtained from monitoring, primarily geophysical monitoring.

The likelihood that a volcano receives magma is discussed in later chapters. Here we mention this topic mainly to emphasise that there is no clear-cut answer to the question: 'How many active volcanoes are there in the world?' Similarly, we cannot answer the question: 'How many seismically active fault zones are there in the world?' However, we can and do assess the probability of, say, a 10-million-year-old volcano in Iceland erupting as being very small indeed.

1.3 Volcanotectonics – the Scope and Aims

Volcanotectonics is a scientific field that uses the techniques and methods of tectonics, structural geology, geophysics, and physics to collect data on, analyse, and interpret physical processes in volcanoes. More specifically, volcanotectonics covers the following topics:

- Collecting and analysing structural-geology, tectonic, and rock-physics data, as well as geodetic, seismic, and stress data, from active volcanoes.
- Collecting and analysing structural-geology, tectonic, and rock-physics data from deeply eroded volcanoes, many of which may actually be, or for practical purposes regarded as, extinct. Such data can also be collected from caldera walls, drill holes, and other sections into active volcanoes.
- Interpreting these data within the framework of volcanology, structural geology/tectonics, geophysics, and classical physics, particularly continuum physics – such as solid mechanics (including rock physics and fracture mechanics) and fluid mechanics.
- Developing models and theories for understanding the processes that generate the data with a view of being able to make reliable forecasts as to the likely course of events or scenarios during unrest periods in volcanoes.

The connection between volcanotectonics and various related scientific fields are indicated schematically in Fig. 1.4. The basic field techniques used are derived from **structural geology**, for the outcrop scale, and from **tectonics**, for the regional scale, and are connected to the general field of **volcanology**. For active volcanoes, geodetic studies, as a part of **geodesy**, are very important and provide data for understanding some of the processes that

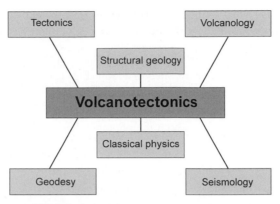

Fig. 1.4   Volcanotectonics uses principles and methods from many scientific fields, the main ones being indicated here. Many of the data from active volcanoes are obtained through the methods of seismology (volcano or volcanotectonic earthquakes), geodesy (geodetic measurements of volcano deformation), and volcanology. Data on extinct and eroded volcanoes (and active volcanoes as well, partly from seismotectonics) are primarily obtained through the methods of tectonics, structural geology, and volcanology. The interpretations of the data in terms of models and theories rest on principles from classical physics (solid mechanics, fluid mechanics, statistical mechanics) and more recent derived fields (fracture mechanics, materials science, and rock physics).

occur inside the volcanoes during unrest periods. Similarly, **seismological data** related to physical processes inside and at the surface of a volcano can be used to monitor fracture development and the propagation of potential feeder-dikes towards the surface of the volcano.

Data, however, are of little use if they cannot be interpreted and understood within the framework of a plausible model or theory of volcano behaviour. Quantitative and testable **models** must, in the end, be based on physical theories and thus on the principles of physics. In volcanotectonics, like in solid-earth sciences in general, the main physical theories used are those that are derived from **classical physics** (physical theories that pre-date relativity and quantum mechanics), particularly continuum and classical mechanics. For solid-earth sciences, the theories are mainly from solid mechanics, which provides the foundation of the much more recent scientific fields of rock mechanics, fracture mechanics, and tectonophysics, as well as from fluid mechanics, including the more recent field of fluid transport in rock fractures (data on physical properties of rocks and crustal fluids, including magmas and lavas, are provided in **Appendices A–F** at the end of the book). Below are given more details on the basic techniques used in volcanotectonics.

1.4   Structural-Geology Techniques and Definitions

All the standard field methods of structural geology and tectonics are useful and are used in volcanotectonic studies. These include the measurements of the strike and dip of strata and structures, measurements of the dimensions and volumes of rock bodies, and related studies. Most of these topics are covered in textbooks on general field techniques in structural geology, but a brief summary is in order for easy reference.

• The **strike** of a rock body or a structure is the direction or trend of the plane of the rock body or the structure, such as a fault (Fig. 1.5). Alternatively, the strike is the direction of the intersection, which forms a line, between the plane of the body or structure and the horizontal plane. In the photographs, such as Fig. 1.5b, the **view** indicated in the captions is the direction in which the photographer is looking. Thus, view southwest (Fig. 1.5b)

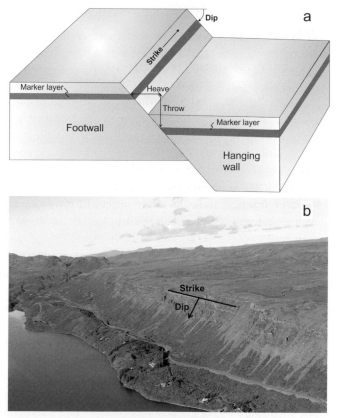

Fig. 1.5   (a) The attitude of a rock fracture is defined by its strike and dip. For a dip-slip fault like the one here (a normal fault) the wall above the fault plane is known as the hanging wall, whereas the wall below the fault plane is known as the footwall. The vertical displacement, here measured using a marker layer (a layer that is easily recognised on both sides of the fault plane), is known as throw and the horizontal displacement is known as heave. (b) Part of a normal fault in the Hengill Volcanic System of the rift zone in Southwest Iceland, with the strike and dip indicated schematically. When the horizontal black line (strike) is extended along the entire fault, it measures the fault strike-dimension. Similarly, when the black arrow (dip) is extended to the lower end of the fault (deep inside the crust), it measures the dip-dimension of the fault. View southwest, the normal fault forms the western boundary of the main graben of the rift zone in Southwest Iceland (Gudmundsson, 2017). The strike and dip dimensions of the segmented fault are in the order of 10–20 km and the vertical displacement or throw about 200 m.

means that the photographer was looking (facing) southwest when the photograph was taken.

- The **dip** is the acute angle between the plane of a structure and the horizontal plane (Fig. 1.5). The dip is measured in an imaginary vertical plane whose direction is at 90° to the plane of the structure. Please note that the dip is defined as an angle; it is therefore tautology to refer to the 'dip angle'.

- The **strike-dimension** (the dimension along the strike) of a body or structure is the length or dimension of the body or structure in the direction of its strike.

- The **dip-dimension** (the dimension along the dip) of a body or structure is the length or dimension of the body or structure in the direction of its dip.

- Together, the strike and dip define the **attitude** (sometimes named the orientation) of the structure or rock body.

- Mechanically, there are two basic types of rock fractures: extension fractures and shear fractures. An **extension fracture** propagates or advances in a direction that is parallel with the maximum compressive principal stress, σ_1 (and parallel with the intermediate principal stress, σ_2), and perpendicular to the minimum compressive (maximum tensile) principal stress, σ_3. The fracture **opening** displacement is thus parallel with the direction of σ_3. There is no relative movement parallel with the fracture walls, only perpendicular to the walls (opening). When an extension fracture is opened or driven by internal fluid overpressure it is known as a **hydrofracture** (a fluid-driven fracture), but when it is opened by tectonic tensile stress, that is, by negative σ_3, it is known as a **tension fracture.** A **shear fracture** forms by shear stress and shows clear evidence of fracture-parallel relative movement of the fracture walls. In geology, shear fractures are normally referred to as **faults.** Thus, all faults are shear fractures.

- For an extension fracture, the opening is of fundamental importance and should be measured whenever possible. For a tension fracture, that is, a fracture opened by tensile stresses, the opening is referred to as an **opening displacement** (or just opening) and is the shortest distance between matching notches and jogs on the fracture walls (Fig. 1.6).

- For a solidified or 'frozen' magma-filled fracture, namely an intrusive sheet such as a **dike**, an **inclined sheet**, or a **sill**, the measured opening is the palaeoaperture, which is measured as the **thickness** of the intrusive sheet (Fig. 1.7). Strictly, the opening of the magma-filled fracture when the magma was fluid may have been somewhat larger than the present thickness of the solidified intrusion. This follows because when a magma body solidifies it contracts or shrinks. However, the density difference between fluid and solidified magma is generally small, suggesting that the shrinkage is normally less than 10%. For mineral-filled extension fractures, known as **mineral veins**, the opening displacement is also the palaeoaperture and is measured as the vein thickness (Fig. 1.8).

- The **aperture** means simply the opening (usually the maximum or average opening) of a fracture. For an extension fracture, such as a tension fracture (Figs. 1.6 and 1.9) and a dike (Fig. 1.7), the aperture corresponds to the opening displacement and can be explained in terms of the theory of fracture mechanics (Chapter 5). For a shear fracture such as a fault (Figs. 1.5 and 1.10), however, the aperture has normally no relation to the displacement on the fault and cannot be forecast or explained in terms of the theory of fracture mechanics. Apertures of both extension fractures and shear fractures are

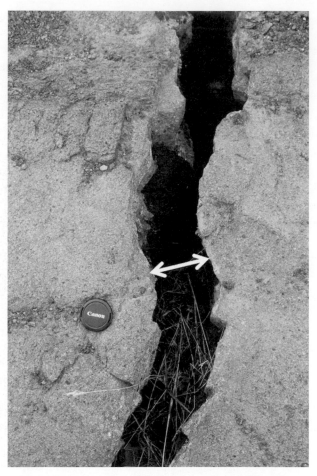

Fig.1.6 Extension fractures are of two main types: tension fractures and fluid-driven fractures, referred to as hydrofractures. This one is a tension fracture, with matching jogs and notches on the opposite fracture walls. The opening (displacement) or aperture is indicated by the white arrow. The diameter of the camera lens cap is about 6 cm.

important for understanding the permeability and general fluid transport in volcanoes, not only of magma but also of geothermal water (Fig. 1.8) and groundwater (Fig. 1.9).

- For **faults** and other shear fractures, the **displacements** can commonly be measured. Fault displacement is a measure of the relative movement of the fault walls. There are two main types of displacement measured in the field. One is the cumulative or total displacement, the other is the slip (Fig. 1.10).
- The total fault displacement or just the **fault displacement** is the maximum (although some use the mean) relative fracture-parallel movement of the fracture walls. In this book, displacement is normally the total cumulative displacement; not the co-seismic slip in individual earthquakes.
- The co-seismic fault slip or simply **fault slip** is the displacement associated with a single (earthquake) rupture. It is either measured at the surface, particularly for a large

Fig. 1.7 Dike thickness is measured as indicated by the black horizontal line. This dike, in the caldera wall of the island of Santorini, Greece, is about 1.5 m thick (cf. Browning et al., 2015). A black and white version of this figure will appear in some formats. For the colour version, please refer to the plate section.

earthquake (many small earthquakes do not rupture the surface), inferred from the inversion of geodetic data, or both. The fault slip does not strictly have to be co-seismic; many faults and fault segments slip **aseismically** (creep), that is, without measured earthquakes – hence the parentheses around the word 'earthquake'.

- The **fault length** is the along-strike or strike-dimension of the fault/extension fracture, as seen at the surface (Figs. 1.5 and 1.10) or as inferred or calculated for the subsurface part of the fault from (mostly geodetic and seismic) data.
- The **fault width** is the dip-dimension of the fault (Fig. 1.10) as observed in the field (for very small faults) or as inferred or calculated from (mostly geodetic and seismic) data. Notice that the use of the term 'fault width' in this sense is very common in seismology and earthquake mechanics, but less so in structural geology and tectonics where 'width' sometimes refers to fracture opening or thickness (of a dike). The latter use should be

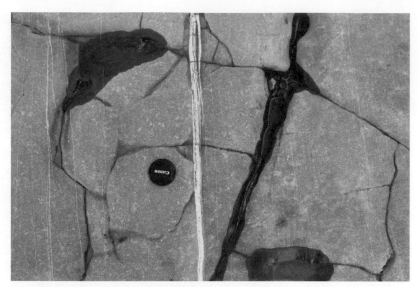

Fig. 1.8 Mineral vein of calcite in limestone in the Bristol Channel, Britain. The vein is multiple (composed of many thinner veins). The diameter of the camera lens cap is about 6 cm.

Fig. 1.9 Tension fracture with the aperture (opening displacement) shown. The direction of the strike-dimension (the horizontal fracture length) is indicated. The maximum fracture opening is about 15 m (cf. Gudmundsson, 2017).

avoided. We do not speak about the 'width' of a lava flow when we mean its thickness; neither should we speak about the width of a dike when we mean its thickness.

- The co-seismic **rupture length**, or simply rupture length (Fig. 1.10), refers to the strike-dimension of the part of an active fault (or fault zone) that ruptures during a particular slip

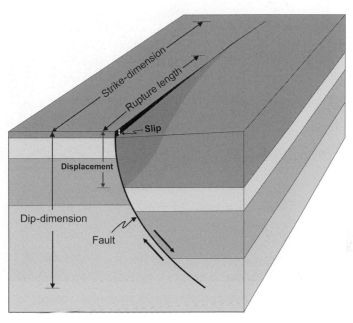

Fig. 1.10   Illustration of some geometric parameters associated with active fault zones. Strike-dimension (length), dip-dimension (width or height), displacement (cumulative fault displacement), rupture length (co-seismic rupture length), and slip (co-seismic slip). Here the fault zone is a listric (curved) normal fault. The slip is recent and seen at the surface, while much of the cumulative displacement is buried (we are supposed to see into the uppermost part of the crust and thereby see the cumulative displacement). The recent fault slip adds to the earlier displacement, so that the displacement does not refer to the same marker layer in the footwall as in the hanging wall. The surface layer is much thicker in the hanging wall (right) of the fault zone than in its footwall (left). This difference is common in active volcanotectonic rift zones, where, for example, lava flows tend to become thicker in the hanging walls of normal faults (inside the graben if the fault forms a boundary of a graben). The approximate scale is so that the maximum displacement is about 100 m and the maximum slip is about 8 m. Only parts of the strike-dimension and the rupture length are shown. The rupture length is considerably shorter than the total length of the fault (cf. Gudmundsson et al., 2013).

and (commonly) an associated earthquake. The rupture length is normally much shorter than the total length of the fault/fault zone within which the rupture (and earthquake) occurs.

- The co-seismic **rupture width**, or simply rupture width, refers to the dip-dimension (Fig. 1.10) of the part of an active fault (or fault zone) that slips during a particular (co-seismic) rupture. For large faults/fault zones, the total width is the thickness of the seismogenic layer (some 10–20 km at many plate boundaries). The rupture width of small to moderate earthquakes in large fault zones is normally much smaller than the total width of the fault/fault zone within which the rupture and earthquake occur.

- The **dike length** normally refers to the strike-dimension of the dike (Fig. 1.11); the same applies to other extension fractures, such as tension fractures (Fig. 1.9) – length mostly means strike-dimension. When modelling fluid flow along a dike, the dike length can, however, refer to the distance from the source magma chamber to the surface. This

Fig. 1.11 Parts of the strike- and dip-dimensions of a well-exposed dike in the island of Santorini, Greece (cf. a close-up of the dike in Fig. 1.7)

distance may correspond neither to the strike-dimension nor the dip-dimension of the dike (Fig. 1.11). This difference in the meaning of the term 'dike length' is clear from the context and explained as such when needed.

- The **dike thickness** is the thickness of the dike as measured directly (using a measuring tape) in the field (Fig. 1.7). Alternatively, the dike thickness may be (crudely) inferred from measurements made – during the dike emplacement – through geodetic (e.g. InSAR and GPS) techniques, such as are explained below.
- The **dike aperture** is the opening of a magma-filled dike (cover photograph and Fig. 7.19). The aperture may be somewhat larger than the eventual thickness of the solidified dike, because the dike material contracts or shrinks as it solidifies. As indicated above, the shrinkage is normally about 10% or less, which is similar to common error estimates for dike-thickness measurements.
- The **controlling dimension**, as used in fault/dike/tension fracture modelling, is the smaller one of the strike- and dip-dimensions. It is the dimension that largely controls the displacement on the fracture for given rock properties and driving stresses/pressures.

1.5 Geophysical Techniques and Definitions

The significant improvement in **monitoring** volcanic unrest periods and eruptions in the past decades is largely attributable to the advancement in monitoring instrumentation, primarily geophysical techniques. All the standard geophysical methods are used for volcano monitoring. In addition, geochemical methods, such as monitoring changes in the chemistry and volumetric flow rates of fluids (gas, groundwater, streams), are routinely being used. While geochemistry is largely beyond the scope of this book, brief outlines are given of the main geochemical monitoring methods. More details on volcano deformation and seismicity and the associated techniques are provided in Chapter 3 and Chapter 4 as well as by Zobin (2003), Dzurisin (2006), Janssen (2008), Segall (2010), and Lu and Dzurisin (2014). What follows here is just a brief summary of some of the main techniques and definitions.

1.5.1 Volcano Deformation

When a volcano becomes subject to **loading** (stress, strain, pressure, or displacement) the rock units and layers constituting the volcano become deformed. The deformation is partly elastic, that is, it is recoverable once the loading is relaxed or removed (Chapter 3). Partly, however, the deformation is inelastic or permanent, that is, it results in the formation of fractures such as tension fractures, faults, and dikes. The permanent part of the deformation is not recoverable; when the load is relaxed the permanent part of the deformation is maintained, as discussed below. The most common large-scale deformation of active volcanoes is either inflation or deflation of the volcano (Fig. 1.12).

Inflation refers to the uplift or doming of the surface of a volcano (or a volcanic system or a volcanic zone). Inflation is generally due to increased fluid pressure within the volcano (Fig. 1.12). The pressure increase may be related to the accumulation or expansion of water (increase in pore-fluid pressure), particularly geothermal water in a reservoir, but it is most often attributed to magma being added to the associated magma chamber (Fig. 1.12). Other causes of inflation include tectonic stresses, particularly when the volcano happens to be

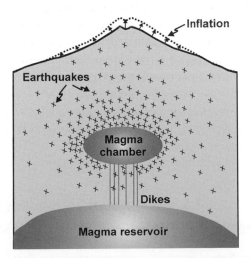

Fig. 1.12   Schematic illustration of an inflation (much exaggerated) and associated earthquakes (each cross indicates the location or focus of an earthquake). The inflation is due to magma-chamber expansion, which, in turn, is related to the shallow chamber receiving new magma (through dikes) from a deeper and much larger source reservoir.

within an area of compression. Inflation is also a measure of the elastic energy (Chapters 7 and 10) stored in the volcano before eventual magma-chamber rupture, dike injection, and (sometimes) eruption.

Deflation refers to the lowering or subsidence of the surface of the volcano as the fluid pressure in the associated magma chamber (or other pressure sources) decreases. Subjecting the volcano to horizontal extension, such as commonly happens in volcanoes in rift zones, may also cause deflation. When the deflation is associated with a pressure decrease in the magma chamber, the standard interpretation is that magma has left the chamber. Normally, the magma leaving the chamber is injected as a dike (or an inclined sheet) into the host rock of the chamber (Figs. 1.2 and 1.11). It does not necessarily follow that the dike reaches the surface to feed an eruption; many, probably most, dike injections do not result in eruptions – the dike propagation stops, the dike becomes **arrested**, at some depth below the surface (Chapter 7). But the magma volume that goes into forming the dike and, if the dike reaches the surface, sustaining the eruption results in a decreasing magma volume in the chamber and thus in a pressure decrease, magma-chamber contraction or shrinkage, and deflation. Even if there is a magma flow from a deeper reservoir into the shallow chamber at the same time (Fig. 1.12), the inflow is normally at a much lower rate than the outflow through the dike and cannot sustain the pressure in the chamber, hence the deflation.

The crustal deformation associated with inflation and deflation is, to a large degree, **elastic** and thus essentially recoverable (Chapter 3). However, part of the deformation, particularly during inflation and dike injection, is **permanent**. The permanent deformation, sometimes referred to as **plastic** or **inelastic**, is exemplified by dikes, inclined sheets, and sills as well as by faults (primarily normal faults and grabens but also strike-slip faults), tension fractures, and volcanic fissures, as are commonly seen at the surface of the volcano. Large-scale inelastic deformation can be studied by the structural geological methods and techniques discussed above. On currently active volcanoes, crustal deformation, both elastic and inelastic, can also be studied by various geophysical methods, some of which are briefly described below.

Volcanic unrest (sometimes called volcano unrest) is the general name used for an increase in various detected geophysical, geochemical, and geological signals indicating changes in the rates or styles of associated physical processes within the volcano. Common signals are changes in seismicity, ground deformation (inflation/deflation), emission of volcanic gases (fumarole activity), flow and chemistry of groundwater and geothermal water, melting of snow/ice, and magnetic and gravity fields. Volcanic unrest is often partly caused by new magma being injected into the magma chamber of the volcano (Fig. 1.12), resulting in its expansion and volcano inflation. The unrest may result in magma-chamber rupture and dike injection, but most unrest periods do not result in a volcanic eruption.

1.5.2 Geophysical Techniques

The main geophysical techniques and methods used to monitor volcanoes are volcano seismology and ground deformation or geodetic studies. Volcanoes and their unrests are also monitored using geochemical methods, such as through gas and hydrological monitoring, as well as gravimetric and magnetic studies. These latter are outside the scope of the book, but geochemical monitoring is briefly discussed for the sake of completeness.

Volcano Seismicity Monitoring

When the magma pressure in a chamber increases, the local stresses in the volcano also change and generally increase, resulting in earthquakes. The seismicity is normally widely distributed within the volcano, but tends to be most concentrated around the magma chamber (Fig. 1.12). When, however, the magma chamber ruptures and a dike (or an inclined sheet) is injected and begins to propagate up into the roof of the chamber, the seismicity becomes more local, generating a **swarm** that can often be used to trace the propagation of the dike within the volcano (Fig. 1.13; Chapter 4). Many of the dike-induced earthquakes occur at and above the propagating tip of the dike, whereas others – presumably the majority – occur along existing fractures and weaknesses (joints, faults, contacts) in the host rock on either side of the dike. Volcanic gases also increase the fluid pressure within certain parts of the volcano, namely along the paths of the migrating gases; the increased pore-fluid pressure commonly triggers earthquakes.

Seismometers are employed for detecting and monitoring the earthquakes. Commonly, during unrest periods, many seismometers are installed on the volcano and in the surrounding area. The seismic network thus established allows the typically small earthquakes associated with the inflation and dike propagation to be detected and their focal mechanisms determined. Volcanic earthquakes are normally small, mostly less than magnitude M3.

The earthquakes can often be used to determine the depth to the magma chamber, and also the depth to a propagating dike (Figs. 1.12 and 1.13). If, eventually, the dike reaches the surface, or close to it, there will be a **volcanic tremor** (Chapter 4). The tremor is related to low-frequency earthquakes, generated during continuous vibration or shaking of the ground and thought to be caused by subsurface movement of magma and gas. While a tremor is generally accepted as an indication of an imminent eruption, some tremors occur without an

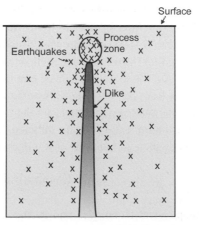

Fig. 1.13 Schematic illustration of the distribution of induced earthquakes (indicated by crosses) around a propagating dike. The process zone, indicated schematically by an ellipse above the tip of the dike, is where the most intense microfracturing and, partly, plastic deformation takes place during the dike (or any fracture) propagation (Gudmundsson, 2011).

eruption eventually happening. Thus, when a volcanic tremor happens there is a high likelihood, but no certainty, of an eruption.

Ground-Deformation Monitoring

Besides monitoring volcano seismicity, measuring the ground deformation is today perhaps the most commonly used technique for monitoring active volcanoes. There has been great progress in this technique in the past decades, primarily through the use of satellites that provide data for GPS and InSAR techniques (the acronyms are explained below). In addition, there are various other ground-deformation methods used, such as levelling, tiltmeters, electronic distance meters, and strainmeters. These techniques are also used to monitor active fault zones as well as to general plate movements – topics that are largely outside the scope of this book.

GPS stands for **Global Positioning System**, which is composed of a total of 32 satellites orbiting the Earth at a distance of about 20 200 km above the surface. The system was developed by the United States Department of Defense, but is currently used for many purposes, including monitoring and measuring volcanic and earthquake deformation. Using five–eight satellites in view (anywhere from the Earth's surface), the position of measurement points or **benchmarks** (metal nails fastened into the surface of solid rocks, such as outcrops at the surface of a volcano) can be estimated very accurately. For maximum accuracy, where changes in position (displacements) of the benchmarks are of the order of centimetres or less, data are collected from the satellites over periods of 8–24 hours. Using various corrections (such as for atmospheric delay of the signals from the satellites), the measurements provide means of detecting slight inflations and deflations of volcanoes. From these it is possible to infer whether magma (or other fluids) are accumulating and building up pressure within the magma chamber of the volcano or, alternatively, leaving the magma chamber (through a dike), thereby decreasing the pressure. The latter is common, for example, during lateral dike propagations, many of which do not result in eruptions.

InSAR stands for **Interferometric Synthetic Aperture Radar** and is a technique used worldwide to measure surface deformation, such as on volcanoes and in earthquakes zones, but also general ground subsidence and ice flow, among other processes. As the name implies, it is a radar technique where use is made of synthetic aperture radar images to map the surface deformation. The basic data are derived from the difference in the known phase of the wave or signal sent by the satellite (the outgoing wave) and the phase of the return wave or signal, reflected from the Earth's surface. Like GPS data, InSAR data can be used to measure surface deformation as small as centimetres, over periods from days to years, in volcanoes and earthquake zones.

Strainmeters are also used for monitoring volcanic (and earthquake) deformation. All these meters measure strain, that is, changes in a dimension such as length divided by the original dimension. The measured strain therefore has no unit (Chapter 3). For surface deformation, the most common strainmeter is an **extensometer**, where the changes in the distance between two points are measured and divided by the original distance. There are various types of these meters, which use either optical methods or the stretching of wire to measure the change in distance. The simplest meters, however, use metal rods, such as have been used to measure the dilation or extension across fractures during rifting events. Fluid-filled **boreholes** (wells) are also sensitive strainmeters. When they are subject to compression, the water level in the wells rises, and when subject to extension, the water level falls. These have been particularly useful in active seismic areas, which often contain numerous groundwater or geothermal wells.

Tiltmeters (also known as **inclinometers**) are useful for measuring the surface deformation of volcanoes. Tiltmeters measure very small changes in the slope angle of the free surface of the volcano at certain specific points. The tilts are measured in radians. Recall that a radian is equal numerically to an arc whose length is the same as the radius of the circle, and thus equal to an arc of about $57.3°$. Because the tilts on volcanoes during inflation and deflation periods are usually very small, the measurements are normally given in microradians, that is, in units of 10^{-6} radians. One microradian is about 6×10^{-5} degrees. For tiltmeters on volcanoes, the smallest tilt that can be measured in a day is around 10^{-3} microradians.

Electronic distance meters are also used to monitor active volcanoes. These meters send and receive electromagnetic signals. By comparing the phase of the sent and received (reflected) signal, the changes in distances between points at the surface of the volcano can be measured with an accuracy of less than a centimetre. The distance changes can then be related to lateral and/or vertical displacements of the surface of the volcano.

Geochemical Techniques

Active volcanoes and their internal processes can also be monitored by geochemical methods, such as through gas and hydrological monitoring. Changes in **gas composition** or the rate of flow, primarily of CO_2 (carbon dioxide) and SO_2 (sulphur dioxide), from magma, is an indication of processes taking place inside the volcano. The gases can be measured directly on site, that is, at fumaroles or vents; or remotely, either from the ground or from aircrafts. Changes in the gas flux (the rate of flow of gas) can be interpreted as being related to changes in the volume of magma in its chamber and/or to changes in the associated geothermal system. Gas monitoring is currently routinely made on many active volcanoes.

There may also be changes in the **chemistry of the springs and rivers** associated with active volcanoes. These are commonly monitored; for example, in glacial rivers associated with some active (largely ice-covered) volcanoes in Iceland.

1.6 Interpreting the Data

1.6.1 General Interpretation

The first aim of volcanotectonic data collection is to provide information for developing a theoretical framework for understanding volcano behaviour. Such an understanding is a necessary condition so as to be able to forecast with any accuracy what is likely to happen during an unrest period. A second aim is to test existing or new theoretical models, the cornerstones of the theoretical framework, on volcanotectonic processes in volcanoes. Here, field data, primarily structural and geophysical, play a crucial role. It is easy to come up with ideas or conceptual models on volcanotectonic processes and principles. But to be useful in the sense of advancing our understanding of the real processes, the ideas must be testable, that is, make some predictions or forecasts that can be checked against existing or, preferably, new data.

Data analysis involves all the standard **statistical techniques** and modes of presentation used in structural geology and tectonics. These include histograms, rose diagrams, and stereoplots of the strike and dip (attitude) of the measured structures, particularly faults, dikes, inclined sheets, volcanic fissures (for which dip measurements do not normally exist), and tension fractures and joints (Chapter 2), as well as histograms/bin plots of dimensions and displacements (for faults) and thicknesses and other dimensions (for dikes).

For an earthquake, the **focus** or **hypocentre** and the **focal mechanism** or **fault-plane solution** (for earthquakes occurring on faults) are normally calculated and analysed (Chapter 4). The focus is simply the site or place within the crustal segment (or the volcano) where the fault rupture begins. The focus thus indicates the site of earthquake nucleation and initiation, which is normally the location of the greatest shear-stress (and strain) concentration prior to the earthquake. Focal mechanisms indicate the type of fault that generates the earthquake, that is, whether the earthquake occurs on a strike-slip or a dip-slip (normal, reverse, thrust) fault. Many earthquakes, of course, occur on faults that are oblique-slip, that is, partly strike-slip and partly dip-slip, and some may also have a large opening component (extension-fracture component). Such faults are referred to as mixed-mode (Gudmundsson, 2011). Some earthquakes, particularly in volcanoes, are not generated by slip on faults but rather by vibration or shaking of the ground, such as in volcanic tremors. Earthquakes related to slip on faults (as most are) are also known as **double-couple**, whereas those that are not so formed are known as **non-double-couple**.

Geophysical deformation data are usually **inverted** so as to infer the depth to the main source or sources. The process is analogous to finding the 'best fit' to a data set on a plot using **regression** analysis, the best known of which is the linear regression. When the geodetic data are inverted the aim is to find the 'best-fit source' of the deformation. For a double-couple tectonic earthquake rupture, the source is a fault, whereas for volcano deformation the **source** is normally a magma chamber and/or a dike (or an inclined sheet or a sill).

The first thing to keep in mind when interpreting volcano deformation is that there are **no unique solutions** as regards the deformation sources (Chapter 3). Several sources with various geometries, fluid pressures, and depths in crustal segments whose mechanical properties are poorly known may fit the deformation data equally well. The second thing to keep in mind is that even when the 'best-fit' source has been selected, its physical characteristics (such as geometry, depth, fluid pressure) are normally poorly constrained.

As regards the lack of a unique solution, the similarity with fitting a line through data is instructive. For a reasonably large set of data points, an **infinite number** of functions can theoretically be fitted to the data. Well-known possible functions include various polynomials. One drawback of using polynomials as functions is that such a fit rarely has much physical meaning. That is, while the polynomial fit may be very good, its connection with the underlying physics of the processes that generate the data is commonly obscure or absent (cf. Hamming, 2004). Fourier series can also fit many data sets. The interpretation of such a fit in terms of physics of the processes producing the data is normally much better than for a polynomial fit.

We tend to choose **straight lines** to fit data, partly because of their perceived 'simplicity'. There are, however, no accepted rigorous simplicity criteria for selecting functions that fit the data. Consider, for example, the following equations:

$$y = ax + b \tag{1.1}$$

$$y = ax^c \tag{1.2}$$

where a, b, and c are constants. Which of these two equations, both of which may fit the same data set, is the simpler one? Many would say that Eq. (1.1) is simpler because it is linear. Both equations contain the same number of terms, however, so that there is no clear-cut reason as to why Eq. (1.1) should be regarded as simpler than Eq. (1.2).

Coming back to the inversion of deformation data, there are many reasons why there are no unique solutions as regards the deformation sources. The main ones may be listed as follows:

- Surface deformation data are essentially two-dimensional. This means that they are data in a plane or a surface. To infer a three-dimensional source such as a magma chamber or a dike from two-dimensional data cannot result in a unique solution. For example, the commonly used 'Mogi model', a nucleus of strain in an infinite elastic half-space (Chapter 3; Dzurisin, 2006; Segall, 2010), cannot distinguish between the size and overpressure of the source magma chamber. **Half-space models** assume that the crustal segment is isotropic and homogeneous, that is, that it contains no layers with contacts and different mechanical properties.
- The crustal segment hosting the volcano and its magma chamber, and the volcano itself, are composed of **numerous layers**, with contacts as well as faults and other discontinuities (Figs. 1.7 and 1.11; Chapter 7). The layers and contacts commonly have widely different mechanical properties, so that the stresses and strains (or displacements) in these layers vary widely, even when the loading, such as the magmatic overpressure in the associated shallow magma chamber, may be constant. When using simple elastic half-space models, the surface deformation thus commonly gives a poor indication of the actual geometry and depth of the source chamber (Chapters 3 and 10; Masterlark, 2007).
- The half-space models are also poor for determining the **geometries and paths** of injected dikes. Dikes commonly open up joints, contacts, and other discontinuities ahead of the dike tips, resulting in surface deformation and stresses that do not allow simple inversion of the geodetic data to determine the dike depth or geometry (Chapters 3, 7, and 10). Simple elastic dislocation models normally use uniform elastic properties for the entire elastic half-space. For a layered crust, the estimated dike thicknesses from such models may easily be in error of 50–100%.

While inversion models are, necessarily, inaccurate as to the determination of the depth and geometry of magma chambers and injected dikes and inclined sheets, they are currently the most widely used models we have for such determinations. One theme that will be explored in the book is how to improve our knowledge of the interiors of volcanoes so as to make our understanding of the real processes and structures that generate the surface deformation more reliable. We will briefly introduce these topics below (Section 1.6.3), but first we turn to a general classification and description of models, particularly those used in volcanotectonics.

1.6.2 Models

The interpretation and use of scientific data is normally through scientific **models**. Existing and new data are then used to construct, develop, and test the models. Generally, we make models to help us understand the natural processes or principles with which we are working. For example, in volcanotectonics we make various types of models to improve our understanding of the physical processes that happen inside volcanoes. In physical science in general, and in earth sciences in particular, there are five basic types of models, namely conceptual, statistical/probabilistic, analogue, analytical, and numerical. Let us now briefly describe and discuss the main characteristics of these models with application to earth sciences in general and volcanotectonics in particular.

1. Conceptual Models. These are basically ideas about some structures or processes that are expressed through words, illustrations, or physical models. Examples include geological maps, cross-sections, and physical models of volcanoes (Fig. 1.14). Other examples include general sketches of geological structures and the processes that generate them.

2. Statistical/Probabilistic Models. As applied in volcanotectonics, these models indicate the likelihood that a randomly selected structure or event will have a certain value. For the strike or orientation of volcanic fissures in a particular volcanic zone or a single volcano, such a model would indicate the purely probabilistic likelihood that the fissure that formed during the next eruption would strike between, say, N20°E and N30°E. That likelihood would depend on the available data, using the frequency theory of probability. If we have additional information, such as the inferred state of stress in the volcano, its mechanical layering, and the location of the most recent volcanic fissure in the volcano/volcanic zone, then the Bayesian probability approach (Stone, 2013) might be warranted. Similar models can be made of the likelihood of values falling within certain class limits or range, such as regarding the length of the next volcanic fissure, its number of crater cones, and so forth. These types of models apply to any types of fractures – including dikes, normal faults, and tension fractures (Chapter 2).

3. Analogue Models. These are also referred to as **scale models** since they draw an analogy between small-scale processes or structures in the laboratory and large-scale processes or structures in nature. Among the earliest analogue models in tectonics were those that used wet clay under tension to develop fracture patterns similar to those seen in rift zones and grabens. In current analogue models, use is made of sand, powder, clay, silicone, jelly, or other materials which have properties that, through scaling factors, can be correlated with the materials (rocks, crustal fluids) which we want to model. Analogue models have been used extensively in volcanotectonics in recent years, primarily to model the emplacement of dikes and sills (Kavanagh et al., 2006, 2015; Menand et al., 2010; Daniels and Menand, 2015) as well as the formation of collapse calderas and related structures (Chapter 5).

4. Analytical Models. These are essentially equations or formulas that express certain relations between volcanotectonic parameters and constants, loading, and associated processes. For example, the variation in aperture (opening displacement), Δu_I, along the tension fracture in Fig. 1.9 can be modelled analytically as a mode I crack through the formula:

$$\Delta u_I = -\frac{4\sigma(1 - v^2)}{E}(a^2 - x^2) \tag{1.3}$$

where σ is the tensile stress applied to the rift zone during fracture formation, v is Poisson's ratio, E is Young's modulus, a is the half length (half strike-dimension) of the fracture which coincides with the x-axis of the coordinate system. The minus sign is because tensile stress is recognised as **negative** (compressive stress as **positive**) so that, in order to get a positive or zero aperture ($\Delta u_I \geq 0$), as it should be, there must be a minus sign before the stress symbol (Chapter 3). In Eq. (1.3) it is assumed that the tension fracture is a through crack, that is, it extends from one **free surface** (rock surface in contact with fluid, for example air, water, or magma) to another free surface. There are many analytical volcanotectonic models presented in the book, some of which are derived and discussed in greater detail by Gudmundsson (2011). In analytical models we normally find the general solution to a problem 'by hand', although there are available several computer programs for finding analytical, closed-form solutions. Analytical models apply to many specific boundary conditions so that the effects of changes in the conditions can be evaluated to provide solutions at any point in the rock. However, analytical solutions can be found only for simple geometries and are normally restricted to small strains and essentially homogeneous and isotropic rock properties.

5. Numerical Models. These are used for volcanotectonic structures and processes where the geometries, mechanical properties, and boundary conditions make the problem too complex or tedious to solve analytically. The volcanotectonic problem is then rewritten as mathematical statements that can be solved numerically. In numerical models we divide or 'discretise' the problem into an equivalent system of small units or 'elements', solve simultaneous algebraic equations (resulting in numerical approximations) for each element, and then combine them into a solution for the entire body or process under consideration (cf. Reddy, 2005; Deb, 2006). These are specific solutions to a particular set of conditions and provide solutions only for the specified points in the body which may, for example, be a crustal segment or a collapse caldera. Numerical solutions can be found for any complex geometries and boundary conditions that we wish to analyse, and are particularly suitable for analysing large strains as well as bodies with

Fig.1.14 Physical model of the volcanoes Eyjafjallajökull (to the left, erupted in 2010) and Myrdalsjökull (to the right, erupted in 1955) in south Iceland. The main volcano in Myrdalsjökull is Katla. The white tops denote ice caps and glaciers. This is a part of a general physical model of Iceland in the City Hall of Reykjavik.

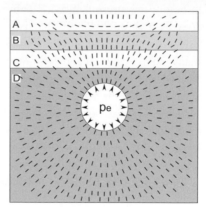

Fig. 1.15 Simple two-dimensional numerical model of the local stress field around a shallow magma chamber of a circular cross-section. The stress field is generated by an excess fluid pressure (p_e), that is, pressure above the lithostatic pressure (Chapter 3) of 10 MPa in the magma chamber. Layer D has a stiffness or Young's modulus of 10 GPa; layer C, 100 GPa; layer B, 1 GPa, and layer A, 100 GPa. The short lines (the ticks) show the directions (the trajectories) of the maximum compressive principal stress σ_1, along which ideal sheet intrusions (dikes, inclined sheets, sills) injected from the chamber would propagate (cf. Gudmundsson and Brenner, 2005).

heterogeneous and anisotropic mechanical properties. Numerical models, especially when presented as animations, are also referred to as **numerical simulations.** An example of a very simple numerical modelling result is the stress field around a magma chamber for given boundary conditions and host-rock properties (Fig. 1.15).

1.6.3 Improving the Models and Monitoring of Volcanoes

The fundamental aim of modelling volcanotectonic processes is to be able to forecast with accuracy the most likely course of events during unrest periods in volcanoes. Volcanic unrest normally includes some or all of the following factors and processes: inflation or deflation, earthquakes, changes in water wells and composition of the water in nearby rivers, as well as changes in gas flux (the rate of flow of gas). Given the sophisticated and dense networks of various geophysical and geochemical instruments on many active volcanoes, one might perhaps expect that we could routinely forecast how an unrest period is going to end – in particular, if it is going to result in an eruption – but normally we cannot. Here, I briefly explain this lack of success, and suggest ways to improve the present modelling and eruption forecasting methods. The points raised and discussed here are elaborated in the subsequent chapters of the book.

All scientific models, including those in volcanotectonics, aim at providing results that improve our understanding of the world and allow us to forecast events. In volcanotectonics, the events of concern include large landslides (lateral collapses, Chapter 5), caldera formation (vertical collapses, Chapter 5), volcanotectonic deformation (Chapter 3), volcanic earthquakes (Chapter 4), and eruptions (Chapters 7, 8, and 10). The forecasts can be either deterministic or probabilistic. **Deterministic** forecasts specify the location, time, and magnitude of an event, whereas **probabilistic** forecasts specify the probability of these parameters within certain ranges. In much of classical physics and derived engineering fields, deterministic forecasting is the rule and has been very successful. Modern physics, however, is largely statistical in nature (quantum mechanics, statistical mechanics), and so are most scientific fields that deal with complex structures and large numbers of entities where stochastic (random) elements play a significant part in processes and events. The same applies to volcanotectonics. Not only are the internal structures and properties of volcanoes complex, but there are always stochastic elements that play a role in the unrest periods, making probabilistic forecasting the only viable method.

So how can we refine our volcanotectonic models with a view to improving the reliability of forecasts? There are several aspects that should and could be bettered at this stage, including the following:

1. Use **numerical models** rather than the nucleus-of-strain models (such as the Mogi model). This follows because the nucleus-of-strain models assume homogeneous, isotropic elastic half-spaces and therefore can never provide accurate depth estimates of the sources (magma chambers). Furthermore, the nucleus-of-strain models cannot distinguish between the size of the source and its internal fluid (magma) overpressure. And these models provide no information about the state of stress in the host rock. Nucleus-of-strain models thus cannot be used to infer the likely paths of injected dikes, their chances of reaching the surface as feeders, or the likelihood of lateral or vertical collapses during unrest. By contrast, as is discussed in later chapters, numerical models can, provided the appropriate data are available, determine reasonably accurately the likely **depth**, **size**, and **shape** of the

source, the state of stress in the host rock, and the likely paths of injected dikes following magma-chamber rupture.

2. Take the **heterogeneity** and **anisotropy** of the volcano and the hosting crustal segment into account. This implies using mechanical layering in the models. There are a lot of data available on mechanical layering in volcanoes through drilling and direct field observations (Fig. 1.16). Taking layering into account refines the models in two principal ways. First, the interpreted surface deformation then reflects much more accurately the depth, size, and shape of the source (a chamber or a dike, for example). Second, reasonably accurate forecasts as to likely paths of injected dikes (or inclined sheets or sills) during the unrest period depend on taking layering into account. This has been demonstrated in many eruptions in well-monitored volcanoes such as in the 2010 Eyjafjallajökull eruption in Iceland (Sigmundsson et al., 2010; Gudmundsson et al., 2012), the 2011 El Hierro eruption in the Canary Islands, Spain (Becerril et al., 2013; Marti et al., 2013), the 2013 eruption in Etna, Italy (Falsaperla and Neri, 2015), and the Bardarbunga 2014–2015 eruption in Iceland (Gudmundsson et al., 2014; Sigmundsson et al., 2015). In all these eruptions, the comparatively complex paths of the feeder-dikes were partly the result of the heterogeneity and anisotropy (mainly layering) of the volcano and the hosting crustal segment.

3. Use data on the geometry of fossil magma chambers and dikes, inclined sheets, and sills in **eroded volcanoes** to provide constraints on typical and likely geometries of chambers and paths of sheet-like intrusions. Some fossil magma chambers are particularly well exposed, so well, in fact, that the roof and the dikes dissecting the roof can be explored in

Fig. 1.16   South slopes of the volcano Eyjafjallajökull in South Iceland (Fig. 1.14) are composed of a variety of layers with different mechanical properties. During the 2010 eruption of Eyjafjallajökull, several sills, presumably similar to the one seen here, were emplaced at great depths within the volcano (Sigmundsson et al., 2010; Tarasewicz et al., 2012; Gudmundsson, 2017).

Fig. 1.17 Part of the exceptionally well-exposed fossil shallow magma chamber (now a pluton) of Slaufrudalur in Southeast Iceland (located in Fig. 3.15). The walls and the roof of the chamber are exposed, and many dikes (as extension fractures) cut the roof. The pluton is made of granophyre and is hosted by a pile of basaltic lava flows. A black and white version of this figure will appear in some formats. For the colour version, please refer to the plate section.

great detail (Fig. 1.17). Similarly, many dikes can be traced in the field (Fig. 1.18), thereby providing constraints on and an understanding of the dike paths that can be inferred from earthquake swarms and deformation during unrest periods in active volcanoes. Vertical dikes are commonly seen to change into inclined sheets or horizontal sills along parts of their paths. In addition, dikes, sheets, and sills show great variations in thicknesses along their paths, commonly being thicker where dissecting soft (low Young's modulus or compliant) layers than when dissecting stiff (high Young's

Fig. 1.18 Deflection of a dike into a sill along part of its path. The deflection occurs at the contact between mechanically dissimilar rocks, with the contact itself being composed of scoria. The vertical dike (1) changes into a thin sill for about 8 m (2), and then back to vertical dike (3).

modulus) layers (Chapter 2; Geshi et al., 2010). Data on **actual dike paths** should be incorporated into the numerical models used to interpret the surface deformation during unrest periods, as well as the models on the seismicity associated with the dike propagation.

1.7 Summary

- An active volcano is a vent through which magma and volatiles are transported to the Earth's surface. The volcanic materials around the vent pile up into a small volcano – a crater cone, normally a cinder/scoria cone or a spatter cone. During many eruptions through one or many nearby vents, the erupted material builds up into a volcanic edifice, most commonly a stratovolcano or a basaltic edifice.

- An active volcano is commonly defined as one that has erupted at least once during the Holocene, that is, during the past 11 700 years. This definition should be treated with caution, however. There are many volcanoes, particularly large collapse calderas, which have not erupted for tens of thousands of years but are still active and need to be monitored. Generally, a volcano is potentially active so long as it has a chance of receiving magma from a source.
- A volcano that is thought to be unable to erupt is regarded as extinct or fossil. When a fossil volcano is also deeply eroded, its fossil shallow magma chamber (or, at least, its top part), a pluton, is commonly exposed.
- Volcanotectonics is the scientific field that combines tectonics, structural geology, volcanology, geodesy, and seismology with theories from classical physics and their modern extensions (including rock mechanics and fracture mechanics) to analyse and interpret processes in active and fossil volcanoes. More specifically, volcanotectonic studies include collecting tectonic, rock-physical, geodetic, seismic, and stress data from active and fossil volcanoes. These data are then interpreted, using particular theories from physics, to provide reliable deterministic and probabilistic models (as in statistical physics) for understanding processes inside volcanoes and for forecasting scenarios and events at active volcanoes.
- Tectonic and structural-geology techniques used in volcanotectonics include all the standard methods such as measurements of strike, dip, length, aperture, and displacement of tension fractures and faults, as well as the thicknesses and other dimensions of dikes, sheets, and sills.
- Geophysical techniques used in volcanotectonics include measurements of volcano deformation (inflation, deflation) and seismicity using methods/equipment such as GPS, InSAR, strainmeters, tiltmeters, distance meters, and seismometers. The data are inverted using standard statistical techniques; for example, to infer the dimensions and location of a subsurface fault, or the dimensions and location of an active magma chamber.
- For understanding, interpreting, and forecasting volcanotectonic processes and events, including eruptions, models are used. The basic types of models are conceptual, statistical/probabilistic, analogue, analytical, and numerical.

1.8 Main Symbols Used

a	half length (half strike-dimension) of a fracture
a	constant
b	constant
c	constant
E	Young's modulus
Δu_I	maximum aperture (total opening displacement) of a mode I (extension) crack
v	Poisson's ratio
σ	normal stress

1.9 Worked Examples

Example 1.1

Problem

The fracture in Fig. 1.19 is at the surface of Holocene lava flow in a rift-zone volcanic system in Iceland.

(a) What type of fracture is this?
(b) Estimate the aperture of the fracture.
(c) If its strike-dimension (length) is 600 m, the dip-dimension or depth 300 m, and the fracture is located in a 20-km-thick crustal segment, which would be its controlling dimension?

Fig. 1.19   Tension fracture in the Holocene lava flows of the rift zone in Southwest Iceland (Gudmundsson, 2017).

Solution

(a) The type of fracture is an extension fracture. This follows because the only displacement is an opening (there is no apparent fracture-parallel movement of the walls). More specifically, since the fracture is at the surface of a Holocene lava flow in a rift zone and shows no evidence of being formed as a result of internal fluid pressure, it is classified as a tension fracture.
(b) Using the scale provided on the photograph, the aperture or opening of the tension fracture is about 10 m.
(c) Since the dip-dimension is smaller than the strike-dimension and the fracture does not extend through the entire crustal segment, the controlling dimension is the dip-dimension. The controlling dimension is the one that mainly controls the displacement on a fracture, here the opening displacement (the aperture).

Example 1.2

Problem

The orthogonal structures in Fig. 1.20 dissect a pile of Tertiary (Neogene) lava flows in Iceland.

(a) What are the general geological names of the structures/fractures?
(b) To which mechanical types of fracture do the structures belong?
(c) Which of the fractures is the youngest one?
(d) What are the thicknesses of the structures?
(e) How do the measured thicknesses of the structures relate to the original apertures of the fractures?

Fig. 1.20   Orthogonal structures/fractures dissecting the Tertiary lava pile in southeast Iceland.

Solution

(a) The vertical structure on the photograph is a dike, the horizontal structure a sill (or, possibly, an inclined sheet, depending on the dip, which cannot be inferred from the photograph).

(b) Both dikes and sills are extension fractures. Since they are driven open by magma pressure, that is, the pressure of the magma breaks or ruptures, the host rock, dikes, and sills are the type of extension fracture referred to as hydrofractures.

(c) The sill clearly dissects (cuts through) the dike and is therefore the younger fracture. The dike is thus the older fracture.

(d) Using the person (about 1.7 m) as a scale, the sill is about 1.4 m thick. The dike thickness changes from about 0.9 m in the lower part (close to the person) to about 0.6 m at its contact with the sill.

(e) The thickness as seen in the field is normally somewhat less than the original aperture of the magma-filled fractures during emplacement. This follows because, on solidification of the magma, the dike/sill volume reduces or shrinks by about 10%.

Example 1.3

Problem

The walls seen in Fig. 1.21 form a part of a large north–south striking river canyon, of Holocene age, in North Iceland. View northeast, the top part of the canyon dissects Holocene layers; the lower part, late Pleistocene layers. Name and describe briefly the structures marked A to E in the photograph.

Solution

(a) Structure A is a cinder/scoria cone, several tens of metres in diameter, and similar to (though smaller than) the one seen in Fig. 1.1b.

(b) Structure B is a Holocene spatter cone that forms part of the surrounding lava flow. The lava flow is mostly about 20 m thick, but reaches about 30 m where the spatter cone is located. These cones form part of a 75-km-long and 8000-year-old crater row or a volcanic fissure (Gudmundsson et al., 2008).

(c) Structure C is a Pleistocene (interglacial) lava flow, 15–20 m thick. It has particularly well-developed sets of columnar joints, mostly vertical.

(d) Structure D is the feeder-dike to the spatter cone (B) and the cinder cone (A). The dike thickness is about 13 m at its contact with the Holocene lava flow (and the spatter cone, B) but 4.5 m in the lowermost parts of the cliff section, close to the river.

(e) Structure E is a reverse fault. Using the Pleistocene lava flow (C) as a marker layer, the fault displacement is about 5 m. The reverse fault is most likely a reactivated normal fault, forming part of a 20-km-long and 0.5–1-km-wide graben. The reactivation is attributed to the horizontal compressive stress generated by the magmatic pressure of the feeder-dike (D) at its time of emplacement (cf. Gudmundsson et al., 2008).

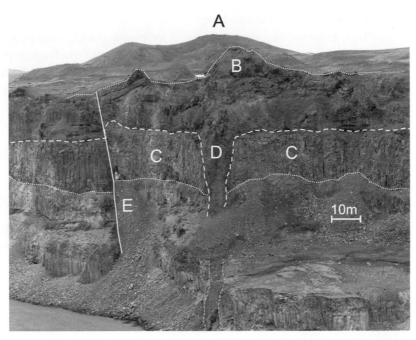

Fig. 1.21　Various volcanotectonic structures and units (A to E) in a canyon of a glacier river in Northeast Iceland.

1.10 Exercises

1.1　Define an active volcano and a potentially active volcano.

1.2　Define an extinct or fossil volcano.

1.3　Define a monogenetic volcano and a polygenetic volcano.

1.4　Define volcanotectonics as a scientific field.

1.5　What is meant by the attitude of a rock structure?

1.6　In what way is the relationship between aperture and displacement for extension fractures different from that for shear fractures?

1.7　What terms are used in volcanotectonics and structural geology for the terms fault length and fault width in seismology?

1.8　Define the term inflation as used in volcanotectonics and provide a standard interpretation of the processes giving rise to inflation.

1.9　Define the term deflation as used in volcanotectonics and provide a standard interpretation of the processes giving rise to deflation.

1.10　Is the volcanotectonic deformation seen in Figs. 1.7 and 1.9 elastic or permanent? Explain why.

1.11　Explain the acronyms GPS and InSAR and discuss briefly the use of the associated techniques in volcanotectonic studies.

1.12 What is a volcanic tremor?

1.13 What are the main geochemical techniques used to monitor active volcanoes?

1.14 Define the terms earthquake focus (hypocentre) and focal mechanism.

1.15 Define and explain the use of analogue models in volcanotectonics.

1.16 Define and explain the use of probabilistic (statistical) models in volcanotectonics.

1.17 Discuss the main difference between using an analytical model for a magma source (magma chamber) and a numerical model.

1.18 What is a crater row?

1.19 What is a feeder-dike?

1.20 Define volcanic unrest.

References and Suggested Reading

Becerril, L., Galindo, I., Gudmundsson, A., Morales, J. M., 2013. Depth of origin of magma in eruptions. *Scientific Reports*, **3**, doi:10.1038/srep02762.

Browning, J., Drymoni, K., Gudmundsson, A., 2015. Forecasting magma-chamber rupture at Santorini volcano, Greece. *Scientific Reports*, **5**, doi:10.1038/srep15785.

Daniels, K. A., Menand, T., 2015. An experimental investigation of dyke injection under regional extensional stress. *Journal of Geophysical Research*, **120**, 2014–2035.

Deb, D., 2006. *Finite Element Method: Concepts and Applications in Geomechanics*. Upper Saddle River, NJ: Prentice-Hall.

Dzurisin, D., 2006. *Volcano Deformation: New Geodetic Monitoring Techniques*. Berlin: Springer Verlag.

Fagents, S. A., Gregg, T. K. P., Lopes, R. M. C. (eds.), 2013. *Modeling Volcanic Processes: The Physics and Mathematics of Volcanism*. Cambridge: Cambridge University Press.

Falsaperla, S., Neri, M., 2015. Seismic footprints of shallow dyke propagation at Etna, Italy. *Scientific Reports*, **5**, doi:10.1038/srep11908.

Geshi, N., Kusumoto, S., Gudmundsson, A., 2010. Geometric difference between non-feeder and feeder dikes. *Geology*, **38**, 195–198.

Gudmundsson, A., 2011. *Rock Fractures in Geological Processes*. Cambridge: Cambridge University Press.

Gudmundsson, A., 2017. *The Glorious Geology of Iceland's Golden Circle*. Berlin: Springer Verlag.

Gudmundsson, A., Brenner, S. L., 2005. On the conditions for sheet injections and eruptions in stratovolcanoes. *Bulletin of Volcanology*, **67**, 768–782.

Gudmundsson, A., Friese, N., Galindo, I., Philipp, S. L., 2008. Dike-induced reverse faulting in a graben. *Geology*, **36**, 123–126.

Gudmundsson, M. T., De Guidi, G., Scudero, S., 2013. Length–displacement scaling and fault growth. *Tectonophysics*, **608**, 1298–1309.

Gudmundsson, M. T., Lecoeur, N., Mohajeri, N., Thordarson, T., 2014. Dike emplacement at Bardarbunga, Iceland, induces unusual stress changes, caldera deformation, and earthquakes. *Bulletin of Volcanology*, **76**, doi:10.1007/s00445-014-0869-8.

Gudmundsson, M. T., Thordarson, T., Hoskuldsson, A., et al., 2012. Ash generation and distribution from the April–May 2010 eruption of Eyjafjallajökull, Iceland. *Scientific Reports*, **2**, doi:10.1038/srep00572.

Hamming, R. W., 2004. *Methods of Mathematics Applied to Calculus, Probability, and Statistics*. Mineola, NY: Dover.

Janssen, V., 2008. *GPS-Based Volcano Deformation*. Saarbrücken: VDM Verlag.

Kavanagh, J., Menand, T., Sparks, R. S. J., 2006. An experimental investigation of sill formation and propagation in layered elastic media. *Earth and Planetary Science Letters*, **245**, 799–813.

Kavanagh, J., Boutelier, D., Cruden, A. R., 2015. The mechanics of sill inception, propagation and growth: experimental evidence for rapid reduction in magmatic overpressure. *Earth and Planetary Science Letters*, **421**, 117–128.

Lu, Z., Dzurisin, D., 2014. *InSAR Imaging of Aleutian Volcanoes: Monitoring a Volcanic Arc from Space*. Berlin: Springer Verlag.

Marti, J., Pinel, V., Lopez, C. et al., 2013. Causes and mechanisms of the 2011–2012 El Hierro (Canary Islands) submarine eruption. *Journal of Geophysical Research*, **118**, 823–839.

Masterlark, T., 2007. Magma intrusion and deformation predictions: sensitivities to the Mogi assumptions. *Journal of Geophysical Research*, **112**, doi:10.1029/2006JB004860.

Menand, T., Daniels, K. A., Benghiat, P., 2010. Dyke propagation and sill formation in a compressive tectonic environment. *Journal of Geophysical Research*, **115**, doi:10.1029/2009JB006791.

Reddy, J. N., 2005. *An Introduction to the Finite Element Method*. New York, NY: McGraw-Hill.

Ritchie, D., Gates, A. E., 2001. *Encyclopedia of Earthquakes and Volcanoes*. New York, NY: Facts on File.

Segall, P., 2010. *Earthquake and Volcano Deformation*. Princeton, NJ: Princeton University Press.

Sigmundsson, F., Hreinsdottir, S., Hooper, A., et al., 2010. Intrusion triggering of the 2010 Eyjafjallajökull explosive eruption. *Nature*, **468**, 426–430.

Sigmundsson, F., Hooper, A., Hreinsdottir, S., et al., 2015. Segmented lateral dyke growth in a rifting event at Bardarbunga Volcanic System, Iceland. *Nature*, **517**, 191–195.

Sigurdsson, H., Houghton, B. F., McNutt, S. R., Rymer, H., Stix, J. (eds.), 2000. *Encyclopedia of Volcanoes*. New York, NY: Academic Press.

Stone, J. V., 2013. *Bayes' Rule: A Tutorial Introduction to Bayesian Analysis*. Berlin: Sebtel Press.

Tarasewicz, J., White, R. S., Woods, A. W., Brandsdottir, B., Gudmundsson, M. T., 2012. Magma mobilization by downward-propagating decompression of the Eyjafjallajökull volcanic plumbing system. *Geophysical Research Letters*, **39**, doi:10.1029/2012GL053518.

Williams, H., McBirney, A. R., 1979. *Volcanology*. San Francisco (California): Freeman.

Zobin, V. M., 2003. *Introduction to Volcanic Seismology*. London: Elsevier.

2 Volcanotectonic Structures

2.1 Aims

Field studies of volcanotectonic structures offer a way of understanding the processes that take place inside volcanoes before eruptions. Collapse calderas and some other large-scale structures are treated separately (Chapter 5), and here the focus is on sheet intrusions, sills, inclined (cone) sheets, and, in particular, dikes. Since they supply magma to most eruptions, it is important to make detailed and **accurate observations and measurements** of sheet intrusions in eroded sections of active and inactive (extinct) volcanoes. All the techniques described here apply equally well to inclined sheets, so that the term '**dike**' in the present context also includes inclined sheets. Most of the techniques also apply to sills; the special aspects of field studies of sills are discussed at the end of the chapter. The observations and measurements provide a better understanding of how dikes propagate, the field conditions that encourage **dike arrest**, as well as the conditions that encourage their propagation to the surface to feed volcanic eruptions. The field data, when combined with geodetic and seismic monitoring data, can be used to test analytical, analogue, and numerical models on internal processes in volcanoes. Well-tested models provide the theoretical framework for understanding unrest periods and, more specifically, for forecasting volcanic eruptions. The main aims of this chapter are to:

- Explain and illustrate the main methods and techniques used to obtain high-quality field data on dikes (and inclined sheets and sills).
- Illustrate the main geometric characteristics of dikes, including segmentation, linkage, and tips.
- Describe the characteristic internal structure of dikes, including criteria to infer magma-flow and propagation directions.
- Describe the characteristics of non-feeder-dikes.
- Describe the characteristics of feeder-dikes.
- Explain the field methods used for studying sills inasmuch as they differ from those used for studying dikes and inclined sheets.

2.2 Field Observations and Measurements

One principal aim of volcanotectonic field studies of dikes is to understand the conditions for their formation and propagation; in particular, the factors that control their geometries and propagation paths. We also want to know the contribution of dikes to crustal dilation, that is, crustal extension, such as in palaeo-rift zones and at plate boundaries. The main discussion in this chapter is on dikes (and, by implication, inclined sheets since these occur often in the same local swarms associated with shallow magma chambers). However, almost all the general field observations and most of the techniques apply as well to studies of sills. Sections for measuring sills and dikes are of course somewhat different; the sections or profiles for dike studies tend to be subhorizontal whereas those for sill studies are mostly subvertical or inclined. Few sill swarms, however, have been measured in traverses in the same way as dike and sheet swarms have been measured. These and other differences as regards field studies of sills versus those of dikes and inclined sheets are discussed in Section 2.9. Otherwise, the observational principles and field methods described here apply to all sheet-like intrusions: to dikes, to inclined sheets, and to sills. But, as indicated, the word mostly used for sheet intrusions in the present description is 'dike'.

Dikes are normally measured along traverses, profiles, sections, or scan-lines. All these words have similar meaning, namely sections along cliffs, river channels, the coast, road-cuts, quarries, or other well-exposed outcrops. A '**traverse**' and a '**profile**' may be of any length but are commonly kilometre-long measurement lines along parts of the coast, roads, or river channels. **Sections** are normally somewhat shorter, and scan-lines tend to be the shortest. More specifically, the word **scan-line** is normally used in connection with measuring smaller structures, such as mineral veins and joints, where the length of the measurement line at each locality is commonly from metres to tens of metres.

A typical cliff section for measuring dikes is shown in Fig. 2.1. The dikes measured in this section form a part of a major dike swarm which, in turn, is a part of the **palaeo-rift zone** that was active in this part of Iceland some 11 million years ago. The measurement line seen here is a part of a measured 10-km-long profile across the total width of the dike swarm. The section itself is located at about 1200 m below the top of the lava pile, that is, below the surface of the active rift zone at the time of dike emplacement.

During measurements along such a profile, every single dike, every fault, and every large-scale tension fracture (if found) is recorded and located. Columnar or cooling joints are only recorded if the aim is to assess the overall mechanical properties of the host rock at the time of dike injection or fault formation. High-quality **aerial photographs** – not Google Earth images, but aerial photographs that allow a three-dimensional view (using a hand stereoscope or a mirror stereoscope) – can be used to locate each measured dike accurately. However, the accuracy of kinematic GPS data has improved much in the past decade and dikes can also be accurately located using **coordinates**. Many geologists use images from Google Earth for mapping. These images are of course very useful, but they are not as accurate for the location of structures in the field as aerial photographs, partly because the Google images mostly lack a three-dimensional view.

When studying the dikes in profiles, the measurements should be made as much as possible at a **similar stratigraphic level** in the host rock. In a lava pile, for example, one would try to have the profile along the same lava flow for as long distances as possible, and then shift up or down to the next well-exposed lava flow to continue the profile (Fig. 2.1). All the measurements should be **accurate** – the strike and dip measurements should be

Fig. 2.1 Typical well-exposed cliff section for measuring regional dikes. The measurement line is marked, and most of the dikes are indicated by arrows. The aim is to keep the measurement line as far as is possible at the same stratigraphic level. In order to follow a continuous exposure, parts of the measurement line, however, may have to be shifted occasionally to higher or lower stratigraphic levels. The height of the cliffs in the centre of the photograph is about 100 m. The conspicuous and somewhat wavy dike in the centre (across which the measurement line is shifted) is 4.5 m thick. View north, the measurement line, here a few hundred metres long, is a part of about 10-km-long traverse (profile) dissecting close to 100 dikes that belong to a dike swarm in East Iceland.

made using a compass (corrected for declination if necessary) and the thickness using a measuring tape. Field studies of dikes often require much physical effort through walking up steep mountain slopes, river channels, climbing, and similar activities. After such an effort – say walking for several hours in a mountain to reach the desired profile elevation – it is not worth saving the few minutes needed for accurate measurements by making instead quick and crude 'estimates' of, for instance, dike strike, dip, or thickness. Also, the general attitude (strike and dip) of the profile itself should be estimated (accurately measured in the case of short scan-lines). Field measurements of dikes should commonly include the following:

Strike. The strike or direction of every dike should be measured. This is normally easy to do (Fig. 2.2), but should not be done on the dike rock itself. Many dikes, particularly basaltic dikes, have strong **magnetic effects**, so that a compass reading on the dike rock itself may yield a large error. The measurements are best made at a certain distance from the dike exposure. Some dikes generate a local topography, either negative or positive (Figs. 1.11 and 2.2). Other dikes, such as are exposed in steeply dipping walls of cliffs or calderas, neither project out from the wall nor form a depression (Fig. 1.20). The strike may then be difficult to measure. Normally,

Fig. 2.2 The strike and dip of a dike. View west, this regional dike in Northwest Iceland, about 5 m thick, strikes parallel with the coast (roughly east–west) and dips steeply to the north.

however, the contacts are somewhat fractured and eroded and the small depressions thus generated can be used to measure the dike strike. If the dike strike is highly variable in the outcrop, several measurements should be made and the average used. The accuracy of a strike measurement is generally 1–2°.

Dip. The dip of every dike should be measured accurately (Figs. 1.11 and 2.2). Many dikes are essentially straight in vertical sections (Figs. 1.7 and 2.2), in which case it is easy to measure the dip. Some dikes, however, show significant variation in dip in the outcrop. The principal method then is to focus on the segment/part of the dike where the strike was measured and use the dip of that part as representative of the dike as a whole. An alternative method, particularly if the dike is curved over a short vertical distance, is to take several dip measurements and use the average as the dike dip. The margin of the dike, its contact with the host rock (Figs. 1.7, 1.11, 1.20 and 1.21), is generally good for measuring the dip. Normally, the accuracy of a dip measurement is 1–2°.

Thickness. The thickness of every dike should be measured accurately using, when possible, the same part of the dike that is used for the strike and dip measurements (Fig. 1.7). This may not always be possible. For example, the best part of the dike for attitude measurements may be high up in a sea cliff and thus not accessible for direct thickness measurements. Then an attempt should be made to make the thickness measurement as close to the attitude measurements as possible. All dike parts where there is no host rock in-between the parts, that is, all the columnar rows (Fig. 2.3), are regarded as a single dike. The dike parts may converge to a single dike and then diverge to several dikes along their strike- and dip-dimensions. Thus, in one outcrop a dike may appear as a single multiple dike (Figs 2.3 and 2.4), whereas in another outcrop it appears as several dikes, with clear host rock in-between the parts/dikes. The rule is, however, that in the outcrop where the strike and dip are measured, the thickness of a multiple or composite dike is the **cumulative (total) thickness** of all its parts (columnar rows). The accuracy of dike-thickness measurements is normally 5–10% of the thickness. Thus, for a dike thickness given as 1 m, the actual value is 1 m $\pm$ 0.05m (or $\pm$ 0.1 m).

It is important and of value to measure each of the columnar rows of a multiple dike (Figs. 2.3 and 2.4), as well as the parts of a composite dike (Fig. 2.5). But in the statistical data for the entire dike set, a multiple/composite dike should be regarded as **one dike**. The dike-thickness accuracy of 5–10% is also roughly the error involved when the thickness of a dike measured in the field is taken as proxy to the opening/aperture of the **volcanic fissure** at the location of the dike when it was transporting magma. This latter follows from simple considerations of the density change from magma to solid rock; on magma solidification, the volume decreases by about 10%.

Fig. 2.3   Multiple basaltic dike from North Iceland. The dike is composed of seven parts or columnar rows (indicated), with a total thickness of 20 m. View northeast, the dike strikes northeast and dips steeply to the southeast. The person provides a scale.

Multiple basaltic dike in North Iceland, composed of at least three parts, with a total thickness of 54 m. Here the dike is seen as a single multiple dike, with no host rock in-between the parts. However, when the dike is followed along strike, the three parts separate (that is, have host rock in-between them), so that in other outcrops this dike would be regarded as three dikes. View north, the white arrow points to a person for scale.

Length. Dike length is most commonly obtained from images, such as aerial photographs and Google Earth images. The total length is normally difficult to measure because of lack of continuous exposures (Figs. 2.2 and 2.5). However, in desert areas, such as in North Africa, Namibia, the Middle East, and parts of the United States, accurate dike lengths can often be measured. Similarly, in other areas with no trees and generally little vegetation, such as in parts of Antarctica, Canada, Greenland, Iceland, and Scotland, dikes can be traced along parts of (Fig. 2.2) or their entire lengths. All dikes are **segmented** (Fig. 2.6), as applies to tectonic fractures in general. The rule for length measurements of all segmented fractures is the same: the fracture (here the dike) is counted as one so long as there is a **physical connection between the segments** (hard-linked). If there is no such connection between segments, then the segments are regarded as separate dikes in the length measurements. Mechanically, closely spaced (soft-linked), comparatively long segments (in relation to the distances between the segment tips) may function as a single fracture (Gudmundsson, 2011). However, calculations are needed to determine if disconnected segments function as a **single dike/fracture**, so that the operational definition used here is preferable for general work on dikes/fractures. For exceptionally well-exposed dikes, one can walk along their lengths and measure their thickness and structural variations, including segmentation and segment connections, variation in thermal effects on host rock, contacts, lithology, vesicles, flow direction, number of columnar rows (parts of a multiple dike), and other aspects related to the mechanics of emplacement. Reported lengths of dikes are often minimum lengths because the exact lateral ends/tips are commonly not exposed. When the lateral ends are exposed, the accuracy is normally 5–10% of the measured length.

Fig. 2.5 Composite dike in East Iceland. View northeast along the strike, the total measured thickness of the dike is 25.5 m. The thickness of the rhyolite part is 13 m, that of the western basalt part is 7.5 m, and that of the eastern basalt part is 5 m. The central acid part of the dike is more easily eroded than the host rock (a basaltic lava pile) and forms a depression, whereas the basaltic parts form ridges (and are thus more resistant to erosion than the basaltic host rock). The dike can be traced laterally for about 14 km, its strike changing from N20°E here to N14°E towards its northern end. The dike extends to the top of the 700-m-high mountain on the other side of the fjord (seen here), but the basaltic parts drop out in the middle part of the mountain so that at the top the dike is purely acid and 35 m thick. A 120-m-thick multiple sill (composed of numerous columnar rows) dissects the dike, and is thus younger than the dike. A black and white version of this figure will appear in some formats. For the colour version, please refer to the plate section.

Lithology. The rock type of every dike should be recorded (Figs. 1.7, 2.4, and 2.5). For field observations of dikes, this normally includes the determination of whether the rock is mafic, intermediate, or felsic. Most dikes, however, are mafic and, more specifically, basaltic. Commonly, it is possible to distinguish between the more primitive **types of basalt**, such as olivine tholeiite, and the more evolved types, such as tholeiite. As lithological terms, these were introduced by Walker (1959) and do not strictly correspond to petrological classifications, but they may be used as an indication of how primitive and hot/fluid the magma was at the time of dike emplacement. The **grain size** of the dike rock is normally recorded.

Host Rock. The lithology of the host rock of the dike at the location of the outcrop should be recorded (Figs. 1.7, 1.18, 1.20, and 1.21). In a volcanic area, the host rock is commonly composed of lava flows or pyroclastic layers, except inside the sheet swarms of central volcanoes and in sheeted dike swarms or complexes of ophiolites. In the latter, the host rock is primarily made of other dikes/inclined sheets. Some dikes also intrude sedimentary and metamorphic rocks. The host rock responds to dike injection in different ways depending on the rock properties. Some sedimentary rocks develop dike-induced fractures in zones parallel to the dike margins, similar to damage zones in fault zones. Igneous rocks, such as lava flows,

Fig. 2.6 All dikes are segmented. Here six segments of the 5-m-thick dike in Fig. 2.2 are numbered. This dike is more resistant to erosion than the host rock, and therefore stands as a segmented ridge above the surroundings. A black and white version of this figure will appear in some formats. For the colour version, please refer to the plate section.

rarely develop such damage zones, but fractures – commonly presented as mineral veins – are common in the host rock next to dikes. The dike **thickness may vary abruptly** between host-rock layers with widely different Young's moduli or stiffness. At the contact between the dike and the host rock, the dike rock may be glassy – forming a so-called **chilled selvage** – if the temperature difference between the magma and the host rock was very great at the time of dike emplacement and the margin of the dike cooled down very rapidly (cf. Gudmundsson, 2017). While the dike-rock grain size is normally finer near its contact with the host rock, a chilled selvage is commonly missing from dikes/sheets in ophiolites and sheet swarms.

The strike, dip, thickness, lithology, and host rock **should be recorded** in any systematic dike study. They provide the basic statistical data that help us understand the process of dike emplacement and dike-path formation. When possible, the length should also be recorded but, as indicated above, the length is commonly difficult to measure because of lack of continuous exposures. There are several additional items that are commonly measured in detailed studies of dikes. Some of these, such as dike tips, should always be measured, but are less commonly observed.

Does the Dike Form a Ridge or a Depression? Is the dike more resistant or less resistant to erosion than the host rock? Many dikes form a topographic depression, indicating that they are **less resistant** to erosion, that is, more easily eroded, than the host rock (Figs. 2.5 and 2.7). Such dikes are often recognised on aerial photographs and other images since they commonly form (unusually straight-line) parts of river channels (Fig. 2.7). Alternatively, dikes may be of the same resistance to erosion as the host rock (Figs. 1.20 and 1.21) or be **more resistant** to erosion and thus form ridges (Figs. 1.11, 2.1, and 2.2). There are many factors that determine the relative resistance to erosion of a dike and its adjacent host rock.

Fig. 2.7   Many dikes are more easily eroded than the host rock and are marked by topographic depressions such as river channels. When the host rock and the dike are similar in composition, like here for a regional basaltic dike in a basaltic lava pile in Northwest Iceland, the most easily eroded parts are commonly at the fractured (due to stress concentration) dike margins. Here the central part of the dike forms a cliff while the marginal parts are eroded and constitute the river channel.

These include the fracture patterns (such as the columnar joints), the alteration (particularly geothermal alteration), and the lithologies of the dike rock and the host rock.

Columnar Joints. Columnar (cooling) joints (Figs. 2.6 and 2.8) form perpendicular to the cooling surfaces (the contacts of the dike with the host rock, the dike-fracture walls) and are thus horizontal in vertical dikes. A detailed study of the joints is important for several reasons. First, the columnar joints commonly form '**rows**' ('sheets' might be a better description, but could be confused with the generic meaning of sheet intrusions and, in particular, with inclined sheets). Each row may correspond to a single **magma injection**, and together they constitute a **multiple dike** (Figs. 2.3 and 2.4). The columnar rows indicate how many magma injections formed the dike (Fig. 2.8). The chilled selvage at the contacts between the rows, or the absence of such a selvage, indicates the total time it took to form the multiple dike (cf. Chapter 6). If there are no chilled selvages or glassy margins between rows, the time between magma injections must have been very short, suggesting that the multiple dike may have formed in a **single rifting event** or episode (during months or years, but not decades or centuries). Second, the frequency and the configuration of the joints, including cross-cutting joints, are an indication of the potential of the dike to transport fluids and to act as a fractured **fluid reservoir**. Dikes, particularly thick ones (Fig. 2.4), can store and conduct various crustal fluids such as groundwater and geothermal water, and the same applies to sills (Gudmundsson and Lotveit, 2012; Barnett and Gudmundsson, 2014). The ability of the intrusions to store crustal fluids depends to a large degree on the configuration of the columnar joints.

Fig. 2.8 Columnar rows in dike segment 6 in Fig. 2.6. The dike has three rows, suggesting a maximum of two main magma injections. The upward bending of the central columns in this segment may indicate late-stage movement of the (by then partly solidified and thus 'plastic') magma (cf. Gudmundsson, 1986).

Vesicles. The size, shape, and location of the vesicles in the dike rock give important information on dike emplacement. Since vesicles are '**frozen gas bubbles**' and are the result of gas escaping from the flowing magma in the dike, they can be used as a measure of degassing of the magma. Just as in lava flows, the shapes of the vesicles are a measure of the viscosity of the magma. Highly **elongated** and, especially, **angular** vesicles indicate comparatively high-viscosity magma, whereas **circular** or slightly **elliptical** vesicles indicate low-viscosity magma (the latter being common in primitive basaltic dikes, for example). Vesicles sometimes form one or more **bands** or zones in dikes, particularly in feeder-dikes close to the surface (Galindo and Gudmundsson, 2012; Gudmundsson, 2017). The configuration and arrangement of vesicles can be used to infer **magma-flow directions** (Marinoni and Gudmundsson, 1999). Still more detailed information on flow directions can be obtained by various other kinematic indicators (Baer, 1995) and by magnetic anisotropy studies (Kissel et al., 2010; Eriksson et al., 2011). Such methods, however, require special techniques and are not among the standard field techniques used in dike studies.

Secondary Minerals. These are generated through geothermal activity within and around the dike following its emplacement. The secondary minerals are of two main mechanical types: amygdales and veins (Fig. 2.9). Most of the mineral **veins** are extension fractures and thus are good indicators of the state of stress in the dike and its vicinity subsequent to the dike emplacement. The veins are also an indication of the **permeability** of the dike and host rock while they were a part of an active geothermal field (strictly palaeopermeability). The **amygdales** are primarily filled vesicles. The mineral type, which can often be identified in the field, is an indication of the composition of the dike rock (and the host rock). Thus,

Fig. 2.9 Mineral veins and amygdales seen at about 1.5 km depth of erosion (in a basaltic host rock) in the Husavik–
Flatey Fault, and oceanic transform fault exposed on land, in North Iceland (cf. Gudmundsson, 2011).

zeolites tend to occur mainly in primitive basaltic rocks, such as olivine tholeiite, whereas quartz amygdales and veins, for example, tend to occur in more evolved basalts (quartz tholeiites) as well as in intermediate and acid rocks. The **zonation** of secondary minerals, particularly zeolites, can be used to indicate the depth in the lava pile at which the minerals formed. This method, developed by Walker (1960), has been used extensively in basaltic lava piles to estimate the depth of erosion. Thus, the estimated **depth of erosion** in Iceland, reaching its maximum of about 2 km in Southeast Iceland, is partly based on the secondary minerals found in the lava pile and in the dikes in this area. This method is of great importance for knowing the depth of dike outcrops studied and to infer the depths of the many plutons (interpreted as the tops of shallow magma chambers) exposed in Southeast Iceland.

Phenocrysts, Xenocrysts, and Xenoliths. For any detailed lithological or subsequent petrological studies, phenocrysts, if they occur, should be recorded. **Phenocrysts** (large crystals, distinct from the groundmass) can be used as one measure of the **viscosity** of the magma (Takeuchi, 2004). Some phenocrysts, particularly high-density crystals such as olivine and pyroxene, can also be used to infer aspects of the fluid dynamics of magma during the emplacement of the dike. Xenoliths and xenocrysts give information about the host rock through which the dike passed on its path to the present outcrop. The host-rock fragments, the **xenoliths**, can also indicate aspects of the fluid dynamics of the magma transport in the dike. **Xenocrysts** are similar to phenocrysts but the crystal is foreign to the magma in the dike. The main criteria used to determine the direction of flow of magma in dikes are summarised by Marinoni and Gudmundsson (1999).

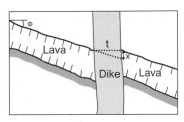

Fig. 2.10 Apparent vertical displacement of a lava flow across a vertical dike gives the false impression of the dike being, or following, a shear fracture. An apparent displacement of this type is common when dikes dissect lava flows (or other layers or intrusions) at angles less than 90°. The apparent vertical displacement x is then given by $x = t \tan\varphi$, where t is the dike thickness and φ the dip of the lava (or other structure dissected by the dike). For example, if the lava flow is dipping 8° and cut by a 5-m-thick vertical dike, x would be 0.7 m.

Faults and Dikes. Does the dike occupy a fault? Is the dike itself faulted? These questions are very important for establishing the age relationships between faults and dikes in the area. Most importantly, however, the question of whether or not the dike occupies a fault relates to the mechanics of dike and sheet emplacement in general. Some dikes use faults for **parts of their paths,** partly for the reason that active or recently active faults have close to zero tensile strengths and thus were easily opened by the magmatic pressure of the dike. When the fault is steeply dipping, as many normal faults are close to the surface in rift zones, then the dike may find it easier to follow the existing fault, at least along part of the dike path (Chapter 10). Statistical data on how common this is contribute to a better understanding of the **propagation paths** of dikes (Chapters 7 and 10) and the related hazards during unrest periods. It is important, however, to distinguish between dikes (and sheets in general) that occupy real faults, and those that only **appear** to occupy faults. It is common for all structures that meet at angles of less than 90°, particularly cross-cutting dikes and dikes cutting lava flows and pyroclastic/sedimentary layers, to give an **appearance of displacement** where there is none (Fig. 2.10). Much of the minor displacement of lava flows or other marker layers across dikes can be explained in this way – by the dikes dissecting the lava flows/pyroclastic layers at less than 90°. However, dikes that occupy real faults, normal, reverse, or strike-slip, should be recorded as the data is important for understanding better the general mechanics of dike emplacement and path formation (Chapter 10).

Contact Characteristics. Many dikes conduct **fluids** along their contacts with the host rock. It is thus important to record whether the contacts are completely sharp and welded to the host rock (Fig. 1.7) or, alternatively, fractured (Fig. 2.7). One should also look for evidence of small vein-like intrusions from the dike at the contact (and into the host rock) and describe and record the overall geometry of the contacts. For example, are the contacts straight, wavy, or do they show abrupt changes (dike thickening or thinning) when the dike passes from one bed/layer to the next one (cf. Fig. 1.18)?

Dike Tips and Feeder-Dikes. Detailed studies and measurements should be made of all observed dike tips, in lateral sections and, in particular, in vertical sections. Most vertical tips seen in the field belong to **non-feeder-dikes**, that is, dikes that did not reach the surface to supply magma to an eruption but rather became arrested at a depth in the rift zone or volcanic edifice (Geshi et al., 2010). While some dikes taper away in vertical sections, most dikes become arrested at contacts between mechanically dissimilar layers, such as soft or compliant pyroclastic layers and stiff lava flows (Fig. 2.11). The **geometry** of the tip – rounded, blunt, or a narrow fracture – indicates the mechanical properties of the host rock as well as the magma overpressure and the state of stress in the host rock at the time of arrest. The **proportion** of dikes seen arrested in a dike swarm exposed in a vertical section is particularly important for volcanic hazard assessments since it is a measure of the probability of an injected dike reaching the surface. All observed **feeder-dikes** should be studied in detail and measured, particularly their connection with the unit or layer (lava flow, intrusions, pyroclastic layer) to which they supplied magma. Well-exposed feeder-dikes are rare and worth great effort to study in detail (Fig. 1.21).

Photographs. Photographs should be taken of all dikes seen as important and unusual. Certainly, all **dike tips** and all **feeder-dike** connections should be photographed in detail. It is usually best to take several photographs of every dike deemed worthy of a photograph in the first place. As a minimum, an overview photograph, showing the relationship between the dike and the surrounding host rock (and perhaps other dikes and structures) should be taken together with several close-ups of the main features of interest. All photographs should have a scale. For detailed features of interest, such as dike tips or dike connections with lava flows, **measuring tapes** are the best scales. The direction in which the photographer is looking when taking the photograph should always be indicated. In this book that direction is indicated by the phrase '**view**' north (or east, or south, or west, as the case may be).

2.3 Basic Data Analysis and Presentation

While in the field, it is important to make some basic data analysis. This follows for several reasons. First, simple statistical plots indicate the **general attitude** of the dikes in the area. The strike distribution, for example, may indicate the lack of profiles/sections in certain directions or, in general, gaps in the data set. Second, a histogram showing the frequency, for example, of the **thicknesses** of dikes may indicate systematic errors in the measurements. If the dike-thickness histogram indicates a normal distribution, a bell-curve, then there is likely a systematic error in the thickness measurements. This follows because all thickness measurements of dikes made so far follow heavy-tailed or, in general, **power-law** distributions, rather than normal distributions. (These distributions are discussed in greater detail below.) Third, the statistical plots could indicate some very interesting features regarding the strike, dip, or thickness, or any other quantities related to the measured dikes that would be

Fig. 2.11 An arrested dike. This basaltic dike is arrested about 5 m below the surface of the active Holocene rift zone in Southwest Iceland. The dike is subvertical and strikes northeast, that is, parallel with the rift zone itself. The dike cuts through a layer of comparatively soft tuff (a pyroclastic layer) and becomes arrested at the contact between the soft tuff and the stiff Holocene basaltic lava flow forming the surface of the rift zone. The dike thickness decreases from about 0.34 m at the bottom of the exposure to about 0.1 m at the tip. The feeder-dike to the lava flow itself is a short distance to the west of this arrested dike (cf. Gudmundsson, 2017).

worthwhile to explore while still in the field. Fieldwork is often done in remote areas that are difficult, time-consuming, and costly to access. Thus, basic statistical analysis of the measured field data while in the field can save time, effort, and money.

Most scientists who do fieldwork today carry with them computers, usually laptops. Field computers should contain basic statistical and geological programs. For the type of statistical analysis described here, simple Microsoft Excel spreadsheets are the starting points. These can be used for **histograms** (binning) as well as for finding best lines or curves through data sets (linear and non-linear **regression analyses**). Specifically for dike and other fracture studies, programs that generate **stereograms** and **rose diagrams** are indispensable. What follows is a brief description of some of the main techniques used for analysing the different types of data obtained in the field.

Strike. It is best to present the data first as a histogram, to **bin** the data (Fig. 2.12). There are various ways of recording strike. Some use the upper half of the circle, so that the strike ranges from 270° (or –90°, or N90°W) to 90° (or + 90° or N90°E). Some use the right half of the circle, so that the strike ranges from 0° to 180°. Still others, particularly in Scandinavia, use the so-called '**right-hand rule**'. Then the dip of every dike (every structure) is regarded as being to the right, in whichever direction the person is looking when measuring the dike. It follows that the recorded strike then ranges over the whole circle, that is, from 0° to 360°. Whatever method is used to record the measurements, a histogram, usually with class or bin limits or widths of either 5° or 10°, is a very clear graphical presentation of the strike distribution (Fig. 2.12).

The favoured presentation of strike among earth scientists is the **rose diagram** (Fig. 2.13a,b) or, alternatively, **circular histograms**. Again, the class sizes or widths of the bins or sectors of the rose are chosen depending on the data set, but normally using either widths of 5° or 10°. The roses can represent complete circles or semicircles. The rose diagrams can be used to present not only the strike of structures such as dikes but also the **direction of processes**, such as the flow of subsurface magma in dikes or on the surface as lava flows, as well as the direction of flow in rivers and the various wind directions. The roses can be presented in different ways, depending on the type of data and the aim of the presentation. These include the following:

- Roses showing **directional data**. For directional data, one end of the structure/process is **distinguished** from the other end. For example, in a lateral magma-flow direction in a dike to the north from a magma chamber, the north end or tip (where the magma flow stops) is distinguished from the south end (at the chamber where the magma flow starts). Similarly, in the flow of lava in a tube or, alternatively, from a fissure or a central crater downslope a volcanic edifice, one end (at the source crater or fissure) can be distinguished from the other end (where the flow stops). Roses for directional data are unidirectional or

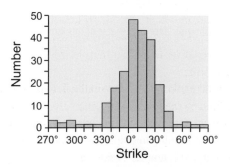

Fig. 2.12   Histogram for the frequency distribution of dike strike. Here, the bin width (the class limits) is 10°, whereas for other data sets different widths, such as 5° or, rarely, 20° might be more suitable. In this case, most of the dikes strike north to north–northeast, and the distribution is crudely normal.

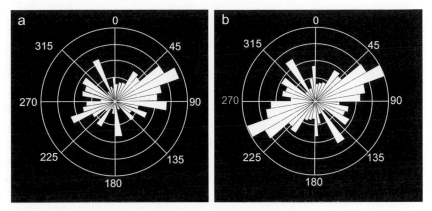

Fig. 2.13 Rose diagrams are of two basic types: asymmetric (unidirectional) and symmetric (bidirectional). (a) An asymmetric rose is for directional data where one end of the structure or process can be distinguished from the other end. (b) A symmetric rose is for oriented data where there is no distinction between the ends of the structure or process. Symmetric roses are used for most crustal fractures, including dikes and faults.

asymmetric; opposite classes or sectors (180° apart) of the rose are normally different and the complete circle must be used (Fig. 2.13a).

- Roses showing **oriented data**. For oriented data there is **no directional distinction** between the ends (Fig. 2.13b). This applies to most crustal fractures, including dikes. When measuring a dike striking, say, N45° E, it normally does not matter if we measure it looking northeast (view northeast) or southwest (view southwest). The strike will be the same, while the exact methods of recording the strike may differ, as explained above. The only exception to this would be if the origin of the dike could be clearly determined as being in an active or fossil magma chamber. Then it might be useful to distinguish one end (the origin) from the other (the lateral tip) so as to determine the propagation direction and/or the magma-flow direction (these may differ widely on a local scale, however). Roses for oriented data are bidirectional or **symmetric** so that the opposite classes or sectors (180° apart) are equivalent. Either the complete circle or a semicircle may be used to present the rose.
- Roses showing **unweighted** (non-normalised) data (Fig. 2.14a). This is the most common presentation, in which case the strikes are shown without taking into account the lengths of the dikes. Thus, for unweighted presentation, every dike (or fracture) counts the same irrespective of its length so that the rose represents the true frequencies of dikes with different strikes.
- Roses showing **weighted** (normalised) data (Fig. 2.14b). Dike lengths can be normalised or weighted by the length of the shortest dike, which is then regarded as the **unit length**. It follows that more weight is given to the longer dikes because they are regarded as being composed of many short segments. The unit segment length is the length of the shortest dike so that, for example, a dike that is 10 km long counts 10 times more than a dike that is 1 km long, and so on. If the main aim of the study is to infer the regional stress field, then the weighted presentation of the dike strike is normally better. This is because long dikes (or long tension fractures or long normal faults, if these are being studied) indicate better

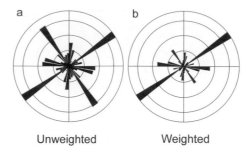

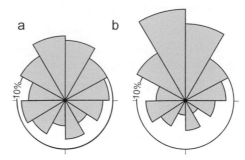

Fig. 2.14   Rose diagrams of lineaments can also be presented as unweighted and weighted. (a) When unweighted (non-normalised) the strike distribution is directly based on the number of data, irrespective of their sizes (e.g. the lengths of fractures). (b) When weighted (normalised), the strike distribution takes the size, normally the length, into account so that long fractures weight more than short ones. More specifically, the length is then normalised by the shortest fractures, regarded as of unit length, so that each long fracture counts in proportion to how many times longer it is than the shortest one.

Fig. 2.15   Rose diagrams are sometimes made so that (a) the sector frequency is proportional to the sector area. (b) More commonly, however, the sector frequency is proportional to the sector radius.

the time-averaged direction of the minimum compressive principal stress, σ_3 (the direction of the local **spreading vector** at divergent plate boundaries), than short dikes (or tension fractures or normal faults).

- Roses showing the **area of the sector** or class as proportional to the class frequency (Fig. 2.15a). In analogy with ordinary histograms or bin plots, this is the logical choice (Nemec, 1988). Here the increasing area of the sector with distance from the centre of the circle is taken into account so that the area shown reflects the number of data that fall within that sector. However, it commonly makes the main orientations poorly presented and the entire rose somewhat unclear.
- Roses showing the **radius of the sector** or class as proportional to the class frequency (Fig. 2.15b). This presentation is not 'area correct' but emphasises the main orientations. This is the more common type of rose and is used in **this book** unless otherwise stated. This presentation is fine so long as it is clear – for example, stated in the text or in the captions – that it is not the area but rather the radius of the sector that represents the data frequency.

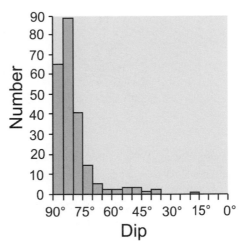

Fig. 2.16   Histogram for the frequency distribution of dike dip. Here, the bin width (the class limits) is 5°, whereas for other data sets different widths, such as 10°, may be more suitable. Most dikes here are steeply dipping, between 80 and 90°, the arithmetic mean dip being about 81°.

Dip. Dips can be well presented by **histograms** or bin plots (Fig. 2.16). In contrast to histograms for the strike, where the range can be 0–360°, the range of the dip is always 0–90°. There are no dips less than horizontal (0°) and no dips greater than vertical (90°). The dips, however, may be in various **geographical directions**. Theoretically, it is then possible to show the dips in different directions on separate histograms. For example, in a study of a north-striking dike swarm, one histogram could show all the dikes dipping in an easterly direction, that is, to southeast, east, and northeast. For the same swarm, another histogram could show all the dikes dipping in a westerly direction, that is, to southwest, west, and northwest.

The dips and the **dip directions** can also be presented as rose diagrams. Then each sector shows the frequency of dikes dipping within that class. For example, the sector 80–90° would show the number of dikes dipping in these directions, that is, close to east. Similarly, the sector 180–190° would show the number or frequency of dikes dipping in these directions, that is, close to south. However, these types of rose diagrams are not much used. It is more common to show the strike and dip together, namely the attitude of the dikes, when one wants to present the dip directions. The attitude is best presented using **stereoplots.**

Attitude. Stereographic projections using stereonets are the most common presentation of attitudes of structures such as beds/layers and fractures of every type (including, of course, dikes). Sometimes the planes themselves are presented; for example, the plane of a dike or an inclined sheet. It is more common, however, to plot the poles to the planes, such as the poles to dip-slip fault planes, inclined sheets, and dikes (Fig. 2.17). The plot involves presenting the **poles** to all the fracture planes (here of dikes and sheets and sills) and plotting them as points at the surface of a half-sphere, namely the **lower hemisphere**. While this presentation requires some three-dimensional geometric visualisation on behalf of the reader, it provides a very clear picture of both **the strike and the dip**, that is, the **attitude** of the fractures such as dikes and inclined sheets.

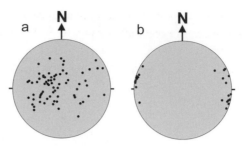

Fig. 2.17   Pole plot for sheet intrusions. (a) Plot for local inclined (cone) sheets, showing that most of them have shallow dips (the points, that is, the poles, are close to the centre of the plot). (b) Plot for regional dikes, most of which are subvertical, as is indicated by their poles being mostly close to the margin of the plot (where the dip is 90°). The poles of the regional dikes also show that their strike is mostly north–northeast.

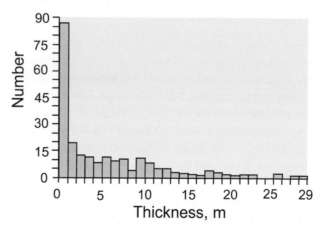

Fig. 2.18   Histogram for a dike-thickness frequency distribution. Here, the bin width is 1 m, but for inclined sheets, for example, a bin width of 0.5 m would be more suitable – because their thicknesses and thickness range are smaller than those for regional dikes. Here the dikes range in thickness from less than 1 m to a maximum of 29 m, with an arithmetic average of 5.4 m. The frequency distribution follows approximately a power law.

Thickness. Thicknesses of dikes (and sheets and sills) as well as openings (apertures) and displacements (slips) of fractures are best represented as histograms or bin plots (Fig. 2.18). The histograms are either presented using the data frequency directly or a cumulative frequency plot. When using the data directly, the classes are selected based on the range in thickness. For example, when analysing regional dikes that range in thickness from tens of centimetres to tens of metres, a suitable class limit or bin width would be 1 m or 2 m. By contrast, when analysing local swarms of inclined sheets that range in thickness from centimetres to perhaps 10 metres, a suitable class limit would be 0.5 m.

In a **cumulative plot**, the data are reorganised so that the classes or bins represent dikes with thicknesses in **excess of** a certain limit. For example, when analysing the thickness data of a regional swarm, the first class or bin would include all thicknesses in excess of 0 m (so all the data), the next bin could include all thicknesses in excess of, say, 1 m, the next bin, all thicknesses in excess of 2 m, and so on. Similar reorganisation of the data is commonly used for fracture apertures and fault displacements (slips), as well as for fracture lengths.

One primary reason for reorganising the data into cumulative presentations is that they are easier to handle in that way when the data are presented as **bi-logarithmic plots**. The latter, also referred to as **log–log plots**, are much used for analysing fracture lengths, fracture openings (dike thicknesses), bed/layer thicknesses, as well as many other data (Gudmundsson and Mohajeri, 2013). It turns out that data of these types commonly show heavy-tailed cumulative size distributions, many of which are approximately **power-law** size distributions. Power-law size distributions also apply to sizes or volumes of eruptions, magnitudes of earthquakes, sizes of floods, and many other processes and hazards in geology (cf. Gudmundsson, 2017).

Length. Lengths of dikes, including those of feeder-dikes and associated volcanic fissures, are also well presented by histograms (Fig. 2.19). For dikes that range in lengths from several hundred metres to about 10 kilometres, the class limits or bin widths of 200 m or 500 m would be suitable. Some dikes are much longer, reaching hundreds of kilometres, but these are rare in relation to the most common lengths, which tend to be hundreds of metres to, at most, several tens of kilometres. **Cumulative plots** are very commonly made for fracture lengths, including dike lengths, although comparatively few detailed and systematic dike-length studies have been made. This is partly because such studies require exceptionally good exposures of dikes over large regions, such as are mostly found in desert areas. The lengths generally show heavy-tailed distributions and allow for tests to be made of whether or not they follow **power-law** size distributions.

It is important to understand and be able to forecast length and thickness distributions of dikes, particularly feeder-dikes. This is because these are direct measures of the lengths and apertures of volcanic fissures during eruptions (Fig. 2.19). More specifically, the horizontal cross-sectional area of a feeder-dike largely controls the **volumetric flow rate** (effusion rate for effusive eruptions) through the volcanic fissure that the dike feeds. The maximum effusion rate, as well as the overall lava or pyroclastic volume, depends strongly on the geometry of the feeder-dike, a topic to which we turn now.

2.4 General Shape

Dikes show a variety of geometric forms. All dikes are **segmented** and the segments vary in shape (Fig. 2.6). Some segments are straight and with parallel contacts, so that the thickness is essentially constant over a distance of tens or hundreds of metres (Figs 1.7, 1.11, and 2.6). Other segments are wavy but with parallel margins. Still other dikes have

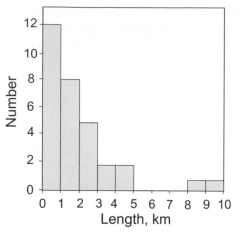

Fig. 2.19 Histogram for the frequency distribution of the lengths of volcanic fissures. Here, the bin width is 1 km. The fissures, which are the surface expressions of feeder-dikes, range in length from a few hundred metres to about 10 km. These Holocene fissures from Iceland have an arithmetic average length of about 2 km.

geometry akin to a pinch-and-swell structure, or change their thicknesses abruptly at contacts between mechanically dissimilar host-rock layers (Fig. 1.18). Apart from the general geometry of the dikes and dike segments, there are many smaller geometric features associated with the dikes. Here, we list and describe briefly some of the main geometric characteristics of dikes.

Most dike segments, and dikes in general, have the overall approximate shape of a flat ellipse. This applies to their shape in lateral sections and, to a lesser extent, in vertical sections (Fig. 2.20). While there are understandably many exceptions (Kavanagh and Sparks, 2011; Daniels et al., 2012; Rivalta et al., 2015), detailed measurements confirm that the **flat ellipse** is a common approximate shape of many dikes, particularly in lateral sections (Fig. 2.21; Pollard and Muller, 1976; Delaney and Pollard, 1981; Pollard and Segall, 1987). Shape variation is greater in vertical sections, primarily because of layering. Thus, at layer contacts, the thickness of the dike commonly changes abruptly, either becoming thinner or thicker, depending primarily on the change in the Young's modulus of the layers across the contact. Non-feeders may be crudely flat ellipses, particularly within individual layers or units (Geshi et al., 2010; Kusumoto et al., 2013). Many feeder-dikes become **thicker** (the aperture increases) on approaching the Earth's free surface (Geshi et al., 2010; Geshi and Neri, 2014), while others become thinner on approaching the Earth's surface (Galindo and Gudmundsson, 2012). Further discussion of the variation in dike thickness with depth is provided in Section 2.8.

A flat ellipse is the shape expected of any fluid-driven fracture hosted by rocks whose elastic properties are **constant** along the dip- and strike-dimensions of the fracture, and where the fluid overpressure of the fracture is also constant. The elliptical shape applies to other fluid-driven fractures, such as mineral veins, formed under the similar mechanical conditions (Gudmundsson, 2011; Philipp, 2012; Kusumoto et al., 2013; Kusumoto and

Gudmundsson, 2014). Deviations from the shape of a flat ellipse, however, are common among dikes. Most such deviations relate to variations in the mechanical properties, particularly in the Young's modulus, along the strike- and dip-dimensions of the dike, as well as variations in the dike overpressure. Some of these aspects are discussed and illustrated further below, and also in Chapter 7.

Dike segments are of very different lengths, but the general **length–thickness** (or strike-dimension/thickness) **ratio** shows a comparatively small variation. Measurements of regional dikes in Iceland indicate a length–thickness ratio from about 300 to about 1500. The most common value is about **1000**. All these dikes were measured in the same type of host rock, a basaltic lava pile. From the length–thickness ratio of a dike and the elastic properties of the host rock, the magmatic overpressure (driving pressure) at the time of the dike emplacement

Fig. 2.20   Most dike segments have the two-dimensional shapes of flat ellipses in cross-sections, as seen here, and as flat oblate ellipsoids in three dimensions. This basaltic dike segment, from Santorini (Greece), has a maximum thickness of about 1 m. The segment is a part of a larger, segmented dike. The top part of another, offset segment is indicated.

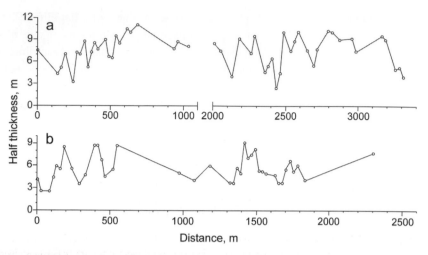

Fig. 2.21 Thickness variation (shown as half thickness) along parts of two regional dikes in Iceland. (a) The dike segment dissects a basaltic lava pile with many soft soil and scoria layers in-between the lavas. Many of the thickness changes are related to stiffness differences between the lava flows and the soil and scoria layers. (b) This dike segment dissects mostly basaltic lava flows and thus shows less variation in thickness (Gudmundsson, 1983).

can be calculated (Example 2.4; Chapter 7). Since basaltic lava flows tend to be stiff (with a high Young's modulus; Chapter 3) in comparison with many sedimentary and pyroclastic rocks, the length–thickness or **aspect ratio** is generally lower for dikes of a given over-pressure in sedimentary and soft pyroclastic rocks (Becerril et al., 2013).

The observed length–thickness aspect ratio depends also on other factors. One factor is the **depth** of the dike exposure below the original surface of the volcanic zone/edifice within which the dike was emplaced. For example, the dikes in Iceland with a typical aspect ratio of about 1000 are mostly exposed at crustal depths of around 800 m below the original top of the lava pile. In a given volcanotectonic regime, dikes tend to become **thinner and longer** at greater crustal depths (Fig. 2.22). This follows partly because the average Young's modulus of crustal layers or units tends to increase with depth. Thus, for a given overpressure, the dikes tend to be somewhat thinner with depth and, to conserve volumetric flow rate for a vertical flow of magma, correspondingly longer.

A second factor is that dikes propagate in different ways, which may be reflected in their measured length–thickness ratios (Fig. 2.23). Dikes (and inclined sheets and sills) commonly become offset at contacts between mechanically dissimilar rocks, sometimes on meeting stress barriers (Chapters 5 and 7). For feeder-dikes, or dikes that become arrested or deflect into sills at mechanically weak (with low tensile strength) or open contacts, the strike dimension – which is normally the measured length in the field or on images –controls the thickness of the dike (Example 2.4; Chapter 7). More specifically, when the upper and lower ends of the dike, that is, the tips, are in contact with free surfaces, then the length is the **controlling dimension**.

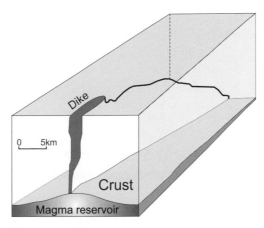

Fig. 2.22 Schematic illustration of the variation in regional dike thickness and length with depth in the crust (below the uppermost 1–3 km of the crust, where the variation is more irregular). Only half of the dike is shown here. Partly because the average Young's modulus gradually increases with increasing crustal depth, dikes tend to become longer and thinner with depth. The scale is approximate for dike length and crustal thickness, and the dike thickness is much exaggerated (Gudmundsson, 1990a,b).

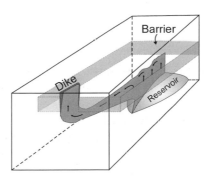

Fig. 2.23 When a dike that is injected vertically from a magma reservoir meets a layer with unfavourable local stress, a stress barrier, the dike may become arrested, or change into a sill. Alternatively, the dike may propagate primarily laterally for a while beneath the stress barrier till the end of the barrier where the dike may, again, propagate primarily vertically. Only part of the dike and the source reservoir are shown here. In this example, the thickness of the dike would be controlled by its dip-dimension (height). Normally, dikes propagate both laterally and vertically in any given layer or rock unit (Gudmundsson, 1990a,b).

(Recall that here a **free surface** is a solid rock surface in contact with a fluid. The contact is then with the atmosphere or water at the Earth's surface or an open fluid-filled discontinuity in the subsurface in general, and with the magma of a chamber or a sill in particular.) For other dikes, however, the controlling dimension is not the strike-dimension but rather the dip-dimension, that is, the height of the dike above its magma source (chamber or reservoir). Many dikes exposed at depths in the crust are non-feeders (Geshi et al., 2010; Kusumoto et al., 2013) and were not in contact with two free surfaces at the time of their emplacement. For these dikes, the dip-dimension is the controlling dimension. Thus, in any one swarm of dikes, the thicknesses of some dikes are controlled by their strike-dimensions whereas the thicknesses of other dikes in the same swarm are controlled by their dip-dimensions. It follows that, even if there is correlation between dike thickness and length (strike-dimension), this correlation is rarely very strong.

2.5 Segmentation and Linkage

All dikes are segmented. This is seen in lateral and vertical sections (Figs. 2.6 and 2.20). In **vertical sections**, the segmentation is commonly at **contacts** between layers, particularly between mechanically dissimilar layers. Some **offsets** are small in relation to the dike thickness, primarily a slight displacement or irregularity in the dike path (Fig. 1.7). There is normally a clear igneous linkage between the offset parts. Other offsets are large in relation to the dike thickness, and commonly without clear igneous linkage between the offset parts (Fig. 2.20). In such cases the linkage between the segments may be located away from the outcrop (for example, inside the cliff where the dike segments are observed) and thus not seen. Although it is possible for a dike to propagate as completely disconnected segments or 'fingers' for long distances, the dike segments are usually linked somewhere inside the rock mass.

There are principally two mechanisms or fracture types by which dike segments are linked. One is the hook-shaped fracture, the other is the transfer fault. **Hook-shaped fractures** are primarily extension fractures that form curved links between the offset segments (Fig. 2.24). The theoretical reason for their formation has been studied extensively in fracture mechanics, as well as in structural geology. They are not limited to dikes. In fact, hook-shaped (primarily) extension fractures are well known for linking various types of fractures in extensional regimes such as at divergent plate boundaries and in rift zones. For example, hook-shaped fractures link large tension fractures and normal faults in rift zones. Also, on a much larger scale, hook-shaped fractures as long as tens or hundreds of kilometres link ocean-ridge segments through structures known as '**overlapping spreading centres**'.

Some hook-shaped parts of dikes form direct links with the main dike segments, while others do not make that direct connection, as seen in the outcrop (Fig. 2.24). Again, inside the rock body the physical links may be complete between the hook-shaped part and the main segments of the dike. This type of linkage is commonly known as '**soft-linked**' in structural geology, in particular in relation to the geometries of normal faults. By contrast, in structural geology, the transfer faults are referred to as '**hard-linked**'. Note, however, that most hook-shaped linkages are so close to the nearby segment (the segment towards which the hook-shaped fracture is propagating) that the segments function mechanically as a **single fracture** even if there is no (or apparently no) physical connection (Fig. 2.6). This follows from general stress analysis of segmented extension fractures (Gudmundsson, 2011).

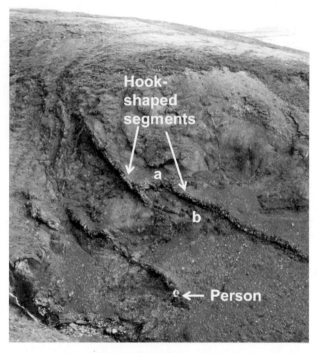

Fig. 2.24 Hook-shaped linkage between segments of an inclined basaltic sheet in Southwest Iceland. The upper segment (a), dipping 28°S, has a thickness of 0.8 m, whereas the lower segment (b), dipping 52°S, has a thickness of 0.9 m. At the location of label a the linkage is complete between segments (hard-linked), but at b, the linkage is incomplete (soft-linked) and the segment ends as a horn (cf. Fig. 2.26). The host rock is basaltic breccia. The person provides a scale.

The second mechanism of linkage of dike segments (and segments of extension fractures and normal faults) is through **transfer faults**. As the name implies, these are faults. However, most are not only shear fractures (faults) but also extension fractures. That is, they are hybrid or **mixed-mode** fractures. Transfer faults are well known for generating linkages between segments of large and small normal faults in rift zones. But transfer faults occur also on a much larger scale, namely with lengths of tens or hundreds of kilometres, linking offset ridge segments. Such large transfer faults, which primarily but not exclusively occur on the ocean floor, are named **transform faults**. Transform faults, like many other transfer faults, contain a large component of extension. In fact, most transform faults and their fracture-zone extensions are large graben structures (Fox and Gallo, 1986).

Fig. 2.25 Dike changing into a sill along part of its path in Iceland. The change occurs at the contact between mechanically dissimilar rocks, the contact itself being composed of scoria and comparatively soft (along which the sill is deflected), whereas the layer above (above the present sill) is stiff basaltic lava flow. The horizontal length of the sill is about 8 m. The vertical dike segments are about 0.8 m thick (cf. Fig. 1.18). The person provides a scale.

Many transfer faults, like dike segments, form along existing discontinuities, primarily contacts. When the magma propagates along the contact and links the two dike segments, a sill forms (Fig. 2.25). There is a clear analogy between transform faults linking ridge segments and, on a much smaller scale, transfer faults linking segments of an extension fracture. In all cases, the movement is initially primarily shear, hence the term transfer fault. However, if the sill expands, as described in Chapter 5, the main movement along the transfer fault may cease to be shear and becomes an **extension**. Such a development may occur during dike linkage (Fig. 2.25) and is common during the development of tension fractures in rift zones.

All hook-shaped and transfer linkages between dike segments (and between segments of tension fractures and segments of normal faults) are partly due to stresses generated by a **crack–crack interaction.** This interaction results in two main changes as regards the stress field (Melin, 1983; Pollard and Aydin, 1984; Gudmundsson, 2011). First, shear and tensile stresses **concentrate** in a zone between the nearby tips of the offset segments. This stress concentration may favour shear fractures (transfer faults), hook-shaped fractures, or both. Second, the principal stresses become **rotated** (change their orientations) in these shear zones and around the nearby tips of the offset segments. This rotation is the main reason why hook-shaped fractures strike differently from adjacent dike segments and yet are primary extension fractures. Extension fractures, by definition, form in a direction that is perpendicular to that of the local minimum compressive principal stress σ_3. It follows that the direction of σ_3 changes gradually between the nearby tips of the offset segments of the dike, so that the hook-shaped fractures connecting the segments are everywhere roughly perpendicular to the local direction of σ_3.

Fig. 2.26   Horns are failed dike paths. More specifically, curved horns, such as this one, are failed hook-shaped linkages between dike segments. The host rock is basaltic lava flow. The dike is in Iceland. The measuring tape provides a scale.

In addition to the development of hook-shaped and transfer-fault linkages of dike segments, there are many other geometric features associated with the general development of dike paths. Among the most common geometric features are '**horns**' (Fig. 2.26). Some horns are curved and parts of short, or failed, hook-shaped linkages. Others are straight (Fig. 2.27). All horns are primarily '**failed magma paths**'. They indicate paths tried by the magma but given up when 'easier' paths were found. By 'easier', we mean here less energy-demanding (Chapter 10).

2.6 Tips (Ends)

Dike tips form where the local stress field becomes unfavourable for further dike propagation. Tips thus indicate that the dike became **arrested** so that the dike propagation **at that location** stopped. The arrest may have been **temporary** during dike propagation, as exemplified by horns (Figs. 2.26 and 2.27). Alternatively, it may have been a **permanent** arrest (the true vertical end of the exposed segment of the dike). Measurements and studies of **lateral tips** are important for understanding dike segmentation and linkage in lateral sections. These have been discussed in Section 2.5. For understanding dike propagation to the surface and the successful forecasting of dike-fed volcanic eruptions, however, **vertical dike tips** are much more important than lateral tips. In this section, the focus is on vertical tips.

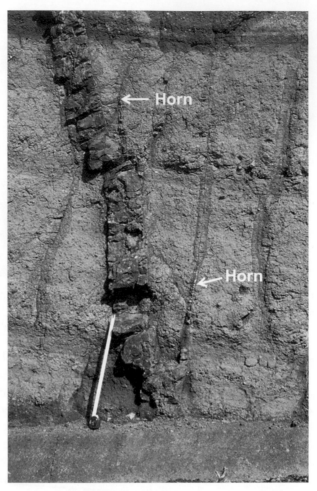

Fig. 2.27 Straight horns extending from slightly offset parts of a dike in Tenerife (Canary Islands). The horns are failed dike paths. The dike paths eventually selected in these two instances are to the left of the horns. The measuring tape is 1 m long.

There are basically three main geometric types of dike tips: the thin crack-like or wedge-like tip, the blunt tip, and the rounded tip. The tip geometries depend on the stress conditions at the tip at the time of arrest, which in turn are related to the mechanical properties of the host rock. Here, we give a brief description of these tips and indicate the mechanical conditions responsible for them. The mechanics of the tip advancement, and dike propagation in general, are given in Chapters 7 and 10.

Wedge-Like Tips. This is the crack-like type where the dike simply **thins out** or tapers away in a vertical section (Fig. 2.20). Commonly, these tips occur within thick layers, such as lava flows, or welded or otherwise relatively stiff pyroclastic layers. Many, perhaps most, of these tips are not the true vertical ends of the dikes, but rather are an indication of the dike segmentation in a vertical section (Fig. 2.20). There are three basic mechanical reasons for the development of wedge-like tips.

- The first reason is **overlap** with another existing or developing segment of the same dike. The magma overpressure in this segment causes horizontal compressive stress in the host-rock layer, say a lava flow, so that the magmatic overpressure in the adjacent, advancing segment gradually decreases to zero as it enters the layer (here the lava flow) of the existing segment. As seen in the field (Fig. 2.20), both segments increase their thickness from the overlapping zone to a full thickness away from that zone. This type of tip is thus not an indication that the dike did not propagate to shallower levels in the crust, and possibly to the surface, but rather an indication of the way dikes become segmented. The tip, in this sense, is just a **temporary halt** in the dike propagation.
- The second reason is that the overpressure driving the dike propagation gradually decreases until it approaches zero, at the tip of the dike. The **gradual decrease** in overpressure can have several causes. These include that the dike is entering layers of gradually lower density (this applies particularly to basaltic dikes). Alternatively, the excess pressure in the chamber (or the source; for some thin dikes the source is a small sill or another dike) was very small when rupture and dike injection took place and rapidly diminishes to zero as the dike propagates into the host rock. This type of tip represents a **real arrest** of the dike/segment propagation in this particular section, and is thus a statistically valid indication of the proportion of arrested dikes in a given dike swarm. This aspect is discussed in more detail below.
- The third reason is that when the tip of a dike segment is hosted by a comparatively compliant layer and propagates towards a contact with a stiffer layer, the tensile-stress concentration at the tip gradually decreases. This is because the dike-induced tensile stress is gradually transferred to the stiff layer ahead (above the contact) as the dike tip approaches the contact, so that the stress at the dike tip itself, located in the compliant layer, decreases.

The first reason discussed in the bullet points is very common and the typical mechanism by which **dikes become segmented** on their vertical paths. Commonly, the offsets are so small that the segments are in contact or close to being in contact with each other (Fig. 2.28). There is normally an **overlap** between the segments where one segment thins out to a wedge-like tip, and the other takes over and continues the upward propagation of the dike.

Blunt Tips. These are almost exclusively confined to tips that are arrested at **discontinuities**. The discontinuities may be joints, faults, or, most commonly, contacts (Fig. 2.29). The blunt ending is particularly common when the dike ends at a contact where the layer above the contact is **stiff**, such as a stiff lava flow or an intrusion (Fig. 2.29). The conditions for such a dike arrest involve one or more of three main mechanisms, namely the Cook–Gordon delamination, the stress barrier, and the elastic mismatch. All these mechanisms are discussed in detail in Chapter 5.

Fig. 2.28 Offset and somewhat overlapping dike segments (A and B) in Tenerife (Canary Islands). This type of segmentation is typical for dikes. The maximum thickness of the lower segment is about 1.3 m. The segments are hosted by pyroclastic rocks.

Rounded Tips. There is a gradual change from completely blunt tips to rounded tips. Rounded tips also tend to occur at discontinuities, particularly at contacts between dissimilar layers. The main difference in geometry between the blunt and the rounded tips is that the former occur mostly at contacts with stiff layers (for example, lava flows) whereas the latter are mostly at contacts with soft or compliant layers (for example, soft pyroclastic rocks or sediments).

An excellent example of a rounded dike tip at contact within pyroclastic rock is seen in Fig. 2.30. The tip is rounded because the layers that constitute the pyroclastic unit were comparatively soft, **compliant**, at the time of dike emplacement. Thus the host rock deforms in a quasi-ductile manner rather than in an entirely brittle manner. While the principles controlling the arrest are basically the same as for a blunt-ended dike, it is perhaps easiest to think of the arrest in terms of material toughness or

Fig. 2.29 A vertically arrested basaltic dike in Tenerife (Canary Islands). The dike is hosted by comparatively soft pyroclastic rock and becomes arrested (with a blunt tip) at its contact with much stiffer (higher Young's modulus) inclined sheet. The maximum thickness of the dike is about 0.8 m.

energy needed to propagate the dike-fracture through the soft materials. Since soft sediments and pyroclastics normally have zero tensile strength, it is essentially impossible to propagate an extension fracture through such materials. They can fail in shear, generating shear fractures or faults, but most dikes are extension fractures. All these factors contribute to dike arrest. Rounded tips are common in soft sedimentary and pyroclastic layers within stratovolcanoes, indicating how fracture-resilient the volcanoes are to dike propagation (cf. Chapter 7).

Measurements and Observations. When accessible, all dike tips, particularly those that mark the **real vertical ends** of dikes as seen in a particular section, should be studied in detail. The main aim of the study is to understand better the conditions that contributed to the dike/segment arrest at that particular site. This understanding relates to the overpressure of the magma at the time of dike arrest, as well as the energy budget. It is important to quantify as much as possible how the **energy** available to drive the dike propagation was partly **dissipated** through deformation of the host rock at the dip, and by other means (Chapter 10). The following is a list of some items to look for at arrested dike tips.

- **Radius of curvature** of the tip. The radius of curvature is the smallest circle that can be fitted into the very tip of the dike. When the dip- or strike-dimensions of the dike are known, Eq. (5.6) can be used to make a crude estimate of the stress concentration at the dike tip at the time of arrest. In the absence of such knowledge (as is commonly the case), the radius of curvature can sometimes be used for the crude stress-concentration estimate through Eqs. (2.1) and (2.3); Example 2.1). This

Fig. 2.30 A vertically arrested dike (highlighted) in the caldera wall of Santorini (Greece). The dike tip is rounded. The dike is hosted by layered pyroclastic rock and the dike arrest occurs at the contact between mechanically dissimilar (partly due to different grain size) layers in the pyroclastic unit. The maximum dike thickness is about 2 m.

estimate, which indicates the **theoretical potential** stress magnitude (not the stress actually reached in nature), makes it possible to infer how the tip stress and the related strain energy were **dissipated** when the dike tip became arrested. Also, the radius of curvature may provide a very crude indication of the size of the non-measureable dimension (strike- or dip-dimension) of the dike.

- **Shear movement** along the arresting discontinuity. Part of the energy associated with the dike tip may be dissipated through shear movement along the discontinuity at which the dike tip is arrested (Fig. 2.29). Look for evidence of slickensides (striations indicating a fault slip) along the discontinuity as well as kinematic indicators (such as Riedel shears) for determining the sense of slip. If these are found, check if the sense of the slip is in agreement with the fault movement along the contact being caused by the dike emplacement.

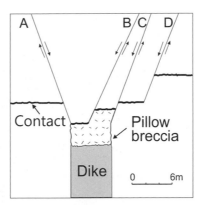

Fig. 2.31 Vertically arrested dike in a graben in Iceland. The graben, here composed of four normal faults denoted A, B, C, and D, existed when the dike was emplaced, as indicated by the top part of the dike meeting fault-transported water and changing into pillow breccia. The slip on the graben faults changes the stress field so that the horizontal principal stress temporarily changes from σ_3 (the minimum compressive principal stress) to the maximum compressive principal stress σ_1, favouring dike arrest or deflection into a sill.

- **Fractures** around the dike tip. Part of the energy associated with an arrested tip is normally dissipated through the formation or reactivation of fractures around the tip. These may be tension fractures or faults. But existing fractures, such as columnar joints, along the potential path of a dike may also have contributed to its arrest. Furthermore, many dike tips are captured and arrested by **existing grabens** (Fig. 2.31). This is because when a graben forms or subsides during a rifting episode, it generates a local **stress barrier** to dike propagation (the maximum compressive principal stress σ_1 commonly becomes temporarily horizontal and perpendicular to the axis of the graben/the rift axis inside the graben, following its formation or subsidence). Thus, on entering a graben, a dike tends to become arrested. It is therefore important to distinguish between fractures and faults **induced** by the dike and those **existing** before the dike emplacement.
- **Dike-rock texture** at the dike tip. Are there large **vesicles** in the dike rock at the tip? For basaltic dikes these indicate easy degassing and normally a shallow depth of arrest. Is the contact between the dike rock and the host rock sharp and **chilled** (chilled selvage, grain-size reduction)? Alternatively, is there **breccia** at the dike tip? The former indicates arrest under essentially dry conditions, whereas the latter indicates that considerable groundwater was in the host rock at the location of the dike-tip arrest. The breccia then forms during explosions when the hot magma comes into contact with the groundwater. If the dike tip is brecciated and there are also faults ahead of the tip, one possible interpretation is that the faults acted as conduits for the groundwater which was responsible for the breccia formation. That interpretation would suggest that the faults existed, perhaps as part of a graben, before the dike was emplaced and contributed to the dike-tip arrest (Fig. 2.31).

2.7 Direction of Propagation and Magma Flow

Knowing the direction in which a dike propagates during an unrest period is of fundamental importance for reliable **hazard assessment**. This direction is normally inferred from seismic data as well as from geodetic data. More specifically, the primary source of information as to the dike-path development is from the migration of **earthquake swarms** associated with the dike propagation (Chapter 4). However, to understand and interpret the seismic and geodetic data so as to be able to make more reliable forecasts for the likely dike paths and the **probability of eruption** we must understand the mechanical conditions that control the dike-path formation inside volcanic zones and edifices (Chapter 10). For this, again, we rely on studies of actual dike paths within eroded volcanoes and rift zones.

Before we discuss some of the criteria used when analysing dike propagation, it is important to understand that the direction of **magma movement** in a dike is not the same as the direction of propagation of the **dike-fracture.** These may differ significantly on a local scale. This difference is best illustrated by feeder-dikes which supply magma to volcanic fissures. Most volcanic fissures start from very small segments, commonly metres to tens of metres in length (Figs. 2.32 and 7.19). From these initial points, the volcanic-fissure segments propagate **laterally**, many linking into longer segments in the manner described for dike-segment propagation above (Section 2.5). At a flat surface, the fissure-segments can only propagate laterally, so that the dike-fracture propagation there must be lateral. By contrast, the magma flows **vertically** up the feeder-dike to the surface. Thus, the magma-flow direction is, close to and at the surface, at roughly **right angles** to the main propagation direction of the volcanic fissure.

Fig. 2.32   Small volcanic-fissure segments marking the initiation of the July 1980 eruption in the volcano Krafla in Iceland. These segments (here tens to hundreds of metres in length) subsequently propagated laterally so as to form a segmented, 4-km-long, volcanic fissure. Photo: Aevar Johannsson.

Generally, dike-fractures propagate in **all directions**. A comparison with joints, many of which are fluid-driven fractures, is instructive. There are several indicators on the surfaces of joints which show that they propagate in all directions. These are primarily the '**ribs**', which indicate where the fluid-driven joint stopped for a while and waited for the fluid pressure to build up again so that the joint could advance its front. Similarly, dikes propagate as '**fingers**' or short segments into layers or units (Chapter 7) and do so in all directions within that unit until the dike-fracture reaches an equilibrium size – as determined by Eq. (7.13). Magma fills the void generated by the propagating fracture front and thus **follows the front**. But the direction of magma flow can, as said, be locally very different from that of the fracture-propagation direction.

Several criteria have been used over the years to infer the direction of dike emplacement. Most of these infer the direction of **magma flow** rather than the direction of dike-fracture propagation. A review of these is given by Baer (1995), and recent applications of some of the methods are by Kissel et al. (2010), Eriksson et al. (2011), and Urbani et al. (2015). The methods used include the following; all refer to the dike rock unless stated otherwise (Marinoni and Gudmundsson, 1999):

- Orientation of xenocrysts and phenocrysts.
- Magnetic fabric (magnetic anisotropy).
- Macroscopic lineations on dike walls and host-rock walls.
- Distribution and geometry of vesicles.
- Dike offsets.
- Earthquake migration (earthquake swarms).

All these methods, except the last one, apply to dikes exposed in eroded areas. They are therefore used to infer **palaeo-flow directions** in dikes. Briefly, these kinematic methods indicate that the magma flow is most commonly inclined, that is, neither perfectly vertical nor perfectly horizontal. The indicators suggest that regional dikes, at least in Iceland, are dominantly formed in **subvertical flow** of magma (e.g. Kissel et al., 2010; Urbani et al., 2015), while local radial dikes are partly formed in subvertical and partly in subhorizontal flow (Eriksson et al., 2011; Urbani et al., 2015). These results are in agreement with earlier results from field studies and mechanical modelling of dikes and inclined sheets (Gudmundsson, 1990a,b; Gautneb and Gudmundsson, 1992). The flow directions in individual dikes may change abruptly along the strike-dimension. For example, the flow direction may change markedly over distances of several metres to tens of metres in the same dike. In some dikes, the inferred magma-flow direction changes from steeply upwards to steeply downwards over short distances. In other dikes, the lineations indicate magma flow vertically downwards.

Some indicators, such as the **magnetic fabric**, primarily reflect the last movement of the magma at the sample location in the dike. The **last movement** of the magma, however, is known to be sometimes different from the main direction of magma flow during dike emplacement. For example, **drain-back** is common in volcanic fissures at the end stages of eruptions, when the pressure in the dike falls. Thus, for many feeder-dikes which, close to the surface, initially had a vertical upward flow of magma, the flow direction recorded by some of the indicators, because of drain-back, suggests a vertical downward flow.

Anisotropy of magnetic susceptibility (AMS) studies may be difficult to use successfully for feeder-dikes where drain-back has occurred at the end of the eruption. While most of the indicators are from the dike rock itself, there are also indicators as to the flow direction from the wall (host) rock. Some studies show a difference between the inferred magma-flow directions from the dike rock and from the host rock next to the dike. Thus, while indicators are certainly useful and worth exploring, the results are sometimes somewhat divergent, even for the same dikes, making it difficult to determine the most likely dominating magma-flow direction.

Earthquake Migration. Earthquake swarms are commonly associated with and induced by dike propagation. This is to be expected since the magmatic pressure **changes** the local **stress field** so that, temporarily, the maximum compressive principal stress σ_1, normally vertical in rift zones and volcanic edifices, becomes horizontal in the vicinity of the dike during and following its emplacement. Earthquakes are mostly associated with **faulting**, that is, slip on shear fractures. The dike-fracture itself is primarily an **extension fracture**. Earthquakes produced during dike propagation are partly related to stresses induced ahead of the dike tip (in the **process zone**), and are partly related to shear failure, that is, faulting, in the compressed host rock on either side of the propagating dike (Fig. 1.13). The compression occurs because of the magma overpressure in the dike. There may also be local small-scale shear failure in the process zone ahead of the dike tip, but the advancement of the tip as an extension fracture is not likely to produce much shear failure.

Earthquake swarms generated during dike propagation in Iceland, Hawaii, Japan, Canary Islands, and elsewhere have been studied in detail (Fig. 4.9; Brandsdottir and Einarsson, 1979; Klein et al., 1987; Peltier et al., 2005; Uhira et al., 2005; Grandin et al., 2011; Wright et al., 2012; Rivalta et al., 2015; Sigmundsson et al., 2015; Marti et al., 2017; Townsend et al., 2017). The general results indicate that dike propagation, as inferred from earthquake data, is a complex process. Some of the main results, particularly as regards earthquake migration in Kilauea in Hawaii (up to 1987) and Krafla in Iceland (1975 to 1984), may be summarised as follows (cf. Brandsdottir and Einarsson, 1979; Klein et al., 1987).

1. Earthquake migration is in various directions. Some swarms start far away from the shallow magma chamber (and associated volcano; the calderas of Kilauea in Hawaii and Krafla in Iceland) and migrate towards the chamber/volcano. Some earthquake swarms thus migrate '**uprift**' rather than '**downrift**' and away from the chamber/volcano.
2. Other swarms starting far from the magma chamber migrate away from the chamber/volcano. These swarms thus migrate downrift. In Krafla, lateral migration is most commonly away from the chamber/volcano, that is, downrift, but in the East Rift Zone of Kilauea, uprift and downrift migration are about equally likely to occur.
3. **Upward migration** of earthquakes is locally common, whereas **lateral migration** is the most common. Lateral migration is normally faster than vertical migration.
4. In Kilauea, the earthquakes are mostly at 2–4 km depth, but some occur down to at least 7 km. Below 4 km depth, much of the rifting is aseismic. In Krafla, the earthquakes are also mostly shallow, but occur down to a maximum depth of about 9 km.

5. In some swarms in Krafla, the lateral migration of earthquakes **stopped**, with downward migration occurring instead. Uprift migration occurred in Krafla, but was much less common than downrift migration.
6. One earthquake swarm in Krafla migrated obliquely **out of** the Krafla Fissure Swarm for a while and then back again into the swarm.
7. In both Kilauea and Krafla, flow of lava **down** into open tension fractures or normal faults is common. The resulting widening and lateral propagation of the tectonic fractures produces earthquake swarms very similar to those associated with dike propagation at depth.
8. **Aseismic** (non-seismic) magma transport is also common.

The general results of these detailed studies indicate that dike propagation, as inferred from earthquake data, is complex. This complexity has been confirmed in other areas of dike propagation and rifting. In particular, during the rifting episode and dike injection in the East African Rift in the past decade (Wright et al., 2012) and the Bardarbunga–Holuhraun episode in 2014–2015 (Gudmundsson et al., 2014; Sigmundsson et al., 2015), the dike injections (vertical and then lateral) in Piton de la Fournaise (Reunion, France; Peltier et al., 2005), magma injection and dike and sill formation in the decades prior to and during the 2010 eruptions in the Eyjafjallajökull Volcano in Iceland (Sigmundsson et al., 2010), as well as prior to the caldera collapse in Miyakejima in Japan in 2000 (Uhira et al., 2005), and the eruption in El Hierro (Canary Islands) in 2011 (Marti et al., 2017). This complexity makes it difficult to forecast the paths of injected dikes and, therefore, to distinguish between non-feeders and potential feeder-dikes, a topic to which we turn now.

2.8 Feeder-Dikes and Non-Feeders

Until recently, comparatively few feeder-dikes had been reported in the literature. This was partly because most dikes were studied only at one particular point, in one particular outcrop, so that they were not followed vertically to see if they were connected with lava flows or pyroclastic layers. Partly, however, few feeders were reported because finding the connection between feeders and their eruptive materials (such as lava flows) requires very good exposures, better than those normally encountered during dike studies.

One important statistical result to be obtained from dike studies is the proportion of **feeder-dikes versus non-feeders** in any particular exposed swarm or section in an edifice. The best outcrops for such studies are **vertical cliffs** with little or no vegetation, so that all dikes can been seen, both feeders and non-feeders. Some such cliffs can be found at the sea, but perhaps the best ones are those formed by ring-faults and exposed through recent caldera collapses. One such outcrop was generated by the caldera collapse in the year 2000 in the Miyakejima Volcano in Japan. The resulting cliff-wall exposes many dikes, both feeders and non-feeders, and these have been analysed in detail by Geshi et al. (2010) and Kusomoto et al. (2013) who find great **difference in shape** between feeders and non-

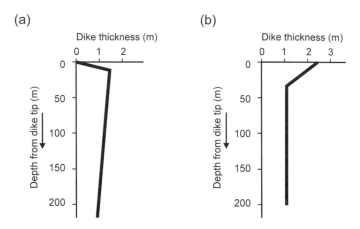

Fig. 2.33 Schematic illustration showing how the thickness of dikes changes with depth below their vertical tips. (a) The non-feeder (arrested dike) tapers away (thins) towards its tip (cf. Figs. 2.11, 2.20, and 2.30). (b) The feeder-dike becomes much thicker towards the surface (cf. Fig. 1.21). Both dikes are from the caldera walls exposed during the 2000 collapse of the Miyakejima Volcano in Japan (modified from Geshi et al., 2010).

feeders. Another recent study of feeder-dikes, in Tenerife (Canary Islands) and Iceland, also demonstrates their differences from non-feeders (Galindo and Gudmundsson, 2012). The main results of these studies may be summarised as follows:

1. Many feeders increase their **thicknesses** close to the free surface (Fig. 2.33). This may be partly because of erosion of the dike walls (the conduit walls), but is primarily related to the elastic **free-surface effects** on extension-fracture openings (Gudmundsson, 2011). At the surface, the thickness of the feeder (its aperture while liquid) may be several times greater than the dike thickness some tens or hundreds of metres below the surface (Figs. 1.21 and 2.33).

2. The sizes and numbers of **vesicles** in feeders increase close to the surface. The exact depth at which this increase becomes noticeable, indicating increasing **degassing** of the dike magma, depends on the magma composition; degassing occurs at greater depths for felsic than for mafic magmas (Chapter 6). However, the increase in vesicle size and number in mafic magmas may extend to depths as great as several hundred metres below the surface. This is supported by observations in Hawaii which indicate that significant degassing of basaltic feeder-dikes may occur to a depth of several hundred metres (Greenland et al., 1988).

3. Many feeders contain elongated (prolate ellipsoidal) **cavities** in their central, topmost parts, particularly within 2–3 m of the surface (Galindo and Gudmundsson, 2012). Commonly, there are solidified magma drops on the walls of these cavities. Cavities of this type have not been reported for non-feeders.

4. Some feeders inject oblique, small 'fingers' or **dike-lets** into the planes of faults that cross the feeder paths. Magma injection of this kind may trigger slip on the existing faults. While this triggering mechanism applies to dike–fault interaction in general, it is more likely to occur close to the surface. This follows because normal faults tend to become steeper close

to the surface, which implies that the normal stress on the fault plane becomes smaller and therefore easier for the magma to use the fault as a channel (Chapter 10).

One main difference between feeders and non-feeders is the shape of their upper or **top parts**. While most feeders tend to expand as they reach the free surface, as indicated above, some do in fact become thinner (Galindo and Gudmundsson, 2012). By contrast, non-feeders generally become thinner on approaching their vertical tips (Fig. 2.33; cf. Figs. 1.21 and 2.11). It should be stressed, however, that the exact geometry of all dikes depends strongly on the properties of the mechanical layers that they dissect. If internal magma overpressure is the only driving 'force' or loading of the dike, then the thickness may be comparatively **large** (large aperture) in compliant or **soft layers**, such as sediments and soft pyroclastic rocks. For the same loading, dike thickness tends to be comparatively small in very stiff or high-Young's modulus layers.

The dike-thickness variation, however, depends on other parameters such as the dip of the dike and the external loading during dike emplacement. At divergent plate boundaries, and in rift zones in general, the external loading is commonly **tension** associated with the spreading vector. Under external tension, stiff layers concentrate tensile stress much more than soft layers. This means that the horizontal compressive stress in the stiff layers becomes reduced. This reduction results in low horizontal compressive stresses in such layers at depth and in absolute tension, negative σ_3, at shallow depths; that is, from depths of, normally, a few hundred metres to the surface. From Eq. (5.3) it then follows that the aperture and subsequent thickness of a dike may in fact be **larger** in the **stiff layers** than in the soft layers, exactly opposite to the situation when there is no external tension operating during dike emplacement. These results apply to all fluid-driven fractures and can be confusing since side by side there may be hydrofractures, such as mineral veins, where one has the greatest thickness or aperture in the stiff layers while another has the greatest thickness in the soft layers. But the reason for this difference is the different external loading conditions during hydrofracture emplacement.

Proportion of Feeders versus Non-Feeders. This proportion is of fundamental importance for reliable **hazard and risk assessments** for volcanoes and volcanic zones/fields. In addition, in order to use geodetic and seismic (and chemical) monitoring data during unrest periods as efficiently as possible, full understanding of this proportion and its relation to the mechanics of dike propagation and arrest are needed.

There have been several estimates of the **proportion of dike tips** that should be observed in a section of a given height and depth below the original top of the pile. For the regional dikes of Iceland, measured mostly at crustal depths between 500 and 1000 m below the original top of the lava pile, the mathematical probability of seeing the vertical tips of the dikes depends on the height of the exposure. The exposures for the regional Tertiary (Neogene) and Quaternary dikes in Iceland are mostly 5–15-m-thick lava flows. The probability of seeing the tip of an observed dike increases at shallower depths in the pile. The maximum mathematical probability would be for a section along a 15-m-thick lava flow at 500 m depth, namely

15/500 = 0.03 or 3%. By contrast, the minimum probability of observing the tip of a particular dike would be for a section along a 5-m-thick lava flow at 1000 m depth in the lava pile, or 5/1000 = 0.005 or 0.5%. In a study of some 5000 dikes in profiles in Iceland, the tips of 1–3% of the dikes were actually observed, in good agreement with these estimates.

Similarly, in a study of regional dikes in Tenerife (the Canary Islands), the average height of the exposure (along road cuts of lava flows and pyroclastic layers) was around 10 m, and their depths below the original surface a few hundred metres. The actual proportion of dike tips seen in the sections is 5–6%, corresponding to crustal depths of about 200 m. Again, the agreement between the observations and the estimated depth of erosion, given the height of the average exposure, is excellent.

Most of the dike tips in Iceland and Tenerife belong to non-feeders. Of about **140 dike tips** observed in these profiles there are very few that belong to feeder-dikes. Although some **11 feeders** have been observed in Tenerife and Iceland (Galindo and Gudmundsson, 2012), none of them belongs to these profiles; they are all found at very shallow crustal depths or at the surface. The results suggest that **most dikes**, both in Tenerife and in Iceland, are **non-feeders** and that the feeder-dikes are no more than several percent of the dikes. These are of course long-term averages. It is known that in some basaltic edifices, in the **short term** (over decades), many, and sometimes nearly all, dikes injected during unrest periods reach the surface, that is, are feeders (Gudmundsson, 2009; Peltier et al., 2010; Poland et al., 2014). Generally, it is easier to fracture a basaltic edifice than a stratovolcano, so that a higher proportion of injected dikes would be expected to reach the surface as feeders in basaltic edifices than in stratovolcanoes (Chapter 7). But these periods are normally short, some decades, and over longer periods of time, even in basaltic edifices, the proportion of feeder-dikes is much lower since many of the dikes become arrested.

It is rare to find steep sections that are high enough to measure the exact **proportions** of feeders and non-feeders. One such section, mentioned above, is in the walls of the collapse caldera of the Miyakejima Volcano in Japan. In the studied 200–450-m-high caldera walls, an estimated maximum of 7% of the dikes are feeders; the remaining 93% being non-feeders (Geshi et al., 2010; Kusumoto et al., 2013). Similar results were obtained in a part of the caldera wall of the volcano on (the island of) Santorini in Greece (Fig. 2.34). The exposed caldera wall of Santorini is mostly 150–200 m high, along which some 60 dikes have been measured. The arrested vertical tips of many dikes were observed (Fig. 2.30), but only two feeders were identified. This suggests that perhaps 3–4% of the dikes are feeders, the remaining 96–97% being non-feeders. The results for the walls of Miyakejima and Santorini are thus very similar, suggesting that 93–97% of dikes in stratovolcanoes become arrested and thus non-feeders, implying that only **3–7% are feeder-dikes.** These results have great implications for **hazard assessments** during unrest periods in stratovolcanoes. However, at this stage the data are very limited. Much more research on this topic is needed to obtain more accurate and larger data sets on the actual proportion of injected dikes reaching the surface to feed eruptions in volcanoes.

Fig. 2.34 Part of the caldera wall of Santorini (Greece). The wall here is about 200 m high. Most of the dikes are non-feeders.

2.9 Sills

All the basic methods described above for field studies of dikes and inclined sheets apply, with proper modifications, to field studies of sills. Sills, however, have characteristics that make certain aspects of field studies of them different from those of dikes and inclined sheets (Fig. 2.35). These differences include the following items:

- It is generally more difficult to **distinguish** sills from the host rock – commonly lava flows – than to distinguish either dikes or inclined sheets from the host rock. This follows because the sills are mostly emplaced parallel (concordant) with the lava flows.
- The main criteria to distinguish sills from lava flows are: (1) sills have chilled selvages at the **upper and lower** margin (lava flows have chilled selvages only at the lower margin, if at all); (2) sills have generally much better developed **columnar (cooling) joints** than lava flows; (3) sills have much smaller and fewer and more regularly shaped (circular to somewhat elliptical) **vesicles** than lava flows; and (4) sills do not normally have scoria layers in contact with their lower and upper margins while lava flows do.
- Because many sills are close to horizontal, that is, have dips of only a few degrees (Fig. 2.35), it is commonly difficult to measure the **attitude** of a sill (the same applies to many lava flows and sedimentary beds). It may be necessary to measure the attitude on either of the margins of the sill itself (commonly the lower margin is the one better

Fig. 2.35 Part of the Whin Sill in northern England. The sill, of basalt (dolerite), has well-developed columnar joints. It was emplaced during the Carboniferous. Here the sill reaches a thickness of some 30 m, whereas its maximum measured thickness is about 70 m. The lower contact of the sill with the host rock is indicated; the upper contact has been eroded away. The person (indicated) provides a scale.

exposed) rather than from a distance (as for dikes and inclined sheets). Since most sills are basaltic, the possible magnetic effects on the measurements must be taken into account.

- Because sills are commonly horizontal, **profiles** (sections) through sill swarms or clusters must generally be steep, preferably vertical.
- In comparison with dikes and inclined sheets, there are comparatively little statistical data available on sill **geometries**. It is known that sills are commonly much thicker than lava flows. Sill thicknesses arc commonly tens of metres and, for many, several hundred metres (Polteau et al., 2008; Cukur et al., 2010; Gudmundsson and Lotveit, 2012; Barnett and Gudmundsson, 2014; Hansen, 2015). Many sills have been traced over long distances, for tens of kilometres and some for one or two hundred kilometres, in the field and on seismic images. But it is not known if they are all formed in single injections. The general ratios between the lateral dimensions and the thicknesses of sills, their **aspect ratios**, are poorly known, but those measured are in the range of several tens to a few hundred.
- Because of their great thicknesses and (normally) lack of scoria at their contacts, sills have **thermal effects** on the host rock that are generally much greater than those of lava flows. These thermal effects – contact metamorphism and related aspects – should be mapped and measured, particularly around thick sills.

Apart from these specific items, the field studies of sills proceed similarly and use the same general techniques as those used for inclined sheets and dikes.

2.10 Summary

- Detailed field studies of sheet intrusions (dikes, inclined sheets, and sills) require good outcrops. Such outcrops are normally found along cliffs, river channels, coastal sections, road-cuts, quarries, and other well-exposed rock bodies. For statistical studies, the dikes are usually measured along sections (also referred to as traverses, profiles, and scan-lines).

- For each dike, the lithology of the dike rock and host rock are recorded and the following dike parameters are measured: strike, dip, and thickness. When possible, and depending on the aim of the study, the following items are also recorded: dike length; dike tip/connection with lava flow/pyroclastic layer; columnar joints; whether the dike forms a topographic high (a ridge) or a topographic low (a depression); the distribution and sizes of vesicles, secondary minerals (amygdales, veins), and phenocrysts; whether the dike occupies a fault or is cut by a fault; and contact characteristics of dike and host rock (chilled selvage, contact metamorphism, eroded contacts).

- While in the field, basic data analysis should be carried out (further analysis is made when back in the office). The basic analysis includes making plots of dike strike (histograms, bin plots, rose diagrams), dip (histograms, bin plots), attitude (stereoplots, normally pole plots), thickness (histograms, bin plots), and length (histograms, bin plots). Plots of these types, done while working in the field, indicate if there are significant errors in the data collection and also provide preliminary results that may suggest where further studies/measurements are needed before completing the fieldwork.

- The general geometry of each dike should be measured, using direct field measurements and images (aerial photographs, Google Earth images). All dikes are segmented, and the segment linkages should be studied, both in lateral and vertical sections. The aim is to understand better the mechanics of dike (and inclined sheet, and sill) propagation.

- All exposed dike tips should be studied. These are of three main types: wedge-like (thin crack-like) tips, blunt tips, and rounded tips. Both lateral and vertical tips should be studied, but the vertical ones are of much greater importance for understanding hazards and are the focus here. Many tips in vertical sections are not the true vertical ends of a dike but simply mark the end of one segment; a new segment then continues to either side of the arrested tip, indicating that the vertical dike propagation continued. Blunt tips are generally associated with discontinuities, most commonly contacts, and rounded tips with soft or compliant host rocks. Several measurements related to tips and the associated fractures provide information on the state of stress at the time of dike arrest, as well as offering general information on the mechanics of dike propagation.

- The direction of the dike-fracture propagation and magma flow should be studied. On a local scale, these need not coincide; that is, the dike-fracture propagation may be partly at a high angle to the main magma-flow direction. During unrest periods, the dike-propagation direction is mainly inferred from the migration of seismic swarms. For dikes in eroded areas, there are various methods to infer (mainly) magma-flow direction, including the orientation of xenocrysts and phenocrysts, magnetic fabric, lineations (on the dike rock as well as on the host rock), distribution and geometry of vesicles, and dike offsets. All the results indicate that dike propagation is a complex process, with abrupt changes in the direction of the fracture propagation and in the direction of the magma flow. Forecasting the direction of dike propagation during unrest periods is one of the major challenges in volcanology.

- All observed feeder-dikes should be recorded and an attempt made to estimate the proportion of feeders versus non-feeders in well-exposed sections. The best sections for such studies are high cliffs, particularly those generated by recent caldera collapses. Few systematic studies of this kind have been made, and there is a strong need to get more information of the percentage of feeders in a given dike swarm. The present results, as far as they go, indicate that feeders may constitute as little as 3–7% of all the dikes in a given swarm, the rest being arrested dikes, that is, non-feeders.

- Most of the field methods used for dikes and inclined sheets, listed above, can be applied to sills. Since sills are normally subhorizontal, studies of sill swarms require steep to vertical profiles.

2.11 Worked Examples

Example 2.1

Problem

The **radius of curvature**, ρ_c, at the end of an elliptical hole, here used as a crude model for a dike, is given by (Gudmundsson, 2011):

$$\rho_c = \frac{b^2}{a} \tag{2.1}$$

where a is the semi-major axis of the elliptical hole and b its semi-minor axis. We also have the maximum tensile stress, σ_{max}, at the tip of the elliptical hole as given by:

$$\sigma_{max} = p_o(2\sqrt{a/\rho_c} - 1) \tag{2.2}$$

where p_o is the magmatic overpressure or driving pressure in the elliptical dike. The measured radius of curvature of a vertical dike tip is 2 cm, while the measured thickness of the dike some 20 m below tip is 3 m. Make a crude estimate of dip-dimension of the dike segment to which the tip belongs.

Solution

As discussed in this chapter, all dikes are segmented, so what we could infer from these calculations is a very crude estimate of the dip-dimension of one segment of the dike. Since dike segments are commonly flat ellipses (Fig. 2.20), any thickness measurement at a considerable distance from the tip may be taken as approximately the maximum thickness – here that distance is 20 m, but it could as well be 10 m or 40 m. With the total dike thickness of 3 m, then $b = 1.5$ m and we also have $\rho_c = 0.02$ m. Rearranging Eq. (2.1) and solving for half the dip-dimension a, we get:

$$a = \frac{b^2}{\rho_c} = \frac{(1.5\,\text{m})^2}{0.02\,\text{m}} = 112\,\text{m} \tag{2.3}$$

This would then be the dimension of one segment of the dike. The segmentation of dikes is well known, but few studies have been made of them, particularly in vertical sections. As observed in cliff sections, many dike segments range in height from tens of metres or less to hundreds of metres (Figs. 2.33 and 2.34), so that the above result is reasonable. Empirical relations suggest that, at shallow depths, strike- and dip-dimensions of basaltic dikes are commonly between 300 and 1500 times the dike thickness, with an average of about 1000. This ratio suggests that the total dip-dimension of this dike might be around 3 km, in which case its source would be a shallow magma chamber (Fig. 1.2).

Example 2.2

Problem

Use Eq. (2.2) to calculate the maximum theoretical tensile stress at the vertical tip of the dike in Example 2.1. Assume a typical overpressure (driving pressure) in the dike of 10 MPa. Explain why the calculated theoretical tensile stress can never occur in nature.

Solution

From Eq. (2.2) and the information in Example 2.1 we get as follows:

$$\sigma_{max} = p_o(2\sqrt{a/\rho_c} - 1) = 10\,\text{MPa}(2\sqrt{112\,\text{m}/0.02\,\text{m}} - 1) = 1487\,\text{MPa}$$

Since the maximum *in situ* tensile strength of rocks is about 9 MPa (Appendix E; Gudmundsson, 2011), no rock can tolerate tensile stresses of hundreds of mega-pascals, let alone nearly 1500 MPa as indicated here. Thus, the result is purely theoretical and indicates how great the theoretical tensile stress can be at the tip of an extension fracture, here a dike, that has a very narrow tip in relation to its dip- and strike-dimensions. Notice that here we used the overpressure in mega-pascals directly in the calculations; there is no risk of error in doing so because the value in the parentheses was calculated separately. Normally, however, all values should be converted to standard SI units in calculations, so that in this case we could have converted mega-pascals to pascals during the calculations, and then back to mega-pascals at the end of the calculations. As said, that was not necessary here, but this is the best procedure in most calculations.

Example 2.3

Problem

A typical dike in Santorini (Fig. 2.34) is about 2 m thick. If the depth to the shallow magma chamber at Santorini at the time of dike emplacement was 4 km (Browning et al., 2015), give a crude estimate of the volume of such a typical dike.

Solution

Here the dip-dimension or height of the dike is assumed to be 4 km, and the measured thickness is 2 m. The missing dimension is thus the strike-dimension or length of the dike. In the absence of better information, and given that the dikes in Santorini are at shallow crustal depths and dissect volcanic rocks, then a length/thickness ratio of about 1000 may be assumed as an approximation – using the field results from Iceland, discussed above. For a 2-m-thick dike, the strike-dimension would then be 2 km. In that case, the dike volume (V) would be:

$$V = 4000 \text{ m} \times 2 \text{ m} \times 2000 \text{ m} = 1.6 \times 10^7 \text{m}^3 = 0.016 \text{ km}^3$$

This is obviously a very crude estimate, but in the absence of accurate data (except for the dike thickness) it is the best that we could come up with. It is important to have an idea of the dike volume when estimating the total volume of magma that is injected from, or flows out of, a shallow chamber during an unrest period, with or without an eruption.

Example 2.4

Problem

Estimate the magmatic overpressure of the dike in Example 2.3 using Eq. (7.13). Assume that the Young's modulus of the rock hosting the dike is 10 GPa and its Poisson's ratio is 0.25.

Solution

From Eqs. (5.4) and (7.13) we have:

$$\Delta u_I = \frac{2p_o(1 - v^2)L}{E}$$

which can be solved for the magmatic overpressure in the dike p_o, thus:

$$p_o = \frac{\Delta u_I E}{2L(1 - v^2)} \tag{2.4}$$

where Δu_I is the dike thickness (strictly the maximum thickness, but taken as the value measured in the field away from the tip), E is Young's modulus of the host rock, and v is Poisson's ratio of the host rock. Using the length and thickness values given for the dike in

Example 2.3, as well as the values for Young's modulus and Poisson's ratio given above, we get:

$$p_o = \frac{\Delta u_I E}{2L(1-v^2)} = \frac{2 \text{ m} \times 1 \times 10^{10} \text{ Pa}}{4000 \text{ m} \times (1-0.25^2)} = 5.3 \times 10^6 \text{ Pa} = 5.3 \text{ MPa}$$

For a comparatively small basaltic dike injected from a shallow magma chamber in a polygenetic volcano, a magmatic overpressure of 3–10 MPa is reasonable, so that the present value of 5.3 MPa is very plausible.

Example 2.5

Problem

Use the estimated source depth for the dike in Example 2.3. Furthermore, assume that the average magma density is 2650 kg m^{-3} (a common value for basaltic magma), and that the average density of the uppermost 4 km of the crust is 2700 kg m^{-3}. Use 3 MPa for the tensile strength of the roof of the shallow magma chamber at the location of its rupture and dike injection, and take the differential stress, σ_d, that is, the difference between the maximum and the minimum principal stress in the host rock where the dike is exposed, as 1 MPa. Using these values and Eq. (5.3), calculate the magmatic overpressure in the dike in Example 2.3 and compare with the results obtained in Example 2.4.

Solution

From Eq. (5.3) we have the overpressure, p_o, in the dike given by:

$$p_o = p_e + (\rho_r - \rho_m)gh + \sigma_d$$

where ρ_r is the average host-rock density, ρ_m is the average magma density, g is acceleration due to gravity, h is the dip-dimension of the dike, and σ_d is the differential stress. Using the values above, and the tensile strength for the excess pressure p_e in the chamber at the time of roof rupture, we get:

$$p_o = 3 \times 10^6 \text{ Pa} + (2700 \text{ kg m}^{-3} - 2650 \text{ kg m}^{-3}) \times 9.81 \text{ } ms^{-2} \times 4000 \text{ m} + 1 \times 10^6 \text{ Pa}$$
$$= 6 \times 10^6 \text{ Pa} = 6 \text{ MPa}$$

Thus, using very reasonable assumptions for magma and crustal densities and *in situ* tensile strength of the chamber roof, the magmatic overpressure obtained by an entirely different method from that in Example 2.4 yields similar results, or 6 MPa (strictly 5.96 MPa) instead of 5.3 MPa. We could argue that the differential stress, σ_d, is poorly constrained in the present calculations. However, as discussed in Chapters 5 and 7, this stress is normally similar to the *in situ* tensile strength, and, at shallow depths in an active volcanic area, σ_d is most likely in the range of 0.5–3 MPa. Thus, even if we took σ_d as 3 MPa, the estimated overpressure of 9 MPa would still be reasonably close to the one obtained in Example 2.4, namely 5.3 MPa. We conclude, therefore, that these independent methods, as used here, yield coherent magmatic overpressure estimates for dikes.

Example 2.6

Problem

Use the information in Fig. 2.10 (including the captions) to solve the following problem. A 6-m-thick dike dipping 85° W dissects a pile of lava flows dipping 10° W. Calculate the apparent vertical displacement across the dike.

Solution

There are several ways of solving this problem. But if you like to use the equation in the captions to Fig. 2.10, then that equation assumes that the dike is vertical. Thus, we first rotate the dike (and the pile) by 5° to the east so as to make the dike vertical. That rotation increases the dip of the pile by 5°, which thus becomes $10° + 5° = 15°$ W. From the equation in the caption to Fig. 2.10 the apparent vertical displacement, x, across the dike is then:

$$x = t\tan\phi = 6 \text{ m} \times \tan 15° = 1.6 \text{ m}$$

Notice: there is no real fault displacement across the dike; the dike is not occupying a fault, a shear fracture. The apparent displacement is simply because the dike dissects the lava flows at an angle that is less than 90°.

2.12 Exercises

2.1 Describe the basic method of selecting and measuring a dike profile (or section or traverse). Where are the best outcrops for measuring dikes normally found – in which landforms? How are the profiles and the dikes that dissect it located – what techniques are used?

2.2 What are the basic geometric measurements made of each dike during dike studies?

2.3 Why is it important to study and measure columnar joints and 'columnar rows' in dikes? What do the rows tell about the mode of emplacement of the dike? What applied implications do the columnar joints have?

2.4 What main information relevant to dike studies can be obtained from detailed studies of the secondary minerals and vesicles in dikes?

2.5 What mechanical types of fracture are most dikes? Do dikes occupy faults? What would be the mechanical implication of a dike occupying a fault?

2.6 List and describe briefly the main techniques for the statistical representation of dike strike.

2.7 List and describe briefly the main techniques for the statistical representation of dike dip and, in general, dike attitude.

2.8 List and describe briefly the main techniques for the statistical representation of dike thickness and dike length.

2.9 What are common length–thickness ratios of dikes? What do these ratios depend on, that is, which mechanical properties?

2.10 Which are the two basic mechanisms by which dike segments link? What types of fractures are generated through these mechanisms? Which large-scale structures at ocean ridges form by the same mechanisms?

2.11 What are the three main types of vertical dike tips? Describe them and discuss their mechanical implications.

2.12 What is the radius of curvature of a dike tip and how can it be used for understanding certain aspects of dike mechanics?

2.13 How can the direction of dike-fracture propagation differ from that of the magma movement in the dike? Provide field examples.

2.14 List and describe the main 'rock fabric' criteria that can be used to infer the magma movement in a dike-fracture.

2.15 What do earthquake-migration studies suggest regarding dike propagation?

2.16 Describe the main geometric differences between non-feeders and feeder-dikes.

2.17 What is thought to be the general percentage of feeder-dikes in a typical volcano?

2.18 In which type of volcano is it normally easier for a dike to reach the surface and become a feeder and why: a stratovolcano or a basaltic edifice?

2.19 Why is the proportion of feeder-dikes in a typical swarm of dikes important for hazard studies?

2.20 List the main criteria used to distinguish in the field between sills and lava flows.

References and Suggested Reading

Acocella, V., Neri, M., 2009. Dike propagation in volcanic edifices: overview and possible developments. *Tectonophysics*, **471**, 67–77.

Baer, G., 1995. Fracture propagation and magma flow in segmented dykes: field evidence and fabric analysis, Makhtesh Ramon, Israel. In Baer, G. and Heimann, A. (eds.), *Physics and Chemistry of Dykes*. Rotterdam: Balkema, pp. 125–140.

Barnett, Z. A., Gudmundsson, A., 2014. Numerical modelling of dykes deflected into sills to form a magma chamber. *Journal of Volcanology and Geothermal Research*, **281**, 1–11.

Becerril, L., Galindo, I., Gudmundsson, A., Morales, J. M., 2013. Depth of origin of magma in eruptions. *Scientific Reports*, **3**, 2762, doi:10.1038/srep02762.

Brandsdottir, B., Einarsson, P., 1979. Seismic activity associated with the September 1977 deflation of the Krafla central volcano in NE Iceland. *Journal of Volcanology and Geothermal Research*, **6**, 197–212.

Browning, J., Drymoni, K., Gudmundsson, A., 2015. Forecasting magma-chamber rupture at Santorini volcano, Greece. *Scientific Reports*, **5**, doi:10.1038/srep15785.

Cukur, D., Horozal, S., Kim, D. C., et al. 2010. The distribution and characteristics of igneous complexes in the northern East China Sea Shelf Basin and their implications for hydrocarbon potential. *Marine Geophysical Research*, **31**, 299–313.

Daniels, K., Kavanagh, J., Menand, T., Sparks, R., 2012. The shapes of dikes: evidence for the influence of cooling and inelastic deformation. *Geological Society of America Bulletin*, **124**, 1102–1112.

Delaney, P., Pollard, D., 1981. Deformation of host rocks and flow of magma during growth of minette dikes and breccia-bearing intrusions near Ship Rock, New Mexico. *US Geological Survey Professional Paper*, 1202, 1–61.

Eriksson, P. I., Riishuus, M. S., Sigmundsson, F., Elming, S. A., 2011. Magma flow directions inferred from field evidence and magnetic fabric studies of the Streitishvarf composite dike in east Iceland. *Journal of Volcanology and Geothermal Research*, **206**, 30–45.

Fox, P. J., Gallo, D. G., 1986. The geology of North American transform plate boundaries and their aseismic extensions. In Vogt, P. R. and Tucholke, B. E. (eds.), *The Geology of North America, Volume M: The Western North Atlantic Region*. Boulder, CO: Geological Society of America, pp. 157–172.

Galindo, I., Gudmundsson, A., 2012. Basaltic feeder dykes in rift zones: geometry, emplacement, and effusion rates. *Natural Hazards and Earth System Sciences*, **12**, 3683–3700.

Gautneb, H., Gudmundsson, A., 1992. Effect of local and regional stress fields on sheet emplacement in West Iceland. *Journal of Volcanology and Geothermal Research*, **51**, 339–356.

Geshi, N., Neri, M., 2014. Dynamic feeder dyke systems in basaltic volcanoes: the exceptional example of the 1809 Etna eruption (Italy). *Frontiers in Earth Science*, **2**, doi:10.3389/feart.2014.00013.

Geshi, N., Kusumoto, S., Gudmundsson, A., 2010. The geometric difference between non-feeders and feeder dikes. *Geology*, **38**, 195–198.

Grandin, R., Jacques, E., Nercessian, A., 2011. Seismicity during lateral dike propagation: Insights from new data in the recent Manda Hararo–Dabbahu rifting episode (Afar, Ethiopia). *Geochemistry, Geophysics, Geosystems*, **12**, doi:0.1029/2010GC003434.

Greenland, L. P., Okamura, A. T., Stokes, J. B., 1988. Constraints on the mechanics of the eruption. In Wolfe, E. W (ed.), *The Puu Oo Eurption of Kilauea Volcano, Hawaii: Episodes Through 20, January 3, 1983 Through June 8, 1984. US Geological Survey Professional Paper, 1463*. Denver, CO: US Geological Survey, pp. 155–164.

Gudmundsson, A. 1983. Form and dimensions of dykes in eastern Iceland. *Tectonophysics*, **95**, 295–307.

Gudmundsson, A., 1986. Formation of dykes, feeder-dykes and the intrusion of dykes from magma chambers. *Bulletin of Volcanology*, **47**, 537–550.

Gudmundsson, A., 1990a. Dyke emplacement at divergent plate boundaries. In Parker, A. J., Rickwood, P. C. and Tucker, D. H. (eds.), *Mafic Dykes and Emplacement Mechanisms*. Rotterdam: Balkema, pp. 47–62.

Gudmundsson, A., 1990b. Emplacement of dikes, sills and crustal magma chambers at divergent plate boundaries. *Tectonophysics*, **176**, 257–275.

Gudmundsson, A., 2009. Toughness and failure of volcanic edifices. *Tectonophysics*, **471**, 27–35.

Gudmundsson, A., 2011. *Rock Fractures in Geological Processes*. Cambridge: Cambridge University Press.

Gudmundsson, A., 2017. *The Glorious Geology of Iceland's Golden Circle*. Berlin: Springer Verlag.

Gudmundsson, A., Lotveit, I. F., 2012. Sills as fractured hydrocarbon reservoirs: examples and models. In Spence, G. H., Redfern, J., Aguilera, R, et al. (eds.), *Advances in the Study of Fractured Reservoirs. Geological Society of London Special Publications, 374*. London: Geological Society of London, pp. 251–271.

Gudmundsson, A., Mohajeri, N., 2013. Relations between the scaling exponents, entropies, and energies of fracture networks. *Geological Society of France Bulletin*, **184**, 377–387.

Gudmundsson, A., Lecoeur, N., Mohajeri, N., Thordarson, T., 2014. Dike emplacement at Bardarbunga, Iceland, induces unusual stress changes, caldera deformation, and earthquakes. *Bulletin of Volcanology*, **76**, 869, doi:10.1007/s00445-014-0869-8.

Hansen, J., 2015. A numerical approach to sill emplacement in isotropic media: do saucer-shaped sills represent 'natural' intrusive tendencies in the shallow crust? *Tectonophysics*, **664**, 125–138.

Jerram, D., 2011. *The Field Description of Igneous Rocks*. Oxford: Wiley-Blackwell.

Kattenhorn, S. A., Watkeys, M. K., 1995. Blunt-ended dyke segments. *Journal of Structural Geology*, **11**, 1535–1542.

Kavanagh, J. L., Sparks, R. S. J., 2011. Insights of dyke emplacement mechanics from detailed 3D dyke thickness datasets. *Journal of the Geological Society of London*, **168**, 965–978.

Kissel, C., Laj, C., Sigurdsson, H., Guillou, H., 2010. Emplacement of magma in eastern Iceland dikes: insights from magnetic fabric and rock magnetic analyses. *Journal of Volcanology and Geothermal Research*, **191**, 79–92.

Klein, F., Koyanagi, R. Y., Nakata, J. S., Tanigawa, W. R., 1987. The seismicity of Kilauea's magma system. *US Geological Survey Professional Paper*, 1350, 1019–1185.

Kusumoto, S., Gudmundsson, A., 2014. Displacement and stress fields around rock fractures opened by irregular overpressure variations. *Frontiers in Earth Science*, **2**, doi:10.3389/feart.2014.00007.

Kusumoto, S., Geshi, N., Gudmundsson, A., 2013. Inverse modeling for estimating fluid-overpressure distributions and stress intensity factors from arbitrary open-fracture geometry. *Journal of Structural Geology*, **46**, 92–98.

Maley, T., 1994. *Field Geology Illustrated*. Troutner Way, Boise (United States of America): Mineral Land Publications.

Marinoni, L.B., Gudmundsson, A., 1999. Geometry, emplacement, and arrest of dykes. *Annales Tectonicæ*, **13**, 71–92.

Marti, J., Villasenor, A., Geyer, A., Lopez, C., Tryggvason, A., 2017. Stress barriers controlling lateral migration of magma revealed by seismic tomography. *Scientific Reports*, **7**, doi:10.1038/srep40757.

McClay, K.R., 1991. *Mapping of Geological Structures*. Oxford: Wiley-Blackwell.

Melin, S., 1983. Why do cracks avoid each other? *International Journal of Fracture*, **23**, 37–45.

Nemec, W., 1988. The shape of the rose. *Sedimentary Geology*, **59**, 149–152.

Peltier, A., Ferrazzini, V., Staudacher, T., Bachelery, P., 2005. Imaging the dynamics of dyke propagation prior to the 2000–2003 flank eruptions at Piton de la Fournaise, Reunion Island. *Geophysical Research Letters*, **32**, doi:10.1029/2005GL023720.

Peltier, A., Staudacher, T., Bachelery, P., 2010. New behaviour of the Piton de la Fournaise volcano feeding system (La Réunion Island) deduced from GPS data: influence of the 2007 Dolomieu caldera collapse. *Journal of Volcanology and Geothermal Research*, **192**, 48–56.

Philipp, S. L., 2012. Fluid overpressure estimates from the aspect ratios of mineral veins. *Tectonophysics*, **581**, 35–47.

Poland, M. P., Miklius, A., Montgomery-Brown, E. K., 2014. Magma supply, storage, and transport at shield-stage Hawaiian volcanoes. In Poland, M. P., Takahashi, T. J. and Landowski, C. M. (eds.), *Characteristics of Hawaiian Volcanoes. US Geological Survey Professional Paper, 1801*. Denver, CO: US Geological Survey, pp. 179–234.

Pollard, D. D., Aydin, A. 1984. Propagation and linkage of oceanic ridge segments. *Journal of Geophysical Research*, **89**, 10 017–10 028.

Pollard, D. D, Muller, O., 1976. The effect of gradients in regional stress and magma pressure on the form of sheet intrusions in cross section. *Journal of Geophysical Research*, **81**, 975–984.

Pollard, D. D., Segall, P., 1987. Theoretical displacements and stresses near fractures in rocks: with applications to faults, joints, veins, dikes, and solution surfaces. In Atkinson, B. K. (ed.), *Fracture Mechanics of Rock*. London: Academic Press, pp. 277–349.

Polteau, S., Mazzini, A., Galland, O., Planke, S., Malthen-Sorensen, A., 2008. Saucer-shaped intrusions: occurrences, emplacement and implications. *Earth and Planetary Science Letters*, **266**, 195–204.

Rivalta, E., Taisne, B., Bunger, A. P., Katz, R. F., 2015. A review of mechanical models of dike propagation: schools of thought, results and future directions. *Tectonophysics*, **638**, 1–42.

Rubin, A. M., 1995. Propagation of magma-filled cracks. *Annual Reviews of Earth and Planetary Sciences*, **23**, 287–336.

Sigmundsson, F., Hreinsdottir, S., Hooper, A., et al., 2010. Intrusion triggering of the 2010 Eyjafjallajökull explosive eruption. *Nature*, **468**, 426–430.

Sigmundsson, F., Hooper, A., Hreinsdottir, S., et al., 2015. Segmented lateral dyke growth in a rifting event at Bardarbunga Volcanic System, Iceland. *Nature*, **517**, 191–195.

Takeuchi, S., 2004. Precursory dike propagation control of viscous magma eruptions. *Geology*, **32**, 1001–1004.

Thorpe, R. S., Brown, G. C., 1985. *The Field Description of Igneous Rocks*. Maidenhead: Open University Press.

Tibaldi, A., 2015. Structure of volcano plumbing systems: A review of multi-parametric effects. *Journal of Volcanology and Geothermal Research*, **298**, 85–135.

Townsend, M., Pollard, D. D., Smith, R., 2017. Mechanical models for dikes: a third school of thought. *Tectonophysics*, **703–704**, 98–118.

Uhira, K., Baba, T., Mori, H., Katayama, H., Hamada, N., 2005. Earthquake swarms preceding the 2000 eruption of Miyakejima volcano, Japan. *Bulletin of Volcanology*, **67**, 219–230.

Urbani, S., Trippanera, D., Porreca, M., Kissel, C., Acocella, V., 2015. Anatomy of an extinct magmatic system along a divergent plate boundary: Alftafjordur, Iceland. *Geophysical Research Letters*, **42**, doi:10.1002/2015GL065087.

Walker, G. P. L. 1959. Geology of the Reydarfjordur area, eastern Iceland. *Quarterly Journal of the Geological Society of London*, **114**, 367–393.

Walker, G. P. L., 1960. Zeolite zones and dike distribution in relation to the structure of the basalts of eastern Iceland. *Journal of Geology*, **68**, 515–527.

Wright, T. J., Sigmundsson, F., Pagli, C., et al., 2012. Geophysical constraints on the dynamics of spreading centres from rifting episodes on land. *Nature Geoscience*, **5**, 250.

3 Volcanotectonic Deformation

3.1 Aims

Polygenetic volcanoes, to a first approximation, behave as if they are **elastic.** When subject to loading such as magmatic excess pressure in a chamber or overpressure in a dike, the volcano deformation is, so long as the loading is small, roughly linear elastic. When related to pressure changes in the source chamber, the measured deformation is referred to as **inflation** when the volcano surface rises (during magma-pressure increase) and as **deflation** when the surface falls or subsides (during magma-pressure decrease). If the loading generates stresses that reach the strength of the rock, then fractures form or reactivate. Slip on shear fractures, that is, faults, commonly trigger earthquakes, which can be used to monitor the state of stress in the volcano as well as magma movement through dike or sheet propagation. Some stresses are sufficiently large to form or reactivate the boundary faults of grabens or the ring-faults of collapse calderas. Similarly, the stresses may result in lateral or sector collapses, that is, landslides. The earthquake activity in volcanoes is treated in Chapter 4, and vertical and lateral collapses in Chapter 5. Here, the focus is on measuring and modelling (mostly) elastic deformation in volcanoes with a view to better understanding the associated unrest processes, some more technical aspects of which are elaborated upon in Chapter 10. The main aims of this chapter are to:

- Describe the main types of volcano-deformation data obtained and used in modelling.
- Review the elementary theory of stress and strain.
- Explain how linear elasticity theory is applied to volcano deformation.
- Present and discuss examples of typical surface deformation in volcanoes.
- Describe and discuss the Mogi model of a magma chamber.
- Present and discuss stresses around magma chambers in layered rocks.
- Describe and discuss stresses around dikes and inclined sheets.
- Discuss the effects of rock heterogeneity and anisotropy on volcano deformation.

3.2 Basic Definitions and Data

Volcano-deformation studies include detailed and accurate measurements as well as modelling, using analytical, analogue, and numerical methods. As elsewhere in the book, the term volcano here means a **polygenetic volcano** (rather than a monogenetic volcano), that is, a central volcano; namely, a stratovolcano, a basaltic edifice, or a collapse caldera. Many collapse calderas form parts of larger central volcanoes, such as stratovolcanoes and basaltic edifices, and the term volcano covers all of these. Most of the volcano deformation that occurs before eruption and/or intrusions of dikes and sheets is due to fluid-pressure changes in the shallow source chamber of the volcano. Similar changes commonly take place in the associated deeper magma reservoir, and may also be detected by measurements. The main techniques and instruments used to obtain the data were discussed in Chapter 1. Those used for obtaining the geodetic data, especially as applied to volcanoes, are also explained in detail in many monographs (e.g. Dzurisin, 2006; Janssen, 2008; Segall, 2010; Lu and Dzurisin, 2014). Thus, in this chapter we focus on the interpretation and modelling of these data.

First, we recall, and add to, some of the basic definitions and assumptions used in relation to volcano-deformation studies, namely the following:

- **Inflation** and **deflation** refer to rise (uplift or doming) and fall (lowering or subsidence), respectively, of the surface of a volcano. Until major rock fractures, such as extension fractures and faults, begin to form or propagate, inflation and deflation on most volcanoes can be modelled approximately as being linear elastic.
- **Linear elastic** behaviour of rock (and other solids) is characterised by a linear relationship between the stress and strain in the rock. The ratio between the stress and the resulting strain is the **Young's modulus** (or stiffness) of the rock.
- When the stress in a brittle rock reaches its strength, the rock fractures. Fracture is a **permanent deformation** and non-elastic – sometimes referred to as **inelastic** or **plastic** – behaviour or deformation. Thus, tension-fracture opening or a normal-fault slip, both common during inflation, are examples of inelastic deformation. Large-scale graben subsidence and caldera collapse are also examples of permanent, but comparatively rare, deformation.
- Almost all models of volcano deformation during inflation and deflation assume **linear elastic** rock behaviour. There exist models assuming plastic or viscoelastic behaviour of the rock, but these are few and somewhat difficult to interpret in terms of volcanotectonic processes. All the models in this book assume linear elastic behaviour.
- The **main models** used for explaining and forecasting volcano deformation are of two types: analytical and numerical. In addition, analogue models are widely used to model large-scale permanent deformation, such as lateral and vertical collapses and graben formation.
- **Analytical models** provide comparatively simple equations to calculate pressures, depths, and sometimes dimensions of magma bodies from geodetic data. For example, such models may be used to calculate the depth to a dike generating surface deformation in a volcano or a rift zone. The best-known analytical model for magma chambers is the nucleus-of-strain model, the Mogi model (Fig. 3.1), which is discussed in detail below.

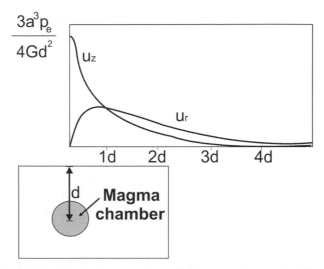

Fig. 3.1 Vertical (u_z) and horizontal (u_r) surface displacements above a shallow magma chamber, located at depth d below the surface and modelled as a Mogi model (point source). The horizontal axis (x-axis) shows the displacements as function of distance (in terms of the depth to the chamber, d) from the centre of the doming or uplift above the chamber. The vertical axis (the y-axis) shows the distance along the surface with the depth d to the point source as a unit.

Analytical models normally assume the rock to be homogeneous and isotropic; in particular, the Mogi model assumes that the modelled crustal segment is a homogeneous, isotropic elastic half-space and is primarily used to calculate the depth to the magma chamber from a given surface deformation.

- **Numerical models** give approximate results for the size and shape of a magma chamber, or a dike, for example, based on surface-deformation data (Fig. 1.15). Alternatively, numerical models are used to calculate the condition for magma-chamber rupture and likely paths of injected dikes and inclined sheets for magma chambers of various shapes, sizes, depths, and loading conditions. Numerical models can deal with any anisotropy and heterogeneity in the volcano and the associated crustal segment, including layering and existing faults, such as boundary faults of grabens and ring-faults, and any shape and size and depth of a magma chamber.

- **Analogue models** use materials that are thought to behave similarly to crustal rocks at the scale, loading and strain rate of the model. A common material used in volcanotectonic studies is sand. A simple analogue model for a collapse caldera uses an inflated balloon buried in sand. The balloon is then assumed to present a shallow magma chamber; and the sand, the crustal rocks, including the volcano. When air is released from the balloon, supposedly corresponding to magma leaving the chamber during an eruption and/or dike intrusion, the sand-layer above the balloon commonly develops a circular fracture, which is supposed to correspond to the ring-fault of a collapse caldera.

The main data obtained in volcano-deformation studies are **displacements of points** at the surface of the volcano. In early geodetic studies, these were measured using a standard **precision levelling technique**, which gives the relative elevation or height of points. This is the oldest technique used for measuring volcano deformation, dating back to the early twentieth century for volcano studies in Japan. Other early techniques include trigonometric levelling and trilateration to measure the displacements (primarily horizontal) of points in a network (at the surface of the volcano). Such methods were, for example, used in an attempt to measure the rate of dilation or spreading across the rift zones of Iceland in the 1930s (Niemczyk, 1943). **Electronic distance meters** are currently used for volcano deformation studies, as are **tiltmeters** of various types, for measuring small changes in the slope angle at specific points at the surface of the volcano.

The main techniques used for obtaining volcano-deformation data today are GPS and InSAR, both of which rely on the use of satellite signals to measure the displacements of points over time. **GPS** is an acronym for **Global Positioning System**, where benchmark points are measured over periods of 8–24 hours and the changes in their location, that is, their displacements, over the periods are recorded. The method is based on access to signals from 32 satellites, of which 5 to 8 are in view during the measurement at each particular location. The accuracy as regards the displacements of the points is **centimetres**, particularly for horizontal displacements (the vertical displacements are less accurate). The overall movement of the points at the surface of the volcano determined with this accuracy is always relative to a certain point, preferably outside the volcano, that is assumed to be stationary.

InSAR, an acronym for **Interferometric Synthetic Aperture Radar**, is based on radar images used to map surface changes that occur over a certain time interval. The deformation data are obtained from the difference in the known phase of the outgoing wave or signal sent by the satellite and the phase of the return wave or signal, reflected from the Earth's surface. The method can be used to measure surface deformation as small as **centimetres** over periods of days to years, in volcanoes and other deforming surface areas such as fault zones.

There are several **physical processes** inside the volcano that generate the main deformation data. The main ones are summarised here with reference to a double magma chamber and dike intrusion (Figs. 1.12, 1.13, and 3.2). Not all polygenetic volcanoes are supplied with magma from double magma chambers; some, for example, may have triple magma chambers, or even more complex magma-storage systems (e.g. a quadruple chamber). However, a double magma chamber is presumably the most common magma-storage system, and is thus used as a basis for the present discussion. The processes inside volcanoes that give rise to surface deformation include the following (cf. Dzurisin, 2006; Segall, 2010):

1. **Magma** becomes added to the shallow chamber, resulting in its expansion and inflation. This is a common process and, if it continues, eventually results in magma-chamber rupture and dike or sheet injection.
2. The shallow chamber is subject to an external **extension**, which eventually results in magma-chamber rupture and dike or sheet injection. This, again, is a very common process and is, in fact, universal at divergent plate boundaries and in rift zones. During this process, magma may or may not be added to the chamber. The surface deformation is more complex than in the first process.
3. Magma or melt becomes added to the reservoir, but not to the chamber. The result is **doming** of the crustal segment above the reservoir and therefore a rise of the surface.

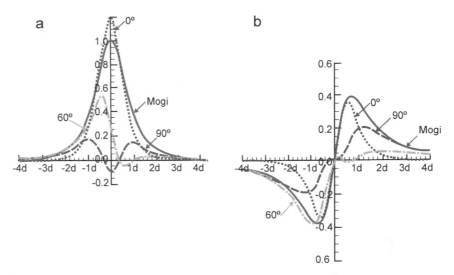

Fig. 3.2 Schematic illustration of the theoretical (a) vertical and (b) horizontal surface displacement induced by a dike with a tip (top) at a shallow depth. The dike/sheet dips are 0° (a sill), 30° and 60° (an inclined sheet), and 90° (a dike). Also shown is the displacements induced by a Mogi (point-source) magma chamber. There is subsidence at the surface right above the dike tip (Dzurisin, 2006).

However, if little or no new magma is added to the chamber, it may encourage ring-fault formation, particularly if the chamber has a sill-like geometry. This is probably the easiest mechanism for generating ring-faults (Gudmundsson, 2007).

4. Magma becomes added to both the chamber and the reservoir. This is a common process and results in **doming** or **inflation** of the surface above the shallow chamber, and also in doming of a much wider area, reflecting the inflation effects of the deeper and larger reservoir.

5. A vertical **dike** approaches the surface (Fig. 3.2). If the crust behaves as roughly isotropic, then there will be symmetric tensile-stress and displacement peaks at the surface on either side of the projection of the dike tip to the surface (the hypocentre of the dike), generating a bimodal displacement and stress distribution (cf. Chapter 10). In many models, there is no stress change at the surface immediately above the tip of the dike (at the hypocentre). In a layered crust with mechanically weak (low tensile strength) or open contacts, however, the surface deformation and stresses induced by the dike can vary widely and there is no simple way of relating the surface deformation with the dike dimensions or depth (Section 3.8; Chapter 10).

6. An **inclined sheet** approaches the surface. In this case the surface deformation is asymmetric, the main uplift or vertical displacement of the ground being in the area above the dip direction of the sheet. For a crustal segment composed of layers with contrasting mechanical properties and weak or open contacts, the surface deformation and stress can be complex. This is particularly so because sheets commonly change their paths abruptly on their way to the surface. For example, many inclined sheets and dikes become deflected along contacts between layers, thus for a while propagating as sills,

only to change into inclined sheets or vertical dikes higher up in the pile (Fig. 1.18). In a layered crust, however, the induced stresses and deformation may vary widely so that it is not straightforward to relate the surface deformation with the sheet dimensions, depth, or dip (Section 3.8).

7. A **sill** being emplaced a certain depth below the surface. If the sill is straight (planar) and the crust above it is not too far from being approximately homogeneous and isotropic, the surface deformation becomes symmetric. However, sills have different geometries and commonly change their elevation levels in the crust during their propagation, moving up or down between contacts. If, in addition, the crustal segment is hetero-geneous and layered, as it generally is in volcanoes, then the sill-related surface deformation may be much more complex and difficult to interpret.

The above list suggests that the deformation associated with volcanotectonic processes is commonly complex, particularly in volcanoes composed of layers with contrasting mechanical properties and mechanically weak or open contacts (Fig. 1.16). It follows that the interpretation of the surface deformation in terms of processes taking place inside the volcano is normally far from simple (cf. Chapter 10). Realistic interpretations require high-quality geodetic and seismic data and also detailed understanding of the processes them-selves. Such an understanding is best obtained through comparison between **active** and **inactive** (and deeply eroded) volcanoes. Once that understanding it obtained, modelling of the processes that give rise to the surface deformation can be made with the help of basic elasticity theory, to which we turn now.

3.3 Stress, Strain, and Elasticity

Here, we provide a very brief and simple presentation of the basic concepts of stress, strain (and displacement), and linear elasticity as a background for analytical and numerical modelling of volcano deformation. A more extensive discussion of these concepts in the earth-science context is given by Gudmundsson (2011a), to which the reader is referred for details.

3.3.1 Stress

Stress is a measure of the intensity of force per unit area. The unit of stress is the **pascal**, Pa, where $1 \text{ Pa} = 1 \text{ N m}^{-2}$. Using this simple definition, the formula for stress, σ, is:

$$\sigma = \frac{F}{A} \tag{3.1}$$

where F is the normal force in newton (N) and A is the area in square metres (m^2) on which the force acts. We use the symbol σ for the stress generated by a force that is normal (is at right angles) to the plane or area – for example, a fault plane or a dike – on which the force acts. Thus, σ is **normal stress**, in geology commonly denoted by σ_n when presenting the normal stress on a particular structure (Fig. 3.3). When a force operates parallel to the plane

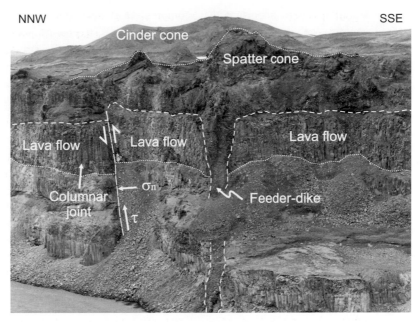

In this section (two dimensions), the stresses acting on a fault plane are the normal stress, σ_n, acting perpendicular or normal to the plane, and the shear stress τ, acting parallel with the fault plane and driving the eventual fault displacement or slip. The fault seen here, dissecting Pleistocene lava flows (with columnar joints) in a river canyon in North Iceland, was originally a normal fault that was subsequently reactivated as a reverse fault (as indicated by the direction of the arrow denoting the shear stress) because of horizontal compressive stress induced by the nearby 4–13-m-thick basaltic feeder-dike that supplied magma to the Holocene lava flow and associated cinder cones (cf. Fig. 1.21 and Gudmundsson et al., 2008).

or area of interest, such as a fault plane, it is a shear force and generates shear stress. We use the symbol τ for **shear stress** (Fig. 3.3), so that:

$$\tau = \frac{F}{A} \tag{3.2}$$

where the F is here the shear force (or shear-force component) acting parallel or tangential to a plane whose area parallel with the direction of the force is A. Stress as defined in Eqs. (3.1) and (3.2) is really a **vector** and referred to as the **traction,** the traction vector, or the **stress vector.** The state of **stress at a point** in the crust or inside a volcano, for example, is the three-dimensional collection of all stress vectors at that point; it defines a matrix with nine components, and is referred to as **stress tensor.** The matrix components are nine because they represent three stress vectors, each with three components.

For any state of stress state at a point, there are three mutually perpendicular planes free of shear stress. For example, the solid Earth's surface in contact with fluid (air or water), as in Fig. 3.3, is free of shear stress. These planes are referred to as principal stress planes or **principal planes of stress.** The normal stress components or normal stresses that act on these planes are known as the **principal stresses** and they act along the **principal axes.** The principal stresses are denoted by σ_1, σ_2, and σ_3. In physics and engineering, tensile stresses are much more important than compressive stresses and are regarded as positive. In geology, geophysics, rock mechanics, and soils mechanics, compressive stresses are much more common than tensile stresses – even if tensile stresses are very important when they occur, such as around magma chambers during unrest periods – and they are considered positive. For this reason **compressive** stresses are regarded as **positive** and **tensile** stresses as **negative** in this book. More specifically, σ_1 denotes the maximum compressive principal stress, σ_2 the intermediate compressive principal stress, and σ_3 the minimum compressive (or maximum tensile) principal stress, that is:

$$\sigma_1 \geq \sigma_2 \geq \sigma_3 \tag{3.3}$$

When modelling magma chambers, we normally assume that before unrest starts – as a result of magma being added to the chamber or other similar processes as listed above – the state of stress around the chamber is lithostatic. A **lithostatic** state of stress is isotropic and depends only on the overburden pressure. **Isotropic** (also named hydrostatic) means that the stress at a given point in the crust is the same in all directions. While both isotropic compression and tension are possible stress states, isotropic tension rarely if ever occurs in the Earth's crust, so that here we consider only isotropic compression. Then all possible planes are subject to equal compressive stress and all the principal stresses are compressive and equal, that is:

$$\sigma_1 = \sigma_2 = \sigma_3 \tag{3.4}$$

The magnitude of the lithostatic stress depends only on the **overburden pressure** (geostatic pressure), which means that the lithostatic stress increases with depth according to the equation:

$$\sigma_v = \rho_r g z \tag{3.5}$$

where we use the vertical stress, σ_v, as a general expression for the lithostatic stress. Here, g is the acceleration due to gravity (9.81 m s^{-2}) and z is the depth, measured as positive down from the free surface of the Earth. Equation (3.5) assumes that the rock density is uniform (constant) and given by the average value ρ_r. However, the equation can easily be extended to crustal segments or volcanoes composed of layers with different densities (Figs. 1.7 and 1.16) simply by using the integral:

$$\sigma_v = \int_0^z \rho_r(z) g \, dz \tag{3.6}$$

where $\rho_r(z)$ is the density of the rock layers as a function of depth z. Using Eqs. (3.5) and (3.6) we can, for example, calculate the vertical stress at the bottom of the canyon in Fig. 3.3.

Many other states of stress may exist in the crust. For example, **general stress** is one where all the principal stresses are different, that is, instead of Eq. (3.3) we have $\sigma_1 > \sigma_2 > \sigma_3$. Also, commonly in the crust the stress field is **heterogeneous** or inhomogeneous, which means that the components of the stress tensor differ between points in the crustal segment or volcano. By contrast, if the stress field is **homogeneous** then all the components of the stress tensor are the same at each point in the crustal segment or volcano. Similarly, as indicated, a state of stress is **isotropic** if the stress components at a given point are the same in all directions, and **anisotropic** if the components differ depending on the direction.

With reference to modelling of volcanotectonic structures and processes, recall that the principal stresses are at right angles to each other, that is, they are mutually **orthogonal**. Generally, close to the Earth's surface outside areas with Alpine or mountainous landscape, one of the principal stresses is normally vertical and the other two horizontal. Because compressive stress is positive and tensile stress negative, then under general geological conditions, particularly those of importance in volcanotectonics, we have as follows:

- σ_1 is **always positive**. Even at the surface of the Earth, the maximum compressive principal stress is positive and equal to the atmospheric pressure (1 bar or about 0.1 MPa, as at the surface in Fig. 3.3). For example, the roofs of most shallow crustal chambers are at crustal depths between 1 km and 5 km. For typical densities, and assuming σ_1 to be the vertical stress (as is common at divergent plate boundaries and in rift zones), then σ_1 would normally be from 20–25 MPa (1 km depth) to 100–130 MPa (5 km depth).
- σ_2 is **normally positive.** Generally, under some temporary conditions at very shallow depths or at the surface, the intermediate principal stress may become tensile, that is, negative. These conditions may be satisfied close to and at the surface during doming, such as above shallow magma chambers subject to inflation (Fig. 1.12). Also, during magma-chamber expansion as a result of increasing excess pressure, σ_2 may become tensile at the chamber boundary. In all these cases where σ_2 is tensile, σ_3 is, by definition, also tensile (Eq. (3.3)). At the Earth's surface, σ_1 is 0.1 MPa, thus close to zero, in which case the resulting stress field may approximate that of **biaxial tension**.
- σ_3 can be either **positive** (the minimum compressive principal stress) or **negative** (the maximum principal tensile stress). In the absence of fluids, σ_3 can be negative (absolute tension) on an outcrop scale only close to or at the Earth's surface (Fig. 1.9). Outcrop-scale tensile stresses are unlikely to exist, in the absence of fluid pressure, at crustal depths exceeding 1 km, and normally reach maximum depths of only several hundred metres. During magma-chamber expansion as a result of increasing magmatic excess pressure, however, σ_3 becomes tensile and eventually reaches the tensile strength of the chamber host rock.

The principal stresses form the basis for analysing stresses around magma chambers, dikes, inclined sheets, sills, faults, ring-faults, and other structures generated by volcano-tectonic processes. For example, all extension fractures, such as essentially all dikes, inclined sheets, and sills, propagate in a direction that is, on average, perpendicular to σ_3 and, since the principal stresses are mutually orthogonal, parallel with σ_1 and σ_2. In many of the models below and in other chapters (cf. Chapter 10) we forecast crudely the propagation paths of these sheet intrusions by showing the variation in the trend or orientation of σ_1 – and less commonly the variation in the orientation of σ_3. Most of the models are two-dimensional so that only the orientations of σ_1 and σ_3 are shown. The orientations or trends of the principal stresses are referred to as **stress trajectories.** These are curves or, more commonly, line segments or **ticks** whose orientations at any point show the directions of the principal stress axes at that point (Fig. 3.4). The **magnitudes** of the principal stresses shown

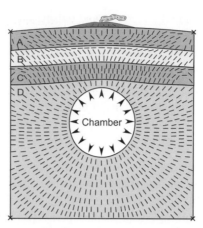

Fig. 3.4 Stress trajectories, shown here as short-line segments (ticks), indicate the directions of the principal stress axes in the rock (or other solids). Here the ticks show the direction of the maximum compressive principal stress σ_1 around a magma chamber of a circular cross-section, subject to internal magmatic excess pressure as the only loading. This numerical model shows the trajectories in the 'deformed state' of the crustal segment (due to the loading), so that the magma chamber and the layers A, B, C, and D (and the surface) are somewhat deformed (cf. Gudmundsson and Philipp, 2006).

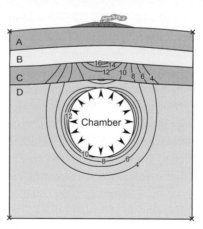

Fig. 3.5 Stress contours show the magnitudes (sizes) of stresses, given in pascals or mega-pascals. Here, the contours, in mega-pascals, show the magnitude of the minimum compressive principal stress σ_3 around a magma chamber (the same as in Fig. 3.4) of a circular cross-section, subject to internal magmatic excess pressure as the only loading. This numerical model shows the stress magnitudes in the 'deformed state' of the crustal segment (due to the loading), so that the magma chamber and the layers (and the surface) are somewhat deformed (cf. Gudmundsson and Philipp, 2006).

in this book are mostly those of the maximum principal tensile stress σ_3. The magnitudes or sizes of the stresses are normally given in mega-pascals, that is, as MPa (1 Pa = 1 N m^2; 1 MPa = 10^6 Pa), and are shown as **stress contours** (Fig. 3.5). It is important to remember that failure, such as a fault slip, occurs only if the stress magnitudes are high enough (indicated by the stress contours) and the orientation of the principal stresses is favourable for that type of slip (indicated by the stress trajectories).

3.3.2 Displacement and Strain

Displacement refers to the change in the position of **particles** or **material points** and so does strain. Strain and displacement are thus closely related. **Strain**, however, involves changes in the internal **configuration** of the body; the distances between the particles/material points that constitute the strained body change. (Notice: many use the term particles or particle points primarily for discrete systems, such as sediments or other loose materials, and material points primarily for continuous systems, such as solid rocks; cf. Chapter 10.) Rigid displacement of a rock body, such as the lava pile far outside plate boundaries, where there is no change in the internal configuration of the body, is **deformation** but not strain. Such ideal displacements of a body are referred to as **rigid-body translation** and **rigid-body rotation.** A rigid body strictly has infinite stiffness, that is, its **Young's modulus** is infinite, as is discussed below. Since the Young's modulus of common solid rocks is generally in the range of 1–100 GPa, it is clearly not infinite and no rocks (and, in fact, no known materials) are strictly rigid. But the assumption of the rock being rigid is commonly made in solving problems in geology; for example, in hydrogeology, volcanology, structural geology and tectonics and, in particular, plate tectonics.

Most geodetic measurements of volcano surface deformation report **displacements** (Fig. 3.2). The same applies to the deformation in active fault zones and at plate boundaries in general; the measurements recorded are normally displacements. For volcanoes, the displacements measured using the various techniques discussed above are as small as centimetres, both in horizontal directions and vertical directions. Similarly, the measured displacements of the tectonic plates, the plate movements, are commonly from a centimetre or less to, for the fastest movements, 10–18 cm per year. All these displacement measurements, however, can be transformed into strain measurements. If we measure the distance between two points or benchmarks today and then again in a year or several years, the change in distance between the points divided by the original distance gives the strain. The time it took the strain between the two points to change by the recorded amount gives the **strain rate**.

A simple (one-dimensional) quantitative definition of strain is as follows. The fractional change in a dimension of a rock body subject to loads is referred to as **strain**. For an elongated body, such as a bar-shaped piece of rock, subject to a tensile or compressive force, the strain is the ratio of the change in length or extension of the body ΔL to its original length L. Denoting the **normal strain** by the symbol ε we have, for one-dimensional tensile strain:

$$\varepsilon = \frac{\Delta L}{L} \tag{3.7}$$

Since Eq. (3.7) represents one length divided by another, the strain ε is dimensionless; it is a pure number and thus has no unit. The strain is often expressed as a percentage, such as:

$$\varepsilon\% = \frac{\Delta L}{L} \times 100 \tag{3.8}$$

Equation (3.7) gives one-dimensional normal strain. For a shear force F, the **shear strain** γ is given by:

$$\gamma = \frac{\Delta x}{y} = \tan\psi \tag{3.9}$$

where Δx is the deflection or movement of the upper face of the body relative to the lower face, with y being the vertical distance between the faces (height of the body) and ψ the angle generated by the deflection (cf. Gudmundsson, 2011a). Now that we have a quantitative definition of strain, we can proceed to linear elasticity theory, the basis of which is Hooke's law.

3.3.3 Hooke's Law

The principal equation of linear elastic behaviour of rocks and other solids is Hooke's law, the one-dimensional version of which is:

$$\sigma = E\varepsilon \tag{3.10}$$

where σ is the **normal stress** applied to the rock (for example, a specimen in a laboratory), ε is the **normal strain** resulting from the applied normal stress, and E is Young's modulus. **Young's modulus**, that is, the ratio $E = \sigma/\varepsilon$, is a material property of rocks and other solids. More specifically, the Young's modulus is the slope or gradient of the stress–strain curve or graph (Fig. 3.6). The steeper the slope, the higher is the Young's modulus of the rock.

Equation (3.10) is the currently used one-dimensional presentation of Hooke's law. As originally stated (by Hooke himself in the seventeenth century), the extension or displacement of a spring or bar is proportional to the applied force. (Notice that **bar** in solid mechanics is an elongated solid of uniform cross-sectional area, normally with a circular or rectangular, sometimes a hexagonal, cross-section.) In this original formulation, Hooke's law may be presented by the following equation:

$$F = k\Delta L \tag{3.11}$$

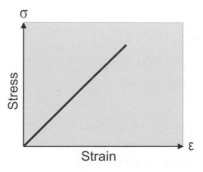

Fig. 3.6 Young's modulus of a solid such as rock is the slope of the stress–strain (σ–ε) curve, here shown as a straight line. The steeper the slope, that is, the closer the line is to being parallel with the σ-axis, the higher is the Young's modulus and the stiffer the rock.

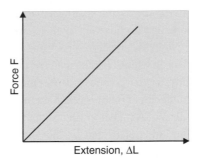

Fig. 3.7   Stiffness k of a solid such as rock is the slope of the force–extension ($F - \Delta L$) curve, here a straight line.

where F is the normal force (usually tensile in solid-mechanics experiments), ΔL is the extension due to the tensile force, and k is a constant, known as **stiffness.** The stiffness constant is the slope or gradient of the force–extension curve or graph (Fig. 3.7), namely the ratio $k = F/\Delta L$, and it has the unit N m^{-1}. By contrast, Young's modulus has the unit N m^{-2} (Pa). Also, k depends on the size of the loaded body (the body subject to the force F). If the size of a rock body subject to a given force F is increased, then the corresponding extension ΔL of that body increases while its measured stiffness k decreases. Since stress is independent of the size of the body considered, it is normally more suitable than force for analysing rock response to loading. Generally, the Young's modulus as measured in laboratories does not depend on the size of the body. However, the field or *in situ* Young's modulus – the modulus of outcrop rocks or, generally, of rocks in nature – depends on the number and distribution of pores, fossils, fractures, and other inhomogeneities and discontinuities that a rock body contains. The same applies to the effective Young's modulus of any solid that contains flaws or cracks. For rocks, pores, fractures, and other discontinuities tend to increase in number with increasing size of the rock body considered. It follows that, up to a certain limit, for rocks in the upper (brittle) part of the crust, Young's modulus generally **decreases** with increasing size of the rock body being considered.

These considerations indicate that k and E are strictly not the same. However, both are used as a measure of rock stiffness, that is, the resistance of the rock to strain. So commonly we refer to rocks with high Young's moduli, such as many lava flows, dikes, and sills, as **stiff**. By contrast, we refer to rocks with low Young's moduli, such as many sedimentary and pyroclastic rocks, as **soft** or, more correctly, as **compliant.** Stiffness is, however, not directly related to **strength**, which is a measure of the rock resistance to failure. For brittle rocks, the main strengths considered are the shear strength (for shear fractures such as faults) and tensile strength (for extension fractures such as tension fractures and dikes). Thus, we can have stiff rocks with low tensile strength; for example, many lava flows are stiff but have low tensile strength because of their numerous columnar joints.

When using the simple version of Hooke's law (Eq. (3.10)), there are several assumptions implied as to the rock body. The main ones, some of which are already defined, are as follows:

- The rock body, say a volcano hosting a magma chamber, is **homogeneous**. This means that the rock properties do not vary with location within the volcano – they are the same everywhere within the volcano. A rock body that is not homogeneous is inhomogeneous or **heterogeneous**, in which case the material properties change with position within the body.
- The rock body, say a volcano, is **isotropic.** This means that its properties are the same regardless of the direction of measurement – they are the same in all directions at a point within the volcano. A rock body that is not isotropic is **anisotropic**, in which case the rock properties are directional, that is, they have different values in different directions at a given point within the body.
- The relation between stress and strain in the rock body is **linear** (Fig. 3.6). This is a basic assumption for a Hookean body. An elastic body that does not show a linear behaviour is **non-linear** (not a straight stress–strain graph but rather a curved one). If the behaviour of the body is not elastic, either linear or non-linear, then it is referred to as **inelastic**.
- The strains are **infinitesimal.** This means that the strains are so small that the second and higher-order powers of strain may be neglected. For some materials this assumption may be valid up to about 10% strain.
- When the rock body becomes loaded (subject to stress, pressure, or displacement) it **instantaneously** becomes strained. And when the load is removed, the strain and the strain energy disappear instantaneously and all the stored strain energy is released. Instantaneous implies that there is no time lag between the loading or unloading of the body and its straining or strain recovery (strain and strain-energy disappearance).

It is well known that rocks in general, and the layers that constitute volcanoes in particular, strictly do not meet all these assumptions. In particular, the rock properties within a volcano are normally **heterogeneous and anisotropic**. The anisotropy and heterogeneity, however, are much more pronounced in stratovolcanoes than in basaltic edifices. The assumption of linear elastic, or Hookean, behaviour is normally a good first approximation when modelling inflation and deflation and many other volcanotectonic processes. Equation (3.10) is, however, one-dimensional, and most volcanotectonic models are either two-dimensional or three-dimensional – and occasionally four-dimensional (when time-dependence is taken into account, and then they are no longer purely elastic).

There are several three-dimensional versions of Eq. (3.10). These are derived and discussed in a geological context by Gudmundsson (2011a). Here, the simplest version is presented, namely the following:

$$\varepsilon_1 = \frac{1}{E}[\sigma_1 - v(\sigma_2 + \sigma_3)] \tag{3.12}$$

$$\varepsilon_2 = \frac{1}{E}[\sigma_2 - v(\sigma_3 + \sigma_1)] \tag{3.13}$$

$$\varepsilon_3 = \frac{1}{E}[\sigma_3 - v(\sigma_1 + \sigma_2)] \tag{3.14}$$

The following notation is used. $\varepsilon_1 \geq \varepsilon_2 \geq \varepsilon_3$ are the **principal strains** in the rock; they are parallel with the principal stresses $\sigma_1, \sigma_2, \sigma_3$, in that order. E is Young's modulus, and v (nu) is **Poisson's ratio**, defined as the negative of the ratio of transverse strain to longitudinal strain (discussed in detail below). Equations (3.12)–(3.14) present the three-dimensional Hooke's law for an isotropic rock body where the principal strains are given in terms of the principal stresses. Other versions of the three-dimensional Hooke's law include (a) the same equations but using an arbitrary coordinate system with the axes x, y, and z rather than the principal stress axes, and (b) the principal stresses (σ_1, σ_2, and σ_3) given in terms of principal strains (ε_1, ε_2, and ε_3) thus:

$$\sigma_1 = \lambda\Delta + 2G\varepsilon_1 \qquad (3.15)$$

$$\sigma_2 = \lambda\Delta + 2G\varepsilon_2 \qquad (3.16)$$

$$\sigma_3 = \lambda\Delta + 2G\varepsilon_3 \qquad (3.17)$$

where λ is Lamé's constant, Δ is the dilation or volume strain, and G is shear modulus or modulus of rigidity. In these equations we have introduced several elastic constants, the meaning of which and use need some further discussion.

3.3.4 Elastic Constants

For isotropic elastic materials there are **two independent** elastic constants or moduli. In addition, several constants are used for convenience in various relations between stress and strain and for formulating and solving different types of problems. We usually refer to **five constants**, that is, Young's modulus, Poisson's ratio, shear modulus, bulk modulus, and Lamé's constant. In volcanotectonics, however, the reciprocal of the bulk modulus, namely the compressibility, is of fundamental importance in that it largely controls how magma chambers and reservoirs respond to changes in the magma volume (the added or subtracted volume of magma) and, most importantly, how much magma can flow out of a chamber or reservoir during an ordinary eruption (Chapters 6 and 8). We have already used the compressibility and many of the other constants in earlier chapters and in the equations above, but here we summarise the basic definitions of and relations between these and related moduli (cf. Appendix D).

Young's Modulus

Young's modulus is normally denoted by E (some authors use Y) and also referred to as the modulus of elasticity. It is the ratio of normal stress to normal strain (σ/ε) and has the unit of stress, N m^{-2} or pascal (Pa). Young's modulus is a measure of stiffness, and is commonly referred to as the stiffness. Thus, rocks with a high or large Young's modulus are referred to as stiff; those with a low or small Young's modulus are referred to as compliant, floppy, or soft. For a **linear elastic** material, Young's modulus is the slope of the stress–strain diagram (Fig. 3.6). Many rocks, however, are somewhat non-linear (that is, the normal stresses and strains are not linearly proportional), in which case the **tangent modulus** may be used. It represents the instantaneous ratio of stress to strain, as given by the slope of the tangent to the stress–strain curve. For such a non-linear material, however, the (tangent) Young's modulus changes within the elastic range, that is, along the curve, since the slope of the curve is changing.

Young's modulus for **fluids** is zero. This is useful when making numerical models: a fluid-filled cavity, such as a totally molten magma chamber, can then be modelled as an empty cavity (in three dimensions) or a hole (in two dimensions), with the internal pressure on the cavity surface or walls given as zero if the chamber is in lithostatic equilibrium or positive (non-zero) if the chamber is subject to an **excess fluid pressure** (fluid pressure in excess of the lithostatic stress). The values of Young's moduli for solid rocks show a very large range, commonly from 1 GPa to 100 GPa. However, very compliant or floppy rocks, such as some pyroclastic rocks, breccias, and clays, may have Young's moduli as low as 0.01 GPa; and unconsolidated sands, gravels, and clays, as low as 0.003 GPa. By contrast, the stiffest rocks, as measured in the laboratory, reach Young's moduli of 150 GPa (Appendix D). In a basaltic edifice, Young's moduli may vary by one or two orders of magnitude, and in a typical stratovolcano easily by three and possibly four orders of magnitude. This variation in Young's moduli is one reason why the local stress in a stratovolcano is so much more heterogeneous and anisotropic than in a basaltic edifice. The heterogeneity of the local stress field also explains partly why stratovolcanoes arrest a higher proportion of their injected dikes and inclined sheets (and fractures in general), that is, why they are stronger (more resistant to brittle failure) than basaltic edifices (Gudmundsson, 2009).

Poisson's Ratio

Poisson's ratio and Young's modulus are the most commonly used elastic constants in volcanotectonic studies. **Poisson's ratio** is denoted by v and is the negative of the ratio of transverse strain to longitudinal strain. It is defined as the negative of the ratio to ensure that the ratio itself is positive. For example, during axial extension in a uniaxial test, Poisson's ratio is the (negative of the) ratio of the lateral contraction to the axial extension. Because Poisson's ratio measures a change in one dimension over another dimension (both measured in the same length unit), it follows that Poisson's ratio itself has no unit and is a pure number. The reciprocal of Poisson's ratio is used in some equations. It is referred to as **Poisson's number**, denoted by m, and defined as:

$$m = \frac{1}{v} \tag{3.18}$$

The range in the values of Poisson's ratio is much more limited than the range in the values of Young's modulus. When Poisson's ratio reaches 0.5, the material behaves as a fluid at rest, that is, it is not subject to any shear stress. Some plant tissues approach this value, and so does rubber. By contrast, if uniaxial extension produces no lateral contraction, then Poisson's ratio is 0.0: a well-known material approaching this value is cork. For most solid rocks, however, Poisson's ratios fall within a very narrow range, between 0.15 and 0.35, with most values between 0.2 and 0.3. Thus, for typical solid rocks, Poisson's ratios are unlikely to vary by more than a factor of three, and usually by a factor of two or less.

This limited range of values for Poisson's ratios of rocks, in comparison with the range of Young's moduli, has important implications for volcanotectonic modelling. For example, when modelling layered rocks, such as in basaltic edifices and stratovolcanoes, Poisson's ratios are often assumed to be uniform whereas Young's moduli are assumed to vary by two or three (or more) orders of magnitude between layers.

For a rock body subject to loading, Poisson's ratio measures the body's tendency to change its **volume** in proportion to its tendency to change its **shape**. As this ratio increases and approaches 0.5, the rock body increasingly responds to the loading through a change in **shape** rather than a change in volume. When the Poisson's ratio reaches the theoretical value of 0.5, there will be no change in volume regardless of loading. Such a material has a very high bulk modulus K and is referred to as **incompressible.** No rocks are incompressible, and neither are any volcanic fluids, but rubber is close to being incompressible, with a typical Poisson's ratio of 0.43. Under load, the shape changes of a soft rubber are comparatively large but the volume changes are minimal. Fluids such as water have Poisson's ratios close to 0.5 and are therefore often assumed to be incompressible. However, no materials are strictly incompressible, and the **compressibility of fluids** must, for example, normally be taken into account when modelling magma chambers and reservoirs (Chapters 6 and 8). Alternatively, we can say that Poisson's ratio measures the tendency of the loaded body to respond through dilation rather than through shear or distortion. Small Poisson's ratios indicate that the body is comparatively resistant to **shear** and distortion but less resistant to **dilation**.

Shear Modulus

The shear modulus is also referred to as the **modulus of rigidity** and is denoted by G. It relates shear stress and shear strain and has the unit of stress. It is a measure of the resistance to **shape changes**. Shear modulus is related to Young's modulus E and Poisson's ratio v through the equation:

$$G = \frac{E}{2(1+v)} \tag{3.19}$$

Bulk Modulus and Compressibility

The bulk modulus is also known as **incompressibility** and is denoted by K. It is a measure of the change in rock volume, that is, the dilation (at a constant shape), when the rock is subject to a given mean or spherical stress. It has the same unit as stress. Bulk modulus is the ratio of the mean stress $\bar{\sigma}$ to the dilation or volume strain Δ, namely

$$K = \frac{\bar{\sigma}}{\Delta} \tag{3.20}$$

Bulk modulus is related to Young's modulus and Poisson's ratio through the equation:

$$E = 3K(1 - 2v) \tag{3.21}$$

For volcanotectonics, particularly magma-chamber and magma-reservoir modelling, a more important constant is the reciprocal of the bulk modulus, namely **compressibility** which is denoted by β. It is defined as:

$$\beta = \frac{1}{K} \tag{3.22}$$

Compressibility has the unit of 1/stress, that is, Pa^{-1}. The relative compressibilities of the fluid, say magma or gas, in a reservoir or a chamber and the host rock (the matrix) partly determine (1) how much fluid/magma the reservoir/chamber can contain before rupture, and (2) when rupture occurs with fluid/magma flow, how the chamber/reservoir responds mechanically to reduced fluid/magma volume (Chapter 6). Compressibility also affects the response of the reservoir/chamber when magma is added to it. Thus, the reservoir/chamber expansion when magma is added to the reservoir/chamber, and the contraction or shrinkage when magma is subtracted from the reservoir/chamber, are partly determined by the fluid/magma and matrix compressibilites (Chapter 6).

Lamé's Constant

Lamé's constant is denoted by λ and has the unit of stress. It has no simple physical meaning but is a mathematical convenience. Its main use is to make some equations more elegant and easier to handle. The constant is defined in terms of Young's modulus and Poisson's ratio as:

$$\lambda = \frac{vE}{(1+v)(1-2v)} \tag{3.23}$$

Various relations between the five main constants are presented in Appendix D at the end of the book. For **isotropic rocks**, only two of the constants are independent and thus are sufficient to set up and solve simple analytical and numerical volcanotectonic models. Many rocks are, of course, **anisotropic** because of the preferred orientation of the crystals. Others, however, may be regarded as approximately isotropic on a small scale, that is, the scale of individual layers or units. This follows because the grains in many sedimentary rocks and of crystals in many igneous rocks (and metamorphic rocks) are essentially randomly oriented so that their individual anisotropies cancel out. Whole volcanic edifices, or large parts of them, however, are normally regarded as being anisotropic because of mechanical layering. As indicated above, the most commonly used elastic constants are Young's modulus and Poisson's ratio. These, together with compressibility, are the constants most often used in the volcanotectonic models and the problems solved in this book.

Lamé's constant λ and the shear modulus G are often referred to jointly as **Lamé's constants.** They are related through Poisson's ratio by the equation:

$$G = \frac{\lambda(1-2v)}{2v} \tag{3.24}$$

It follows from Eq. (3.24) that when Poisson's ratio $v = 0.25$, as is common for many solid rocks, then $G = \lambda$. For this case, referred to as **Poisson's relation**, the above five constants for isotropic rocks reduce to **four** constants. The values of the constants E, G, K, and λ range from 0 to very large but finite values ($<\infty$), whereas v ranges from -1 to approximately 0.5, but is mostly between 0 and less than 0.5.

3.4 Magma Chambers as Nuclei of Strain – the Mogi Model

Traditionally, in surface-deformation studies of volcanoes, the source magma chambers have been modelled in two basic ways: as a pressure source (a nucleus of strain without finite size) and as a finite-size cavity. Both types of model normally assume that the crustal segment hosting the chamber behaves as linearly elastic, although there are also some cavity models assuming viscoelastic or plastic crustal behaviour (e.g. Bonafede et al., 1986; Segall, 2010).

The most common pressure-source model is the **point source**, which, as the name implies, does not really have any finite-size geometry. This model is still the most popular one to explain surface deformation associated with inflation and deflation in volcanoes, as obtained from geodetic measurements during periods of unrest. The other basic model type regards the magma chamber as a finite-size body – as a **hole** for two-dimensional models and as a **cavity** for three-dimensional models. These basic types of model can both be presented through analytical solutions. The crustal segment is then normally assumed to be a homogeneous and isotropic elastic body so that all the layers, faults, and other variations in lithology and mechanical properties are ignored. For realistic modelling of stress fields and surface deformation associated with a composite volcano, however, the crustal segment hosting the chamber must be regarded to be layered and therefore anisotropic. For such models, numerical methods are used.

During periods of unrest in a volcano its surface deformation is traditionally explained in terms of a pressure change in the associated magma chamber, modelled as a **point pressure** or a **nucleus of strain**. This is the so-called '**Mogi model**', a widely used model in volcanology (Mogi, 1958). In this model, the chamber is regarded as a concentrated (point) force of an infinitesimal (very small, a point) volume (Fig. 3.1). The stresses and displacements produced by a nucleus of strain located at a certain distance below the surface of a semi-infinite elastic body or an elastic half-space can be obtained through analytical solutions. These solutions were initially derived by Melan (1932) and Mindlin (1936) and used in geology by Anderson (1936).

More specifically, using a nuclei-of-strain model, Anderson (1936) was the first to explain, in formal terms, the attitude of dikes and inclined sheets injected from a shallow magma chamber, located in a homogeneous, isotropic **elastic half-space**. (In volcanotectonics, an elastic half-space is a model where there is only one surface – the Earth's surface – and all the other dimensions are practically infinite. It is thus analogous to an infinite body or space cut in half, hence the name. A magma chamber, a dike, or a fault, for example, subject to a given loading, is then introduced into the half-space and the stresses and displacements calculated.) Since dikes and sheets are mostly extension fractures (Chapters 1 and 2) they follow the trajectories of the maximum compressive principal stress σ_1, and are perpendicular to the trajectories of the minimum compressive principal stress σ_3 (Chapters 7 and 10). From the nucleus-of-strain model, the orientations of the principal stresses can be calculated and, thereby, the attitudes of ideal dikes and inclined sheets far away from the magma chamber (the strain nucleus) itself (Chapters 7 and 10). Using a nucleus of strain referred to as the **centre of compression** (Love, 1927) for a magma chamber, Anderson (1936) also provided a model on the formation of collapse calderas and ring-dikes, assuming that both result from a magma-chamber underpressure, that is, the compression or contraction of an associated magma chamber.

The basic equations of the nucleus-of-strain or Mogi model can be presented in various ways. Here, we focus on the simplest versions. The vertical (u_z) and horizontal (u_r) displacements at the surface above the magma chamber (Fig. 3.1) can be expressed by the following equations:

$$u_z = \frac{(1-v)p_e R_1^3}{G} \frac{d}{(r^2 + d^2)^{3/2}} \tag{3.25}$$

$$u_r = \frac{(1-v)p_e R_1^3}{G} \frac{r}{(r^2 + d^2)^{3/2}} \tag{3.26}$$

where v is Poisson's ratio, p_e is the magmatic excess pressure in the chamber, R_1 is the radius of the chamber, G is shear modulus of the host rock, d is the depth to the centre of the chamber below the surface of the Earth, and r is the radial coordinate at the surface. Notice that the surface displacements at the free surface may be larger than those on a plane at the same distance from the chamber but inside the crustal segments. This is a well-known **free-surface** effect, which is also taken into account when measuring the apertures of fractures intersecting a free surface. (Recall that a surface is regarded as free if it is not subject to any shear forces or stresses, only normal stresses or forces. A solid surface in contact with a fluid, such as rock in contact with magma or water, is a free surface.) As discussed later, the relation between the **dome** or **depression** at the surface induced by **inflation** or **deflation** of an associated magma chamber is complex, particularly when trying to infer the volume of magma leaving the chamber from the depression or subsidence at the surface.

Equations (3.25) and (3.26) can be written in different forms. For example, the Poisson's ratio for rocks is commonly 0.25. Substituting that value for v into Eqs. (3.25) and (3.26), they simplify to the following equations:

$$u_z = \frac{3p_e R_1^3 d}{4G(r^2 + d^2)^{3/2}} \tag{3.27}$$

$$u_r = \frac{3p_e R_1^3 r}{4G(r^2 + d^2)^{3/2}} \tag{3.28}$$

In Eqs. (3.25) to (3.28), the term $p_e R_1^3$ is a **single entity**, so that the excess pressure p_e cannot be separated from the radius R_1 of the chamber. This follows partly because the chamber is a nucleus of strain and thus not a cavity of a finite size but rather a '**point energy source**'. In the Mogi model it is always assumed that the chamber radius is much smaller than the chamber depth below the free surface, so that $R_1 \ll d$. Since the depth to a shallow chamber is commonly only a few kilometres, this assumption implies that the chamber radius should be very small, generally a fraction of a kilometre. (Some modellers, however, assume the chamber radius to be 1 km – Dzurisin (2006) – for reasons that are not entirely clear.) That assumption, however, is known to be untrue for most shallow magma chambers. The term $p_e R_1^3$ has the unit of energy (J) and is a measure of the **strain energy** accumulation in and around the chamber prior to its rupture and dike/sheet injection.

Because the term $p_e R_1^3$ is a single entity it follows that it is not possible to estimate independently the size (the radius R) of a Mogi chamber. The change in chamber volume ΔV of a Mogi chamber, however, can be crudely estimated. The volume ΔV added to the chamber to generate an excess pressure p_e is equal to the radial displacement of the surface of chamber u_R times the surface area of the spherical chamber, $A = 4\pi R_1^2$, so that:

$$\Delta V = 4\pi R_1^2 u_R \tag{3.29}$$

For a spherical cavity, used here for the Mogi chamber, it is known (e.g. Jaeger et al., 2007) that the radial displacement u_R for a full-space approximation to the half-space model is given by:

$$u_R = \frac{p_e R_1^3}{4Gr} \tag{3.30}$$

Combining Eqs. (3.29) and (3.30), the volume change ΔV (for $r = R_1$) becomes:

$$\Delta V = \frac{\pi p_e R_1^3}{G} \tag{3.31}$$

Using Eq. (3.31), Eqs. (3.27) and (3.28) may be rewritten in terms of volume change ΔV of a Mogi chamber as follows:

$$u_z = \frac{3d\Delta V}{4\pi(r^2 + d^2)^{3/2}} \tag{3.32}$$

$$u_r = \frac{3r\Delta V}{4\pi(r^2 + d^2)^{3/2}} \tag{3.33}$$

If the assumption of Poisson's ratio being 0.25 is regarded as unwarranted, then the more general Eqs. (3.25) and (3.26) may be similarly rewritten in terms of magma-chamber volume change and used for matching surface deformation during volcano unrest. The volume change ΔV in Eqs. (3.31) to (3.33) is not strictly equivalent to the magma volume added to the chamber. This follows partly because in these equations the compressibility of the magma (and gas) in the chamber, as well as the pore compressibility, are not considered (Chapters 6 and 8).

When the nucleus-of-strain methods for elastic half-spaces were initially developed in the 1930s, analytical methods of this kind were the only ones available to relate magma-chamber pressure changes to the local stress and displacement fields. Similarly, at the time when these solutions were introduced to match the measured surface deformation associated with volcanic unrest periods in the 1950s, computers were still in their infancy and numerical programs for modelling volcano deformation were not available. At that time, the Mogi model was a clear step forward and had a very great and lasting impact on volcano-deformation studies. In the past decades, however, many easy-to-use numerical software programs have become readily available. These programs allow finite-sized magma chambers to be modelled at any crustal depth and in crustal segments with very sophisticated internal structures in terms of heterogeneity and anisotropy, including contacts and other discontinuities such as faults. Some numerical models of magma chambers are discussed below, but here we summarise the main advantages and disadvantages of the Mogi model in comparison with numerical models for volcanotectonic studies. Some of the **advantages** of the Mogi model are as follows:

- The Mogi model is conceptually and mathematically **simple** and easily understood. In particular, the analytical solutions are comparatively simple (e.g. Eqs. (3.27) and (3.28)) and their relation to the surface deformation or displacements is clear (Fig. 3.1).
- The Mogi model provides a clear relation between the term $p_e R_1^3$, with the unit of joule, and **strain energy**. While this term is not unique to the Mogi model – the basic form relates to all magma chambers modelled as cavities and subject to internal fluid excess pressure – it is particularly clear in the Mogi-model formulation (Eqs. (3.25)–(3.28)).
- Inversion of surface deformation data so as to fit a Mogi model normally gives a crude indication of the **depth** to the pressure change associated with a particular unrest period. This is, in fact, the main current use of the Mogi model. It is of interest and importance to know, for example, if the pressure source inducing surface deformation during an unrest period is a shallow chamber at a depth of, say, 2–4 km below the surface or, alternatively, a deep-seated reservoir at a depth of 15–20 km.

Some of the **disadvantages** of using the Mogi model in volcanotectonic studies are as follow:

- Because of the many assumptions made in the Mogi model, the chamber/reservoir depth estimate, which is the main result of the model, is commonly **poorly constrained**. For example, even where reasonably good GPS or InSAR data exist, the estimated depths of shallow chambers associated with a given unrest period may vary by factors of two to three or more. When, for instance, the magma-chamber depth is estimated as somewhere between 3 km and 10 km (as happens) the usefulness of the results is limited as regards assessing the likely time needed for a vertical dike to propagate to the surface from the chamber, and related hazards.
- For most volcanoes, many of the Mogi assumptions are poor. These include the following: (a) The assumption that the chamber **radius** R_1 is much smaller than its depth d below the surface. At the time of collapse-caldera formation the radius of the chamber is normally similar to that of the caldera. Given that many calderas have radii of 5–10 km or larger (Chapter 5), it is clear that the radius of the chamber is commonly of the same order of magnitude as, or larger than, the depth to the chamber. (b) The assumed chamber **shape**, such as it is, in the Mogi model is that of a sphere, whereas many, perhaps most, real magma chambers are oblate ellipsoids or sill-like for much of their lifetimes (Chapters 6 and 8). The stress and displacement fields around sill-like chambers are very different from those around spherical chambers. (c) Direct field studies of active and extinct volcanoes, as well as drill holes into many volcanoes, show that the assumption of an elastic half-space is a poor one. Volcanoes are heterogeneous and anisotropic

structures and commonly have many faults, numerous mechanically weak contacts, and other discontinuities. More specifically, volcanoes are composed of layers, often with widely different mechanical properties which may have great effects on the induced stresses and surface deformation.

• The main disadvantage of the Mogi model from the point of view of volcanotectonics, however, is that its results, even if the depth estimates were reasonably accurate, throw hardly any light on the **volcanotectonic processes** involved. Basically, the model can only account for the displacement and stresses far away from the chamber, because the model substitutes actual cavity-like magma chambers with vanishingly small point sources. Thus, the **stress concentration** around the chamber itself – which determines if and where the chamber rupture and magma injection takes place during an unrest period – cannot be determined using a Mogi model. Thus, the model cannot answer the question of if, and then where, dike injection is going to happen during an unrest period. And because of its unrealistic chamber shape, size, and elastic half-space assumptions, the Mogi model is of no help in forecasting the likely propagation path of an injected dike.

3.5 Magma Chambers as Cavities – Analytical Models

All magma chambers have finite size, commonly with a radius of several kilometres (Fig. 1.17). Shallow magma chambers in many areas have roofs at depths of 1–5 km below the surface (Chapter 6). It follows that the chamber radius is commonly similar to or larger than the depth to the chamber. Magma chambers are thus best modelled as cavities, of shapes and depths below the surface in accordance with the available data for any particular chamber. Depending on the aim of the modelling, the chamber may be treated as a two-dimensional **hole** or as a three-dimensional **cavity**. Cavity models of chambers, particularly where crustal anisotropy is taken into account, give much more accurate information about the associated local stresses than nucleus-of-strain models and, in addition, allow us to model the physical processes associated with magma-chamber rupture and dike/sheet injection and propagation.

The models discussed in this section are of two types: three-dimensional cavities and two-dimensional holes. Both types are discussed in detail elsewhere as a part of elasticity theory (e.g. Savin, 1961; Soutas-Little, 1973; Saada et al., 2009) and as part of magma-chamber modelling (Gudmundsson, 2006, 2011a), but the main results are summarised here. First, I present the cavity results, and then the hole results.

3.5.1 Cavities

In analogy with the Mogi model, consider first a spherical magma chamber with a radius R_1 much smaller than the distance d to its centre from the free surface or the hosting crustal segment (volcanic zone or volcano). For coordinate convenience, the elastic crustal segment containing the chamber is also assumed to be a sphere with a radius R_2. The solution is for a **full space** (not a half-space), so that the radius of the crustal 'sphere' hosting the

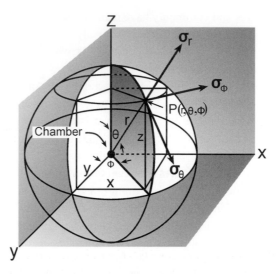

Fig. 3.8 Coordinates of a spherical magma chamber (indicated) in the crust. Spherical coordinates of a given point in the crust $P(r, \theta, \varphi)$ as well as the radial compressive stress σ_r and the circumferentical tensile stresses σ_θ and σ_φ associated with the magma chamber are indicated. The chamber is subject to fluid excess magmatic pressure p_e as the only loading.

chamber is much larger than that of the chamber, that is, $R_2 \gg R_1$. Using spherical polar coordinates $(r, \theta, \varphi) - r$ is the radius vector (distance), θ is the angle between the radius vector r and a fixed axis z, and φ is the angle measured around this axis (Fig. 3.8) – the radial stress σ_r away from the chamber and due to the chamber's excess pressure p_e is given by:

$$\sigma_r = p_e \left(\frac{R_1}{r} \right)^3 \tag{3.34}$$

Similarly, the two other principal stresses σ_θ and σ_φ, which must be equal because of spherical symmetry, are given by:

$$\sigma_\theta = \sigma_\varphi = -\frac{p_e}{2} \left(\frac{R_1}{r} \right)^3 \tag{3.35}$$

In case the sphere becomes very small so that the radius is a tiny fraction of the depth to the chamber, that is, $R_1 \ll d$ and $R_1 \to 0$ while the product $p_e R_1^3$ is still finite, the intensity of the point-pressure cavity is given by:

$$U_n = p_e R_1^3 \tag{3.36}$$

which has the unit of energy (J). More specifically, U_n is then the **strain-energy nucleus** used in the Mogi model (cf. Gudmundsson, 2012a). Mathematically, such a tiny strain nucleus makes perfect sense as the source of stress and displacement in the crust. Physically, however, as a model of a shallow magma chamber, the sense is less obvious. This follows because a tiny strain nucleus would, in geological terms, be with a radius of, say, tens of metres or less, and thus would be unable to function as a magma chamber. For

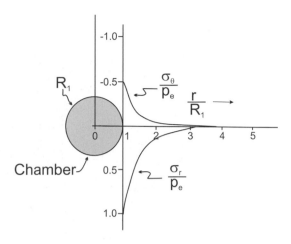

Variation (decrease) in radial σ_r and circumferential σ_θ stress around the spherical magma chamber in Fig. 3.8. The chamber radius is R_1 and its only loading is excess magmatic pressure p_e. The radial distance as measured from the chamber r is made dimensionless through dividing by the chamber radius R_1. The stresses generated by the excess pressure in the chamber decrease as the cube of the distance from the margin of the chamber (cf. Gudmundsson, 2006).

example, a spherical 'magma chamber' with a radius of 5 m would have a volume of just over 500 m³, which is far too little to generate a dike, let alone feed an eruption. To have the chance of functioning as a magma chamber, the intrusion has to be at least several hundred metres, and normally kilometres, in diameter.

Inspection of Eqs. (3.35) and (3.36) indicates that the intensity of the stress field around a spherical chamber, with p_e as the only loading, falls off inversely as the cube of the distance (radius vector) r from the surface of the chamber. Substituting $r = R_1$ in these equations, the compressive stress at the surface of the magma chamber is $\sigma_r = p_e$ and the tensile stress $\sigma_\theta = \sigma_\varphi = -0.5\, p_e$ (Fig. 3.9). The results show that the effect of the chamber on the surrounding stress field is **local**. At distances of between one or two radii of the chamber, the **local stress** effects become vanishingly small and the **regional stress** field takes over from the local field. An example of the change from a local to a regional stress field is documented from an extinct central volcano (a collapse caldera) in West Iceland (Fig. 3.10). There, the thickness and attitude of the sheet intrusions change abruptly at a distance of about 11 km from the centre of the caldera (Fig. 3.11). Thus, at about 11 km from the centre of the caldera, or about **one radius** from the margin of the magma chamber (assumed at one stage to be equal in diameter to that of the caldera), the local stress field (controlling the attitude and, partly, the thickness of the local inclined sheets) vanishes and the regional rift-zone stress field takes over, resulting in dominatingly vertical, thick regional dikes being emplaced.

Other loadings than internal excess pressure p_e may also be responsible for unrest periods, magma-chamber rupture, and dike injection. One such loading is external regional **tensile stress** $-\sigma$ in the direction of the spreading vector σ_3. Using the coordinate system defined in Fig. 3.12, the uniaxial tensile loading $-\sigma$ is parallel with the z-axis. In a rift zone,

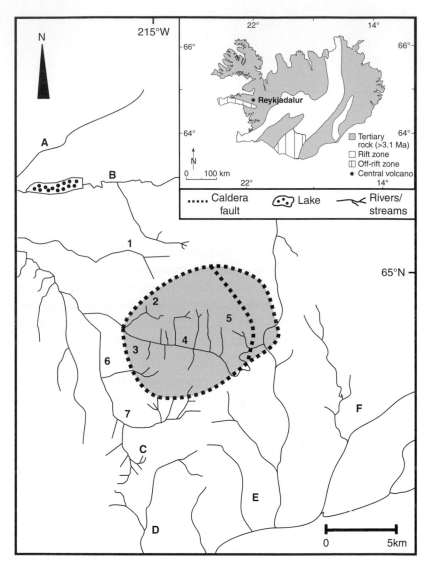

Fig. 3.10 Location of the Reykjadalur Central Volcano, active from about 6 Ma to 4 Ma, in West Iceland. The numbers and letters denote the stations where the sheet intrusions were measured: 1–7, primarily inclined sheets and radial dikes of the local sheet swarm; and A–F, primarily the dikes of the regional swarm (Gautneb and Gudmundsson, 1992).

the spreading vector, and thus $-\sigma$, is horizontal, but the results presented here are completely general and apply to any spherical cavity subject to tensile loading. It thus does not matter with which particular axis of the coordinate system the direction of the tensile loading coincides. For a rift zone, however, it may be helpful to imagine that the sphere in Fig. 3.8 is rotated by 90° so that the z-axis becomes horizontal and parallel with the spreading vector.

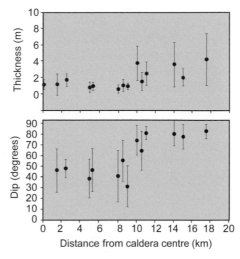

Fig. 3.11 Changes in arithmetic average thickness and dip of sheet intrusions with distance from the centre of a caldera (shallow magma chamber). The vertical error bars indicate the range in values at each station in the fossil Reykjadalur Central Volcano in West Iceland (Fig. 3.10). The thickness and dip increase abruptly at a distance of about 10–11 km from the centre of the caldera/chamber, indicating a change from a local sheet swarm to a regional dike swarm at that distance (Gautneb and Gudmundsson, 1992).

Before loading by $-\sigma$, the chamber is assumed to be in lithostatic equilibrium with its host rock, in which case the vertical stress due to overburden pressure (Eq. (3.6)) is **balanced** by the magma pressure and may be ignored in the analysis. With reference to the coordinates in Fig. 3.12, the stresses at the surface of the sphere are (Timoshenko and Goodier, 1970; Soutas-Little, 1973):

$$\sigma_\theta = -\frac{\sigma(27 - 15v)}{2(7 - 5v)} + \frac{15\sigma}{(7 - 5v)}\cos^2\theta \tag{3.37}$$

$$\sigma_\varphi = -\frac{\sigma(15v - 3)}{2(7 - 5v)} + \frac{15v\sigma}{(7 - 5v)}\cos^2\theta \tag{3.38}$$

With $\theta = 90°$ the second terms in Eqs. (3.37) and (3.38) becomes zero, so that the tensile stress is maximum at the equatorial plane of the spherical chamber and given by:

$$\sigma_\theta = -\frac{\sigma(27 - 15v)}{2(7 - 5v)} \tag{3.39}$$

$$\sigma_\varphi = -\frac{\sigma(15v - 3)}{2(7 - 5v)} \tag{3.40}$$

At the top and bottom of the spherical chamber, the loading by $-\sigma$ generates a compressive stress, the magnitude of which is:

$$\sigma_\theta = \sigma_\varphi = \frac{\sigma(15v + 3)}{2(7 - 5v)} \tag{3.41}$$

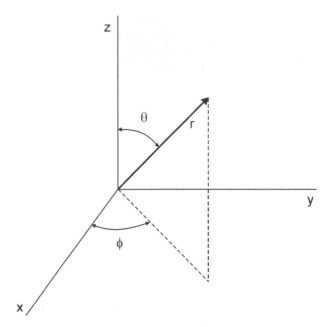

Fig. 3.12 Coordinate system for a magma chamber. Here, spherical coordinates (r, θ, φ) are used, as in Fig. 3.8.

Thus, external tensile loading generates not only tensile stresses around a spherical magma chamber but also compressive stresses. More specifically, the compressive stress obtained from Eq. (3.41) concentrates at the equatorial plane of the magma chamber whereas the tensile stresses (Eqs. (3.39) and (3.40)) concentrate at the top and bottom of the chamber.

Most magma chambers have shapes different from that of a sphere. Some chambers are highly **irregular** in shape, but are unlikely to remain so except over periods which are short in comparison with the lifetimes of the chambers (Chapters 6 and 8). The **long-term shape** of most chambers can be approximated through cavities of a general ellipsoidal form (Fig. 3.13). Simple closed-form solutions such as we obtained above for spherical chambers are, unfortunately, not available for stress concentrations around chambers of general ellipsoidal form. There are stress-concentration results, however, that can be used for general **ellipsoidal magma chambers** (Fig. 3.13) located at comparatively great depths below the surface. In a general **triaxial** ellipsoidal chamber, all the semi-axes of the ellipsoid are of dissimilar lengths, that is, $a \geq b \geq c$ (Fig. 3.13); in a **two-axial** (biaxial) ellipsoidal chamber, two of the semi-axes or dimensions of the ellipsoid are of the same length. For a prolate ellipsoidal chamber (prolate means that the 'polar dimension' is larger than the other two dimensions) with horizontal width $2c$, height (dip-dimension) $2b$, and horizontal length (strike-dimension) $2a$ (Fig. 3.13), the fraction c/b is referred to as its shape ratio. The focus is on the uppermost part of the chamber where rupture and dike or sheet injection is most likely to take place. The tensile-stress concentration at the top of the prolate magma chamber, for host rocks with Poisson's ratio of 0.25, is presented in Fig. 3.14.

For the special case $c/b = 1$ and $a = 1$, the chamber becomes spherical and the tensile-stress concentration is as calculated from Eqs. (3.39) and (3.40).

In the case of a triaxial magma chamber that is elongate parallel with the axis of a volcanic rift zone, two-dimensional (hole) models may be used to calculate the stress concentration around the vertical cross-section of the chamber. For example, the chamber strike-dimension (parallel with the rift-zone axis) is much greater than its dip-dimension and width (so that $2a \gg 2b$ and $2c$), the maximum theoretical tensile stress σ_{max} at point A at

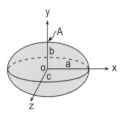

Fig. 3.13   Magma chamber modelled as a three-dimensional general ellipsoidal cavity where all the axes – being double the lengths of the semi-axes, that is, $2a$, $2b$, and $2c$ – have different lengths. The tensile stress concentration at point A is calculated and presented in Fig. 3.14.

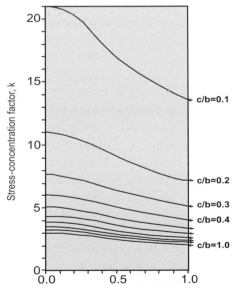

Fig. 3.14   Tensile-stress concentration, presented by the stress-concentration factor k, at point A on the surface of the ellipsoidal magma chamber in Fig. 3.13. Here, the only loading is the remote applied tensile stress σ. The tensile stress presented by the stress-concentration factor $k = \sigma_{max}/\sigma$, where σ_{max} is the maximum tensile stress (data from Sadowsky and Sternberg, 1947, 1949; cf. Gudmundsson, 2006, 2011a). The tensile-stress concentration is given for various shape ratios (c/b) and aspect ratios or elongation (b/a) for the ellipsoidal magma chamber.

the chamber top, is obtained approximately from the 'elliptical hole' (two-dimensional) formula:

$$\sigma_{max} = -\sigma\left[\frac{2b}{c} + 1\right] \tag{3.42}$$

Notice that the theoretical σ_{max} is generally much higher than the actual maximum tensile principal stress σ_3 at the magma-chamber roof rupture and dike/sheet injection . Here, $-\sigma$ is the remote tensile stress related to divergent plate movements, that is, plate pull. As an example, when $c/b = 0.1$ and a very elongate magma chamber (so that $b/a \approx 0.0$), Eq. (3.42) gives $\sigma_{max} = -21\sigma$, in agreement with Fig. 3.14. This brings us to general two-dimensional models of magma chambers.

3.5.2 Holes

To get rough ideas about the stress field around a magma chamber, it is often helpful to make simple two-dimensional hole models. The basic assumption in such models is that one of the dimensions of the chamber is **much larger** than the other dimensions. For example, if the ellipsoidal reservoir in Fig. 3.13 is such that $2a \gg 2b = 2c$ (a **two-axial** ellipsoid), the stress concentration around a vertical cross-section not too close to the lateral ends of the chamber can be calculated from Eq. (3.42) on the assumption that the only loading is external tensile stress $-\sigma$ in a direction perpendicular to the direction of the long axis $2a$. If the only loading is internal excess pressure p_e in the chamber (and thus no external tensile stress), then Eq. (3.42) becomes:

$$\sigma_{max} = -p_e\left[\frac{2b}{c} - 1\right] \tag{3.43}$$

Some deep-seated reservoirs beneath volcanic zones are presumably very **elongated** (Chapters 6 and 8), in which case the maximum stress concentration in cross-sections may be calculated using Eqs. (3.42) and (3.43). In addition, some shallow crustal chambers are elongated in the direction of the volcanic zone within which they are located. An example is the well-exposed fossil magma chamber of Slaufrudalur in Southeast Iceland (Fig. 3.15; cf. Fig. 1.17). For this particular fossil magma chamber, the aspect (a/c) ratio is about 4.2, so that the hole approximation would be justified.

By analogy with Eqs. (3.34) and (3.35), the variation in the radial compressive stress σ_r with distance from the two-dimensional cross-section through the chamber with an excess magmatic pressure p_e is given by:

$$\sigma_r = p_e\left(\frac{R_1}{r}\right)^2 \tag{3.44}$$

and the variation in the circumferential tensile stress σ_θ is given by:

$$\sigma_\theta = -p_e\left(\frac{R_1}{r}\right)^2 \tag{3.45}$$

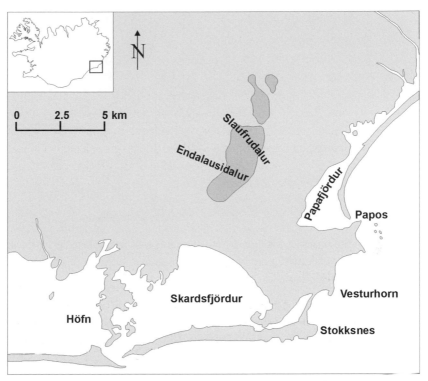

Fig. 3.15 Fossil magma chamber in Southeast Iceland. The elongated magma chamber, now exposed as a granophyre pluton at depths of as much as 2 km below the original surface of the associated central volcano, was an active chamber some 7–10 million years ago. The names of the two main valleys where the magma chamber is best exposed (Slaufrudalur and Endalausidalur) are indicated (cf. Gudmundsson, 2011b).

While both the radial and the circumferential stresses diminish with distance from the two-dimensional magma chamber (Fig. 3.16), they do so **more slowly** than in the case of the three-dimensional chamber (Fig. 3.9). More specifically, while for a three-dimensional spherical chamber the stress magnitudes decrease in proportion to the third power of the distance from the margin of the chamber, the stress magnitudes for the two-dimensional cylindrical chamber decrease in proportion to the second power of the distance from the margin of the chamber. Thus, the local stress effects would normally extend to a **shorter distance** from the margin of a spherical chamber than from a penny-shaped (sill-like) or generally a cylindrical (either with the long axis horizontal or vertical) magma chamber.

When the magma chamber is at a comparatively shallow depth, so that its lateral dimensions are similar to or larger than its depth below the Earth's surface, then the local stresses around the chamber become modified because of the free-surface effect. We consider again a chamber elongated in a direction parallel with the volcanic zone, with a depth to its centre d, and with a top at a depth that is similar to or less than the radius R_1 of

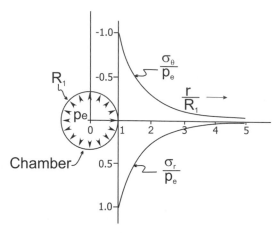

Fig. 3.16   Decrease in radial σ_r and circumferential σ_θ stresses with distance from the surface of a laterally elongated magma chamber with a circular cross-section of radius R_1 subject to an excess magmatic pressure p_e as the only loading (cf. Gudmundsson, 2006).

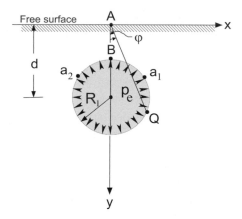

Fig. 3.17   Elongated magma chamber/reservoir with a vertical cross-section located with a centre at depth d below the Earth's free surface. The radius of the chamber is R_1, which is subject to an excess magma pressure p_e as the only loading. The tangent to the boundary of the magma chamber at points a_1 and a_2 is indicated by the line AQ (cf. Gudmundsson, 2006).

its circular cross-section (Fig. 3.17). If the only loading is the excess magma pressure p_e, then the stress stress σ_x at the surface of the hosting rift zone is (Savin, 1961):

$$\sigma_x = 4p_e \left[\frac{R_1^2(x^2 - d^2 + R_1^2)}{(x^2 + d^2 - R_1^2)^2} \right] \tag{3.46}$$

The maximum tensile stress σ_t at the surface of the volcanic zone or volcano hosting the magma chamber occurs at point A (at $x = 0$), the magnitude of which is:

$$\sigma_t = -\frac{4p_e R_1^2}{d^2 - R_1^2} \tag{3.47}$$

This tensile stress may either form or widen existing tension fractures and normal faults at the surface of the volcanic zone where, however, the tensile stress changes to compressive stress at points $x = |(d^2 - R_1^2)^{1/2}|$. The largest compressive stress σ_c occurs at points $x = |3(d^2 - R_1^2)^{1/2}|$ and its magnitude is:

$$\sigma_c = \frac{p_e R_1^2}{2(d^2 - R_1^2)} \tag{3.48}$$

its absolute value being $1/8\ \sigma_t$.

Normally, rupture and dike injection is most likely to occur at the chamber boundary where, theoretically, the **tensile stress** reaches its maximum value. To find this value, the general equation for the circumferential stress at the boundary of the chamber σ_θ is first presented as (Savin, 1961):

$$\sigma_\theta = -p_e(1 + 2\tan^2\varphi) \tag{3.49}$$

the angle φ being defined in Fig. 3.17. The upper part of the magma chamber is normally more likely to inject dikes than the lower part, so that the focus is on the upper part. From Eq. (3.49) the highest value of σ_θ, denoted by σ_b, occurs at the points at a_1 and a_2 where the angle φ is maximum, that is, where the line AQ is the tangent to the boundary of the magma chamber (Fig. 3.17), and it is given by:

$$\sigma_b = -\frac{p_e(d^2 + R_1^2)}{d^2 - R_1^2} \tag{3.50}$$

From Eqs. (3.49) and (3.50) we conclude the following:

- When $d > 1.73R_1$, the maximum chamber-induced tensile stress occurs at its boundary (perimeter), at points a_1 and a_2 (Fig. 3.17; Eq. (3.50)). Under these conditions, the maximum tensile stress occurs at the chamber boundary and favours the injection of dikes or inclined sheets.
- When $d < 1.73R_1$, the maximum chamber-induced tensile stress occurs at the surface of the volcanic zone, at point A (Fig. 3.17; Eq. (3.47)). Under these conditions, the stress field favours the formation of tension fractures (Fig. 1.6) or the widening of existing tension fractures (Fig. 1.9) and normal faults at or close to the surface.
- When $d = 1.73R_1$, the maximum chamber-induced tensile stresses at the surface of the volcanic zone and at the chamber boundary are equal, the magnitude of the stress being $\sigma_b = \sigma_t = 2p_e$.

As regards the local stress field at the boundary of the chamber, at point B (Fig. 3.17) the tensile stress is equal to p_e, whereas at points A, a_1, and a_2 the tensile stress is larger than p_e. At points A, a_1, and a_2 the stress depends on the difference between the depth to the centre of the magma chamber d and the chamber radius R_1. In Eqs. (3.47) and (3.50) the denominator is $d^2 - R_1^2$. It follows that when $R_1 \to d$, that is, when the depth to the point B at the top of the magma chamber decreases, then σ_b and σ_t may, theoretically, become many times greater than p_e. The *in situ* tensile strength of the host rock forming the boundary of the chamber is normally 0.5–9 MPa, so that the actual tensile stresses generated are limited to these values. When the condition of rupture is reached at points a_1 and a_2, there will be an injection of inclined sheets. One or more such injections result in compressive stress being generated in the roof of the chamber, and this stress relaxes the tensile stress in its vicinity, so that, for a while, no new sheet injections will take place.

3.5.3 Sills

Many, and perhaps most, shallow magma chambers have **sill-like geometry** (Chapters 6 and 8). Some may be described as oblate ellipsoids, but not all have circular lateral cross-sectional area. Sills, like other sheet intrusions, can be modelled using Volterra's dislocations. This approach is used by many and is summarised and discussed by Okada (1985), Dzurisin (2006), and Segall (2010). In particular, Dzurisin (2006) presents simplified versions of the equations so as to calculate the surface displacements induced by sills.

Fracture-mechanics principles, however, provide better ways for static and dynamic modelling of dikes, inclined sheets, and sills, and their emplacement, than dislocation models. For example, imposed displacements, such as are used in dislocation models, do not match the boundary conditions well – in particular, the magmatic pressure and its variation along the sill/sheet/dike and the relation to magma transport – whereas hydraulic-fracture-mechanics models do.

Consider a totally molten sill-like chamber of circular lateral cross-section, the volume of which increases during an unrest period and inflation. This model is the so-called '**penny-shaped crack**', which has been widely used in fracture mechanics (Sneddon and Lowengrub, 1969). The mechanics of sill emplacement, and the relevant equations, are discussed in detail in Section 5.5, and by Gudmundsson (2011a,b, 2012a) and Barnett and Gudmundsson (2014). Here we merely state the main results relevant for the present discussion. The volume increase ΔV (fractional volume change) generated during the expansion of a sill-like chamber as magma is added to it is given by (cf. Eqs. (5.7)–(5.9)):

$$\Delta V = \frac{4}{3}\pi a^2 u \tag{3.51}$$

where a is the radius of the circular chamber and u is the vertical displacement, uplift, or doming of the roof of the chamber. Thus, u is half the total expansion or 'extra opening' of the chamber Δu_I and is given by (cf. Eq. (5.5)):

$$u = \frac{4p_e(1 - v^2)a}{\pi E} \tag{3.52}$$

where v is the Poisson's ratio of the host rock and E is its Young's modulus. Here we use p_e, which is the magmatic excess pressure generating the chamber expansion, rather than the overpressure for the sill formation used in Eq. (5.5). Combining Eqs. (3.51) and (3.52), the volume increase ΔV due to the inflation of the sill-like chamber becomes:

$$\Delta V = \frac{4}{3}\pi a^2 \left[\frac{4(1 - v^2)p_e a}{\pi E}\right] = \frac{16(1 - v^2)p_e a^3}{3E} \tag{3.53}$$

which is analogous to Eq. (5.7).

The spherical Mogi chamber expansion has the unit of energy, and it is useful to know the **strain energy** associated with the inflation of a sill-like chamber as well. We know that work done ΔW is equal to force times the displacement or distance in the direction of the force. The excess pressure p_e has the unit of force per unit area. Here the area is the magma-chamber boundary, such as the roof of the sill-like chamber. The work done in expanding (inflating) the sill-like chamber by the volume ΔV is positive (but would be negative when shrinking (deflating) the chamber) and it is given by:

$$\Delta W = p_e \Delta V \qquad (3.54)$$

The total strain energy in a body in equilibrium (here the crustal segment or volcano hosting the sill-like chamber) is equal to half the work done by the forces acting on the body through the associated displacements (e.g. Love, 1927; cf. Gudmundsson, 2012a). It follows that the strain energy through the work in Eq. (3.54) is half that work. Substituting Eq. (3.53) for ΔV in Eq. (3.54) and multiplying by 1/2, the strain energy U_0 generated in the crustal segment or volcano hosting the expanding sill-like chamber is given by:

$$U_0 = \frac{8(1 - v^2)p_e^2 a^3}{3E} \qquad (3.55)$$

This equation allows us to calculate the strain energy stored in the crustal segment or volcano hosting a sill-like magma chamber as a result of expansion or inflation of that chamber. During magma-chamber rupture and dike injection, the strain energy stored in the volcano is one of the main **driving 'forces'** available for the **dike** propagation and, in case an eruption happens, for '**squeezing' magma** out of the chamber to the surface.

The maximum surface uplift may correspond roughly to the vertical expansion of the sill-like chamber. Sill-induced surface deformation has been analysed by Sun (1969) – for hydraulic fractures, 'water-sills' – and by Fialko et al. (2001), and summarised by Dzurisin (2006) and Segall (2010). Generally, sill-like chambers induce proportionally larger maximum vertical surface displacements and smaller maximum horizontal displacements than corresponding Mogi spherical chambers. Also, the volume of the surface deformation (doming for inflation, depression for deflation) is roughly equal to that of the volume change of the sill-like chamber – in contrast to the common difference between these volumes for a Mogi chamber (Example 3.7).

In all these models, the magmatic excess pressure p_e is regarded as uniform (constant). There are, however, available solutions for sill-like chambers (and sills and dikes in general) where the excess pressure varies along the dimensions of the intrusion. Such overpressure or excess-pressure variations can, for example, be modelled using polynomials or Fourier series (Gudmundsson, 2011a; Kusumoto et al., 2013; Kusumoto and Gudmundsson, 2014).

Some of the models discussed above can be applied to sill-like chambers, and sills in general, that have lateral dimensions greater than their depths below the surface. Such sills, however, are commonly modelled using the model of a circular-plate bend through magmatic pressure applied to it. Eventually, many such sills may develop into **laccoliths**. The details of that development are discussed in Section 5.6, and by Pollard and Johnson (1973) and Gudmundsson and Lotveit (2012).

3.6 Magma Chambers as Cavities – Numerical Models

All polygenetic volcanoes are composed of rock **layers**, many of which have widely different mechanical properties. The **contacts** between the layers also have different mechanical properties. This means that even if the loading is simple, such as uniform excess pressure in an ideally shaped magma chamber as the only loading, the resulting stresses and displacements, and thus the associated deformation, may be complex. In particular, the **surface stresses** and **displacements** for a given loading depend strongly on the layering and contacts of the uppermost part of the volcano. This applies not only to large-scale deformation induced by magma-chamber inflation and deflation but also to the deformation associated with dike emplacement, as discussed in Section 3.7. To understand the tectonic processes inside a volcano during an unrest period, it is therefore important that the models used to interpret the local stress and displacement fields match as well as possible the actual geology. For any close-to-realistic matching of the internal structure of a volcano and how it responds to loading, numerical modelling is normally required.

In the past decade, there have been many numerical studies of the local stresses, displacements, and surface deformations of volcanoes. Here, we give a brief indication of typical results of numerical modelling of, primarily, the local stresses around magma chambers of various shapes in volcanoes with different types of layering. The displacements and, more specifically, the surface deformation can also be obtained from numerical models of this kind. The **local stresses**, however, are more important than either of these two for understanding the processes that occur inside volcanoes during unrest periods. In particular, the propagation, deflection, and arrest of dikes – that is, the dike paths and the likelihood that they reach the surface, resulting in an **eruption** – depend primarily on the local stresses. Similarly, faulting and earthquake activity inside the volcano (Chapter 4) depend much on the local stresses in different layers and, of course, stress concentrations around existing faults and other fractures. Most of the models are two-dimensional, but extensions to three-dimensions are easily made.

3.6.1 Chambers in Homogeneous and Isotropic Crustal Segments

For comparison, consider first a **two-dimensional** magma chamber of a circular vertical cross-section, a circular-hole model, located at comparatively great depth below the free surface of the volcano/volcanic zone. This is then effectively a magma chamber in a homogeneous, isotropic full space, which means that the thickness of the crustal segment hosting the chamber is many times greater (strictly infinite) than the diameter of the chamber. The results (Fig. 3.18) show that the trajectories of the maximum compressive principal stress σ_1 are radial from the chamber. Any injected dikes or inclined sheets will follow these trajectories. How far the dike/sheet propagates from the chamber depends on the injected magma volume (Chapters 6 and 8), overpressure (driving pressure), and, eventually, the solidification of the magma. The geometry of some **local sheet swarms** associated with shallow magma chambers shows certain general similarities with that of stress trajectories in Fig. 3.18 (cf. Chapter 7).

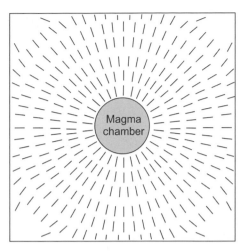

Numerical-model results of the trajectories, shown as short lines or ticks, of σ_1 around a magma chamber of a circular cross-section subject to internal excess fluid pressure p_e of 10 MPa as the only loading. The geometry of the stress trajectories, however, does not depend on the exact magnitude of the excess pressure. Sheet intrusions propagate parallel with the σ_1 trajectories. In contrast to the models in Fig. 3.4, Fig. 3.25, and Fig. 3.26, the results are here, and in the subsequent models in this chapter until Fig. 3.25, shown for the 'undeformed state' (cf. Gudmundsson, 2006).

Consider next the changes that occur when the free surface of the volcano/volcanic zone is taken into account. The **free surface** modifies the stress field associated with a magma chamber of a circular cross-section in a way that is easily understood (Fig. 3.19a). Recall that a free surface is one without any shear stresses (or forces) operating on it. Since fluids do not tolerate any shear stress/force – they flow under the action of shear stress – a free surface in a geological context is normally solid rock in contact with fluid. And that is the condition at the Earth's surface, where the solid surface of the crust is in contact with a fluid, either the air of the atmosphere or the water of lakes or the oceans. Because **principal stresses** define planes of no shear stress, it follows that principal stresses must be either perpendicular or parallel to a free surface. Thus, when the trajectories of σ_1 reach the surface they are no longer inclined as they may be inside the rock (Fig. 3.18) but rather are vertical (Fig. 3.19a).

The local stress field in Fig. 3.19a is for a shallow magma chamber where the only loading is internal excess **magma pressure** p_e. Consider next the same model, that is, a shallow magma chamber in a volcanic zone (semi-infinite elastic plate) but with a **horizontal tension** as the only loading (Fig. 3.19b). This would be a suitable loading for many unrest periods in volcanoes hosted by rift zones. Clearly, the local stress field in Fig. 3.19b is very different from that in Fig. 3.19a. Here the local stress field favours subvertical dikes rather than inclined sheets. Both loadings are presumably common in rift zones. That is, when magma is added to the chamber so that the internal excess pressure reaches the condition for rupture, the local stress field may be generally similar to that in Fig. 3.19a. Alternatively, when little or no new magma is added to the chamber but it

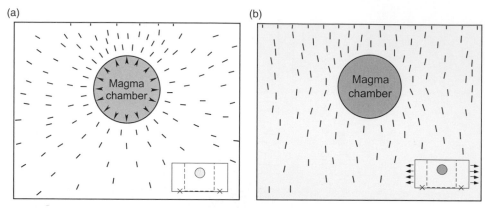

Fig. 3.19 Trajectories, shown as short lines or ticks, of σ_1 around a magma chamber of a circular cross-section located comparatively close to the Earth's surface. In (a) the chamber is subject to an internal magmatic excess pressure of 5 MPa as the only loading, whereas in (b) the chamber is subject to an external tension (plate pull) of 5 MPa as the only loading. The difference between the pattern of the σ_1 trajectories in (a) in this numerical model and the one in Fig. 3.18 is primarily because this chamber is close to the Earth's free surface, which modifies the stress field. The inset indicates that the model is fastened (shown by crosses), with the conditions of no displacement, in its lower parts. Fastening of the lower edges or in the corners are applied to all the numerical models in this chapter (cf. Gudmundsson, 1998).

reaches the condition for rupture as a result of the external tensile loading (due to the spreading vector), the local stress field may be similar to that in Fig. 3.19b. The alternating of stress fields is one major reason why the dips of sheets in local swarms in central volcanoes, such as in the rift zones of Iceland, commonly show two main peaks (Fig. 3.20). One peak, at 75–90°, corresponds to the dikes and the stress field in Fig. 3.19b, whereas the other peak, at about 10–40°, corresponds to the inclined sheets and the stress field in Fig. 3.19a (cf. Chapter 7; Gudmundsson, 1995).

To determine whether the magma chamber ruptures and injects dikes or inclined sheets, we must not only know the **trajectories** of the principal stresses but also the **magnitudes** of the stresses. For dike/sheet injection, the important principal stress is the maximum tensile (minimum compressive) stress, namely σ_3. Clearly, dike/sheet injection occurs only if the tensile stress at the boundary of the chamber reaches the *in situ* tensile strength of the host rock (range 0.5–9 MPa). And the injection is most likely to occur where the theoretical concentration of σ_3 is highest.

Figure 3.21 shows the concentration of σ_3 around the shallow chamber subject to internal excess pressure as the only loading. The main tensile-stress concentration occurs not at the top part of the chamber, but at its upper margins to either side of the top part. The location is favourable for the injection of inclined sheets, in agreement with the stress trajectories (Fig. 3.19a). Thus, when the internal magmatic **excess pressure** is the main loading in a shallow chamber of circular cross-section, the injection of inclined **sheets** is normally favoured. By contrast, the concentration of σ_3 around the same geometric type of a chamber when subject to external **tensile stress** as the only loading is very different (Fig. 3.22). The main tensile-stress concentration then occurs at the top of the chamber and, to a lesser

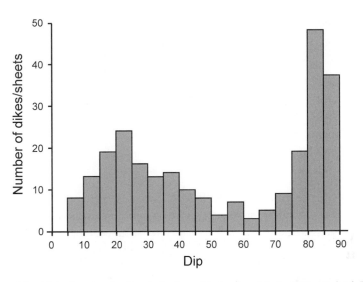

Fig 3.20 Dip distribution of sheet intrusions in a fossil central volcano (stratovolcano) in Southwest Iceland. The dip of the sheets ranges from 5° to 90°, but the frequency distribution shows two peaks. One is at 75–90° and represents steeply dipping (partly radial) dikes whereas the other is at 10–40° and represents gently dipping inclined (cone) sheets. The sheet intrusions range in thickness from several centimetres to 14 m, but most are of a thickness less than 0.5 m (cf. Gudmundsson, 1998).

degree, at its bottom. Clearly, this local stress field favours the injection of subvertical **dikes**, in agreement with the stress trajectories (Fig. 3.19). Thus, when external tensile stress is the main loading, as is common in rift zones, the injection of subvertical dikes is favoured. Again, the results generally fit with field observations, even if the models are for a homogeneous and isotropic crustal segment. In particular, the results show that the two peaks in the dip distribution in a typical sheet swarm in central volcanoes in Iceland coincide with subvertical dikes and much more gently dipping inclined sheets (Fig. 3.20; Chapter 7; Gudmundsson, 1995).

These results are generally applicable to volcanoes worldwide. For some volcanoes, such as at convergent plate boundaries, loading through internal excess pressure may predominate, whereas for others, such as at divergent plate boundaries, both processes are likely to operate. The surface stresses and deformation associated with magma chambers with circular cross-sections are similar to those obtained from the Mogi model (Fig. 3.1). The surface deformation above a **sill-like** (oblate spheroidal) magma chamber subject to internal excess pressure as the only loading is a simple doming of the surface above the chamber (e.g. Barnett and Gudmundsson, 2014). A more interesting case is the sill-like chamber subject to external tensile stress. Here, the surface tensile stress σ_3 concentrates above the **lateral ends** of the chamber (Fig. 3.23). Such a surface tensile stress (and the surface deformation roughly coincides with the σ_3 peaks) might possibly encourage the formation of ring-faults and caldera collapse, although the high tensile stress zone then needs to extend from the surface to the margins of the chamber, which is not always the case (Fig. 3.24). Thus, other types of loading are more likely to encourage ring-fault formation, a topic that will be discussed in connection with chambers in layered rocks, to which we turn now.

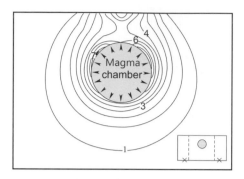

Fig. 3.21 Numerical model of the stress contours, showing the magnitude of maximum tensile (minimum compressive) principal stress σ_3 in mega-pascals around a magma chamber of circular cross-section subject to magmatic excess pressure p_e of 5 MPa. The highest tensile stresses occur at two localities on the upper margin of the chamber (where stresses exceed 7 MPa, as indicated). In this model, the chamber is located in a homogeneous, isotropic crust with a stiffness of 40 GPa and a Poisson's ratio of 0.25. This is the same model as in Fig. 3.19a (cf. Gudmundsson, 1998).

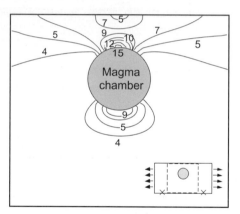

Fig. 3.22 Numerical model of the stress contours, showing the magnitude to maximum tensile (minimum compressive) principal stress σ_3 in mega-pascals around a magma chamber of circular cross-section subject to remote horizontal tensile stress of 5 MPa as the only loading. This is the same model as in Fig. 3.19b. The highest theoretical tensile stresses occur at the upper margin of the chamber, close to the free surface (where the stresses reach 15 MPa).

3.6.2 Chambers in Layered (Anisotropic) Crustal Segments

All crustal segments are heterogeneous and anisotropic, and this applies particularly to volcanoes. Of the two main types of polygenetic volcanoes, basaltic edifices and strato-volcanoes (including their collapsed versions, collapse calderas), the layers of stratovolcanoes have more varied mechanical properties. Many **basaltic edifices** are primarily composed of basaltic lava flows (and intrusions) with basaltic scoria layers in-between, so that the variation in mechanical properties is comparatively modest. By contrast,

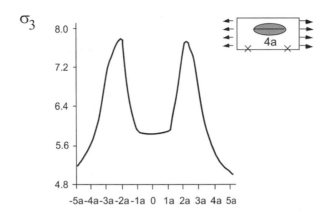

Fig. 3.23 Tensile stress induced by the loading of the shallow sill-like magma chamber in Fig. 3.24. The only loading is a horizontal tensile stress of 5 MPa. The highest theoretical tensile stresses occur above the margin of the chamber. The chamber diameter is 4*a*.

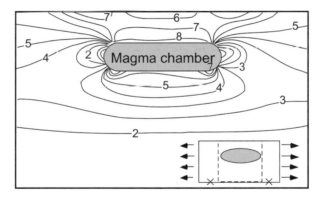

Fig. 3.24 Stress concentration around a shallow sill-like magma chamber subject to a remote horizontal tensile stress of 5 MPa as the only loading, the surface tensile stresses of which are shown in Fig. 3.23. The highest theoretical tensile stresses, indicated by contours of the maximum tensile (minimum compressive) principal stress, σ_3, occur at the upper margin of the chamber next to the free surface (where stresses reach 8 MPa). In this model, the chamber is located in a homogeneous, isotropic crust with a stiffness of 40 GPa and a Poisson's ratio of 0.25. This stress field, as well as the one at the surface, generally favour ring-fault formation (cf. Gudmundsson and Nilsen, 2006).

stratovolcanoes contain not only layers of stiff lava flows and (welded) pyroclastic rocks and intrusions, but also layers of soft (compliant) pyroclastic rocks, sedimentary rocks, as well as contacts with various mechanical properties. Here, again, we focus on two-dimensional models, so what is seen corresponds to observations made in vertical cross-sections through the volcanoes. Such sections are normally the best exposures that we have, and include cliffs (particularly sea cliffs), the walls of erosional and landslide valleys, and the walls of collapse calderas.

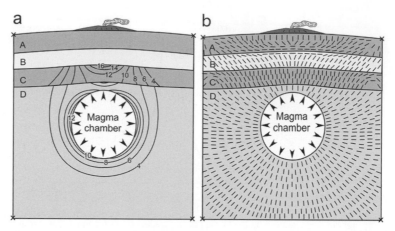

Fig. 3.25 Numerical model of a magma chamber with an excess magmatic pressure of 10 MPa as the only loading. The chamber is located in a layered crust of a stratovolcano volcano/rift zone where layers A and C are very stiff (with high Young's moduli), layer B is very soft (with a low Young's modulus), and layer D is moderately stiff. (a) The magnitude of the maximum tensile (minimum compressive) principal stress σ_3 is given in mega-pascals, the minimum values shown being 4 MPa. (b) The ticks show the trajectories of the maximum compressive principal stress σ_1 along which ideal dikes should propagate. A change in the direction of σ_1 from vertical, favouring dike propagation, to horizontal, favouring dike arrest (or sill formation), occurs at the contact between layers A and B. The model is shown in a 'deformed state' (cf. Gudmundsson and Philipp, 2006).

Consider first models similar to those in the last section, where the chamber has a circular cross-section. But instead of the crustal segment hosting the chamber being homogeneous and isotropic it is now layered (Fig. 3.25; a larger version of this model is in Figs. 3.4 and 3.5). More specifically, the model has four layers: one hosting the chamber and three above the chamber. The layers have different Young's moduli (stiffnesses), but are all welded together, that is, there are no contacts with specific mechanical properties between the layers. In contrast to most of the numerical models in this book, here the deformation of the layers (somewhat exaggerated) as a result of the magmatic excess pressure of 10 MPa is also shown.

In the model in Fig. 3.25, layers or units A and C are very stiff (100 GPa), layer B is very compliant or soft (1 GPa), and layer D (hosting the chamber) is moderately stiff (40 GPa). Each of the layers A–C has a thickness of 0.1 unit, the chamber has a diameter of 0.25 units, and layer D has a thickness of 0.7 units. In a 20-km-thick crust, for example, each of the layers A–C would initially be 2 km thick, and the magma chamber would have a diameter of 5 km. In the (exaggerated) **deformed view** in Fig. 3.25, the chamber diameter has grown, particularly vertically, and become considerably larger than the initial diameter, whereas parts of some of the layers have become compressed and thus thinner than they were originally. Thus, the deformed model suggests that the chamber would **dome** the crust above its top (as seen at the surface) while expanding, particularly along its vertical axis. After the deformation, the cross-section of the chamber is therefore somewhat elliptical. In reality, a single magma injection and excess-pressure increase results in a very small, if any, permanent expansion of the chamber.

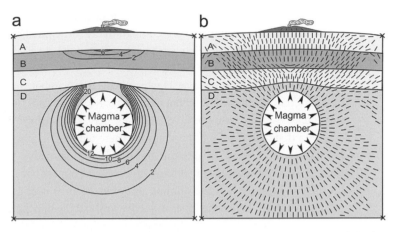

Fig. 3.26 Numerical model of a magma chamber with an excess magmatic pressure of 10 MPa as the only loading. The model is the same as in Fig. 3.25 except that here the surface layer A and layer C are soft (with a stiffness of 1 GPa), whereas layer B is stiff (with a stiffness of 100 GPa). (a) The magnitude of the maximum tensile (minimum compressive) principal stress σ_3 is given in mega-pascals. The lowest values shown are 2 MPa. (b) The ticks show the trajectories of the maximum compressive principal stress σ_1 along which ideal dikes should propagate. Again, there is a change in the direction of σ_1 from vertical, favouring dike propagation, to horizontal, and favouring dike arrest, but it is less marked than in Fig. 3.25 and, here, at the contact between layers B and C. The model is shown in a 'deformed state' (cf. Gudmundsson and Philipp, 2006).

In the model in Fig. 3.26 the arrangement of the stiff–soft layers is different. Here, layers or units A and C are compliant or soft (1 GPa), whereas layer B is very stiff (100 GPa). Layer D, hosting the chamber, has the same stiffness as before (40 GPa). Thus, the layer closest to the top of the magma chamber, layer C, is now soft. This is reflected in that layer becoming more deformed – it has become **thinner** – than in the model in Fig. 3.25.

Apart from the internal deformation of the layers as well as the surface deformation, the results of these simple models highlight some of the main effects of mechanical layering on local stresses around magma chambers. The key points may be summarised as follows:

- The stresses **concentrate** or become raised in the stiff (high Young's modulus) layers. This is reflected in the concentration of σ_3 in layer C in Fig. 3.25a and in layer B in Fig. 3.26a. Here, the concentrated stresses shown are tensile, but the same applies to compressive and shear stresses – these also concentrate in the stiff layers. Thus, earthquakes, driven by shear stresses, would normally be more common in stiff than soft layers if both have suitably orientated fractures.
- The compliant or soft layers **suppress** or reduce stresses. Stresses are thus less easily transmitted to layers or units far from the chamber (or any stress source) where there are very soft layers in-between. This is seen in Fig. 3.25a where in the stress-magnitude range given (lowest contour shown is 4 MPa), there is no tensile-stress concentration in the surface layer A, despite its being stiff. Here, the compliant layer B suppresses the tensile stress. Tensile stresses are transmitted to the stiff layer B in Fig. 3.26a, that is, through the

compliant layer C, but layer B is closer to the chamber than layer A in Fig. 3.25a, and the stresses transmitted to layer B are comparatively low (maximum 6 MPa).

- The **highest stresses** do not have to be in layers or units closest to the magma chamber but depend on the stiffnesses of the layers. In Fig. 3.25a the highest tensile stresses, 16 MPa, occur in layer C and not in layer D which hosts the chamber. This is because the stiffness of layer C is 100 GPa whereas that of layer D is 40 GPa. Thus, in layered rocks the highest (tensile, shear, and compressive) stresses may, depending on the layering, occur at a certain **distance** from the stress source (here the magma chamber).

- For a given layer subject to doming, the highest tensile stresses tend to occur in the **upper part** of the layer whereas the highest compressive stresses occur in the lower part of the layer. This is partly because doming of the layer, as is here generated by the pressured magma chamber, bends the layer in such a way that the upper part is extended and is thus subject to tension whereas the lower part is compressed and thus subject to compression. Thus, in layer C in Fig. 3.25a, the highest tensile stresses, 12–16 MPa, occur in the uppermost part of the layer. Similarly, in layer B in Fig. 3.26a, the highest tensile stresses, 2–6 MPa, occur in the upper part of the layer.

- Similarly, the **deformation** (displacement and strain) in the layers depends on their stiffnesses. For a given loading, soft layers deform more than stiff ones. For example, the soft layer C in Fig. 3.26 deforms much more than the stiff layer C in Fig. 3.25, while both have the same thickness and location in relation to the magma chamber. The difference in deformation depending on the stiffness of the rock adds a source of uncertainty to **inversion** of geodetic data, such as when using the Mogi model.

- The orientation of the **stress trajectories** (here the trajectories of the maximum principal stress σ_1) may change abruptly between the layers. In fact, a 90° flip in the orientation of the stress trajectories is common at the contacts between mechanically dissimilar layers. An abrupt change of this kind occurs in the orientation of the stress trajectories at the contact between layers A and B in Fig. 3.25b and, less extensive, at the contact between layers B and C in Fig. 3.26b. Such an abrupt change in the orientation of the principal stresses at contacts has important implications for the propagation of dikes, inclined sheets, and sills. In particular, the stress change at the contact between layers A and B in Fig. 3.25b will tend to arrest or deflect a dike propagating vertically from the magma chamber through layer D and into the lower part of layer B. If the dike is deflected, it will form a sill, most likely along the contact between layers A and B (Chapter 7).

Three-dimensional modelling gives geometrically generally similar results to those obtained in two-dimensional modelling. As an example, Fig. 3.27 shows a **three-dimensional** model of the local stresses around a spherical magma chamber hosted by a layered crustal segment. A magmatic excess pressure of 10 MPa is the only loading. Layers A and C have a stiffness of 100 GPa, whereas layer B has a stiffness of 1 GPa, and layer D, hosting the chamber, a stiffness of 40 GPa. The main difference between the two-dimensional circular magma-chamber models discussed above and the three-dimensional spherical model is that the maximum tensile stress at the boundary of the chamber is lower in the three-dimensional model. More specifically, for a spherical (three-dimensional) magma chamber in an infinite elastic space subject to magmatic excess pressure p_e as the

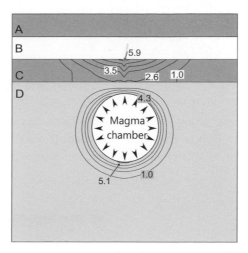

Fig. 3.27 Numerical model of the stress concentration around a spherical magma chamber subject to 10 MPa magmatic excess pressure as the only loading. Layers A and C are stiff (100 GPa), layer B is soft (1 GPa), and layer D, hosting the chamber, is moderately stiff (40 GPa). The contours show the magnitude of the maximum tensile (minimum compressive) principal stress σ_3 in mega-pascals. The model is shown in 'undeformed state'.

only loading, the maximum tensile stress at the chamber boundary is $0.5p_e$, that is, half the magmatic excess pressure. This follows from Eq. (3.35) and is seen in Fig. 3.9. By contrast, the maximum tensile stress at the boundary of an elongated (two-dimensional) magma chamber of circular cross-section subject to excess pressure p_e as the only loading is equal to p_e. The latter follows from Eq. (3.45) and is seen in Fig. 3.16.

In Fig. 3.27 the excess magmatic pressure is 10 MPa, so that the maximum tensile stress magnitude, as presented by the contours of σ_3, is about 5 MPa. Somewhat higher tensile stresses are reached in the stiff layer C close to the chamber. This, again, is because the stiff layers tend to concentrate the stresses and become more highly stressed than the softer adjacent layers. In this case, even if layer C is further away from the chamber than the hosting layer D, layer C is much stiffer than D (100 GPa versus 40 GPa) and thus concentrates somewhat higher tensile stress.

Note also that the stresses discussed here apply only when the loading is excess magmatic pressure p_e. For example, where the loading is external tensile stress of magnitude σ, the maximum tensile stress around the spherical magma chamber, from Eq. (3.39) and with the Poisson's ratio $v = 0.25$, is approximately equal in magnitude to $\sigma_\theta = 2.2\sigma$. Thus, if the external tension is, for example, 5 MPa, the theoretical maximum tensile stress magnitude at the boundary of a spherical chamber, located at great depth below the surface in relation to the chamber diameter, is about 11 MPa.

Most magma chambers, however, are not spherical but rather **oblate ellipsoids** and thus are referred to as sill-like. In particular, such chambers are known to exist at many, perhaps most, intermediate to fast-spreading plate boundaries. Such shapes are also the most likely to initiate or reactivate ring-faults, that is, to general caldera collapses. The most favourable loading for ring-fault formation (Section 3.6.3) is a sill-like chamber supplied with magma from a larger deep-seated reservoir, a double magma chamber, subject to a slight pressure increase in the reservoir, that is, a slight doming of the crustal segment hosting the shallow chamber.

3.6.3 Magma Chambers and Ring-Fault Formation

Here, several magma-chamber deformation models are considered in relation to their potential for caldera-fault or **ring-fault formation**. In the first model (Fig. 3.28) there is an excess pressure of 10 MPa in the deep-seated reservoir, giving rise to a slight **doming** of the crustal

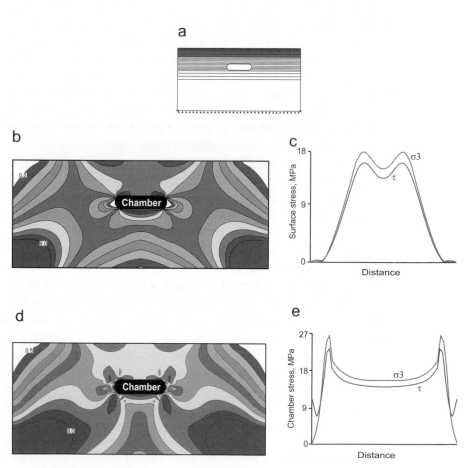

Fig. 3.28 Numerical model of the stress concentration around a sill-like magma chamber located in a homogeneous, isotropic crustal segment subject to a doming stress (due to a magmatic pressure increase in the reservoir) of 10 MPa at the segment bottom. The segment has a stiffness of 40 GPa, and is 20 km thick and 40 km wide. The chamber is located at 5 km depth and is 8 km wide and 2 km thick. All stresses are given in mega-pascals. (a) The general configuration of the numerical model. (b) Contours of the maximum principal tensile stress σ_3 around the magma chamber. (c) Maximum principal tensile stress σ_3 and von Mises shear stress τ at the free surface above the magma chamber (surface of a volcano or a rift zone). (d) Contours of the shear stress τ around the magma chamber. (e) Tensile σ_3 and shear stress τ at the upper boundary of the magma chamber (cf. Gudmundsson, 2007).

segment within which the shallow chamber is contained. In all the following models the crustal segment considered is 40 km wide and 20 km thick, so similar to that of a typical volcanic zone. The surface of the crustal segment is the volcanic zone/volcano associated with the double magma chamber. The shallow chamber is located at 5 km depth; it is 2 km thick and is 8 km wide in a direction perpendicular to the axis of the volcanic zone within which it is located. The results show that the maximum tensile and shear stresses in the crustal segment around the chamber roughly coincide. Both reach their maximum above the **lateral ends** (vertices; the endpoints of the horizontal diameter of the circular chamber) of the sill-like (flat elliptical in cross-section) shallow magma chamber (Fig. 3.28b and c). At the free surface of the crustal segment/volcanic zone the tensile and shear stresses also peak (reach their maximum values) above the lateral ends of the magma chamber (Fig. 3.28 b, c, and d). The tensile and shear stresses are high enough (as much as 18 MPa at the free surface and 27 MPa at the upper boundaries close to the vertices of the chamber) to form or reactivate a ring-fault.

The same loading for a shallow magma chamber of a circular cross-section generates **two peaks** (or a double peak) in the tensile and shear stress at the free surface of the volcanic zone (Gudmundsson, 2007). However, these peaks do not occur at the best location for a ring-fault to form. Furthermore, and most importantly, the tensile and shear stresses at the boundary of the chamber do not peak at its lateral ends but rather at the point on the chamber closest to the free surface. The same applies if the loading of the circular chamber is horizontal tensile stress. Thus, for these loading conditions, a shallow chamber of circular cross-section would tend to inject a dike from its **top** part rather than develop a ring-fault.

The classic model of ring-fault formation is the Anderson's (1936) concept of magmatic **underpressure** in a nucleus-of-strain magma chamber. As discussed above, this analytical solution (using underpressure for deflation and excess pressure for inflation) is the basis of the Mogi model of volcano surface deformation. Here the focus is on the local stress field that such an underpressure could generate. For an underpressure of 10 MPa in the same sill-like chamber as in Fig. 3.28, the results show that there is no tensile stress at the free surface directly above chamber (that part of the surface is subject to compression), but there are small peaks (here about 3 MPa) above the lateral ends of the chamber (Fig. 3.29). Also, the shear stress at the free surface concentrates not above the lateral ends of the chamber but rather right above the centre of the chamber. The shear-stress concentration around and above the chamber (Fig. 3.29d) shows that if any faults were to develop, they would be outward-dipping **reverse faults**. We can infer from this model that when the underpressure is greater than the present one (10 MPa), inward-dipping **normal faults** might also develop at the small peaks of σ_3 at the free surface (Fig. 3.29c), whereas outward-dipping reverse faults could develop in the central part. These results are in agreement with analogue models which suggest that some calderas may develop along two sets of faults: (1) inward-dipping normal ring-faults, where the diameter is somewhat larger than that of the associated chamber; and (2) inward-dipping reverse ring-faults, in which the diameter is much smaller than that of the associated chamber, as discussed below.

So far, in the ring-fault-forming models, the focus has been on homogeneous and isotropic crustal segments. Now we turn to more realistic models, namely **layered** ones. The layered models have the same size and geometry of the shallow chamber as the previous caldera models, 8 km wide and 2 km thick, but the chamber is located at

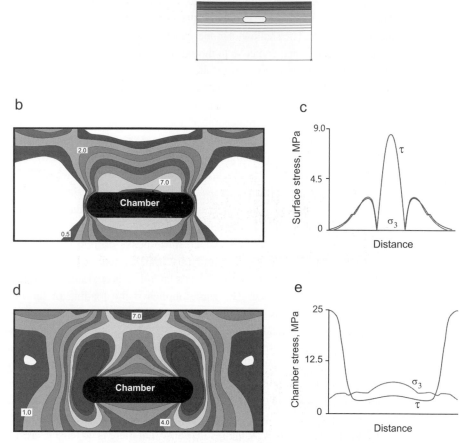

Numerical model of the stress concentration around a magma chamber located in a homogeneous, isotropic crustal segment subject to a magmatic underpressure (pressure less than lithostatic) of 5 MPa. The sill-like chamber, 8 km wide and 2 km thick, is located at 5 km depth in a homogeneous, isotropic crustal segment, 20 km thick and 40 km wide, with a Young' modulus of 40 GPa. All stresses are given in mega-pascals. (a) The general configuration of the numerical model. (b) Contours of the maximum principal tensile stress σ_3 around the magma chamber. (c) Maximum principal tensile stress σ_3 and von Mises shear stress τ at the free surface above the magma chamber (surface of a volcano or a rift zone). (d) Contours of the shear stress τ around the magma chamber. (e) Tensile σ_3 and shear stress τ at the upper boundary of the magma chamber (cf. Gudmundsson, 2007).

a shallower depth, namely 3 km (instead of 5 km) below the free surface. The crustal segment hosting the chamber is, again, 20 km thick and 40 km wide. The layer hosting the chamber has a stiffness of 40 GPa. Between the upper boundary or roof of the chamber and the free surface there are 30 layers (each 100 m thick), alternating in stiffness between 100 GPa (very stiff) and 1 GPa (very soft).

In the first of the layered models, the only loading is a **horizontal tensile stress** of 5 MPa, a reasonable tensile stress in volcanic rift zones. The results (Fig. 3.30) show that the tensile

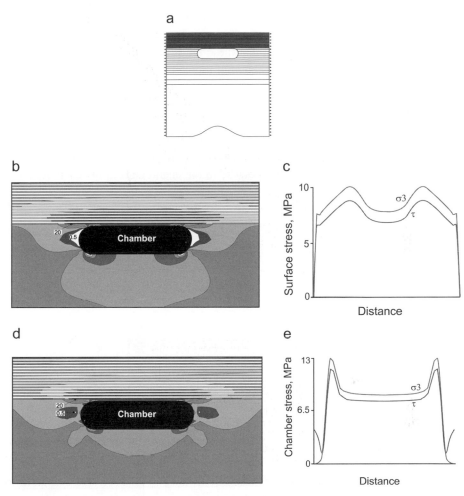

Fig. 3.30 Numerical model of the stress concentration around a sill-like magma chamber, 8 km wide and 2 km thick, located at 3 km depth in a 20-km-thick and 40-km-wide layered crustal segment subject to a horizontal tensile stress of 5 MPa. The lower part of the segment, including the part hosting the chamber, has a Young's modulus of 40 GPa. By contrast, the uppermost 3 km above the chamber consist of 30 layers of alternating stiffness (Young's modulus) of 1 GPa and 100 GPa. All stresses are given in mega-pascals. (a) The general configuration of the numerical model. (b) Contours of the maximum principal tensile stress σ_3 around the magma chamber. (c) Maximum principal tensile stress σ_3 and von Mises shear stress τ at the free surface above the magma chamber (surface of a volcano or a rift zone). (d) Contours of the shear stress τ around the magma chamber. (e) Tensile σ_3 and shear stress τ at the upper boundary of the magma chamber (cf. Gudmundsson, 2007).

and shear stresses at the free surface both peak above the lateral ends (the vertices) of the magma chamber (Fig. 3.30c). Similarly, these stresses peak at the upper boundary of the chamber, close to its lateral ends (Fig. 3.30e). The stress magnitudes, for the given loading, are theoretically high enough (maximum values 10–13 MPa) to initiate a ring-fault. But the layering affects the likelihood of ring-fault formation. In particular, while the tensile and shear stresses are high in most of the stiff layers, they are very **low** in the soft layers in-between the stiff layers. It is thus not clear, even if the stresses in the stiff layers are high enough to initiate faults, that these faults would be able to propagate through all the soft layers so as to form a coherent, if segmented, ring-fault.

In the second layered model, all the geometric parameters and layering are the same as in the first layered model, but here the only loading is a magmatic overpressure, that is, a **doming stress** in the deep-seated reservoir of 10 MPa. The results (Fig. 3.31) show, again, that the tensile and shear stresses at the free surface reach up to 21 MPa and peak above the lateral ends of the shallow chamber (Fig. 3.31b). Similarly, at the upper boundary of the chamber, the stresses peak close to the lateral ends of the chamber and reach as much as

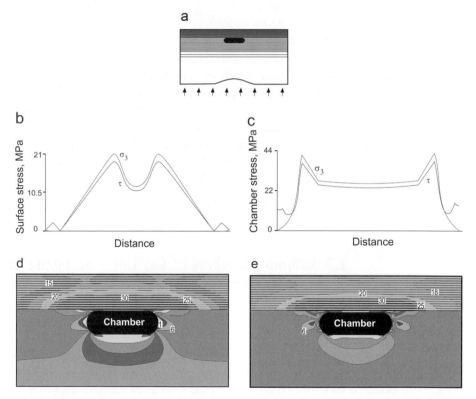

Fig. 3.31 Numerical model of the stress concentration around a sill-like magma chamber, 8 km wide and 2 km thick, located at 3 km depth in a 20-km-thick and 40-km-wide layered crustal segment subject to a doming stress (due to magmatic pressure increase in the reservoir) of 10 MPa at the segment bottom. The lower part of the segment, including the part hosting the chamber, has a Young's modulus or stiffness of 40 GPa, whereas the uppermost 3 km above the chamber consist of 30 layers, with alternating Young's moduli of 1 GPa and 100 GPa. All stresses are given in mega-pascals. (a) The general configuration of the numerical model. (b) Maximum principal tensile stress σ_3 and von Mises shear stress τ at the free surface above the magma chamber (surface of a volcano or a rift zone). (c) Tensile σ_3 and shear stress τ at the upper boundary of the magma chamber. (d) Contours of maximum principal tensile stress σ_3 around the magma chamber. (e) Contours of the shear stress τ around the magma chamber (cf. Gudmundsson, 2007).

44 MPa (Fig. 3.31c). While the same consideration applies as in the model in Fig. 3.30, namely that the soft layers have very **low stresses**, there is a clear difference in the local internal stress field in this model (Fig. 3.31) compared with the previous model (Fig. 3.30). The main difference is that in the present model of slight doming the stiff layers show zones of high tensile (Fig. 3.31d) and shear (Fig. 3.31e) stresses from the upper margin of the chamber and to the free surface. Thus, even if the soft layers tend to arrest the propagation of individual fault segments, the likelihood of the formation of a through-going ring-fault is **high** for this type of loading.

Other loading conditions may operate from time to time in volcanoes. For example, both doming (due to pressure increase in the deep-seated source reservoir) of the crustal segment hosting the shallow chamber and an external tensile stress (associated with spreading) may operate at the same time. For the present geometry of a magma chamber, crustal-segment dimensions, and layering, the results are very similar (Fig. 3.32) to those of tension (Fig. 3.30) and doming (Fig. 3.31) individually. The main difference is that the induced tensile and shear stresses are larger. This applies both to the stresses at the free surface (Fig. 3.32b) as well as the stresses around the upper margin of the magma chamber (Fig. 3.32c). Combined **tension and doming** certainly favours the formation or reactivation of a ring-fault, and thus collapse-caldera formation, in the same way, and with the same limitations as to the effects of the soft layers, as for tension alone (Fig. 3.30) and doming alone (Fig. 3.31). Because of the variation in the mechanical properties, particularly the stiffness, of the layers that constitute typical stratovolcanoes, and associated variations in the local stress field between layers, the attitude of the ring-fault is likely to change between layers (Fig. 3.33). Thus, some fault segments may show outward dips, others inward dips, while the overall dip of the ring-fault is close to vertical (cf. Chapters 5 and 8).

3.7 Magma Chambers as Cavities – Analogue Models

Analogue models are commonly used to advance our understanding of volcano deformation. The basic idea is to find material which behaves similarly to crustal rocks at the scale, loading, and strain rate of the model. Such models have been used for a long time in structural geology and tectonics, for example to understand large-scale crustal deformation, including diapir formation, folding, and fracture development (e.g. Cloos, 1955; Ramberg, 1967; McClay and Ellis, 1987; Fossen and Gabrielsen, 1996; Fossen, 2016). In volcanotectonics, the use of **analogue models** is very common, particularly for modelling caldera collapses, with a focus on the **ring-fault** development and structure (Marti et al., 1994; Acocella et al., 2000; Marti and Gudmundsson, 2000; Cole et al., 2005; Acocella, 2007; Marti et al., 2008; Geyer and Marti, 2014). The material most commonly used in volcanotectonic modelling is sand. One principal aim is that the grain size in the material used is as small as possible. This follows because a sand grain that is, say, 1 mm in diameter in a scale model that is 20 cm thick, would in a 20-km-thick volcanic zone have a diameter of 100 m. Many scale models in general (such as for modelling normal faults) are much thinner than this, and the grain size may be less, but most analogue models of this type, when scaled up

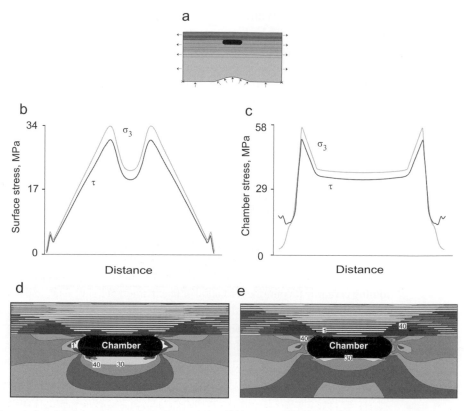

Fig. 3.32 Numerical model of the stress concentration around a sill-like magma chamber, 8 km wide and 2 km thick, located at 3 km depth in a 20-km-thick and 40-km-wide layered crustal segment subject to a doming stress (pressure) of 10 MPa at the segment bottom as well as horizontal tensile stress of 5 MPa. The lower part of the segment, including the part hosting the chamber, has a Young's modulus of 40 GPa, whereas the uppermost 3 km above the chamber consists of 30 layers with alternating Young's moduli of 1 GPa and 100 GPa. All stresses are given in megapascals. (a) The general configuration of the numerical model. (b) Maximum principal tensile stress σ_3 and von Mises shear stress τ at the free surface above the magma chamber (surface of a volcano or a rift zone). (c) Tensile stress σ_3 and shear stress τ at the upper boundary of the magma chamber (d) Contours of tensile stress σ_3 around the magma chamber. (e) Contours of shear stress τ around the magma chamber (cf. Gudmundsson, 2007).

to the natural structures they are supposed to model, end with grains that correspond to rock blocks with diameters of the order of **tens of metres**. Grains or blocks of this size do not occur in volcanoes or crustal segments; in particular, blocks of this size do not occur in the cores of faults, including ring-faults. Thus, the frictional and other effects of the grains in the sand models need to be minimised so as to make the conditions as similar as possible to the actual ones in the crustal segment during volcano deformation.

In analogue models, magma chambers are commonly modelled as **balloons** (Marti et al., 1994). During inflation, the air pressure in the balloon is increased, resulting in doming of the free surface of the volcano/volcanic zone. During deflation, the air is allowed to escape from the balloon, resulting in subsidence of the free surface above. If the deflation is large

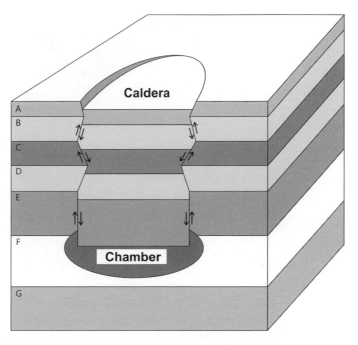

Fig. 3.33   The dip of any fault depends on the local stresses in the layers that the fault dissects. The local stresses, in turn, depend on the mechanical properties, particularly Young's modulus, of these layers and how the properties change at contacts between layers. When the mechanical properties of the layers that constitute a polygenetic volcano change abruptly from one layer or unit to another, the local stresses may also vary abruptly between the layers, affecting ring-fault attitudes. Thus, the local stresses that favour outward-dipping ring-fault segments in some layers may favour vertical or inward-dipping segments in other layers.

enough, a ring-fault forms. More specifically, when the air is allowed to escape from a buried, inflated balloon, commonly two sets of ring-faults are generated (Fig. 3.34). In the upper part of the model, there develops an inward-dipping **normal ring-fault**. Usually, this set has a diameter (caldera diameter) that is considerably larger than the diameter of the associated shallow magma chamber – the balloon. In the central and lower part, by contrast, an outward-dipping **reverse ring-fault** develops. This set is with a much smaller diameter than both the normal ring-fault and the chamber itself. This is in agreement with the results from the numerical models for underpressure (Fig. 3.31).

   Analogue models have thrown much light on the various ways by which ring-faults can develop. Some analogue models indicate that the dip-dimension (height) of the ring-fault can be **much larger** than its diameter (Fig. 3.34; cf. Fig. 3.33). This may seem surprising, but depends in the end on the magnitude of the shrinkage of the magma chamber or, what it would amount to in an analytical or numerical model, the magnitude of the underpressure. If the shrinkage or size reduction of the balloon is large enough, then the mechanical result is essentially equivalent to an empty cavity into which the segment above would subside along a ring-fault. Some interpretations of the structure of the 2000 caldera collapse of Miyakejima in Japan indicate that the dip-dimension of the collapsed piston is larger than the caldera diameter (Geshi et al., 2002).

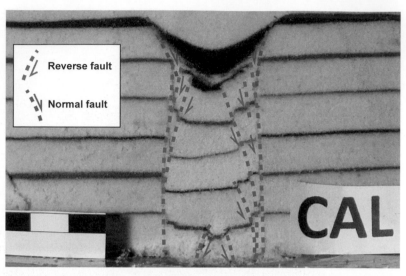

Fig. 3.34 Analogue model providing a section view of an experiment simulating ring-fault formation during the subsidence of a piston at the base of a sand pack. This experiment is characterised by the aspect ratio between the thickness of the sand pack and the diameter of the piston being larger than 1. Multiple vertical to high-angle outward-dipping reverse faults propagate from the bottom and induce the collapse of the overlying wedges, promoting the formation of the inward-dipping normal faults in the upward portion (Ruch et al., 2012).

Models with very large underpressures – **empty cavities** – are very suitable for modelling pit craters (Figs. 1.1a and 3.35). Most **pit craters** form above 'empty cavities' of some sort. For example, a common cause of pit craters is a sudden pressure decrease in the associated feeder-dike. This may happen when the magma finds a new path to the surface during an eruption, that is, when the feeder-dike propagates and reaches the ground at, for example, a lower surface elevation. Then the magma would tend to flow towards the lower ground and, accordingly, stop flowing up through the conduit of an existing crater at a higher elevation. The magma pressure in the feeder-dike below that crater then falls suddenly with the result that it may collapse to form a pit crater. Similarly, the opening of a tension fracture or a normal fault beneath an existing crater cone may result in its collapse into a pit crater. All these collapses can be, and have been, modelled as the result of an abrupt pressure decrease and cavity formation beneath the surface.

3.8 Stress and Deformation Associated with Dikes and Sheets

Dikes and inclined sheets and sills emplaced at comparatively shallow depth all generate measurable **surface stresses and deformation**. As for the magma chambers, analytical, numerical, and analogue models are used to model dikes and sheets. Many results on surface stress and deformation associated with emplaced dikes, sheets, and sills are discussed in Chapters 7 and 10. The focus here is on some general aspects of the stresses and deformation induced by injected sheet intrusions.

Fig. 3.35   Pit crater inside the Erta Ale caldera, northern Afar, East Africa. The crater, approximately 80 m wide and 60 m deep, has been hosting an active lava lake for more than a century. Photo: Valerio Acocella.

3.8.1 Dikes and Sheets in Elastic Half-Spaces

One of the oldest analytical models on the surface effects of fractures is due to Isida (1955), who studied the surface stresses and deformation above elliptical fractures, with their upper tip (or top) at a certain depth below the surface of an elastic half-space. His principal results were that the stresses and displacements above a vertical elliptical fracture, such as a dike, form two peaks, a **double peak**, one to each side of the projection of the dike tip to the surface (Fig. 3.2). By contrast, no stresses occur right above the tip of the buried fracture (dike).

More recent elastic half-space numerical models on the surface effects of arrested fractures generally agree with these results. For example, numerical models by Pollard et al. (1983) show that the dike-induced vertical displacement (and horizontal surface tensile stress) forms a double peak above the projection of the dike to the surface, but no displacement above the tip of the dike itself (Fig 3.36). In this numerical model, the dike-induced vertical displacement at the surface is compared with the measured displacement in one rifting event in the Krafla Volcanic System in North Iceland in the 1980s. In Fig. 3.36a the measured surface displacement is compared directly with the dike-induced displacement, whereas in Fig. 3.36b two normal faults have been added. Since faults (and other fractures) concentrate stresses, adding faults to the model modifies the stresses and the associated surface displacements. Neither model (Fig. 3.36a and b), however, is able to explain the measured graben subsidence of about 1.5 m during the rifting event. Also, the surface-stress peaks and the surface-displacement (uplift) peaks do not normally coincide; the horizontal distance between the stress peaks is much smaller than between the

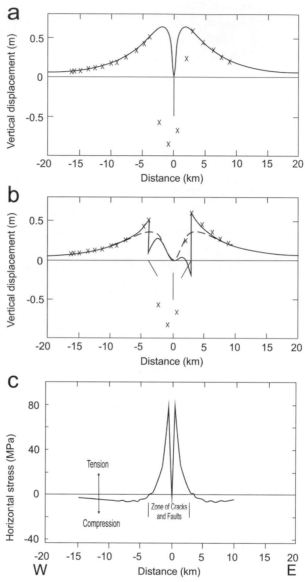

Fig. 3.36 Theoretical displacement/horizontal tensile stresses at the surface of a rift zone compared with actual deformation in the rift zone. (a) Numerical model showing the theoretical vertical displacement induced by a dike extending with a vertical tip (top) at 0.25 km below the surface to a depth of 6 km below the surface. The crosses indicate the actual surface deformation (vertical displacements) measured following the January 1978 rifting event in the Krafla Fissure Swarm in North Iceland (Sigurdsson, 1980). (b) For a better fit with the actual surface deformation two normal faults have been added as the boundary faults of the graben, and the dike now extends from a depth of 2 km (top) to 6 km below the surface. (c) The zones of theoretical surface tension and compression compared with the main zone of surface deformation (fracture formation and reactivation) in the 1978 event. Neither model (a) nor (b) explains the observed maximum relative graben subsidence of about 1.5 m. Modified from Rubin and Pollard (1988).

displacement peaks. If dike-induced grabens form at all, the graben faults would normally form not at the displacement peaks but rather at the stress peaks (Chapter 10).

Subsequently, many numerical models for dikes and inclined sheets in elastic half-spaces have been made. Many of the results are summarised by Okada (1985), Dzurisin (2006), and Segall (2010). The newer results agree with the earlier ones in that the vertical displacement shows two peaks on either side of the projection of the tip of the dike to the surface. Most early models show no displacement at the surface right above the dike tip (Pollard et al., 1983; Segall, 2010), which is one reason why the **graben subsidence** commonly observed during rifting and dike injection could hardly be explained by displacements induced only by the magmatic pressure or opening of a vertical dike in an elastic half-space. More recent half-space models of dikes, however, show some subsidence of the surface above the dike tip (Fig. 3.29; Dzurisin, 2006; Segall, 2010), and so do some numerical models of dikes in layered crustal segments (Chapter 10). Subsidence of the surface, however, is not generally enough to initiate a graben; a concentration of stresses at the right locations so as to form parallel normal faults is also needed.

More specifically, for a **wide graben** to form or reactivate in a rift zone, divergent movement associated with the spreading vector, resulting in a tensile-stress concentration in the layers close to and at the surface above a dike, is normally also needed. (By wide graben we mean one with a surface width of the order of many kilometres.) When a vertically propagating dike approaches the free surface, the vertical displacement peaks normally become higher (the displacement increases) and the distance between the peaks decreases, that is, the peaks become closer to the projection of the tip of the dike onto the surface. If the vertically propagating dike – particularly a feeder-dike – generated a graben, we might then expect a **nested graben** along the lines shown very schematically in Fig. 3.37. While stresses tend to concentrate on existing faults, still when the dike tip is close to the surface, new faults might be expected, resulting in a nested graben. However, such nested grabens related to vertically propagating dikes have rarely, if ever, been observed.

The most popular half-space models for sheet intrusions, however, are neither the analytical models, initiated by Isida (1955) and elaborated in later papers, nor the numerical models initiated by Pollard et al. (1983) and subsequently developed, but rather **elastic-dislocation** models. Basically, all these models derive from the Volterra (1907) dislocation model for calculating the elastic stress and displacement fields. While the basic solution is thus over a century old, it was introduced as a tool for analysing brittle deformation in earth sciences, primarily through its application to earthquake faults, in the 1950s and 1960s (Steketee, 1958a,b; Press, 1965). The application of dislocation theory to fracture mechanics in general has been elaborated by Weertman (1996), but has never really been much used in that field. In materials science, including ductile deformation of rocks, moving dislocations are one of the principal means by which plastic flow occurs. However, materials science dislocation theory is different from the elastic dislocation theory used in brittle deformation studies. In particular, in volcano deformation studies, dislocation models have been used for a long time, primarily for the inversion of surface deformation data to infer the geometric aspects – opening or thickness, strike, dip, and depth – of dikes, inclined sheets, and sills.

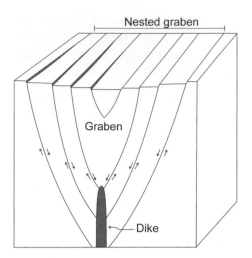

Fig. 3.37   Schematic illustration of the possible effects of surface deformation entirely induced by a propagating dike. As the dike approaches the surface, the distance between the two peaks of vertical surface displacement (and stress), seen in Figs. 3.2, 3.36 and 3.39, would normally decrease (depending on layering) so that a nested graben would be expected. Normally, however, a vertical dike becomes emplaced only if there is already an extension in the crustal segment, which, at shallow depths, gives rise to absolute tension. One primary function of the dike is thus to concentrate this external tensile stress in the zone or layers between the dike tip at any particular time during its propagation and the surface. If there already exist tension fractures and normal faults in the rift zone/volcano hosting the dike, these would normally concentrate the stress and become reactivated. Thus, nested grabens composed of many normal faults are not very common; more commonly, the existing faults become reactivated, partly because of dike-induced stresses and partly because of stress concentration in the layers between the dike tip and the surface.

The application of the dislocation theory of volcano deformation is reviewed in detail by Okada (1985, 1992), where all the relevant equations are given, and also by Dzurisin (2006) and Segall (2010) to which the reader is referred for further details on this method of volcano-deformation analysis. Basically, the aims and methods of the **dislocation modelling** for sheet intrusions and associated induced volcano deformation are as follows:

- **Collect** deformation data during an unrest period in a volcano or volcanic zone. The data may be obtained through any of the methods discussed in Chapter 1, but most commonly vertical and horizontal crustal displacement data come from GPS or InSAR measurements.
- **Invert** the geodetic data using an elastic half-space model. That is, find the best dike, inclined sheet, or sill model (or fault model, in case of an earthquake rupture) that fits with the data. Here the strike, dip, thickness (opening) and other geometric aspects of the sheet intrusion are compared with the data so as to obtain the best fit.
- Try to fit or **model** the data with a single sheet intrusion with constant opening displacement, that is, with constant thickness. This is the most common procedure. The dike or inclined sheet can, however, also be divided into segments that may have different sizes and opening displacements (e.g. Sigmundsson et al., 2015). Then the model is no longer simple and numerical methods are normally better for solving the stress and displacement fields.

- **Compare** the results with other data. The main data useful in this connection are seismic data, since they are the best indication of the likely location and dimensions of the sheet intrusion (Chapter 4). In fact, the joint inversion of geodetic and seismic data is now increasingly common (Segall et al., 2013).

Theoretically, the **inversion method** for finding the attitude and dimensions of the sheet intrusion associated with a given volcano surface deformation is very clear, and all the mathematical aspects are comparatively simple and well understood. In fact, this is a routine method in volcano-deformation studies worldwide. The difficulties arise when we try to use the results to advance our understanding of the actual **volcanotectonic processes** responsible for the deformation and assess the likelihood of an eruption. If the principal aim of the modelling is to understand better the associated volcanotectonic processes, then the elastic half-space dislocation inversion method has several drawbacks. These are of course well known to all experts who work in this field and may be briefly summarised as follows:

1. The geodetic measurements are made in a plane, namely at the Earth's surface. They are therefore basically **two-dimensional** data. By contrast, the geometric structures of dikes, inclined sheets, and sills (and faults) that we are trying to determine using these data are **three-dimensional**. Unfortunately, there is no unique solution when using two-dimensional surface data to infer three-dimensional geometries of intrusions. Thus, there is essentially an infinite number of different intrusion geometries that could fit with the deformation data.

2. The solution adopted is to find the intrusion geometry that gives the **best fit** to the geodetic data. Unfortunately, there are many poorly constrained variables when using the standard half-space models. For example, only one Young's modulus is normally used for the entire crustal segment or volcanic zone/volcano within which the dike or inclined sheet is emplaced. As indicated above, the real host rock is highly heterogeneous and, in particular, anisotropic, that is, layered. The Young's moduli of the layers can easily vary by two orders of magnitude.

3. Most non-feeder-dikes – the most common type of sheet intrusion modelled using half-space dislocation models – become **arrested** at contacts between mechanically dissimilar rocks, particularly rocks with different Young's moduli. As discussed further in the Section 3.8.2, such contacts may completely alter the surface deformation so as to yield highly inaccurate results for dike depth and geometry when interpreted in terms of elastic half-space dislocation models. Furthermore, such models, by their very nature, cannot provide any useful information to help in forecasting sheet- or dike-propagation paths during unrest periods and assessing the likelihood of their reaching the surface to erupt (cf. Chapter 10).

When deformation data are used to infer dike or inclined-sheet geometries from dislocation half-space models, the question arises as to how accurately the model geometry matches the real **intrusion geometry**. In highly anisotropic volcanoes and crustal segments the match is presumably generally poor, but may be better in some basaltic edifices with comparatively little anisotropy. To improve our understanding of the actual volcanotectonic processes, geometric inversion models need to be connected with dynamic models on fluid-driven fracture propagation – a subset of which are models on dike/sheet emplacement.

To explain the needed connection with **fracture mechanics** better, let us take a hypothetical example. Consider a case where the best-fit dislocation model of a vertical dike, arrested at depth of 1 km below the free surface, is such that the dike has the following dimensions: strike-dimension (length) 30 km, dip-dimension (height) 15 km, opening displacement (thickness) 0.2 m. Such a geometry might easily fit the geodetic data better than any other, but the result is, as a model of a real dike, highly implausible and probably impossible. Why? Because there is no known way by which a dike with a controlling dimension (here the dip-dimension) of 15 km in a crustal segment with a reasonable average Young's modulus and with a reasonable driving pressure (overpressure) could be so thin – a dike of these dimensions and typical rock properties and overpressure would always have a thickness or an opening of the order of metres or more. This follows from standard and widely tested equations in fracture mechanics relating elastic properties of the solid (here the host rock) with the fluid overpressure and fracture dimensions (Chapter 5).

So where do we stand as regards the usefulness of dislocation half-space models for dikes and inclined sheets emplaced during unrest periods? Apart from feeder-dikes, for which the models are best suited, a more suitable approach for dikes and inclined sheets arrested at depth is to use **earthquakes** to estimate their dip- and strike-dimensions. Earthquake foci are of course not very accurately known, but the inferred dike dimensions using earthquake distributions should be correct within a factor of two or less. (This uncertainty includes those parts of dikes that may propagate as aseismic.) From the crude estimates of the dike dimensions, the controlling dimension (the smaller of the strike- and dip-dimensions) can be used, using fracture-mechanics equations, to estimate the minimum opening displacement or **thickness** of the dike. This estimate is made using static Young's moduli of (preferably) layered crustal models and a reasonable driving pressure or overpressure in the dike. Likely overpressure values can be estimated from the inferred properties of the magma and the height of the dike, using formulas in Chapter 7 (Eq. (7.5)). The results obtained in this way could then be inserted into standard dislocation half-space models and the fit with the surface deformation data could be tested. The results should also be compared with standard fracture-mechanics models, analytical ones for the elastic half-space approach and numerical ones for layered crustal models, to which I turn now.

3.8.2 Dikes and Sheets in Layered Crustal Segments

When some information is available about the layering of the crustal segment through which dikes or inclined sheets (or sills) are propagating, the normal approach is to use numerical models and compare the results with the geodetic data. This is particularly important where, as is common, the layers that constitute the volcano/volcanic zone have widely different mechanical properties and, furthermore, contain mechanically weak or open contacts. This is discussed in more detail in Chapters 7 and 10, including (in Chapter 10) the internal dike-induced deformation. Here, the discussion focuses on the general effects of layering on the surface deformation and stresses.

Consider first the effect of a mechanically **weak contact** at a shallow depth in the crust on the surface stresses induced by a vertical dike (Fig. 3.38). Here the dike is arrested at a depth of 1 km below the free surface of a volcanic zone. The uppermost part of the crust is composed of the following layers (the stiffnesses are indicated in parentheses): surface layer A (40 GPa), layer B (100 GPa), and layer C (40 GPa), which hosts the dike. There is

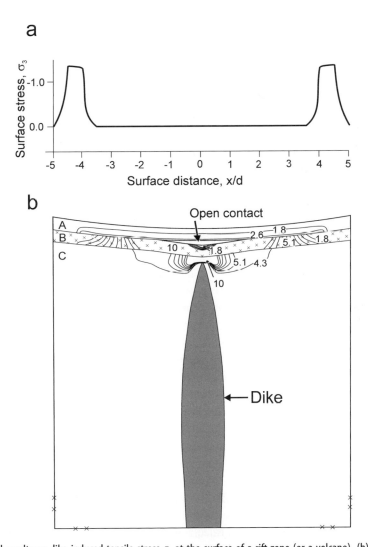

Fig. 3.38 (a) Numerical results on dike-induced tensile stress σ_3 at the surface of a rift zone (or a volcano). (b) The arrested dike tip is at 1 km depth below the surface of the rift zone, and the theoretical tensile stress at the tip reaches 174 MPa (this is entirely a theoretical stress – no rock can tolerate tensile stress in excess of about 9 MPa). But much of this high tensile stress is suppressed at the weak contact between layers A and B, located at 0.4 km below the surface, which opens up because of dike-induced tension perpendicular to the contact, parallel with the dike dip in accordance with the Cook–Gordon mechanism of fracture arrest. The surface tensile-stress peaks outside the lateral ends of the contact (cf. Gudmundsson, 2003).

a mechanically weak 8-km-wide contact between layers A and B at a crustal depth of 0.4 km (400 m), so 600 m above the tip of the arrested dike. More specifically, the contact is modelled as an internal spring with a stiffness of 6 MPa m^{-1}. The contact with this low stiffness could correspond to scoria or tuff layers in-between lava flows. Models where the stiffness is zero, so that the contact is an open discontinuity, yield results similar to those here. The only loading is internal magmatic driving pressure (overpressure), which varies linearly from 10 MPa at the bottom of the dike (here regarded as 9 km tall) to 0 MPa at its tip.

At the **tip of the dike** in Fig. 3.38, the theoretical calculated tensile stress reaches 174 MPa. But much of this high tensile stress is suppressed at the weak contact between layers A and B, which opens up because of dike-induced tension perpendicular to the contact (parallel with the dike dip-dimension). Part of the tensile stress that concentrates in layer B passes to the surface layer through the areas outside the weak and opened contact, namely close to the ends of the layer. This location of the tensile stress at the lateral ends of the contact is reflected in the surface stresses (Fig. 3.38a). There are no tensile surface stresses at all above the contact itself, that is, in the 8-km-wide central part of the volcanic rift zone above the dike, but tensile-stress peaks of 1.3 MPa just outside the lateral ends of the contact. Thus, the surface stresses and the surface deformation, as measured by GPS or InSAR studies, for example, in this case yield data that, if inverted, will give incorrect information as to the geometry and depth to the top of the dike.

The **effects of layering** on the local dike-induced stress field are further illustrated through comparison of the models in Figs. 3.39 and 3.40. In Fig. 3.39 the vertical dike has a constant overpressure of 10 MPa as the only loading. The model is 3 units high, whereas the dike tip is at 0.5 units below the free surface. Thus, for example, if the dike was injected from a shallow chamber with a roof at 1.5 km depth, then its vertical tip would be at 250 m depth below the free surface. This would be similar to a common dike-injection depth from some shallow chambers in Iceland and elsewhere. By contrast, if the dike was injected from a deep-seated reservoir at, say, 15 km depth, the vertical tip of the dike would be at 2.5 km depth. These latter figures are similar to those estimated for the dike emplaced during the 2014 Bardarbunga eruption, for the part of the dike tip that did not reach the surface (Gudmundsson et al., 2014; Sigmundsson et al., 2015).

In the first model (Fig. 3.39), the dike is located within a **homogeneous and isotropic** crustal segment. The stiffness of the crust is high, with a uniform Young's modulus of 100 GPa, and a Poisson's ratio of 0.25. For the constant overpressure of 10 MPa, the theoretical tensile stress the dike tip reaches 149 MPa. From there the tensile stress decreases to 0 MPa at the free surface right above the tip of the dike. Thus, the model is in agreement with several other models and theoretical calculations that show that there is no tensile stress at the free surface above the tip of the dike (Figs. 3.2 and 3.36). However, high tensile stress forms two inward-dipping zones to either side of the dike tip, and these zones reach the surface. Thus, at the surface, the tensile stress peaks on either side of the surface projection of the tip. At these peaks, the tensile stress reaches a maximum of 4 MPa (Fig. 3.39a). These tensile stresses are similar to the typical *in situ* tensile strength of rocks, so that tension fractures could form, or reactivate, at the location of these peaks. Indeed, when dike tips are very close to the surface, tension fractures are likely to form at the surface (see the book cover and Fig. 7.19).

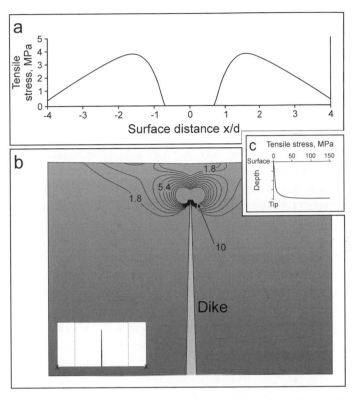

Fig. 3.39 (a) Surface tensile stress induced by an arrested dike. The surface stresses peak at 4 MPa to either side of the projection of the dike tip to the surface. The stress peaks occur at a distance that is 1.6 times the depth to the tip of the dike. (b) The dike is located within a homogeneous and isotropic crustal segment. The only loading is 10 MPa constant overpressure in the dike. The model is 3 units high, whereas the dike tip is at 0.5 units below the free surface. The Young's modulus of the crust is 100 GPa and the Poisson's ratio is 0.25. The theoretical tensile stress at the dike tip reaches 149 MPa but decreases to 0 MPa at the free surface right above the tip of the dike (inset) (cf. Philipp et al., 2013).

As the dike propagates to shallow depths, so that its tip approaches the surface, the two peaks become **taller**, that is, the magnitude of the tensile stress increases, and they move **closer** to the centre, that is, to the projection of the dike tip at the surface (Fig. 3.37). If the tension fractures formed at the peaks reach a critical depth, they could change into normal faults (Gudmundsson, 2011a), thereby forming a graben (Fig. 3.37). But this conclusion rests on the dike being emplaced in what may be approximated as an elastic half-space, that is, a crustal segment that behaves as if it is homogeneous and isotropic and thus with little or no mechanical layering.

Very different results are obtained when **layering** in taken into account (Fig. 3.40). Consider the same model of a dike, subject to an overpressure of 10 MPa as the only loading. The tip of the dike is at 2 units below the free surface. The main difference is that between the free surface and the tip of the dike there are 16 layers: eight stiff (100 GPa) and comparatively thick (0.2 units each) layers, marked by the letter A, and eight soft (1 GPa) and comparatively thin (0.05 units each) layers, marked by the letter B. The layers are welded together in the sense that there are no mechanically weak contacts between them. The dike itself is hosted by layer C whose stiffness is 40 GPa. The dike tip touches the lowermost soft layer B, whose stress-suppressing effects are such that the maximum tensile stress at the tip is only 30 MPa. This contrasts with the 149 MPa maximum tensile stress at the tip of the similarly loaded dike in Fig. 3.39. Although the tensile stress generally decreases in a similar fashion from the dike tip to the surface in Figs. 3.39 and 3.40, the difference in the shape of the stress curve in Fig. 3.40 reflects the layering. More specifically, the tensile stresses in the stiff layers are everywhere at their peaks in the lower parts of the stiff layers A, but very low in all the soft layers B. The location of the maximum tensile stress in the lower parts of each of the stiff layers A is also seen in the stress contours. In contrast to Fig. 3.39, no tensile stress reaches the surface in the model in Fig. 3.40. In fact, the tensile stress in the stiff layers approaches zero already in the middle part of the layered pile above the dike tip.

Further models show the effect of compliant layers in the upper part of the volcanoes/volcanic zones on the dike-induced stresses. All active, and many inactive, polygenetic volcanoes and volcanic zones contain soft or compliant layers at various depths. These layers are commonly either of pyroclastics, scoria, soil, or sediments. In addition, volcanoes and volcanic zones contain numerous stiff lava flows and intrusions, and some stiff, welded pyroclastic layers. In an active volcanic zone/volcano there is thus normally a great variation in stiffness between the layers; variation that can have a very great effect on the internal and, in particular, the surface stresses induced by dikes.

As examples, let us consider a few more models on the tensile and shear stresses induced by an arrested dike in a rift zone (Figs. 3.41–3.44). With the only loading being the overpressure of 6 MPa, the dike, with a dip-dimension of 10 km, is arrested at 500 m below the free surface of the rift zone. The dike is hosted by a thick unit with a Young's modulus of 40 GPa. Above that unit are three layers, in some of the models, and four, in others. Each layer is 100 m thick, and all the layers and the thick unit have a Poisson's ratio of 0.25. All the models are fastened in the corners (indicated by crosses), so as to avoid rigid-body translation and rotation. In each of the figures, part a shows the model set-up including, schematically, the layering (the thickness of the layers and the height of the dike are not to scale), part b the contours of the maximum principal tensile stress σ_3, in megapascals, part c the dike-induced von Mises shear stress τ at the Earth's surface, and part d the dike-induced maximum principal tensile stress σ_3 at the Earth's surface.

For a comparison, the first model has all the layers with equal Young's moduli (Fig. 3.41). Thus, while the layers are shown, they have all the same stiffness of 40 GPa, which is also equal to the stiffness of the thick unit hosting the dike. This is, therefore, effectively an **elastic half-space** model. The results indicate dike-induced surface peak tensile and shear (von Mises) stresses of 10–11 MPa. Given the common tensile (1–6 MPa) and shear (1–12 MPa) strengths of rocks, these stresses are clearly large enough to initiate or reactivate surface tension fractures and (mostly normal) faults.

The Young's moduli of some of the layers in this model, however, are unrealistically high. This applies, in particular, to the surface layer; 40 GPa is far too high. Depending on the rock type, a reasonable dynamic Young's modulus of a surface layer in a volcanic rift zone might be between 5 GPa and 20 GPa, whereas the common static modulus would be

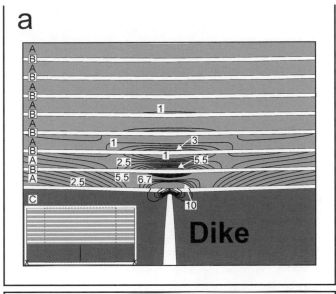

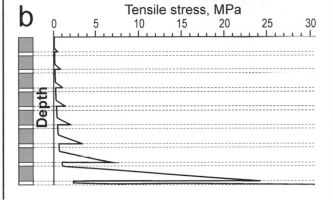

Fig. 3.40 (a) Dike located in a layered crust with 10 MPa overpressure as the only loading. The tip of the dike is at 2 units below the free surface. Between the free surface and the tip of the dike there are 16 layers: eight stiff (100 GPa) and comparatively thick (0.2 units each) layers, marked by the letter A, and eight soft (1 GPa) and comparatively thin (0.05 units each) layers, marked by the letter B. The layers are welded together in the sense that there are no mechanically weak contacts between them. The dike itself is hosted by layer C, with a stiffness of 40 GPa. The dike tip touches the lowermost soft layer B, the stress-suppressing effects of which are such that the maximum tensile stress at the tip is only 30 MPa. (b) The tensile stresses are everywhere at their peak in the lower parts of the stiff layers A, but are very low in all the soft layers B. No tensile stress reaches the surface (cf. Philipp et al., 2013).

between 1 GPa and 10 GPa. But for an elastic half-space model such as this one, there must be a uniform Young's modulus, and 40 GPa is a reasonable average stiffness of the 10-km-thick part of the crustal segment dissected by the dike. The comparatively high surface stresses are largely the result of the unreasonably high stiffnesses assumed for the layers, particularly the surface layer.

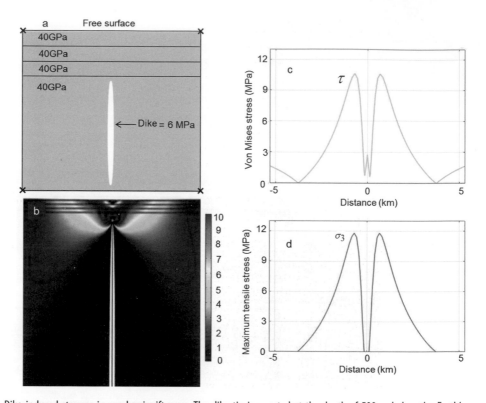

Fig. 3.41 Dike-induced stresses in a volcanic rift zone. The dike tip is arrested at the depth of 500 m below the Earth's surface. The magmatic overpressure of 6 MPa is the only loading. The Poisson's ratio of all the rock layers and units is 0.25. Each of the three top layers is 100 m thick. Here all the layers, as well as the main rock unit hosting the dike, have the same Young's modulus of 40 GPa, making this an elastic half-space model. (a) Schematic set-up of the model. The thickness of the layers and the height of the dike are not to scale. (b) Contours of the maximum principal tensile stress σ_3, in mega-pascals. (c) Von Mises shear stress τ at the Earth's surface. (d) Maximum principal tensile stress σ_3 at the Earth's surface (cf. Al Shehri and Gudmundsson, 2018). A black and white version of this figure will appear in some formats. For the colour version, please refer to the plate section.

In the next model, where somewhat more realistic stiffnesses of the layers are assumed, the dike-induced stresses are much reduced (Fig. 3.42). Here there are three layers above the tip of the dike, the lower-most two with the same stiffness, 27 GPa, whereas the surface layer has a stiffness of 17 GPa. These stiffnesses for the layers are still high, particularly for the surface layer. However, the dike-induced surface stresses reach maximum values of only 3–4 MPa. While 3–4 MPa may be enough to initiate or reactivate some fractures (tension fractures and normal faults), these would tend to be small. Thus, a dike with an overpressure of 6 MPa and with a tip at only 500 m below the surface is barely able to initiate tension fractures and normal faults at the surface, even if the surface layer has a stiffness of 17 GPa. It follows that the same dike with a tip at greater depth would not be able to initiate any surface fractures.

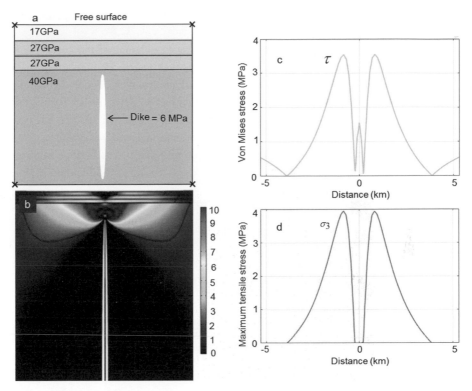

Fig. 3.42 Dike-induced stresses in a volcanic rift zone. Same dike-tip depth, loading, and boundary conditions as in Fig. 3.41 except that here the surface layer has a Young's modulus of 17 GPa, followed by two layers with a Young's modulus of 27 GPa. The unit or layer hosting the dike has a Young's modulus of 40 GPa. (a) Schematic set-up of the model. The thickness of the layers and the height of the dike are not to scale. (b) Contours of the maximum principal tensile stress σ_3, in mega-pascals. (c) Von Mises shear stress τ at the Earth's surface. (d) Maximum principal tensile stress σ_3 at the Earth's surface (cf. Al Shehri and Gudmundsson, 2018). A black and white version of this figure will appear in some formats. For the colour version, please refer to the plate section.

As indicated above, a surface layer in an active volcanic rift zone would normally have stiffness between 1 and 10 GPa. In addition, there are commonly soft sedimentary or pyroclastic layers in-between the stiffer lava flows in active volcanoes and volcanic rift zones. To take these into account, the next model includes an additional layer, so that there are now four layers above the dike tip (Fig. 3.43). The surface layer here has a stiffness of 3 GPa, which is similar to that of some young lava flows or pyroclastic layers. In addition, there is a soft layer, with a stiffness of 1 GPa, below the other three, that is, with a top at the depth of 300 m. As discussed above, each layer is 100 m thick, so that together the four layers are 400 m thick, whereas the dike tip is, as before, at the depth of 500 m below the surface.

The results (Fig. 3.43) show very clearly the great effects of the soft (1 GPa) layer on the rift-zone stresses. The soft layer suppresses much of the stress, so that very low stresses are transmitted across the layer. The peak surface tensile and shear stresses are less than 1 MPa.

It follows from the low surface stresses that neither tension-fracture formation nor normal faulting is likely.

By how much would the surface stresses change if the surface layer were very stiff rather than reasonably compliant? This was tested for the same loading and layering as in the model in Fig. 3.43, except that in this new model the surface layer again has a very high (for a surface layer) stiffness of 17 GPa (rather than 3 GPa). Even for such an unrealistically stiff surface layer, the dike-induced surface stresses remain very low (Fig. 3.44). More specifically, the maximum tensile and shear stresses at the surface reach about 3 MPa. Thus, even for an extremely stiff surface layer and a very shallow dike tip (500 m below the surface in all the models) much dike-induced surface fracturing is unlikely.

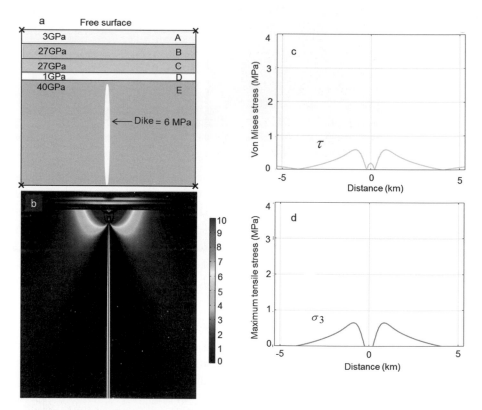

Fig. 3.43 Dike-induced stresses in a volcanic rift zone. Same dike-tip depth, loading, and boundary conditions as in Fig. 3.42, except that here there are four top layers, each 100 m thick. The surface layer has a Young's modulus of 3 GPa, followed by two layers each with a Young's modulus of 27 GPa, and then a compliant layer with a Young's modulus of 1 GPa. The unit or layer hosting the dike has a Young's modulus of 40 GPa. (a) Schematic set-up of the model. The thickness of the layers and the height of the dike are not to scale. (b) Contours of the maximum principal tensile stress, σ_3, in mega-pascals. (c) Von Mises shear stress τ at the Earth's surface. (d) Maximum principal tensile stress σ_3, at the Earth's surface (cf. Al Shehri and Gudmundsson, 2018). A black and white version of this figure will appear in some formats. For the colour version, please refer to the plate section.

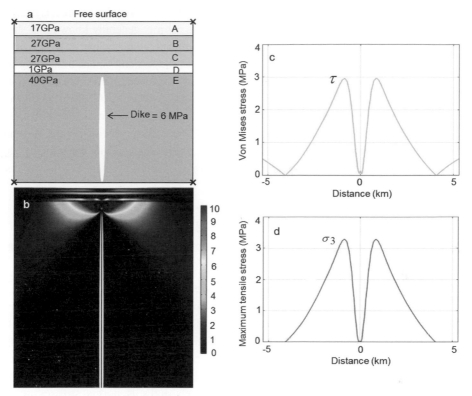

Fig. 3.44 Dike-induced stresses in a volcanic rift zone. Same dike-tip depth, loading, and boundary conditions as in Fig. 3.43 except that the surface layer has a Young's modulus of 17 GPa (rather than 3 GPa). (a) Schematic set-up of the model. The thickness of the layers and the height of the dike are not to scale. (b) Contours of the maximum principal tensile stress σ_3, in mega-pascals. (c) Von Mises shear stress τ at the Earth's surface. (d) Maximum principal tensile stress σ_3 at the Earth's surface (cf. Al Shehri and Gudmundsson, 2018). A black and white version of this figure will appear in some formats. For the colour version, please refer to the plate section.

The results in Figs. 3.38 to 3.44 are in agreement with the models in Chapters 7 and 10 and underline the great effects that **mechanical layering** can have on the local dike-induced stresses and associated surface deformation in volcanoes and volcanic zones. The use of simple elastic half-space models may sometimes be justified, particularly for unrest periods in some basaltic edifices, so as to give a rough indication of the depths and geometries of dikes and inclined sheets inducing surface stresses and deformation during an unrest period. Such models, however, are generally **poorly justified** when dealing with stratovolcanoes, calderas, and volcanic (rift) zones composed of layers with widely different mechanical properties. Most volcanoes and volcanic zones contain compliant or soft layers and contacts with stiffness of the order of 1 GPa or even less. This is well known from drill holes and direct observations of exposed parts (such as in walls exposed through collapses or erosion) of active (and inactive) volcanoes and volcanic zones.

While the exact layering in a given active volcano/volcanic zone is rarely known, the **range in stiffness** of the layers that constitute them is generally known and should be considered when interpreting measured surface deformation induced by dikes (or magma chambers). The results in this chapter show that compliant layers and contacts can have great effects on the surface stresses and associated deformation induced by magma chambers, dikes, and inclined sheets. Further results presented in Chapter 10 and by Al Shehri and Gudmundsson (2018) indicate not only the effects of layering on surface stresses and deformation, but also how the stresses and displacements inside volcanoes during unrest periods depend on the layering. Straightforward half-space-model inversion of the geodetic data to infer magma chamber and dike depths and geometries are thus rarely likely to provide reliable results.

3.9 Summary

- Understanding stresses and deformation inside and at the surface of a volcano during unrest periods is one of the principal tasks of volcanotectonics. Such an understanding relies on knowledge of the basic elements of elasticity theory, particularly stress, strain, and displacement, which is presented in this chapter. Stress is force per unit area with the unit of $N\,m^{-2}$, which has the name pascal (Pa). Normal stress is perpendicular (normal) to the area or plane on which it acts whereas shear stress is parallel with the area or plane on which it acts. For dikes, inclined sheets, and sills, normal stresses are of main importance whereas for faults, including ring (caldera)-faults, shear stresses are of main importance. Displacements are changes in the position of particles or material points in a body (here a rock body) or points at the surface of the body (such as GPS stations at the surface of a volcano). They are given as displacement vectors (with magnitude and direction) in units of metres, for large displacements, or centimetres or millimetres for smaller displacements (such as are commonly measured during volcanic unrest periods). Strain denotes change in distances between particles in a body, a fractional change in the body's dimensions, and has no unit (is a pure number).

- Planes in a rock body on which no shear stresses act are referred to as principal stress planes. When such a plane is in contact with a fluid (such as water, magma, or air) it is known as a free surface. The normal stresses that act on the principal stress planes are referred to as principal stresses and are denoted by σ_1, σ_2, and σ_3. In volcanotectonics and related fields, compressive stresses are the most common and regarded as positive whereas tensile stresses are much more rare – but very important when and where they occur – and regarded as negative. In the present notation, σ_1 is the maximum compressive principal stress, σ_2 is the intermediate compressive principal stress, and σ_3 is the minimum compressive (maximum tensile) principal stress. In volcanotectonics, σ_1 is always, and σ_2 normally, positive, that is, they are compressive stresses. By contrast, σ_3 may be positive (the minimum compressive stress) or negative (the maximum tensile stress). During magma chamber inflation, σ_3 normally becomes increasingly negative until the chamber ruptures and a dike or an inclined sheet is injected.

- Linear elastic behaviour of rock is presented by Hooke's law, which relates stress and strain through elastic constants. For isotropic rock there are two independent elastic constants. The most commonly used constants in volcanotectonics are Young's modulus, which is also referred to as stiffness, and Poisson's ratio. Other constants include the shear modulus (modulus of rigidity), Lamé's constant, and the bulk modulus (incompressibility). All these constants have the unit of stress, Pa, except Poisson's ratio, which is a ratio of strains and is thus a pure number (with no unit). The reciprocal of the bulk modulus is compressibility, which has the unit of 1/stress or Pa^{-1}. The relative compressibilities of magma and its host rock partly determine how much magma a chamber/reservoir can hold before rupture and dike/sheet injection.

- A magma chamber is the shallow part of a double (for some volcanoes, triple) magma chamber, the deep-seated source (normally much larger) being referred to as a magma reservoir. For analysing volcano stress and deformation (strain and displacement), magma chambers can be modelled analytically primarily in three ways: as nuclei of strain, as three-dimensional cavities, and as two-dimensional holes. The nucleus of strain is commonly known as the Mogi model in volcanology. It assumes that the magma chamber may be regarded as very small in diameter in comparison with its depth below the free surface (of the volcanic zone/volcano). The Mogi model cannot separate the chamber excess pressure from the chamber size (assumed very small), and thus does not provide any information about the stresses at the boundary of the chamber; more specifically, the model cannot be used to infer whether or not the magma chamber is going to rupture and inject a dike during the unrest period. The Mogi model is thus mainly used to infer, crudely, the depth to a chamber associated with a particular unrest period and surface deformation.

- The cavity/hole analytical models can be used to analyse the local stress field around the chamber, as well as the surface stresses and deformation during unrest periods. Furthermore, they can be used to infer the size and depth of the chamber associated with the unrest. The three-dimensional cavity model is more accurate for the stresses and displacements, but for elongated magma chambers, such as many sill-like, tunnel-shaped chambers, a two-dimensional hole model gives a reasonably accurate overview of the stress fields and associated deformation. In the limit of a spherical cavity becoming very small in diameter, the cavity model reduces to the Mogi model. While certainly useful, all analytical models suffer from the same disadvantage: they are limited to homogeneous, isotropic crustal segments, commonly known as half-space (or, much more rarely, semi-infinite) models. They cannot, therefore, take into account one of the characteristics of crustal segments in general, and volcanoes/volcanic zones in particular, namely mechanical layering.

- To take mechanical layering, as well as discontinuities such as intrusions, faults – particularly ring-faults – into account in the models, numerical methods are normally used. Numerical methods can of course also be used for chambers in homogeneous, isotropic crustal segments, but their greatest use is for layered and fractured crustal segments. Mechanical layering of the simplest kind – say crustal segments with three or four distinct mechanical layers – can have profound effects on the local stress field around magma chambers during unrest periods. In particular, both two-dimensional hole models and three-dimensional cavity models show that the local stresses in mechanically layered crustal segments and volcanoes tend to arrest/deflect propagating dikes, inclined sheets, and ring-faults. Thus, mechanical layering in general makes fracture propagation of any kind more difficult. This means that the likelihood of a dike injected during an unrest period reaching the surface to feed an eruption is less for a volcano composed of layers of widely different mechanical properties (such as most stratovolcanoes and many calderas) than for a volcano composed of mechanically similar layers (such as many basaltic edifices).

- The tendency to dike/sheet arrest and/or deflection is even greater when there are mechanically weak contacts between the layers that constitute the volcano. In addition, such contacts may have strong effects on the dike-induced surface stresses and deformation. Thus, the layering and associated contacts may alter the dike-induced stress and deformation fields at the surface of the volcanic zone/volcano in such a way that straightforward inversion of the geodetic data, using standard elastic half-space models, yields unreliable results. Layering and contacts make it more difficult to provide reliable inversion results from geodetic data during unrest periods – results which could confidently be used to advance our understanding of the volcanotectonic processes inside the volcano that are responsible for its deformation.

3.10 Main Symbols Used

A	area
a	radius of a sill or a sill-like magma chamber
a	semi-major axis of an elliptical solid or a notch (a hole)
a, b, c	semi-axes of a triaxial ellipsoid (a chamber or a reservoir)
b	semi-minor axis of an elliptical solid or a notch (a hole)
d	depth of a chamber below the Earth's surface
E	Young's modulus (modulus of elasticity)
F	force
G	shear modulus (modulus of rigidity)
g	acceleration due to gravity
h	height or thickness of a sill-like magma chamber
K	bulk modulus (incompressibility)
k	stiffness
L	length, length of a bar
ΔL	extension, change in length of a bar
m	Poisson's number
p_e	magmatic excess pressure
R_1	radius of a magma-filled cylindrical or spherical cavity (a chamber)
R_2	radius of a solid cylindrical or spherical crustal segment (hosting a chamber)
r	radial distance at the surface above a magma chamber
r	radius vector (distance)

u	displacement or half the total opening of a sill
u_R	radial displacement of the surface of a magma chamber
u_r	horizontal surface displacement
u_z	vertical surface displacement
U_n	intensity of a point-pressure cavity (strain-energy nucleus)
U_0	strain energy
ΔV	fractional change in the volume of a magma chamber
ΔV_s	volume change (doming or subsidence) at the surface induced by magma-chamber expansion or contraction
ΔW	work done during expansion or shrinkage of a magma chamber
Δx	deflection or movement of the upper face of a body
y	vertical distance between faces or surfaces of a body
z	depth below the Earth's surface (crustal depth, vertical coordinate)
β	compressibility
γ	shear strain
Δ	dilation, volume strain
ε	normal strain
ε_1	maximum compressive principal strain
ε_2	intermediate compressive principal strain
ε_3	minimum compressive principal strain
θ	angle, a spherical polar coordinate
λ	Lamé's constant
v	Poisson's ratio
π	3.1416
ρ_r	rock or crustal density
σ	normal stress
$\bar{\sigma}$	mean stress
σ_c	maximum compressive stress around a magma chamber
σ_{max}	maximum tensile stress at the tip of a semi-elliptical notch
σ_r	radial stress, a principal stress away from a spherical cavity (a reservoir)
σ_t	maximum tensile stress around a magma chamber
σ_θ	circumferential stress, a principal stress around a spherical cavity (a reservoir)
σ_φ	tangential stress, a principal stress around a spherical cavity (a reservoir)
σ_1	maximum compressive principal stress
σ_2	intermediate compressive principal stress
σ_3	minimum compressive (maximum tensile) principal stress
σ_v	vertical stress
τ	shear stress
φ/ϕ	angle, a spherical polar coordinate
ψ	deflection angle

3.11 Worked Examples

Example 3.1

Problem

A basaltic rock specimen tested in the laboratory has a dynamic Young's modulus of 70 GPa and a Poisson's ratio of 0.25. Calculate the rock:

(a) shear modulus
(b) compressibility
(c) Lamé's constant
(d) Poisson's number.

Solution

(a) From Eq. (3.19) we have:

$$G = \frac{E}{2(1+v)} = \frac{7 \times 10^{10} \text{ Pa}}{2(1+0.25)} = 2.8 \times 10^{10} \text{ Pa} = 28 \text{ GPa}$$

(b) Rewriting Eq. (3.21), we get:

$$K = \frac{E}{3(1-2v)} = \frac{7 \times 10^{10} \text{ Pa}}{3(1-0.50)} = 4.7 \times 10^{10} \text{ Pa} = 47 \text{ GPa}$$

which is incompressibility. To find the compressibility, we use Eq. (3.22), thus:

$$\beta = \frac{1}{K} = \frac{1}{4.7 \times 10^{10} \text{ Pa}} = 2.1 \times 10^{-11} \text{ Pa}^{-1}$$

(c) From Eq. (3.23) we have:

$$\lambda = \frac{vE}{(1+v)(1-2v)} = \frac{0.25 \times 7 \times 10^{10} \text{ Pa}}{(1+0.25)(1-0.50)} = 2.8 \times 10^{10} \text{ Pa} = 28 \text{ GPa}$$

The reason why the result here for λ is the same as for G is that we have used the Poisson's relation, which applies when Poisson's ratio is 0.25. Then, $G = \lambda$ and the number of independent elastic constants for isotropic rocks reduces from 5 to 4.

(d) From Eq. (3.18) we have:

$$m = \frac{1}{v} = \frac{1}{0.25} = 4$$

Example 3.2

Problem

In a volcanic rift zone, the vertical stress is the maximum stress. At a shallow depth in the zone the principal stresses in a thick lava flow are measured as 20 MPa, 15 MPa, and 10 MPa. The Young's modulus of the rock where the stress is measured is 10 GPa and its

Poisson's ratio is 0.25. The average crustal density to the depth of the stress-measurement site is 2400 kg m^{-3}. Calculate:

(a) the likely depth of the lava flow where the stresses are measured;
(b) the principal strains in the directions of the stresses at the measurement site.

Solution

(a) Rewriting Eq. (3.5), the depth z of the lava flow is obtained as follows:

$$z = \frac{\sigma_v}{\rho_r g} = \frac{2 \times 10^7 \text{ Pa}}{2400 \text{ kg m}^{-3} \times 9.81 \text{ m s}{-2}} = 849 \text{ m}$$

(b) The principal strains in terms of principal stresses and the Young's modulus and Poisson's ratio are given by Eqs. (3.12–3.14). For the maximum compressive principal strain we have:

$$\varepsilon_1 = \frac{1}{E}[\sigma_1 - v(\sigma_2 + \sigma_3)]$$

$$= \frac{1}{1 \times 10^{10} \text{ Pa}}[2 \times 10^7 \text{ Pa} - 0.25(1.5 \times 10^7 \text{ Pa} + 1 \times 10^7 \text{ Pa})]$$

$$= 0.00137 = 0.137\%$$

For the intermediate principal strain we have:

$$\varepsilon_2 = \frac{1}{E}[\sigma_2 - v(\sigma_3 + \sigma_1)]$$

$$= \frac{1}{1 \times 10^{10} \text{ Pa}}[1.5 \times 10^7 \text{ Pa} - 0.25(1 \times 10^7 \text{ Pa} + 2 \times 10^7 \text{ Pa})] = 0.00075$$

$$= 0.075\%$$

For the minimum principal strain we have:

$$\varepsilon_3 = \frac{1}{E}[\sigma_3 - v(\sigma_1 + \sigma_2)]$$

$$= \frac{1}{1 \times 10^{10} \text{ Pa}}[1 \times 10^7 \text{ Pa} - 0.25(2 \times 10^7 \text{ Pa} + 1.5 \times 10^7 \text{ Pa})] = 0.00012$$

$$= 0.012\%$$

Example 3.3

Problem

A spherical magma source of 1 km in diameter is at a depth of 4 km below the surface of a volcanic zone. The average Young's modulus of the crust above the magma source is 25 GPa and its Poisson's ratio is 0.25.

(a) Calculate the horizontal and vertical displacements at the surface above the centre of the source induced by an excess magmatic pressure of 5 MPa in the source using a Mogi model.
(b) Would the surface deformation result in fracture formation?

(c) Explain quantitatively why this magma source is unlikely to function as a long-lasting magma chamber.

Solution

(a) Since Poisson's ratio is given as 0.25, it follows that we can use Eqs. (3.27) and (3.28) to calculate the surface displacements. In these equations, the radius of the chamber rather than the diameter is used. Since the diameter is 1 km, we have, in SI units, $R_1 = 500$ m. Also, the shear modulus G rather than the Young's modulus E is used in the equations, so we first have to obtain the shear modulus. From Eq. (3.19) we get:

$$G = \frac{E}{2(1+v)} = \frac{2.5 \times 10^{10} \text{ Pa}}{2(1+0.25)} = 1 \times 10^{10} \text{ Pa} = 10 \text{ GPa}$$

Using this value for G, and the other information given, from Eq. (3.27) the vertical displacement is:

$$u_z = \frac{3p_e R_1^3 d}{4G(r^2 + d^2)^{3/2}} = \frac{3 \times 5 \times 10^6 \text{ Pa} \times (500 \text{ m})^3 \times 4000 \text{ m}}{4 \times 1 \times 10^{10} \text{ Pa} \times [(0 \text{ m})^2 + (4000 \text{ m})^2]^{3/2}} = 0.0029 \text{ m}$$
$$= 2.9 \text{ mm}$$

For the horizontal displacement, from Eq. (3.28) we get:

$$u_r = \frac{3p_e R_1^3 r}{4G(r^2 + d^2)^{3/2}} = \frac{3 \times 5 \times 10^6 \text{ Pa} \times (500 \text{ m})^3 \times 0 \text{ m}}{4 \times 1 \times 10^{10} \text{ Pa} \times [(0 \text{ m})^2 + (4000 \text{ m})^2]^{3/2}} = 0 \text{ m}$$

The first result here shows a very small vertical displacement, 2.9 mm, above the centre of the magma chamber. The second result shows that the horizontal displacement at that location is zero, a well-known result from Mogi models (Fig. 3.1).

(b) The Mogi displacement results as such do not tell us if the displacement is enough to generate fractures at the surface. A rough estimate of the vertical strain, however, suggests that the displacements would be too small. The vertical displacement is 0.0029 m over a vertical distance (depth to the magma chamber) of 4000 m. It is strictly the depth to the centre of the chamber, but the chamber is small, so we calculate the strain from Eq. (3.7) as:

$$\varepsilon = \frac{\Delta L}{L} = \frac{0.0029 \text{ m}}{4000 \text{ m}} = 7.25 \times 10^{-7}$$

This is very small strain, many times smaller than most solid rocks tolerate before brittle failure. The horizontal strain at the deformation or doming centre right above the chamber is zero, but is of the same order of magnitude as the vertical strain at a certain distance from the centre (Fig. 3.1; Example 3.4).

An estimate of the associated stress would also follow from Hooke's law, where Eq (3.10) could be used as an approximate estimate of the stress, yielding:

$$\sigma = E\varepsilon = 2.5 \times 10^{10} \text{ Pa} \times 7.25 \times 10^{-7} = 18125 \text{ Pa} \approx 0.002 \text{ MPa}$$

where the Young's modulus given for the uppermost 4 km, namely 25 GPa, has been used. Clearly, the stresses are far too small to generate any fractures, a conclusion also indicated by the strain calculations. We conclude that no fracture formation would be induced by surface displacements of this magnitude.

(c) The Mogi model assumes the chamber to be very small in comparison with its depth below the surface (in fact, a point source). The chamber radius here is 1/8 of its depth below the surface and may thus approximate crudely the assumptions made in the Mogi model. Using the formula for the volume of a sphere, the volume of a Mogi chamber of radius $R_1 = 500$ m is:

$$V = \frac{4}{3}\pi R_1^3 = \frac{4}{3} \times 3.1416 \times (500 \text{ m})^3 = 5.23 \times 10^8 \text{ m}^3 \approx 0.52 \text{ km}^3$$

From the theory developed in Chapters 6 and 8 it follows that during a typical rupture of a shallow magma chamber, the volume of magma leaving the chamber is of the order of 0.07% for a basaltic chamber (Eq. (6.8)) to a maximum of 4% for a purely felsic magma chamber (Eq. (6.10)). Thus, the largest dike plus eruptive volume such a chamber could produce would be about 3.6×10^{-4} km^3 for a basaltic chamber, and 0.02 km^3 for a felsic chamber, assuming both chambers to be totally molten (fluid). Given that the chamber is at 4 km depth, the volume of a feeder-dike would be of the order of 0.01 km^3 or more, so that most likely this magma source would not erupt and thus not function as a long-term magma chamber. The volume of a shallow chamber is normally at least 10–100 times larger than that of this source (Chapters 6 and 8).

Example 3.4

Problem

Consider a magma chamber at 4 km depth in a crust with an average Young's modulus of 25 GPa, a Poisson's ratio of 0.25, and subject to excess pressure of 5 MPa as in Example 3.3, but now with a radius of 3 km. Although the radius (3 km) is thus still less than the depth to the chamber (4 km), it is clearly too large to fit with the point-source assumptions of the Mogi model. However, magma chambers of dimensions similar to or even greater than their depths are commonly modelled using the Mogi model. McTigue (1987) has suggested that Mogi models may be applicable to spherical chambers only if the following holds:

$$\left(\frac{R_1}{d}\right)^5 << 1 \tag{3.56}$$

where, as before, d is the depth to the chamber and R_1 the chamber radius. Here we have:

$$\left(\frac{3000 \text{ m}}{4000 \text{ m}}\right)^5 = 0.23$$

which is clearly not many times less than 1. By contrast, if the radius of the chamber were 1 km then, for this depth (4 km), we would have $(1000/4000)^5 \approx 0.001 \ll 1$, in which case the assumption of a Mogi model might be more reasonable. In fact, the chamber in a Mogi model is commonly assumed to have a radius of 1 km (Dzurisin, 2006), an assumption that would not be very sound when using a Mogi model for shallow chambers at depths of 1–1.5 km and certainly not valid for chambers with lateral dimensions of many kilometres.

Despite these limitations and reservations, we shall here calculate the surface deformation using the Mogi model. Since we know that the horizontal displacement is zero in the centre of the inflated area, we will calculate only the vertical displacements at the centre, and then both the vertical and the horizontal displacements at a distance from the centre equal to the depth of the chamber. Thus, calculate:

(a) the vertical displacement (actually the maximum displacement) in the centre of the inflation area or dome at the surface
(b) the vertical displacement at a radial distance of 4 km from the centre of the inflation dome
(c) the horizontal displacement at a radial distance of 4 km from the centre of the inflation dome.

Solution

(a) Using the same crustal elastic properties and excess magmatic pressure as in Example 3.3, from Eq. (3.27) we obtain:

$$u_z = \frac{3p_e R_1^3 d}{4G(r^2 + d^2)^{3/2}} = \frac{3 \times 5 \times 10^6 \text{ Pa} \times (3000 \text{ m})^3 \times 4000 \text{ m}}{4 \times 1 \times 10^{10} \text{ Pa} \times [(0 \text{ m})^2 + (4000 \text{ m})^2]^{3/2}}$$

$$= 0.63 \text{ m} = 63 \text{ cm}$$

A maximum vertical displacement of tens of centimetres is common during inflation periods in many volcanoes. For some chambers and inflation episodes, however, the maximum displacements reach many metres or more. Here we are thus modelling a common type of unrest period.

(b) For a radial distance of 4 km, in Eq. (3.27) we no longer have $r = 0$ m but rather $r = 4000$ m, in which case the vertical displacement becomes:

$$u_z = \frac{3p_e R_1^3 d}{4G(r^2 + d^2)^{3/2}} = \frac{3 \times 5 \times 10^6 \text{ Pa} \times (3000 \text{ m})^3 \times 4000 \text{ m}}{4 \times 1 \times 10^{10} \text{ Pa} \times [(4000 \text{ m})^2 + (4000 \text{ m})^2]^{3/2}}$$

$$= 0.22 \text{ m} = 22 \text{ cm}$$

This shows that the vertical displacement declines with distance from the centre of the doming or inflated area, in agreement with the appropriate curve in Fig. 3.1.

(c) For a radial distance of 4 km, Eq. (3.28) gives the horizontal displacement as:

$$u_r = \frac{3p_e R_1^3 r}{4G(r^2 + d^2)^{3/2}} = \frac{3 \times 5 \times 10^6 \text{ Pa} \times (3000 \text{ m})^3 \times 4000 \text{ m}}{4 \times 1 \times 10^{10} \text{ Pa} \times [(4000 \text{ m})^2 + (4000 \text{ m})^2]^{3/2}}$$

$$= 0.22 \text{ m} = 22 \text{ cm}$$

So, at a radial distance from the centre of the inflated area equal to the depth to the chamber ($r = d$) the radial and the vertical displacements are equal. This follows directly from comparing Eqs. (3.27) and (3.28). Both have the same denominator. The only difference in the numerator is that in Eq. (3.28) the radial distance r is substituted for the depth d to the chamber. Thus, when $r = d$, $u_r = u_z$, that is, the curves for u_r and u_z intersect (Fig. 3.1).

The maximum horizontal displacement occurs at the distance of:

$$\pm \frac{d}{2^{1/2}} \approx 0.7d$$

from the centre of the surface deformation (cf. Dzurisin, 2006).

Example 3.5

Problem

(a) Estimate the volume of magma that must be added to the shallow magma chamber in Example 3.4 so as to induce the estimated maximum vertical displacement of 63 cm.
(b) Estimate the excess pressure needed to generate that uplift and comment on whether the estimated excess pressure is realistic.
(c) Estimate the excess pressure needed for a maximum uplift of 2 m, and discuss whether or not the scenario is realistic.

Solution

(a) Rewriting Eq. (3.32) and solving for the added volume (volume change) ΔV, we get:

$$\Delta V = \frac{4\pi u_z (r^2 + d^2)^{3/2}}{3d} \tag{3.57}$$

For the maximum vertical displacement we have $r = 0$, in which case Eq. (3.57) reduces to:

$$\Delta V = \frac{4}{3}\pi d^2 u_z \tag{3.58}$$

For an uplift of 0.63 m, we get from Eq. (3.58):

$$\Delta V = \frac{4}{3} \times 3.1415 \times (4000 \text{ m})^2 \times 0.63 \text{ m} = 4.2 \times 10^7 \text{ m}^3 = 0.042 \text{ km}^3$$

(b) For a Mogi model, the volume added to the chamber and the excess pressure are related through Eq. (3.31), which can be solved for the excess magmatic pressure p_e thus:

$$p_e = \frac{G\Delta V}{\pi R_1^3} \tag{3.59}$$

Substituting the values obtained above into Eq. (3.59), we get the excess pressure associated with the magma volume added to the chamber as:

$$p_e = \frac{1 \times 10^{10} \text{ Pa} \times 4.2 \times 10^7 \text{ m}^3}{3.1416 \times (3000 \text{ m})^3} = 4.95 \times 10^6 \text{ Pa} = 5 \text{ MPa}$$

This is the same excess pressure (5 MPa) as used in Example 3.4, as expected. Since the excess pressure should be similar to the tensile strength of the rock, which is in the general range of 0.5–9 MPa and most commonly 1–6 MPa, this excess pressure is reasonable. This reference to the tensile strength is used even if Eq. (3.35) shows that the maximum tensile stress around a spherical chamber is only half the excess pressure because magma-chamber rupture normally occurs at the sites of the margin of the chamber where irregularities (deviations from a perfect sphere) raise or concentrate the tensile stress.

(c) For a maximum vertical displacement of 2 m, Eq. (3.58) gives the volume of magma added to the chamber as:

$$\Delta V = \frac{4}{3} \times 3.1415 \times (4000 \text{ m})^2 \times 2 \text{ m} = 1.3 \times 10^8 \text{m}^3 = 0.13 \text{ km}^3$$

Using this value for the added magma, Eq. (3.59) gives the associated excess pressure in the chamber as:

$$p_e = \frac{1 \times 10^{10} \text{ Pa} \times 1.3 \times 10^8 \text{ m}^3}{3.1415 \times (3000 \text{ m})^3} = 1.5 \times 10^7 \text{ Pa} = 15 \text{ MPa}$$

This excess pressure is higher than any measured *in situ* tensile strength in the Earth's crust, and is thus unlikely to be reached. We conclude, therefore, that this scenario is unrealistic and that the chamber would be unlikely to be able to accommodate this added volume of magma before rupture and dike injection.

Example 3.6

Problem

As indicated in Section 3.4, the volumetric change in the Mogi model related to magma excess pressure changes is commonly different from that of the volume of the associated surface deformation (inflation dome or deflation depression). The difference is commonly given as (Delaney and McTigue, 1994):

$$\Delta V_s = 2(1 - v)\Delta V \tag{3.60}$$

where ΔV_s is the volume change (doming or subsidence) at the surface induced by the volume change ΔV of the chamber, and v is Poisson's ratio. This applies to a spherical (Mogi) chamber and is based on the assumption that the magma in the chamber is incompressible (Johnson et al., 2000) – whereas real magma is of course compressible. Nevertheless, given this assumption, calculate the difference in the volume change between the surface and the chamber for common volcanotectonic scenarios, and explain under what conditions the volume change at the surface and of the chamber would be equal.

Solution

For common volcanotectonic scenarios, the Poisson's ratio for the host rock is close to 0.25. In fact, that is the assumption we have made (referred to as the Poisson's relation) in deriving Eqs. (3.27), (3.28), (3.32) and (3.33) used in the calculations above. From Eq. (3.60) we then get $\Delta V_s = 1.5 \times \Delta V$. Thus, for a common situation the volume change at the surface would be 50% larger than that of the Mogi chamber which induced the surface deformation. This is assuming an elastic half-space, so that there is only a single layer with constant elastic properties between the chamber and the surface. For a layered crust, the relationship between the volume change in a chamber and the volume change in the layers above it, including the surface, however, depends strongly on the mechanical properties of the layer.

For the volume change at the surface to be equal to that of the magma chamber, based on the above assumptions, we must have:

$$2(1 - v) = 1$$

This is possible only if $v = 0.5$, which means that the material is incompressible. No rocks or magmas are incompressible, and the Poisson's ratio for most rocks is between 0.15 and 0.35, with an average of about 0.25 (Appendix D). Thus, for a Mogi model, the volume change during surface deformation in an elastic half-space is normally different from the volume change of the spherical chamber inducing the deformation.

Example 3.7

Problem

Consider a finite-sized spherical magma chamber of radius 4 km. During an unrest period the chamber is subject to a magmatic excess pressure of 5 MPa. Calculate:

(a) the volume of the chamber and decide if it could act as a long-term magma source to a central volcano (a stratovolcano, a collapse caldera, or a basaltic edifice);
(b) the local principal stresses at a distance of 9 km from the centre of the chamber, that is, 5 km from the boundary of the chamber;
(c) the strain energy associated with the pressured and expanded chamber.

Solution

(a) Using R_1 for the chamber radius, the volume V_c of a spherical chamber is given by:

$$V_c = \frac{4}{3}\pi R_1^3 \tag{3.61}$$

from which we get the chamber volume as:

$$V_c = \frac{4}{3}\pi(4 \text{ km})^3 = 268 \text{ km}^3$$

where we use kilometres rather than metres as the normal unit for measuring the dimensions of a major magma chamber. This is clearly a reasonably large magma chamber and it is easily able to supply magma to numerous dike injections and eruptions, in accordance with the general theory set forth in Chapters 6 and 8. Thus, a chamber of this size could certainly develop, and act as a source for, a central volcano.

(b) Here we can either give the distances in metres or kilometres since they come into the calculations as a ratio. We follow our normal practice and use the SI units, so the distances are in metres and the excess pressure in pascals. From Eq. (3.34) the radial stress (the compressive stress) at a distance of $r = 9000$ m is:

$$\sigma_r = p_e \left(\frac{R_1}{r}\right)^3 = 5 \times 10^6 \text{ Pa} \times \left(\frac{4000 \text{ m}}{9000 \text{ m}}\right)^3 = 4.4 \times 10^5 \text{ Pa} \approx 0.4 \text{ MPa}$$

Similarly, from Eq. (3.35), the tensile principal stresses are:

$$\sigma_\theta = \sigma_\varphi = -\frac{p_e}{2}\left(\frac{R_1}{r}\right)^3 = -2.5 \times 10^6 \text{ Pa} \times \left(\frac{4000 \text{ m}}{9000 \text{ m}}\right)^3$$
$$= -2.2 \times 10^5 \text{ Pa} \approx -0.2 \text{ MPa}$$

Here we give the tensile stress with a minus sign. This is, however, normally not necessary since tensile stress is known to be regarded as negative in geology, so that often we give all the stress magnitudes as absolute values (without signs).

The results show that both the compressive (radial) and tensile (circumferential) local stresses induced by the overpressured (and inflated) chamber have declined to less than 1 MPa at this distance. It is therefore to be expected that at about this distance from the chamber, for the given chamber size and excess pressure, the local stress has diminished so much that its influence on the attitude and dimensions of dikes and sheets is negligible, and the regional stress field takes over as the controlling field. This is, indeed, what is observed in central volcanoes in Iceland (Figs. 3.10 and 3.11). The sheet swarms with thin inclined sheets and radial dikes extend only several kilometres from their shallow magma chambers (exposed as plutons); at greater distances from the chamber/volcano the regional rift-zone stress field, with thick, subvertical, dikes takes over.

(c) When the excess pressure of the chamber gradually increases to 5 MPa, the chamber expands and strain energy becomes stored in its host rock. Eventually, if the chamber roof ruptures and a dike or sheet is injected, part of the stored strain energy is used to generate the surface of the dike-fracture (surface energy), and part to squeeze out the magma in the chamber into the dike-fracture (and to the surface in case of an eruption). Using Eq. (3.36), the strain energy may be calculated as:

$$U_n = p_e R_1^3 = 5 \times 10^6 \text{ Pa} \times (4000 \text{ m})^3 = 3.2 \times 10^{17} \text{ J}$$

This is a large strain energy. For a typical feeder-dike, the energy release rate, and thus the surface energy needed to form the dike-fracture, is of the order of 2×10^7 J m^{-2} (Gudmundsson, 2009). This means that to rupture 1 m^2 of rock during the dike propagation, an elastic energy of the order of 10^7 J needs to be transformed into surface energy – the energy used to rupture the rock and move the rupture surfaces apart (Gudmundsson, 2011a). The elastic energy is here supposed to derive from the strain energy stored in the volcano. Here, the strain energy stored in the host rock around the chamber would be enough to generate a dike-fracture of the order of 1600 km^2 – say a dike with a strike-dimension of 80 km and a dip-dimension of 20 km. Alternatively, the strain energy would be sufficient to generate many smaller dikes and inclined sheets.

This conclusion assumes two things: (i) that all the strain energy is used to overcome the surface energy needed to form the dike-fracture; and (ii) that the magma chamber can be treated as a point source. Neither assumption holds true. First, much of the strain energy is transformed into work needed to squeeze magma out of the chamber into the dike-fracture and, in case of an eruption, to sustain the flow of magma to the surface. Second, the assumption in Eq. (3.36) that the chamber is a Mogi source (a point source) in the strain-energy calculations is clearly not warranted. However, the calculated strain energy is still a useful measure of a part of the elastic energy available to drive the dike propagation and, eventually, the associated eruption (cf. Chapter 10).

Example 3.8

Problem

A magma chamber with a circular vertical cross-sectional area is located at a depth that is somewhat less than the vertical diameter of the chamber. The chamber is elongated parallel with the axis of the volcanic rift zone within which it is located, so that a two-dimensional model may be applied. The magmatic excess pressure in the chamber is 4 MPa. Calculate the theoretical maximum induced tensile stress at the surface of the rift zone and at the boundary of the chamber and decide whether or not failure will happen for the case when:

(a) the chamber radius is 2 km and the depth to its centre is 3 km;
(b) the chamber radius is 2.5 km and the depth to its centre is 5 km.

Solution

(a) From Eq. (3.47) the maximum induced tensile stress σ_t at the surface of the rift zone is:

$$\sigma_t = -\frac{4p_e R_1^2}{d^2 - R_1^2} = -\frac{4 \times 4 \times 10^6 \text{ Pa} \times (2000 \text{ m})^2}{(3000 \text{ m})^2 - (2000 \text{ m})^2} = -1.28 \times 10^7 \text{ Pa} \approx -13 \text{ MPa}$$

Similarly, the maximum theoretical tensile stress at the boundary of the chamber σ_b is obtained from Eq. (3.50) as:

$$\sigma_b = -\frac{p_e(d^2 + R_1^2)}{d^2 - R_1^2} = -\frac{4 \times 10^6 \text{ Pa} \times [(3000 \text{ m})^2 + (2000 \text{ m})^2]}{(3000 \text{ m})^2 - (2000 \text{ m})^2}$$
$$= -1.04 \times 10^7 \text{ Pa} \approx -10 \text{ MPa}$$

Clearly, the maximum theoretical induced tensile stress at the surface of the rift zone above the magma chamber is higher than the maximum theoretical tensile stress at the margin of the chamber. Thus, for this type of loading and magma-chamber geometry and depth, tensile fractures (and presumably normal faults) would form at the surface before any inclined sheets would be injected from the margins of the chamber.

(b) Again from Eq. (3.47) we get the maximum theoretical tensile stress at the surface as:

$$\sigma_t = -\frac{4p_e R_1^2}{d^2 - R_1^2} = -\frac{4 \times 4 \times 10^6 \text{ Pa} \times (2500 \text{ m})^2}{(5000 \text{ m})^2 - (2500 \text{ m})^2} = -5.33 \times 10^6 \text{ Pa} \approx -5 \text{ MPa}$$

Similarly, the maximum theoretical tensile stress at the boundary of the chamber is:

$$\sigma_b = -\frac{p_e(d^2 + R_1^2)}{d^2 - R_1^2} = -\frac{4 \times 10^6 \text{ Pa} \times [(5000 \text{ m})^2 + (2500 \text{ m})^2]}{(5000 \text{ m})^2 - (2500 \text{ m})^2}$$
$$= -6.67 \times 10^6 \text{ Pa} \approx -7 \text{ MPa}$$

For this type of loading and chamber depth and size, the theoretical tensile stress at the boundary of the chamber is greater than that at the surface of the rift zone. Thus, here sheet injections would take place before any tensile fractures or normal faults would develop or reactivate at the surface of the rift zone. In fact, such tensile fractures and normal faults might not develop at all since sheet injection would be likely to start well before the tensile stress at the chamber boundary reached 7 MPa because the rock tensile strength is most commonly 3–4 MPa. As soon as sheet injection begins, the magma excess pressure in the chamber normally falls, in which case the value of 5 MPa at the surface might not be reached.

The results of the calculations in (a) and (b) are in agreement with the conclusions of Section 3.5.2 that when the depth to the centre of the chamber is less than 1.73 times the radius of the chamber (so here 3 km < 1.73 × 2 km = 3.46 km) the maximum induced theoretical tensile stress occurs at the free surface (as in a). By contrast, when the depth to centre of the chamber is greater than 1.73 times the radius of the chamber (so here 5 km > 1.73 × 2.5 km = 4.32 km) the maximum induced theoretical tensile stress occurs at the boundary of the chamber (as in b).

Example 3.9

Problem

The sizes of calderas are an indication of the sizes of the associated shallow magma chambers at the time of ring-fault formation or reactivation. That is, the ring-fault diameter corresponds roughly to the lateral diameter of the associated chamber, many of which have sill-like geometry at the time of caldera collapse. Consider a recently formed circular collapse caldera, a ring-fault, with a diameter of 8 km. Seismic data suggest that the top of the chamber is at about 3 km depth, that the average Young's modulus of the host rock is 10 GPa, and that the chamber is 2 km thick. Calculate:

(a) the volume of the chamber; and
(b) the strain energy stored in the host rock of the chamber if the current magmatic excess pressure is 3 MPa.

Solution

(a) The volume V_c of a circular sill-like chamber is given by:

$$V_c = \pi h R_1^2 \tag{3.62}$$

where h is the height or thickness of the chamber (here 2 km) and R_1 its radius – here assumed the same as the radius of the caldera, that is, 4 km. We can calculate the volume directly in cubic kilometres thus:

$$V_c = \pi h R_1^2 = 3.1415 \times 2 \text{ km} \times (4 \text{ km})^2 = 100.5 \text{ km}^3$$

This chamber is thus reasonably large and clearly within the common volume range of shallow chambers, namely 30–500 km³.

(b) From Eq. (3.55), the strain energy associated with a chamber of this geometry and subject to excess pressure of 3 MPa is:

$$U_0 = \frac{8(1 - v^2)p_e^2 a^3}{3E} = \frac{8(1 - 0.25^2)(3 \times 10^6 \text{ Pa})^2(4000 \text{ m})^3}{3 \times 2 \times 10^{10} \text{ Pa}} = 7.2 \times 10^{13} \text{ J}$$

This strain energy is not large in comparison with the one calculated for the chamber in Example 3.7. For a dike-fracture with a surface energy of the order of $2 \times 10^7 \text{ J m}^{-2}$, this strain energy would be enough to supply energy for a dike 3 km tall (dip-dimension of a feeder-dike) and 1.2 km long (strike-dimension), or, in general, any dike or inclined sheet with an area of 3.6 km². Dikes/sheets of these dimensions are very common in the swarms associated with shallow magma chambers.

Example 3.10

Problem

Consider the same magma chamber geometry as in Example 3.9. During an unrest period, an inflation or maximum uplift of 2 m is measured in the associated volcano (a collapse caldera). Calculate the strain energy stored in the volcano as a consequence of the magma-chamber expansion and surface inflation or doming if the host rock has a Young's modulus of 10 GPa and a Poisson's ratio of 0.25.

Solution

First we need to calculate the volume of the chamber expansion. From Eq. (3.51) the expansion ΔV is:

$$\Delta V = \frac{4}{3}\pi a^2 u = \frac{4}{3} \times 3.1416 \times (4000 \text{ m})^2 \times 2 \text{ m} = 1.3 \times 10^8 \text{ m}^3$$

Here we assume that the measured uplift or doming corresponds to the vertical displacement of the roof of the sill. That is normally a good approximation for a chamber in an elastic half-space, but may be a poor one for a chamber in a crust composed of layers of widely different mechanical properties (cf. Chapter 10). In the latter, the main expansion of the chamber may, in fact, be a vertical displacement or down-bending of the floor of the sill-like chamber if the rocks below the chamber are more compliant or softer than those in its roof.

3.30 In the Mogi model we cannot strictly distinguish between the excess pressure p_e and the radius of the chamber R_1 (Eq. (3.31)). Commonly, the radius is assumed to be 1 km in order to explain the surface displacements. Explain why this assumption for the chamber radius is commonly a poor one and what assumption would be better so as to be able to distinguish between p_e and R_1.

3.31 If the Young's modulus of a rock is 50 GPa and its Poisson's ratio is 0.25, find its shear modulus and compressibility. Use Poisson's relation to estimate Lamé's constant.

3.32 For the models of magma chambers in this book we commonly use Young's moduli of 15–30 GPa and shear moduli of 6–12 GPa. By contrast, the moduli given for common rock in Appendix D are typically a factor of 2 to 4 times higher than these values. Explain why the lower values are used in our models and the justification for doing so.

3.33 A totally fluid shallow magma chamber has an upper boundary (roof) at a depth of 3.5 km. (a) Calculate the vertical stress on the chamber if the average density of the uppermost 3.5 km of the crustal segment hosting the chamber is 2600 kg m^{-3}. (b) Explain why the minimum compressive principal horizontal stress in the roof of the chamber must be similar to the vertical stress; why, for example, the minimum horizontal stress cannot be reduced – through, say, plate-tectonic movement or extension – to 60 MPa. What would happen as the minimum compressive principal horizontal stress is gradually reduced and long before it becomes as low as 60 MPa?

3.34 In a basaltic layer at a shallow depth in a rift zone with an average Young's modulus of 20 GPa, an average density of 2500 kg m^{-3}, and a Poisson's ratio of 0.25, the principal stresses are 25 MPa, 20 MPa, and 15 MPa. Calculate the principal strains at this location in the basaltic layer.

3.35 Consider a volcano with an average shear modulus of 6 GPa and a Poisson's ratio of 0.25 hosting a magma chamber at a depth of 3 km. During an unrest period the maximum vertical displacement or inflation of the surface of the volcano is measured as 1 m. Use the standard Mogi chamber with a radius of 1 km to calculate the excess pressure needed to generate the measured uplift. Discuss whether or not the results are realistic.

3.36 Use the same values for the measured surface uplift, rock properties, and chamber depth as in Exercise 3.35, but now assume the excess pressure is roughly equal to the tensile strength of the host rock of the chamber. What chamber radius would be needed to explain the measured uplift? Discuss whether or not the results are realistic and, if not, what would be a more realistic model for the magma chamber.

3.37 A magma chamber is known to be at 4 km depth. Use the standard assumption of 1 km radius for a Mogi model to calculate (a) the maximum vertical surface displacement, and (b) the horizontal and vertical surface displacements at a radial distance of 2.8 km from the centre of the surface uplift or doming. In what way is this radial distance of 2.8 km of special importance as regards the surface deformation?

3.38 During an unrest period the maximum measured vertical surface displacement is 50 cm and the maximum horizontal displacement is 19 cm. Seismic data suggest the chamber diameter is about 4 km. Using a Young's modulus of 30 GPa and a Poisson's ratio of 0.25 for the host rock, calculate the depth to a Mogi chamber associated with the deformation. Comment on how realistic the model is for this particular case.

3.39 Estimate the volume change of the magma chamber needed to explain the surface deformation in Exercise 3.38. Estimate also the magmatic excess pressure needed in the chamber to generate that expansion and surface deformation.

3.40 An elongated chamber of circular cross-sectional area has a radius of 1.7 km and depth to its centre of 3.5 km. The magmatic excess pressure is estimated at 5 MPa. Using a two-dimensional model, calculate the maximum induced tensile stresses at the surface above the chamber as well as at the boundary of the chamber. Based on the results, which volcanotectonic processes are most likely to take place?

3.41 A circular shallow chamber has a horizontal diameter of 6 km and a thickness of 1.5 km. The chamber's roof is at 3 km depth. The average shear modulus of the host rock is 6 GPa and its Poisson's ratio is 0.25. Calculate (a) the volume of the chamber, and (b) the strain energy stored in the host rock if the chamber excess pressure is 5 MPa. Compare the results with the strain energy needed to propagate dike-fractures.

References and Suggested Reading

Acocella, V., 2007. Understanding caldera structure and development: an overview of analogue models compared to natural calderas. *Earth-Science Reviews*, **85**, 125–160.

Acocella, V., Cifelli, F., Funiciello, R., 2000. Analogue models of collapse calderas and resurgent domes. *Journal of Volcanology and Geothermal Research*, **104**, 81–96.

Al Shehri, A., Gudmundsson, A., 2018. Modelling of surface stresses and fracturing during dyke emplacement: application to the 2009 episode at Harrat Lunayyir, Saudi Arabia. *Journal of Volcanology and Geothermal Research*, **356**, 278–303.

Anderson, E. M., 1936. The dynamics of formation of cone sheets, ring dykes and cauldron subsidences. *Proceedings of the Royal Society of Edinburgh*, **56**, 128–163.

Barnett, Z. A., Gudmundsson, A., 2014. Numerical modelling of dykes deflected into sills to form a magma chamber. *Journal of Volcanology and Geothermal Research*, **281**, 1–11.

Bonafede, M., Dragoni, M., Quareni, F., 1986. Displacement and stress fields produced by a centre of dilation and by a pressure source in a viscoelastic half-space: application to the study of ground deformation and seismic activity at Campi Flegrei, Italy. *Geophysical Journal of the Royal Astronomical Society*, **87**, 455–485.

Cloos, E., 1955. Experimental analysis of fracture patterns. *Bulletin of the Geological Society of America*, **66**, 241–256.

Cole, J. W., Milner, D. M., Spinks, K. D. 2005. Calderas and caldera structures: a review. *Earth-Science Reviews*, **69**, 1–26.

Delaney, P. T., McTigue, D. F., 1994. Volume of magma accumulation or withdrawal estimated from surface uplift or subsidence, with application to the 1960 collapse of Kilauea Volcano. *Bulletin of Volcanology*, **56**, 417–424.

Dzurisin, D., 2006. *Volcano Deformation: New Geodetic Monitoring Techniques*. Berlin: Springer Verlag.

Fagents, S. A., Gregg, T. K. P., Lopes, R. M. C. (eds.), 2013. *Modeling Volcanic Processes: The Physics and Mathematics of Volcanism*. Cambridge: Cambridge University Press.

Fialko, Y., Khazan, Y., Simons, M., 2001. Deformation due to a pressurized horizontal circular crack in an elastic half-space, with applications to volcano geodesy. *Geophysical Journal International*, **146**, 181–190.

Fossen, H., 2016. *Structural Geology*, 2nd edn. Cambridge: Cambridge University Press.

Fossen, H., Gabrielsen, R. H., 1996. Experimental modelling of extensional fault systems by use of plaster. *Journal of Structural Geology*, **18**, 673–687.

Gautneb, H., Gudmundsson, A., 1992. Effect of local and regional stress fields on sheet emplacement in West Iceland. *Journal of Volcanology and Geothermal Research*, **51**, 339–356.

Geshi, N., Shimano, T., Chiba, T., Nakada S., 2002. Caldera collapse during the 2000 eruption of Miyakejima volcano, Japan. *Bulletin of Volcanology*, **64**, 55–68.

Geyer, A., Marti, J., 2014. A short review of our current understanding of the development of ring faults during collapse caldera formation. *Frontiers in Earth Science*, **2**, doi:10.3389/feart.2014.00022

Gudmundsson, A., 1995. Infrastructure and mechanics of volcanic systems in Iceland. *Journal of Volcanology and Geothermal Research*, **64**, 1–22.

Gudmundsson, A., 1998. Magma chambers modeled as cavities explain the formation of rift zone central volcanoes and their eruption and intrusion statistics. *Journal of Geophysical Research*, **103**, 7401–7412.

Gudmundsson, A., 2003. Surface stresses associated with arrested dykes in rift zones. *Bulletin of Volcanology*, **65**, 606–619.

Gudmundsson, A., 2006. How local stresses control magma-chamber ruptures, dyke injections, and eruptions in composite volcanoes. *Earth-Science Reviews*, **79**, 1–31.

Gudmundsson, A., 2007. Conceptual and numerical models of ring-fault formation. *Journal of Volcanology and Geothermal Research*, **164**, 142–160.

Gudmundsson, A., 2009. Toughness and failure of volcanic edifices. *Tectonophysics*, **471**, 27–35.

Gudmundsson, A., 2011a. *Rock Fractures in Geological Processes*. Cambridge: Cambridge University Press.

Gudmundsson, A., 2011b. Deflection of dykes into sills at discontinuities and magma-chamber formation. *Tectonophysics*, **500**, 50–64.

Gudmundsson, A., 2012a. Strengths and strain energies of volcanic edifices: implications for eruptions, collapse calderas, and landslides. *Natural Hazards and Earth System Sciences*, **12**, 2241–2258.

Gudmundsson, A., 2012b. Magma chambers: formation, local stresses, excess pressures, and compartments. *Journal of Volcanology and Geothermal Research*, **237–238**, 19–41.

Gudmundsson, A., Nilsen, K., 2006. Ring-faults in composite volcanoes: structures, models and stress fields associated with their formation. In Troise, C., De Natle, G., Kilburn, C. R. .J. (eds.), *Mechanism of Activity and Unrest at Large Calderas. Geological Society of London Special Publications, 269.* London: Geological Society of London, pp. 83–108.

Gudmundsson, A., Philipp, L., 2006. How local stress fields prevent volcanic eruptions. *Journal of Volcanology and Geothermal Research*, **158**, 257–268.

Gudmundsson, A., Lotveit, I. F., 2012. Sills as fractured hydrocarbon reservoirs: examples and models. In Spence, G. H., Redfern, J., Aguilera, R, et al. (eds.), *Advances in the Study of Fractured Reservoirs. Geological Society of London Special Publications, 374.* London: Geological Society of London, pp. 251–271.

Gudmundsson, A., Friese, N., Galindo, I., Philipp, S. L., 2008. Dike-induced reverse faulting in a graben. *Geology*, **36**, 123–126.

Gudmundsson, A., Lecoeur, N., Mohajeri, N., Thordarson, T., 2014. Dike emplacement at Bardarbunga, Iceland, induces unusual stress changes, caldera deformation, and earthquakes. *Bulletin of Volcanology*, **76**, 869, doi:10.1007/s00445-014-0869-8.

Isida, M., 1955. On the tension of a semi-infinite plate with an elliptic hole. *Scientific Papers of the Faculty of Engineering, Tokushima University.* **5**, 75–95.

Jaeger, J. C., Cook, N. G. W., Zimmerman, R. W., 2007. *Fundamentals of Rock Mechanics*, 4th edn. Oxford: Blackwell.

Janssen, V., 2008. *GPS-Based Volcano Deformation*. Saarbrücken: VDM Verlag.

Johnson, D. J., Sigmundsson, F., Delaney, P. T., 2000. Comment on "Volume of magma accumulation or withdrawal estimated from surface uplift or subsidence, with application to the 1960 collapse of Kilauea Volcano" by T. T. Delaney and D. F. McTigue. *Bulletin of Volcanology*, **61**, 491–493.

Kusumoto, S., Gudmundsson, A., 2014. Displacement and stress fields around rock fractures opened by irregular overpressure variations. *Frontiers in Earth Science*, **2**, doi:10.3389/feart.2014.00007

Kusumoto, S., Geshi, N., Gudmundsson, A., 2013. Inverse modeling for estimating fluid-overpressure distributions and stress intensity factors from arbitrary open-fracture geometry. *Journal of Structural Geology*, **46**, 92–98.

Love, A. E. H., 1927. *A Treatise on the Mathematical Theory of Elasticity*. New York, NY: Dover.

Lu, Z., Dzurisin, D., 2014. *InSAR Imaging of Aleutian Volcanoes: Monitoring a Volcanic Arc from Space*. Berlin: Springer Verlag.

Martì, J., Gudmundsson, A., 2000. The Las Canadas caldera (Tenerife, Canary Islands): an overlapping collapse caldera generated by magma-chamber migration. *Journal of Volcanology and Geothermal Research*, **103**, 161–173.

Marti, J., Ablay, G. J., Redshaw, L. T., Sparks, R. S. J., 1994. Experimental studies of collapse calderas. *Journal Geological Society London*, **151**, 919–929.

Marti, J., Geyer, A., Folch, A., Gottsmann, J., 2008. A review on collapse caldera modelling. In Gottsmann, J., Marti, J. (eds), *Caldera Volcanism: Analysis, Modelling and Response*. Amsterdam: Elsevier, pp. 233–283.

McClay, K. R., Ellis, P. G., 1987. Analogue models of extensional fault geometries. In Coward, M. P., Dewey, J. F., Hancock, P. L. (eds), *Continental Extensional Tectonics. Geological Society of London Special Publications*, *28*. London: Geological Society of London, pp. 109–125.

McTigue, D. F., 1987. Elastic stress and deformation near a finite spherical magma body: resolution of the point source paradox. *Journal of Geophysical Research*, **92**, 12931–12940.

Melan, E., 1932. Point force at internal point in a semi-infinite plate. *Zeitschrift fur Angewandte Mathematik und Mechanik*, **12**, 343–346 (in German).

Mindlin, R. D., 1936. Force at a point in the interior of a semi-infinite solid. *Physics*, **7**, 195–202.

Mogi, K., 1958. Relations between eruptions of various volcanoes and the deformations of the ground surfaces around them. *Bulletin of the Earthquake Research Institute University of Tokyo*, **36**, 99–134.

Niemczyk, O. (ed.), 1943. *The Fractures of Iceland* (Spalten auf Island). Stuttgart: Wittwer (in German).

Okada, Y., 1985. Surface deformation due to shear and tensile faults in a half-space. *Bulletin of the Seismological Society of America*, **75**, 1135–1154.

Okada, Y., 1992. Internal deformation due to shear and tensile faults in half-space. *Bulletin of the Seismological Society of America*, **82**, 1018–1040.

Philipp, S., Philipp, S.L., Afsar, F., Gudmundsson, A., 2013. Effects of mechanical layering on the emplacement of hydrofractures and fluid transport in reservoirs. *Frontiers of Earth Science*, **1**, doi:10.3389/feart.2013.00004.

Pollard, D. D., Fletcher, R. C., 2005. *Fundamentals of Structural Geology*. Cambridge: Cambridge University Press.

Pollard, D. D., Johnson, A. M., 1973. Mechanics of growth of some laccolithic intrusions in the Henry mountains, Utah, II. Bending and failure of overburden layers and sill formation. *Tectonophysics*, **18**, 311–354.

Pollard, D. D., Delaney, P. T., Duffield, W. A., Endo, E. T., Okamura, A. T., 1983. Surface deformation in volcanic rift zones. *Tectonophysics*, **94**, 541–584.

Press, F. 1965. Displacements, strains, and tilts at teleseismic distances. *Journal of Geophysical Research*, **70**, 2395–2412.

Ramberg, H., 1967. *Gravity, Deformation and the Earth's Crust*. Cambridge, MA: Academic Press.

Rubin, A. M., Pollard, D. D., 1988. Dike-induced faulting in rift zones of Iceland and Afar. *Geology*, **16**, 413–417.

Ruch, J., Acocella, V., Geshi, N., Nobile, A., Corbi, F., 2012. Kinematic analysis of vertical collapse on volcanoes using experimental models time series. *Journal of Geophysical Research*, **117**, doi:10.1029/2012JB009229.

Saada, A. S., 2009. *Elasticity Theory and Applications*, 2nd edn. London: Roundhouse.

Sadowsky, M. A., Sternberg, E., 1947. Stress concentration around an ellipsoidal cavity in an infinite body under arbitrary plane stress perpendicular to the axis of revolution of cavity. *Journal of Applied Mechanics*, **14**, A191–A201.

Sadowsky, M. A., Sternberg, E., 1949. Stress concentration around a triaxial ellipsoidal cavity. *Journal of Applied Mechanics*, **16**, 149–157.

Savin, G. N., 1961. *Stress Concentration Around Holes*. New York, NY: Pergamon.

Segall, P., 2010. *Earthquake and Volcano Deformation*. Princeton (New Jersey): Princeton University Press.

Segall, P., Llenos, A. L., Yun, S. H., Bradley, A. M., Syracuse, E. M., 2013. Time-dependent dike propagation from joint inversion of seismicity and deformation data. *Journal of Geophysical Research*, **118**, doi:10.1002/2013JB010251.

Sigmundsson, F., Hooper, A., Hreinsdottir, S., et al., 2015. Segmented lateral dyke growth in a rifting event at Bardarbunga Volcanic System, Iceland. *Nature*, **517**, 191–195.

Sigurdsson, H., Houghton, B. F., McNutt, S. R., Rymer, H., Stix, J. (eds.), 2000. *Encylopedia of Volcanoes*. New York, NY: Academic Press.

Sigurdsson, O., 1980. Surface deformation of the Krafla Fissure Swarm in two rifting events. *Journal of Geophysical Research*, **47**, 154–159.

Sneddon, I. N., Lowengrub, M., 1969. *Crack Problems in the Classical Theory of Elasticity*. New York, NY: Wiley.

Soutas-Little, R. W., 1973. *Elasticity*. New York, NY: Dover.

Steketee, J. A., 1958a. On Volterra's dislocations in a semi-infinite elastic medium. *Canadian Journal of Physics*, **36**, 192–205.

Steketee, J. A., 1958b. Some geophysical applications of the elasticity theory of dislocation. *Canadian Journal of Physics*, **36**, 1168–1198.

Sun, R. J. 1969. Theoretical size of hydraulically induced horizontal fractures and corresponding surface uplift in an idealized medium. *Journal of Geophysical Research*, **74**, 5995–6011.

Timoshenko, S., Goodier, J. N., 1970. *Theory of Elasticity*, 3rd edn. New York, NY: McGraw-Hill.

Volterra, V., 1907. On the equilibrium of multiply connected elastic bodies. *Annales scientifiques de l'École Normale Supérieure*, **24**, 401–517 (in French; English translation).

Weertman, J., 1996. *Dislocation Based Fracture Mechanics*. London: World Scientific.

Zobin, V. M., 2003. *Introduction to Volcanic Seismology*. London: Elsevier.

Volcanic Earthquakes

4.1 Aims

Volcanic (or volcanotectonic) earthquakes are those that occur inside volcanoes or close to them. The study of these earthquakes in volcanoes is commonly referred to as **volcano seismology**. Most earthquakes associated with volcanoes occur at comparatively shallow depths, normally of less than 10 km. They differ from other earthquakes at plate boundaries partly in that volcanic earthquakes commonly occur in **swarms**, that is, clusters in time and space of many comparatively small and similar-sized earthquakes. Volcanic earthquakes provide information about the state of stress in the volcano and of eventual magma-chamber rupture during unrest periods (e.g. Massa et al., 2016). They also indicate the location and, crudely, the size of magma chambers. When a dike (or an inclined sheet) is injected from a magma chamber, the formation of its propagation path (dike-fracture path) produces earthquakes. Thus, accurate monitoring of earthquakes during such events can be used to map out the propagation path of the dike and help assess the likelihood of the dike (or inclined sheet) reaching the surface, causing an eruption. The main aims of this chapter are to:

- Explain briefly earthquakes and their magnitudes.
- Summarise the information earthquakes yield about the state of stress.
- Discuss the main types of volcanic earthquakes.
- Describe and explain high-frequency (A-type) earthquakes.
- Describe and explain low-frequency (B-type) earthquakes.
- Describe and explain explosion earthquakes.
- Describe and explain volcanic tremors.
- Discuss and explain seismic monitoring of volcanoes.

4.2 Seismic Waves

Earthquakes, as the name implies, are **vibrations** in the Earth. Earthquakes therefore refer to seismic events, that is, events (mostly) in the Earth's crust and mantle that generate seismic waves. The vibrations are due to the passage of seismic waves of energy – that is,

waves carrying energy. The waves result from a sudden release or transformation of energy in the Earth's crust or mantle, the energy sources most commonly being sudden fracturing of the rock or explosion at the site of the earthquake. The seismic wave velocities are proportional to the density of the rock units or layers through which the waves pass. Since the rock density generally increases with depth in the crust and mantle, the velocities are, on average, higher in the lower mantle (as high as 13 km s^{-1}) than in the crust and lithosphere (where the velocities commonly range from 2 km s^{-1} to 8 km s^{-1}). For most scientists, however, an earthquake is the process of **rupture** or **slip** on the associated fault rather than the resulting ground movement or shaking.

There are various types of waves generated by the energy released or transformed during earthquakes. The two basic types are **body waves** and **surface waves**. These two basic types can be subdivided into other classes. As the name implies, body waves propagate through the body – in this case the Earth. Thus body waves propagate through the interior of the Earth, its crust and lithosphere, as well as the mantle and the core. Surface waves, by contrast, travel along the Earth's surface.

4.2.1 Body Waves

Body waves are of two main subtypes: primary waves and secondary waves. Both travel along paths or rays that are refracted or reflected (or both) at contacts between mechanically dissimilar layers or units, that is, layers with different Young's moduli (stiffness) and density. The exact velocities of seismic waves in rocks depend on many factors such as (1) rock density, (2) rock matrix stiffness (Young's modulus), (3) pore-fluid pressure, (4) fracture orientations, shapes, and frequencies in the rock, and (5) depth or overburden pressure. However, most of these factors can be combined so as to yield the effective elastic constants (for example, Young's modulus and Poisson's ratio) which, together with density, provide expressions from which the velocities can be calculated (see Eqs. (4.1)–(4.6)). Some crude average wave velocities for crustal rocks and layers are given below.

Following an earthquake, the **primary waves** (P-waves) are the first to arrive at the seismic stations and be recorded on the seismograms. They are also referred to as compressional, pressure, or push–pull waves. These waves propagate through both **solids and fluids**, so that they can move through geothermal reservoirs and active (fluid) magma chambers. The movement of the particles in the primary wave is push and pull, which, in homogeneous and isotropic rocks, is parallel with the wave-propagation direction (longitudinal). In air, they propagate as **sound waves**, at the velocity of sound, about 340 m s^{-1} in air at 20 °C. Their velocities of propagation generally increase with depth in the Earth and are commonly around twice that of secondary waves.

For solid crustal rocks, common P-wave velocities are 2–7 km s^{-1}, while in sediments such as sand the velocities may be as low as (and sometimes lower than) 1.0 km s^{-1}. When P-waves move through the crust, their velocities are much reduced in totally molten bodies, such as totally molten parts of magma chambers (Fig. 4.1). When the magmatic or melting temperature is reached for a rock of given type, the P-wave velocity may decrease to half its velocity in the solid rock of the same type (Murase and McBirney, 1973). For example, the

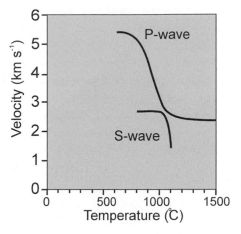

Fig. 4.1 P-wave and S-wave velocities in basalt as a function of temperature (up to 1500 °C). On meeting magma (in a magma chamber) at temperatures of about 800–1200 °C there is a considerable reduction in P-wave velocity, as indicated here, whereas the S-waves drop out entirely at about 1100 °C. The latter is referred to as an S-wave shadow and used to detect magma chambers. Based on data from Murase and McBirney (1973).

P-wave velocity of basaltic magma in a chamber may be about 3 km s^{-1}, whereas in the surrounding basaltic host rock at the same depth it may be about 6 km s^{-1}.

The **secondary waves** (S-waves) are the next ones to arrive at the seismic stations following an earthquake. These waves are also known as transverse, shear, or shake waves. The waves propagate only through **solids**, so that they are commonly used to detect magma fluid chambers – identified through **S-wave shadows.** S-waves thus drop out on meeting totally fluid bodies. It follows that there are no S-waves travelling through the Earth's outer core, which is known to be liquid. The movement of the particles in S-waves is in a direction transverse to the wave-propagation direction. This means that the particles move side to side or up and down in a plane perpendicular to the direction in which the wave itself is travelling.

S-waves propagate at lower velocities than P-waves; commonly at about half that of P-waves. Typical velocities of S-waves in the crust are 2–4 km s^{-1}. However, for sediments such as sand these are as low as 0.5 km s^{-1}.

The velocities of the seismic waves can be presented by various formulas depending on the elastic constants used. Common expressions for the velocities of the P-waves (V_p) are as follows (e.g. Mavko et al., 2009):

$$V_p = \left[\frac{\lambda + 2G}{\rho} \right]^{1/2} \tag{4.1}$$

and, alternatively, as:

$$V_p = \left[\frac{K + \frac{4}{3}G}{\rho} \right]^{1/2} \tag{4.2}$$

Similarly, the S-wave velocity (V_s) is given by:

$$V_s = \left[\frac{G}{\rho}\right]^{1/2} \qquad (4.3)$$

where λ is Lamé's constant, G is the shear modulus, and ρ is the material (fluid or solid) density (cf. Chapter 3). While Eqs. (4.1) to (4.3) are the most common representations of the P-wave and S-wave velocities, for obtaining the dynamic Young's modulus from seismic experiments it is more appropriate to write Eq. (4.1) in the following form:

$$E = \frac{V_p^2(1+v)(1-2v)\rho}{(1-v)} \qquad (4.4)$$

where E is here the dynamic Young's modulus and v is the dynamic Poisson's ratio. Similarly, the dynamic Poisson's ratio may be obtained directly from the expression:

$$v = \frac{V_p^2 - 2V_s^2}{2(V_p^2 - V_s^2)} \qquad (4.5)$$

The ratio between the seismic wave velocities can be found as follows:

$$\frac{V_p}{V_s} = \left[\frac{2(1-v)}{1-2v}\right]^{1/2} \qquad (4.6)$$

For a typical Poisson's ratio for crustal rocks of $v = 0.25$, Eq. (4.6) yields the V_p/V_s ratio of 1.73 and its reciprocal, the V_s/V_p ratio, as 0.57. Thus, given the typical range of Poisson's ratios for common rocks, 0.20–0.30 (Appendix D), the S-wave velocity is commonly 50–60% of the P-wave velocity.

4.2.2 Surface Waves

Surface waves travel along the surface of the Earth. Thus, they occur only in the Earth's crust, but neither in the mantle nor the core. In physics, the general term surface wave refers to waves that propagate at the **interface** (the contact) between different media; for example at the interface between different types of fluids (e.g. water and air) or the interface between a solid and a fluid (e.g. the crust and the air). They are analogous to capillary waves or, as we commonly refer to them, **ripples** such as occur on a pond or a lake when a stone falls into the water. They have lower frequencies and arrive later at the seismic stations than the P-waves.

Most of the **damage** of buildings and other human constructions during an earthquake is due to surface waves, primarily because the particle motion in such waves is much larger than that in body waves. Also, the **amplitude** (one-half the distance between the crest and the trough of the wave, that is, the maximum surface displacement associated with the seismic wave) of surface waves decreases more slowly with distance of wave propagation than that of body waves. More specifically, denoting the distance travelled by the wave by d, the energy of a body wave decays as d^{-2}, whereas that of a surface wave decays as d^{-1}. It follows that the **ground movement** – shaking or disturbance – is dominated by surface waves, particularly far from the earthquake source. In large earthquakes, the surface waves

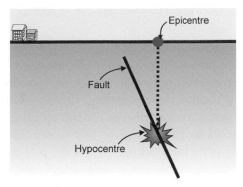

Hypocentre and epicentre. Hypocentre is the place of origin of the slip on the earthquake fault. Epicentre is the point on the Earth's surface right above the hypocentre.

can have amplitudes of several centimetres. For an earthquake of a given magnitude or size, the strength of the surface waves, and thus the damage they cause, decreases with increasing depth of the source (the hypocentre or focal depth; Fig. 4.2).

The main surface waves generated during earthquakes are named after two British scientists, namely Love waves (after Augustus Love) and Rayleigh waves (after Lord Rayleigh). **Love waves** have the greatest velocity of the surface waves. As they travel along the surface the ground movement is horizontal, that is, from side to side, and thus in a direction perpendicular to the propagation of the wave itself. Love waves propagate more slowly than body waves, but are generally the ones that cause the greatest damage to man-made constructions, except perhaps in the **epicentre** area itself (epicentre is the projection of the source point, the focal point, of the earthquake onto the surface above the source, as discussed below; Fig. 4.2). They are thus the waves most easily felt during an earthquake.

Rayleigh waves move more slowly than Love waves and the particle movement is very different. In Rayleigh waves the particles have a rotating movement in the vertical section, moving the surface ground side to side as well as up and down. That is, the particles are subject both to longitudinal and transverse displacements. The particle movement is, for isotropic rocks, in ellipses, oriented so as to be parallel to the wave propagation and perpendicular to the surface. The displacements of the particles are greatest at the surface itself and decrease with depth. In some ways their movement is similar to that of **ocean waves**. Like Love waves, Rayleigh waves commonly produce much **damage** of human-made constructions. Both Rayleigh waves and Love waves propagate at velocities somewhat less than the S-wave velocity (commonly at close to $0.9V_s$).

4.3 Hypocentre, Epicentre, and Focal Mechanism

Earthquake sizes have long been measured using well-defined scales. Currently, there are many scales in use, but the most popular one is the moment-magnitude scale, discussed below. Before explaining magnitudes, however, two concepts of importance for understanding the sizes of earthquakes, should be defined, namely earthquake hypocentre and epicentre.

The earthquake **hypocentre** is also known as the earthquake **focus** (Fig. 4.2). For a typical tectonic earthquake on a fault, the hypocentre is that part of the fault where the rupture or slip begins. In other words, the part along the fault-slip plane where the first movement occurs and elastic energy is first transformed or released during the earthquake. The hypocentre is often regarded as a point (Fig. 4.2), while in reality it is a part of the slip or rupture plane of a fault and has a finite size. The depth to the hypocentre is commonly referred to as the **focal depth**.

The processes that determine whether an earthquake becomes small (most earthquakes) or large (comparatively few earthquakes) are not well understood. Basically, there are two main models. In one model, large earthquakes originate in exactly the same way as small ones but for some unknown reason develop into large ones. In the other model, the build-up process for a large earthquake is somewhat different from the build-up process for a small earthquake. This latter model is favoured here for the simple reason that the total slip (the displacement during the earthquake, the **co-seismic slip**) at any point along a fault is reached almost instantaneously. For this to be possible, the controlling dimension (see below) of the earthquake rupture must be predetermined, possibly because of prior stress-field homogenisation of that part of the fault zone that slips during the earthquake (Gudmundsson and Homberg, 1999) or by other means.

The earthquake **epicentre** is the point at the surface of the Earth that is right above the earthquake hypocentre or focus (Fig. 4.2). More specifically, the epicentre is above that part of the fault where the rupture or slip beings. For smaller earthquakes the epicentre is the location of the greatest, sometimes the only (if any), damage. For large earthquakes (say of magnitude 7 or larger), however, the damage to human-made constructions at the surface may extend for tens of kilometres, sometimes hundreds of kilometres, and thus far beyond the epicentre itself. The observed damage depends not only on the availability of human-made structures that can be damaged, but also on the properties of the near-surface and surface rocks and soils.

Related to these concepts is the **focal mechanism** or fault-plane solution which, for a fault slip, is supposed to yield the orientation of the fault segment in the source region (the hypocentre) as well as the fault-slip direction (Figs. 4.3 and 4.4). When the fault slips, that is, when rupture occurs, the movements of the fault walls generate movements in the host rock, namely **seismic waves**. The push on each side of the fault generates the compressional P-waves, and the shear displacement across the fault generates the shear S-waves. Both the P-waves and the S-waves are generated at the hypocentre, the instant the fault slip or rupture is initiated.

If we present the crust as four quadrants, the fault slip (rupture) results in compression, that is, **transpression**, in two quadrants, and extension, that is, **transtension**, in the other two. The first waves associated with the fault slip (associated, in fact, with the hypocentre – the source) to arrive at the seismic stations are the P-waves. In the two compressional quadrants, the **ground motion** – the displacement of the ground – when the first P-waves arrive will be up, that is, the ground surface will rise. Similarly, in the two extensional quadrants, the ground motion associated with the first P-waves will be down, that is, the

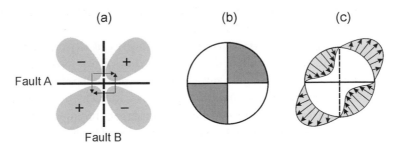

Fig. 4.3   Earthquake focal mechanisms. (a) For any double-couple earthquake, the slip could have occurred of one of two possible fault planes at 90° to each other, namely fault A and fault B. Here, fault A is the real fault plane (solid line) and fault B is the auxiliary plane (broken line). Areas or zones of compression (transpression for strike-slip faults) associated with the slip are indicated by plus (+), and those of extension (transtension for strike-slip faults) or tension by minus (−). (b) Beach-ball diagram (the focal-sphere projection) of the fault slip in (a), with areas of compression in grey (sometimes black) and those of extension in white. (c) Arrows show the direction of initial movement of points or particles around the hypocentre (Fig. 4.2) for the fault slip in (a).

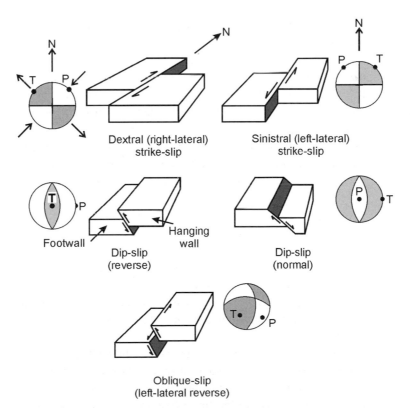

Fig. 4.4   Beach-ball diagrams (focal-sphere projections) for various types of fault slip. Areas of compression are in grey and those of extension in white. Also shown is the direction of minimum compressive principal stress, the axis of tension, T, and the direction of maximum compressive principal stress, the axis of compression, P. Modified from Reiter (1990).

ground surface will subside. More specifically, in the compressional quadrants the movement will be outward, but inward in the extensional quadrants (Fig. 4.3).

The seismic waves thus give an indication of the fault-slip direction as well as the orientation or attitude of the fault (Fig. 4.4). When there is an earthquake due to shear movement on a single fault plane, the event is referred to as being **double couple**. Earthquakes that are not generated in this way, such as earthquakes associated with mode I (extension or opening) fracture movement, are referred to as being **non-double couple**. The symbol used to present the earthquake is the **beach-ball** diagram or symbol (Figs. 4.3 and 4.4). Such a diagram is generated in the following way. The focus of the earthquake is imagined to be spherical. The lower hemisphere (the lower half) of that sphere (the **focal sphere**) is projected onto a horizontal plane (the Earth's surface), showing quadrants of compression and extension. The quadrants of compression are normally shaded (grey or black), while areas of extension are non-shaded (white).

The compressional gray quadrants contain the **axis of tension**, which shows the direction of the maximum tensile (minimum compressive) principal stress and is denoted by T. Similarly, the extensional white quadrants contain the **axis of compression**, denoted by P, which shows the direction of the maximum compressive principal stress (Fig. 4.4). In terms of principal stresses, the P- and T-axes would, for a two-dimensional stress field, correspond to σ_1 and σ_3, respectively. For **normal faulting**, σ_1 is vertical, so that the corresponding P-axis is in the centre of the circle (dipping 90°), indicating that it is vertical. Similarly, for a normal fault, σ_3 is horizontal (horizontal extension, say, parallel with the spreading vector), so that the corresponding T-axis is at the edge of the circle (dipping 0°). For **reverse faulting**, the location of the P- and T-axes is simply swopped. For **strike-slip faulting**, the T- and P-axes are both at the edge of the circle, indicating that σ_1 and σ_3 are both horizontal. Since these are the driving stresses of the fault displacement, the P-axis is located in the area that, following the fault slip, becomes a zone of transtension, and the T-axis is located in the area that, following the fault slip, becomes a zone of transpression.

In theory, the focal mechanisms are straightforward to obtain and apply. In practice, however, things arc somewhat more complex. For example, there are always two possible fault planes, **nodal planes**, corresponding to a single focal mechanism (Fig. 4.3). One is of course the real fault plane on which the earthquake originates; the other is the auxiliary or additional plane, at right angles to the real fault plane. Thus, there are two possible types of fault slip that correspond to a given focal mechanism. Based only on the focal mechanism itself, it is not possible to determine which of the two nodal planes is the real fault-slip plane. To determine the **real fault-slip plane**, a combination of (1) geological data, such as the observed surface rupture (as is normally associated with reasonably large – say larger than M5 – earthquakes), (2) geophysical data, such as geodetic data from InSAR or GPS, or (3) knowledge of the general tectonic regime, including the direction of plate movements at a plate boundary, can be used. The earthquake **velocity model** (based on the assumed mechanical layering of the crust) is commonly crude and sometimes very poor, resulting in an incorrect location (hypocentre) of the earthquake. There are several other potential sources of errors. Nevertheless, earthquake focal mechanisms are of great importance in seismology in general, and in volcano seismology in particular, since they give an indication of the state of stress associated with the earthquakes and, therefore, the likely volcanotectonic processes (such as dike propagation, sill emplacement, magma-chamber inflation) that are generating the earthquakes.

4.4 Earthquake Moments and Magnitudes

There are two main types of earthquake-magnitude scales. One quantifies the surface effects of the earthquake, primarily on human-made constructions. The other quantifies the energy released or transformed during the earthquake. The first is the **Mercalli intensity scale.** This scale has the range from 1 to 12, usually given by Roman numbers, I–XII, based on the effects of the earthquake on humans and their constructions. Thus, an earthquake of intensity I is hardly felt by anyone, except under especially favourable conditions. By contrast, an earthquake of intensity VI is felt by everyone, while damage to constructions is generally small. At the top end of the scale, an earthquake of intensity XI is felt by everyone and very few human-made structures remain standing following the earthquake. For example, rails are then greatly bent and bridges destroyed. An earthquake of the highest intensity, XII, results in what is effectively a **total destruction** of all human-made structures. Table 4.1 provides a description of the Mercalli intensity scale.

Intensity	Magnitude	Description
I	1–2	Detected only by seismometer
II	2–3	Noticed by some people at rest
III	3–4	Felt by many indoors; resembles vibration due to traffic
IV	4	Felt indoors by many, and outdoors by some; standing objects rock
V	4–5	Felt by most; some dishes and windows break.; some unstable objects overturned
VI	5–6	Felt by all; some heavy furniture moved; some chimneys may fall
VII	6	Considerable damage to poorly made buildings, but negligible in well-made ones; some chimneys broken; felt by car drivers
VIII	6–7	Considerable damage to ordinary buildings; chimneys fall
IX	7	Ground fractures; some houses collapse and pipes rupture
X	7–8	Many buildings destroyed; much fracturing of the ground; some landslides; rails become bent
XI	8	Most buildings are destroyed as well as many pipes, bridges, railways; urban areas without water, gas, and electricity
XII	8–9	Total destruction of urban areas; objects thrown in the air, very much ground fracturing and distortion

Table 4.1 Mercalli scale in comparison with earthquake magnitudes. The earthquake magnitudes shown are very approximate. The comparison is partly based on data from the US Geological Survey.

While the Mercalli scale is useful, it is not a measure of the real size of an earthquake in terms of the energy released or transformed. The scale is just a measure of the damage or destruction induced by the earthquake, which does depend strongly on the distance from the earthquake, the type of rock or soil on which the structures stand, and also partly on how strong or earthquake-resistant the structures are. Thus, a comparatively small earthquake with a hypocentre at a shallow depth can cause great damage at its epicentre (Fig. 4.2). By contrast, a large earthquake with a hypocentre at a great depth may cause comparatively little damage at its epicentre.

The earthquake-magnitude scale is a direct measure of the energy released or transformed during an earthquake. While all the energy scales in some way quantify the energy transformed, the exact method of estimating the energy varies. Thus, the early scales were based on the logarithm of the amplitude of the wave trace or signal on the seismograms at a certain distance from the hypocentre. This method of energy estimate is, for example, the basis of the well-known **Richter** local-magnitude (M_L) scale. The same method is used for the subsequently developed scales based on the **surface-wave** magnitude (M_s) and the **body-wave** magnitude (m_b).

In the past decades these amplitude scales have to a large degree been replaced by the **moment-magnitude** (M_w) scale. This replacement applies in particular to all moderate to great earthquakes (earthquakes larger than magnitude M5). The scale is directly connected to the **seismic moment** M_0 of an earthquake which, in the simplest way, may be defined as (Fig. 4.5):

$$M_0 = \Delta u_a A G \qquad (4.7)$$

where Δu_a is the average slip on the fault plane, assumed everywhere to be parallel with the slip surface (thus a non-curving fault plane), A is the total slip or rupture area associated with the seismogenic faulting, and G is the shear modulus. (Notice that in seismology G is

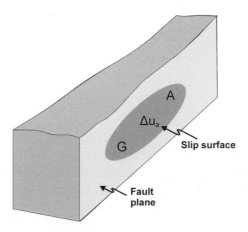

Fig. 4.5   For most earthquakes, which are small to moderate, the slip surface or rupture plane, here the area marked by A, forms only a small part of the much larger fault plane. The slip plane shown here is an interior crack, that is, it does not reach any free surface – such as the Earth's surface. The average slip is denoted by Δu_a and the shear modulus by G. Together with the area, these are used to determine the seismic moment (Eq. (4.7)).

commonly denoted by μ, in which case Eq. (4.7) becomes $M_0 = \mu \Delta u_a A$.) As defined in Eq. (4.7), the seismic moment is a scalar and applies to a single, planar fault. Many earthquake faults, however, are curved, composed of segments, or both, in which case Eq. (4.7) can be generalised so to take these effects into account (Madariaga, 1979; Scholz, 1990).

The seismic moment need not be expressed in terms of the shear modulus G. Using the average slip, the seismic moment can be expressed in terms of the **driving shear stress** τ_d. The driving shear stress is roughly equivalent to the **stress drop** (the difference in stress on the fault before and after the earthquake) and may be defined as:

$$\tau_d = \tau_i - \tau_f \tag{4.8}$$

Here, τ_i is the initial shears stress on the rupture plane of the fault before (earthquake) slip and τ_f is the final shear stress on the plane after slip. If the driving shear stress τ_d is constant and the average fault slip Δu_a is everywhere parallel on the single rupture plane (non-curving and non-segmented fault plane), the seismic moment may be given by:

$$M_0 = k \tau_d \Delta u_a A \tag{4.9}$$

where, as before, A is the slip area. Here k is a constant, the value of which depends on the geometry of the slip plane, that is, whether it is circular or rectangular, and whether it is an interior crack (one that does not intersect any free surface, including the Earth's surface), a part-through crack (one that intersects one free surface), or a through crack (one that interests two free surfaces). For details see Kanamori and Anderson (1975) and Gudmundsson (2011).

Some expressions for the seismic moment do not use the average slip Δu_a. For example, the moment may be presented as:

$$M_0 = q \tau_d A^{\frac{3}{2}} = q \tau_d L^3 \tag{4.10}$$

where q is a constant which, like that of k in Eq. (4.9), depends on the fault geometry and other factors. Here L is the **characteristic dimension** of the earthquake rupture or the slip surface, a dimension which is commonly defined as the square root of the slip area, that is, as $\sqrt{A}$.

Since the seismic (scalar) moment may be expressed in different ways, that is, alternatively in terms of the shear modulus G and the driving shear stress τ_d, we must make sure that the unit of the moment is the same for both expressions. In terms of G (Eq. (4.7)), we have shear modulus [N m^{-2}] × average slip during the rupture [m] × rupture area [m^2], so that the unit of the **seismic moment** is N m. In terms of τ_d (Eq. (4.9)), we have stress [N m^{-2}] × average slip during the rupture [m] × the rupture area [m^2], so the unit is, again, N m. This shows that the unit is consistent and the same as for work or **energy**, that is, N m.

In this book we primarily use fracture mechanics for modelling rock fractures, including seismogenic faults. In this formulation, the relation between moment, fault geometry, size (area), slip, and driving shear stress is somewhat different from that given in the equations above. This difference is partly because, in fracture mechanics, the Young's modulus E and Poisson's ratio v are more commonly used than the shear modulus G, where $E = 2G(1 + v)$ (Chapter 3; Appendix D). Partly, however, the difference is because the

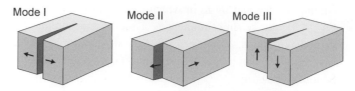

Fig. 4.6 There are three ideal crack-displacement modes, I, II, and III. Dikes are normally modelled as mode I, dip-slip faults as mode II, and strike-slip faults as mode III. However, in the case of a magma reservoir at the bottom of a through-going (surface- and bottom-reaching) fault, dip-slip faults may be modelled as mode III and strike-slip faults as mode II.

fracture-mechanics equations are a direct measurement of the elastic energy transformed or released during an earthquake, rather than a measure of the seismic moment.

There are three main ideal elastic **crack modes**, namely mode I, mode II, and mode III (Fig. 4.6). Strike-slip faults are most commonly modelled as mode III, whereas dip-slip faults are most commonly modelled as mode II. Some strike-slip faults, however, may be modelled as mode II, particularly large faults that extend from one free surface to another. Thus, a large strike-slip fault in a volcanic zone or a volcano that extends from the Earth's surface down to the magma chamber/magma reservoir would be modelled as mode II.

For a fault that is modelled as a mode II crack, the energy transformed or released U_{II} is given, in terms driving shear stress τ_d, by (Gudmundsson, 2011):

$$U_{II} = \frac{\tau_d^2(1 - v^2)\pi aA}{E} \tag{4.11}$$

and in terms of slip Δu by:

$$U_{II} = \frac{E\Delta u_{II}^2 \pi A}{16(1 - v^2)a} \tag{4.12}$$

where a is half the strike-dimension (rupture length) of the slip plane in the case of a strike-slip fault but the total dip-dimension in the case of a dip-slip fault (the latter being regarded as an edge crack). Here, the fault slip Δu (either for mode II or III) is, strictly speaking, the **maximum slip**, but it may often be regarded as the average slip measured anywhere along the fault except near its tips or ends. Eqs. (4.11) and (4.12) are suitable for plane-strain conditions, namely when the crustal segment hosting the fault is of a thickness similar to, or greater than, the lateral dimensions considered. The lateral dimensions cannot be smaller than the strike-dimension (horizontal length) of the fault rupture, and are usually considerably larger than the rupture dimension. When the crustal segment hosting the fault is much thinner than the lateral dimensions considered, then the analogous plane-stress equations are used. The only difference is that for plane stress the term $1 - v^2$ in Eqs. (4.11) and (4.12) drops out (becomes equal to 1).

When a fault is modelled as a mode III crack, the energy transformed or released is given, in terms of driving shear stress τ_d, by (Gudmundsson, 2011):

$$U_{III} = \frac{\tau_d^2(1 + v)\pi aA}{E} \tag{4.13}$$

Similarly, in terms of slip Δu, the energy transformed is given by:

$$U_{III} = \frac{E\Delta u_{III}^2 \pi A}{16(1+v)a} \tag{4.14}$$

where a is half the strike-dimension of a dip-slip fault and the dip-dimension of a dip-slip fault. Equations (4.11) to (4.14) can be used to calculate the energy released during seismogenic slip using the models of either mode II or mode III cracks. Similar, but more complex, equations can be presented for mixed-mode crack models of seismogenic slip; for example, partly mode II and partly mode III. The energy release calculated from these equations can then be compared with the standard seismic-moment calculations from Eqs. (4.7), (4.9), and (4.10) and other similar equations (Kanamori, 1977; Madariaga, 1979; Kanamori and Brodsky, 2004). The seismic-moment results, calculated in this way, can be used to estimate the magnitude of the earthquake, to which we turn now.

The moment magnitude M_w may be defined as follows:

$$M_w = \frac{2}{3}\log M_0 - 6.0 \tag{4.15}$$

where log is the common (decimal) logarithm (to the base 10) and the seismic moment M_0 has the unit N m. In many scientific papers and monographs, however, the seismic moment is still given in dyne-centimetres (dyne-cm) and the constant 6.0 is then given as 4.7. The relationship in Eq. (4.15) can be expressed in different ways. For example, some recent publications give the last constant value as 6.07 (Stacey and Davis, 2008; Udias et al., 2014). Normally, the constant is between 6.0 and 6.1 (Kagan, 2014). However, Eq. (4.15) makes some restrictive assumptions, such as that the driving shear stress (the stress drop) is constant. Furthermore, it is assumed that the ratio between the stress drop and the shear modulus G is 10^{-4} (Udias et al., 2014). Thus, an implied level of accuracy in the constant of three significant figures may be unwarranted, and here we use 6.0, as is common (Kagan, 2014).

Observations show that, for any seismogenic zone, as well as for the Earth as a whole, if the period considered is sufficiently long (commonly a year or, depending on the earthquake frequency, a few decades), small-magnitude earthquakes are much more common than large-magnitude earthquakes. This relationship is referred to as the **Gutenberg–Richter frequency–magnitude relation** or law for earthquakes, which is among the best known of all the **power laws** or heavy-tailed distributions in physical science. Since the fault slip is a function of the fault size (the controlling dimension, which is either the strike-dimension or the dip-dimension), and since earthquake moments (energies) and related magnitudes depend on the slip and the rupture area, it could be argued that when the fault sizes follow power laws, then so must the magnitudes. Whatever the explanation for the Gutenberg–Richter law, it may be expressed as follows:

$$\log N(\geq M) = a - bM \tag{4.16}$$

where, as above, log stands for the common logarithm, N is the cumulative number of earthquakes larger than M, a and b are constants (or, rather, parameters) and M is the earthquake magnitude as measured on any of the magnitude scales used and discussed

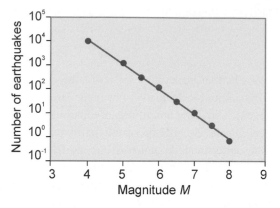

Fig. 4.7 Earthquake frequency (number) versus earthquake magnitudes M. The curve (straight line) shows the magnitude distribution of earthquakes in the world from 1904 to 1980 (modified from Kanamori and Brodsky, 2004). For the Gutenberg–Richter law (Eq. (4.16)) a slope of –1, as here, implies a b-value of 1.

above (cf. Udias et al., 2014). A typical plot is given in Fig. 4.7. More specifically, a and b are constants for specific seismic areas or zones over certain periods, but, as indicated below, they are strictly parameters that vary between tectonic regimes and periods. On such a plot (Fig. 4.7), the data are log-transformed, both on the vertical (y-) axis for the cumulative number of earthquakes above magnitude M, as well as on the horizontal (x-) axis, for the earthquake magnitude. The result is what is called a bi-logarithmic or **log–log plot**, because the cumulative number of earthquakes N is given on a logarithmic scale (the vertical axis), and the magnitude M is proportional to the logarithm of the amplitude, that is, to the energy radiated by seismic waves from the fault during the earthquake.

The details of the relation in Eq. (4.16) can be explained as follows. The solid line in Fig. 4.7 gives the number of earthquakes of a magnitude larger than or equal to the magnitude M in Eq. (4.16). Commonly, the magnitude M is the surface-wave magnitude M_s. Thus, in a given area (in Fig. 4.7 the whole Earth), N is the number of earthquakes over a given time – the time considered is commonly one year. The value of the parameter b, the **b-value**, indicates the slope on the log–log plot of number versus magnitudes of earthquakes (Fig. 4.7), whereas the parameter a, the **a-value**, indicates the intersection of the solid line with the horizontal (x-) axis. The **slope** is $-b$ and is a measure of the proportion of smaller earthquakes to larger earthquakes. For tectonic (non-volcanic) earthquakes b is generally in the range 0.8–1.1, and most commonly around 1.0. However, b tends to be higher for normal faulting, particularly where volcanotectonic events are common, such as at mid-ocean ridges where values of 1.0–2.0 are common. By contrast, in areas of reverse and thrust faulting, b-values of 0.7–1.0 are common, and they are occasionally as low as 0.4–0.7.

In seismic areas with **high b-values**, smaller earthquakes are proportionally more common (in relation to larger earthquakes) than in areas with low b-values. Thus, at divergent plate boundaries with normal faulting and dike injection, smaller earthquakes are more common than at convergent plate boundaries with reverse and thrust faulting. In particular, where dike emplacement is common, such as at mid-ocean ridges, earthquake swarms are common, and in some of these the b-values may be as high as 3.0. Thus, on a typical Gutenberg–Richter plot, earthquakes related to normal faulting, dike emplacement, and eruptions have **steeper slopes** (scaling exponents) than those related to purely tectonic faulting, particularly reverse and thrust faulting at convergent boundaries, as well as to intraplate seismicity.

The other parameter, a, simply marks the intersection of the sloping line or curve on the log–log plot (Fig. 4.7) with the horizontal (x-) axis. When the value of a changes, the Gutenberg–Richter (G–R) curve or line becomes **shifted** laterally. If the **a-value** decreases, the lateral shift of the G–R curve is to the left; if the a-value increases the shift of the G–R curve is to the right. For example, if the G–R curve represents the seismicity in one active area in a year, then, for b = 1.0, a corresponds to the minimum magnitude of the largest earthquake that occurs in that area, on average, every year. If b is larger or smaller than 1.0, then the G–R relation for that seismic area implies that an earthquake of magnitude a/b (or larger) occurs, on average, once every year.

The total energy U_t radiated as seismic waves by a fault can be approximately estimated from the Gutenberg–Richter law and the earthquake magnitude, usually using the surface-wave magnitude M_s. If the **total radiated energy** U_t is given in joules, the energy-magnitude relation may be given as (Stacey and Davis, 2008; Udias et al., 2014):

$$\log U_t = 1.5\, M_s + 4.8 \tag{4.17}$$

Equation (4.17) shows that an increase in earthquake magnitude by a unit, say from M6 to M7, implies that the radiated energy associated with the earthquake increases by a factor of $\Delta U_t = 10^{1.5} = 31.6$. That is, moving up or down the earthquake-magnitude scale by one unit changes the radiated energy by a **factor of about 32**. Similarly, if we move up the scale by two units, say from M6 to M8, then the radiated or released energy increases by a factor of $\Delta U_t = 10^3 = 1000$. Thus, an M8 earthquake radiates 1000 times more energy than an M6 earthquake. Furthermore, an M8 earthquake radiates or releases $\Delta U_t = 10^6$ or **million times** as much energy as an M4 earthquake. M4 earthquakes are classified as light but are normally easily felt. What this relationship thus demonstrates is that numerous light or small earthquakes (M4 or less), while felt and recorded, contribute very little to the total radiated seismic energy from an earthquake area or an active fault zone. Each strong to major (M6 to M8) earthquake, while rare, radiates so much more energy than small earthquakes that the cumulative effect of hundreds or thousands of small earthquakes in the zone contributes very little to the total energy budget (cf. Stein and Wysession, 2002).

Note, however, that the radiated seismic energy (Eq. (4.17)) is normally only a **small fraction** of the total energy, measured by the seismic moment M_0, which is relaxed or transformed during the earthquake. While the exact fraction of the energy that is radiated as seismic waves during an earthquake is not known, it is commonly thought to be **about 10%** of the total transformed energy (Stacey and Davis, 2008). The radiated energy (Eq. (4.17)) and the seismic moment (Eq. (4.7)) do therefore not measure the same thing and normally have a different numerical value. The energy radiated as seismic waves is a measure of the potential of the earthquake to cause **damage** to human constructions. By contrast, the moment measures the total **energy** transformed or released during the earthquake.

Coming back to Eq. (4.16), the parameter a is the intercept of the sloping line with the horizontal (x-) or magnitude axis whereas parameter b is the slope of the line. The value of a depends on the seismic area and the number of earthquakes studied. The gradient of the line is very commonly around -1, which corresponds to a b-value of 1. For example, in the plot in Fig. 4.7 the following relationship holds (Stacey and Davis, 2008):

$$\log N(\geq M) = (8.0 \pm 0.2) - (1.0 \pm 0.03)M \tag{4.18}$$

Different values of a, however, are obtained from different areas and different periods of sampling, and thus different numbers of earthquakes. The b-value tends to be very close to 1.0 for many earthquake areas and periods, as indicated above, in which case there is a **10-fold increase** in the number of earthquakes for each unit drop in magnitude.

Eqs. (4.15) and (4.16) can be combined to get:

$$\log N = \log\alpha - \beta\log M_0 \tag{4.19}$$

where α and β are parameters to be interpreted in a moment. Equation (4.19) can be recast in the form:

$$N(\geq M_0) = \alpha M_0^{-\beta} \tag{4.20}$$

which is a power law. In such a power-law size distribution, α is a constant or parameter and β is the **scaling exponent**. Here, α may be regarded as the number of earthquakes per year with a moment larger than 1 N m, and the moment itself should be interpreted as the ratio $M_0/1$ N m and thus dimensionless. Using Eq. (4.18) and Eq. (4.15), the constants in Eq. (4.20) become $\alpha \approx 10^{14}$ and $\beta \approx 2/3$ (Stacey and Davis, 2008).

Notice that b in Eq. (4.16) is a constant, or rather a parameter, for the earthquake-magnitude distribution whereas β in Eq. (4.20) is a constant or parameter for the seismic-moment distribution. For a 'perfect' power law, such as Eq. (4.20), the relationships between these parameters should be $\beta = b/1.5$ (Kagan, 2014). Theoretical analysis suggests that β should be about 0.5, but measurements show that the parameter varies, while almost all its values are larger than 0.5. Similarly, although b is normally close to 1.0, its value also varies somewhat between different tectonic regions with variable rock properties (heterogeneities) and states of stress, as discussed above. Many possible explanations of the variations in the values of β and b are listed by Kagan (2014).

Since the value of b for tectonic earthquakes is commonly at about 1.0 (and for β about 2/3), values below this are normally regarded as low and values higher as high. We have seen that in volcanic areas the b-values are commonly high, that is, well above 1.0. Since the b-value is the slope on the log–log plot of number versus magnitudes of earthquakes (Fig. 4.7), it follows that high b-values indicate steep slopes and thus an unusually high proportion of **small-magnitude** earthquakes. As mentioned, this is common in volcanic swarms and, generally, in **volcanic areas** and at mid-ocean ridges. By contrast, small b-values characterise seismogenic fault zones in continental areas and at subduction zones. More specifically, **normal faulting**, as is common at divergent plate boundaries in general, and mid-ocean ridges in particular, is mostly associated with small earthquakes and thus tends to have comparatively high b-values. By contrast, **reverse and thrust faulting** at convergent plate boundaries is often associated with large faults, and comparatively larger fault slips, and is commonly associated with smaller b-values.

4.5 High-Frequency Earthquakes

Volcanic earthquakes are classified in various ways. Here, we follow a widely used classification whereby there are four main groups. These are: (1) high-frequency or A-type earthquakes, (2) low-frequency or B-type earthquakes, (3) explosion earthquakes, and (4) volcanic tremors (McNutt, 2000, 2005; Zobin, 2003). We shall now discuss these four main classes and what they imply for volcano unrest, starting with the **high-frequency or A-type** earthquakes.

These are also referred to as **volcanotectonic earthquakes**, a term used more generally in this book for volcanic earthquakes. Most or all these earthquakes are thought to be related to shear failure, that is, fault slips. In this sense they are very similar to ordinary tectonic earthquakes. There are several differences, however, between ordinary tectonic earthquakes and high-frequency (volcanotectonic) earthquakes. The main differences include (1) the stress sources driving the earthquake rupture, (2) the tendency for volcanotectonic earthquakes to occur in swarms, and (3) the differences in b-values between tectonic and volcanotectonic earthquakes.

4.5.1 Stress Sources

One difference between ordinary tectonic earthquakes and volcanotectonic earthquakes is the source of the driving shear stress of faulting. In volcanotectonic earthquakes the proposed sources are many, including tectonic forces, but commonly are somehow related to the fluid pressure of an intrusion, such as a dike or an inclined sheet, or that of a magma chamber. By contrast, for most tectonic earthquakes generated outside volcanoes the stress sources connect directly to regional plate movements. Stress sources suggested for volcanotectonic earthquakes include the following (McNutt, 2005):

1. Regional, mostly plate-tectonic, forces.
2. Pore-pressure changes, such as in geothermal fields (Fig. 4.8).
3. Gravitational loading; loading through (2) and (3) may generate landslides as well.
4. Hydrofracture formation and propagation, including fractures formed by geothermal water (eventually mineral veins) and by magma (dikes, inclined sheets, and sills).
5. Magma-chamber pressure changes, resulting in inflation and deflation.
6. Thermal stresses generated by intrusions.
7. Cooling of intrusions and magma-chamber envelopes.

For many of these earthquakes, there may be two or more stress sources acting simultaneously. For example, during the generation of about 100 000 earthquakes in the Hengill central volcano in Southwest Iceland (Fig. 4.8) in the period 1994–2000 (Feigl et al., 2000), there was a pore-fluid-pressure change in the associated geothermal field but also uplift or doming of the area, which may have been partly the result of a pressure increase (inflation) of the associated magma chamber (Fig. 4.8). Similarly, during rifting episodes in active volcanoes in rift zones, volcanotectonic earthquakes may be partly the result of faulting

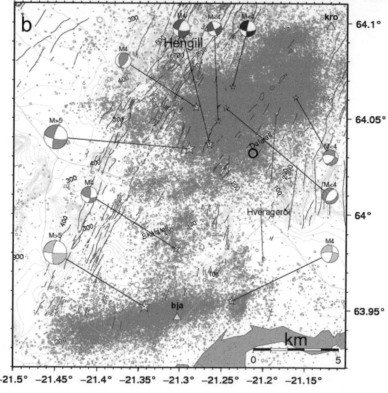

Fig. 4.8 (a) View south, the Hengill central volcano in Southwest Iceland is dissected by numerous normal and strike-slip faults. (b) Hengill experienced doming and intensive seismicity, with about 100 000 earthquakes in the period 1994–2000 (Feigl et al., 2000). The maximum uplift or doming rate during this period was 20 mm a^{-1}, the cumulative maximum being 7.5 cm (Vogfjord et al., 2005; Jakobsdottir, 2008) and occurred partly in the graben area between the faults in (a). Much of the seismicity, however, occurred on strike-slip faults, although there was also normal faulting. The uplift has been explained in terms of a Mogi pressure source at about 7 km depth.

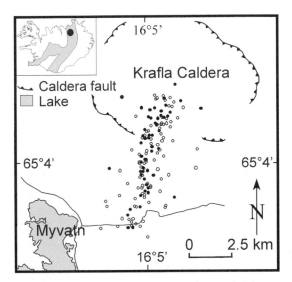

Fig. 4.9 Earthquake swarm in the Krafla Volcanic System in North Iceland. This swarm migrated from the Krafla caldera to about 8 km south of the centre of the caldera. The earthquake activity then migrated southwards along the Krafla fault swarm with a rate of about 0.5 m s^{-1}. Hypocentres (focal depths) were mostly 0–6 km, the focal mechanisms generally suggesting slip on normal faults, which are numerous in the fissure swarm. Modified from Brandsdottir and Einarsson (1979).

generated by the regional plate-tectonic stress field, partly due to fault slips generated by propagating dikes (Figs. 1.13 and 4.9), and partly because of inflation and deflation of shallow and/or deep-seated magma chambers (Fig. 3.1).

4.5.2 Earthquake Swarms

A second difference between volcanotectonic earthquakes and general tectonic earthquakes, as indicated in Fig. 4.9, is that volcanotectonic earthquakes much more commonly occur in specific swarms. A **volcanotectonic swarm** is characterised by a large number of earthquakes at a given location (the area or zone within which the swarm occurs), all of which are of **similar magnitude** or size. This means that in a volcanotectonic swarm there is no clear mainshock (largest earthquake). More specifically, the magnitude difference ΔM between the largest and the second largest earthquake in the swarm is commonly around $\Delta M = 0.5$ or less. This is in contrast to common large tectonic earthquakes that are characterised by foreshock–mainshock–aftershock sequences where $\Delta M \geq 1.0$ (McNutt, 2000). As the name implies, high-frequency volcanotectonic earthquakes have, indeed, a high frequency. The characteristic frequency is 5–15 Hz. Both the P-waves and the S-waves are normally recognised.

For earthquake swarms associated with **eruptions** two distinct processes can be identified. One process operates before the swarm peaks, that is, reaches its maximum frequency of earthquakes per unit time. The other process operates after the swarm peaks (McNutt, 2005). The process controlling the swarm prior to the peak is mechanically most likely caused by the

Fig. 4.12 Segmented basaltic dikes, located in Northwest Iceland, dissecting a basaltic lava pile with many fractures oriented obliquely to the dikes. Oblique fractures are potential slip planes generating earthquakes in the host rock during dike propagation. Only some of the larger oblique fractures are indicated; there are many smaller ones. The dikes, some of which are indicated, form a part of a Tertiary swarm with an average dike thickness of 5–6 m.

stress induced by a vertically propagating dike (Gudmundsson, 2011). Consequently, the joints and contacts function as potential shear fractures, many of which, when loaded through horizontal compressive stresses induced by the dike, slip in shear and generate earthquakes. Much of the earthquake activity in swarms associated with dike emplacement is thus likely to be related to slip on oblique joints and contacts, as well as existing faults (Fig. 1.21) in the surrounding host rock rather than to fracturing at the tip of the dike. Fault slip in the walls of a propagating dike is therefore presumably primarily on existing fractures that are subject to shear stresses induced by the magmatic pressure. All types of fault slip are expected, including reverse and strike-slip faulting even if the dike is emplaced in a rift zone (e.g. Passarelli et al., 2015; Agustsdottir et al., 2016; Bonaccorso et al., 2017).

The **duration** of volcanotectonic earthquake swarms that eventually result in eruptions varies widely (McNutt, 2005). Commonly, the swarms last from several weeks to many months. For example, the earthquake swarm prior to the 1980 Mount St. Helens eruption lasted 2 months. Similarly, the duration of the earthquake swarm associated with the dike emplacement prior to the Bardarbunga–Holuhraun eruption in Iceland in 2014 was about 2 weeks (Gudmundsson et al., 2014; Sigmundsson et al., 2015). Much shorter durations are known, however. For example, the earthquake swarm prior to the 1991 Hekla eruption in Iceland was about half an hour (Gudmundsson et al., 1992).

These and other similar results (e.g. McNutt, 2005) show that it is difficult to **forecast** the time of an eruption based on the likely duration of the earthquake swarm prior to the eruption.

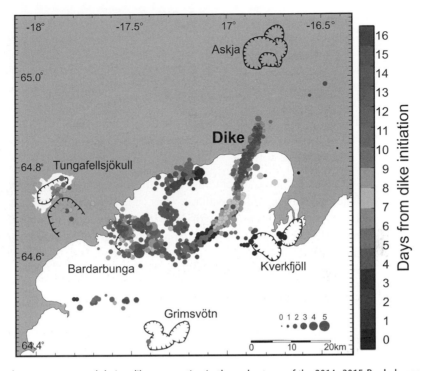

Earthquake swarm generated during dike propagation in the early stages of the 2014–2015 Bardarbunga–
Holuhraun eruption in central Iceland. Most of the earthquakes associated with the dike were generated in the first
2 weeks, when the magma driving pressure was fracturing the crust so as to form a path towards the surface. This
illustration, modified from Gudmundsson et al. (2014), shows the earthquake swarm generated during the first 2
weeks (from 16 August 2014) of the episode. While breaking and compressing the rock around it (Figs. 4.11 and
4.12), the dike reached a strike-dimension of about 45 km and a dip-dimension of at least 15–20 km and
generated some 17 000 earthquakes; which is more earthquakes in 2 weeks than is normal in all the volcanoes and
earthquake zones in Iceland during a whole year (cf. Sigmundsson et al., 2015).

Such forecasts become even more difficult when we take into account that most dike
injections and associated earthquake swarms do not result in an eruption. Most propagating
dikes become arrested and thus never reach the surface to feed an eruption (Chapters 2, 5 and
7). Even in highly active volcanoes, measured surface-deformation periods (unrest periods)
in the majority of cases do not end in an eruption (Biggs et al., 2014). Well-known recent
examples of major earthquake swarms associated with propagating dikes that eventually
became arrested, resulting in **failed eruptions**, include the following:

- Dike injection close to the surface of the Teide Volcano in Tenerife in 2004 (Carracedo
  and Troll, 2006; Garcia et al., 2006; Gottsmann et al., 2006; Marti et al., 2008).
- Dike injection close to the surface of the Harrat Lunayyir, Saudi Arabia, in 2009 (Pallister
  et al., 2010; Al Amri et al., 2012; Xu et al., 2016; Al Shehri and Gudmundsson, 2018).
- Many dike injections during the 20-year period prior to the 2010 Eyjafjallajökull eruption
  in Iceland (Sigmundsson et al., 2010; Gudmundsson et al., 2012; Tarasewicz et al., 2012).

4.5.3 Differences in *b*-Values

As indicated above (Eq. (4.16)), tectonic earthquakes generally have Gutenberg–Richter-law *b*-values close to 1.0. While this applies to most tectonic regions, the variation in *b*-values for volcanic areas is much greater, with values commonly being much higher than 1.0. For example, *b*-values as high as 3.0 have been reported from some volcanic areas (McNutt, 2005). There are several possible explanations for this and the main ones are listed below. Generally, the characteristics of earthquake *b*-values in volcanic regions may be summarised as follows:

- All volcanic areas, including the volcanic edifices themselves, show significant heterogeneity in their *b*-values. Thus, within volcanic zones there are crustal segments or rock volumes with high *b*-values close or adjacent to segments with normal (around 1.0) values. Studies suggest that the high-*b*-value segments remain high for at least years to decades; they are thus, apparently, not just transient features.
- High *b*-values in volcanic zones are normally at shallow depths, ranging to depths of about 7 km from the surface and most commonly at about 3–4 km depth. These are the typical depths for shallow magma chambers (Chapters 5 and 7) and associated dike and inclined-sheet injections.
- These depth ranges lend further support to the suggestion that the high *b*-values are associated with volcanotectonic earthquakes generated by pressure changes in nearby magma bodies, chambers and injected dikes, as well as possible gas exsolution and expansion. It has also been suggested that these depths coincide with those of open fractures in volcanic zones. However, once formed, and unless filled by secondary minerals, fractures may remain open (but normally with openings of millimetres or less at great depths) in the brittle parts of volcanic and non-volcanic areas down to crustal depths of at least 9–10 km (Gudmundsson, 2011). The formation of tension fractures, however, is limited to the uppermost several hundred metres of the crust – at greater depths, tension fractures transform into normal faults (Gudmundsson, 2011).
- Many volcanotectonic swarms associated with eruptions and/or arrested dike intrusions show significant variations in their *b*-values with time (McNutt 2005). Some show two *b*-value peaks, followed by more normal *b*-value periods. It has been suggested that the first peak relates to short-lived high-temperature gradients (associated with the magma emplacement), whereas the second peak may relate to pore-fluid pressure increase. In this model, the first peak lasts for a shorter period because the extra heat is carried away rather quickly through fluids (e.g. geothermal water or gas). The second peak, however, may last longer because it is primarily controlled by diffusion, which takes a longer time.

It is fair to say that abnormally high *b*-values in volcanic areas, while common, are still not well understood. This is an area of active research and the associated processes will, no doubt, be better understood in the coming years.

4.6 Low-Frequency Earthquakes

As the name implies, these earthquakes have low frequencies. The general frequency range is from 1 Hz to 5 Hz, with 2–3 Hz being the most common (McNutt, 2000, 2005). While these earthquakes show P-waves they lack S-waves. These earthquakes are also known as **B-type** earthquakes or **long-period** earthquakes.

One difference between the low-frequency events and the high-frequency events is that the latter are well understood in terms of the actual fracture mechanics causing the earthquake. There is no doubt that most high-frequency events relate to slip on faults, that is, on shear fractures. By contrast, the origin of low-frequency events is less well understood in terms of basic physics. It is thought that many, perhaps most, of the low-frequency earthquakes are due to **fluid-pressure changes**. The associated processes include bubble formation, bubble collapse, faulting (shear failure), and extension-fracture development, all of which are supposed to occur at shallow depths. The energy sources for the earthquakes can be magma-filled fractures or cylindrical conduits that radiate energy.

Assuming that low-frequency earthquakes are mostly generated by fluid-pressure changes then, in theory, they may be used to monitor magma-pressure changes and magma movement within a volcano. However, similar seismicity is recorded from geysers and geothermal areas, suggesting that identifying all or even most low-frequency earthquakes with magma-pressure changes may not be warranted.

Another difficulty in understanding the origin and volcanotectonic implications of low-frequency earthquakes is the comparatively recent recognition that similar earthquakes are apparently common on ordinary tectonic faults. It is now known that **slow earthquakes** (also known as creep) occur frequently on many tectonic faults (Stein and Wysession, 2002; Peng and Gomberg, 2010; Gudmundsson et al., 2013). Some of the slow earthquakes are similar to the low-frequency earthquakes discussed here. There is, however, one significant difference: namely, the true volcanotectonic low-frequency earthquakes lack S-waves, whereas the slow tectonic earthquakes normally show S-waves (McNutt, 2005).

In summary, we still have to learn much more about low-frequency earthquakes in volcanoes before we can maintain that we fully understand them. While it is generally accepted that they are largely, or mostly, related to fluid-pressure changes, it is less clear how these pressure changes originate, and what fluids are involved.

4.7 Explosion Earthquakes

These earthquakes are primarily associated with **explosive eruptions** (McNutt, 2005). The energy released or transformed during the explosion spreads or travels in different ways. Partly, the energy generates ordinary seismic waves propagating through the ground (the solid rock). Partly, however, the energy travels through the air (generating an air-shock phase on the seismograms) as acoustic (sound) waves or air waves.

Most of the explosions originate in the main conduit associated with magma transport to the surface during the explosive eruption. Many explosions originate at very shallow depths in the conduit – some at depths of only a few hundred metres. Other explosions, however, originate at considerably greater depths, with some depths of origin being of the order of kilometres.

4.8 Volcanic Tremors

Volcanic tremors are perhaps the most **characteristic seismicity** of volcanoes. They are continuous (or semi-continuous) ground vibrations or shaking that lasts from minutes to several days or longer. Volcanic tremors are generally thought to be generated in association with the migration of fluids, that is, gas or magma, or both. More specifically, the standard interpretation of a volcanic tremor is that it is caused by **magma movement** in fractures, such as dikes, or boiling and **pulsation** of fluids under high pressure, both at shallow depths within a volcano or a volcanic zone. Some volcanic tremors may also be related to the escape of high-pressure steam and gas from fumaroles.

Volcanic tremors are commonly interpreted as an indication of an **imminent eruption**. This is particularly a standard interpretation when the tremor follows or is a continuation of an earthquake swarm. It should be noted, however, that many recorded volcanic tremors have not eventually resulted in an eruption (McNutt, 2000). Nevertheless, the general idea is that the tremor is largely generated by magma migration in the potential feeder-dike just below the surface (Figs. 3.3 and 4.10). A volcanic tremor is composed of low-frequency events, characteristically 1–5 Hz, with 2–3 Hz being the most common. Thus, the frequency is essentially the same as that characterising low-frequency or B-type earthquakes. However, B-type earthquakes are different from volcanic tremors in other respects. For example, while the magnitude-size distribution of most earthquakes, including B-type earthquakes, follow power laws (Eqs. (4.16) and (4.20); Fig. 4.7), the size distribution of earthquakes in a volcanic tremor follows an exponential law. Strictly, when the tremor is continuous, individual earthquakes cannot be counted, but the different amplitudes and their duration can be recognised and used for this purpose. If the frequencies of evenly spaced spectral peaks in a volcanic tremor vary regularly with time, the process is referred to as gliding.

As indicated, common sources of volcanic tremors are thought to be potential feeder-dikes with tips at shallow depths. For such tremors, an exponential size distribution normally implies that the source, here the dimensions of the feeder-dike, remains essentially constant during the tremor (McNutt, 2000). This is likely to be approximately true for a feeder-dike very close to and approaching the surface (Fig. 4.10). While the strike-dimension (length) and aperture or thickness of a feeder-dike may vary at any depth (Figs. 2.5, 2.33 and 4.11), the strike-dimension in particular is likely to be similar in the uppermost metres or tens of metres just before the dike reaches the surface (Figs. 2.22 and

2.23). This is partly because the feeder-dike is then normally propagating through a single layer or unit with essentially uniform mechanical properties.

Volcanic tremors associated with feeder-dikes (fissure eruptions) tend to be stronger than those associated with cylindrical or somewhat elliptical conduits (McNutt, 2005). This may be because the surface area of a feeder-dike is normally much larger than that of a typical cylindrical conduit. Feeder-dikes also change their openings (apertures) more easily as the magma pressure changes than cylindrical conduits do. Phreatic eruptions commonly generate strong volcanic tremors. These strong tremors are likely to be related to boiling of groundwater that comes in contact with the hot magma in the feeder-dike.

4.9 Earthquakes Triggering of Volcanic Eruptions

It is sometimes observed that volcanic eruptions, particularly explosive ones, occur within a short time of a large, often distant, earthquake. In particular, it is estimated that about 0.4% of explosive eruptions occur within a few days following a large distant earthquake (Hill et al., 2002; Manga and Brodsky, 2006). For example, Watt et al. (2009) suggested that there was a significant increase in the eruption rate in parts of Chile and the Andes following M8 and larger earthquakes. They suggest that the triggering effects may be felt from earthquakes of this magnitude at volcanoes as far as 500 km from the earthquake rupture.

Many mechanisms have been proposed to explain how large and commonly remote earthquakes could trigger eruptions. These **mechanisms** include (1) increased bubble growth within the magma chamber, (2) accumulation of gas bubbles at the top of a chamber, and (3) overturn of magma chambers (Manga and Brodsky, 2006; Namiki et al., 2016). All these mechanism are supposed to result in increased excess pressure in the chamber, primarily through bubble growth and/or migration within the chamber, thereby resulting in magma-chamber rupture. In addition to these mechanisms for volcanic eruptions, mud volcanoes and geysers are also affected or triggered by earthquakes (Manga and Brodsky, 2006). Mud volcanoes are outside the scope of the present discussion. It is well established, however, that earthquakes affect **geysers** and geothermal fields in general. In fact, the fracture-related permeability of geothermal fields worldwide is maintained through earthquake activity. It is well known, for example, that the geysers of the Haukadalur area in South Iceland – which includes the Great Geysir and Strokkur, the only erupting hot springs in Europe – change the frequency and size of their eruptions following strong to major earthquakes in the nearby South Iceland Seismic Zone (Gudmundsson, 2017).

The mechanisms above focus mainly on an excess-pressure increase in the chamber due to earthquake effects. However, an obvious effect of many large earthquakes on nearby volcanoes is through **stress transfer**. Stress transfer from earthquakes and earthquake zones to volcanoes is common (Gudmundsson and Brenner, 2003; Mathieu et al., 2011), as is stress transfer and mechanical interaction between volcanoes (Gudmundsson and

Andrew, 2007; Andrew and Gudmundsson, 2008; Biggs et al., 2016; Elshaafi and Gudmundsson, 2018). Here, stress transfer, from earthquakes to volcanoes or between volcanoes, implies stress concentration around the associated magma chambers. In particular, when stress is transferred from a fault zone or an earthquake to a nearby central volcano, the tensile-stress concentration around its source magma chamber may, depending on the location of the volcano in relation to the earthquake fault/fault zone, either increase or decrease (Gudmundsson and Brenner, 2003). If the tensile-stress concentration around the chamber increases, this may trigger magma-chamber rupture and **dike injection**.

But there is a long way from a magma-chamber rupture and dike injection to a dike-fed eruption. Most dikes become **arrested** and therefore do not erupt (Chapters 5, 7, and 10). Thus, unless the magma-chamber stress or excess-pressure changes triggered by an earthquake also make the stress conditions in the entire crustal segment between the chamber and the surface such that they favour dike propagation – as seems to be the case for the Hekla Volcano in Iceland (Gudmundsson and Brenner, 2003) – there is no reason why an eruption should follow a large earthquake. Given that **only 0.4%** of explosive eruptions can be interpreted as following recent large and 'nearby' earth-quakes – where 'nearby' is regarded as a distance as great as 500 km between the earthquake rupture site and the volcano – the correlation between such earthquakes and eruptions is **weak**. And so it should be because even if an earthquake could trigger excess-pressure/stress changes that favour dike injection, most injected dikes would not feed eruptions. It might thus be a better approach in studying the mechanical effects of earthquakes on nearby volcanoes to look at the frequencies of unrest periods rather than the frequencies of eruptions. Unrest periods are more likely than eruptions to show a reasonable general correlation with the occurrence of nearby earthquakes.

4.10 Seismic Monitoring

One of the main methods for monitoring volcanoes is the use of volcanic earthquakes, that is, seismic monitoring. The principal reason for the importance of seismic monitoring is that most if not all volcanic eruptions are **preceded by earthquakes**. Furthermore, earthquake activity in a volcano is a general indication of its state of stress (Massa et al., 2016), which has implications for the likelihood of an unrest period resulting in an eruption.

To be useful in assessing the likelihood of an eruption during an unrest period with earthquakes, seismic monitoring of a volcano must be compared with the **background seismicity**. This may take years or decades to establish. It is the average number and the characteristics of earthquakes in a given volcano when the volcano is in a period of quiescence, that is, not subject to unrest of any kind. Once the (normally low) background or quiescence seismicity of the volcano has been established, any changes (normally increase) in the frequency (intensity) of earthquakes as well as in their types (e.g. whether the focal mechanisms indicate primarily normal faulting, reverse faulting, or strike-slip faulting) is recorded and interpreted in terms of models of unrest.

Seismic activity almost always **increases** before an eruption and forms a part of the unrest period of the volcano. The earthquakes associated with a typical unrest period that results in an eruption can be crudely grouped into several stages, the main ones being the following:

1. The early stages of increased seismic activity are often related to **inflation**, that is, to magma-chamber expansion, which is reflected in (normally) small (of the order of centimetres to tens of centimetres) doming or rise of the surface of the volcano (Figs. 1.12, 3.5 and 3.25). During some unrest periods in some volcanoes the inflation may be much larger; that is, many metres or more.
2. If magma-chamber **rupture** with dike (or sheet) injection occurs, then an earthquake swarm is expected to occur (Figs. 4.9 and 4.13). If the swarm is reasonably intense (Fig. 4.13), the propagation of the dike can be monitored as a function of time. The earthquake swarm is characterised by high-frequency volcanotectonic earthquakes as the host rock is fractured at the tips of the dike (partly non-double-couple earthquakes) and slip occurs on either side of the dike (mostly double-couple earthquakes generated by slip on existing fractures), both as a result of loading by the driving pressure of the dike (Figs. 1.13, 4.12 and 4.13).
3. If and when the dike propagates very close to the surface (Figs. 2.11 and 4.10), a volcanic **tremor** may start. The tremor is characterised by low-frequency earthquakes, primarily due to vibration of fluid (magma and gas)-filled fractures. Thus, the fluid-filled vibrating fractures generate low-frequency earthquakes in contrast to the high-frequency earthquakes generated as the rock fractures during dike propagation. The tremor normally continues until, eventually, the dike hits the surface to form a volcanic fissure.

Most earthquakes in volcanoes are small, that is, less than M3. However, this depends on the tectonic regime and the driving pressure of the magma and other factors. For example, in the 2014–2015 Bardarbunga–Holuhraun eruption in Iceland, some hundreds of earthquakes of M4–4.9 occurred during the dike propagation, and many tens of M5–5.7 events were associated with fault slips on the Bardarbunga caldera (Gudmundsson et al., 2014; Sigmundsson et al., 2015; Agustsdottir et al., 2016). Because most of the earthquakes triggered during unrest periods in a volcano are associated with the inflation of its shallow magma chamber – and the subsequent rupture or dike injection from the chamber – the unrest-related earthquakes tend to be shallow. More specifically, they are generally at depths of less than 10 km, and many occur at much shallower depths. However, again, the depth depends on the tectonic regime and the exact course of events. Thus, during the Bardarbunga 2014–2015 volcanotectonic episode there were earthquakes as deep as 25 km. Note, however, that the depth estimates of earthquakes are normally not very accurate. For example, for the Bardarbunga 2014–2015 earthquakes the depth estimates have error bars (uncertainty) of at least ± 1–2 km.

While seismic monitoring is one of the most useful methods of volcano monitoring and, in particular, for indicating dike or magma paths, the absence of earthquakes does not necessarily imply the absence of magma transport and crustal deformation. Some volcanoes apparently show **hardly any seismicity** immediately before the eruption. For example, no earthquakes were detected in the hours before the 1993 eruption of the Galeras Volcano in Columbia (McNutt, 2000). For some volcanoes, a lack of seismicity before eruption may indicate that the magma uses the same (already hot and plastic) path or conduit again and again.

Unless the same, still-hot **path is used again and again** for magma migration to the surface, one would normally expect considerable seismicity associated with magma-path formation through dike propagation. An exception to this rule is if the main magma source is very deep or, generally, below the seismogenic layer (essentially, the brittle part of the crust). Then, very little deformation and seismic activity would be associated with magma accumulation in the source prior to rupture and potential eruption (McNutt, 2005). Furthermore, although most volcanoes show considerable seismicity during inflation (expansion of the magma chamber), some show little seismicity during considerable inflation, including the Three Sisters Volcano in Oregon (in the USA), and the Westdahl Volcano and the Ugashik-Peulik Volcano, both in Alaska (USA) (McNutt, 2005). Also, some volcanoes show little seismic activity at the assumed depths of their magma sources (based primarily on petrological evidence) prior to eruptions, while generating seismicity at shallower depths (Roman and Cashman, 2018). This is to be expected if the magma source is located below the bottom of the seismogenic (the brittle) layer of the crust, if the strain rate is very low, or if the magma uses essentially the same path/conduit again and again at comparatively short intervals. Accurate estimates of the tops of the magma sources are also very difficult for some volcanoes.

The general rule, however, is that volcanic unrest periods, particularly with magma-chamber rupture and dike injection, result in **increased seismic activity**. This follows because when a dike (or an inclined sheet) is injected into the crust, the fluid pressure in the crust changes and tends to bring the surrounding host rock to failure. While some of the failure is through extension, much of it, particularly in the dike walls, is through brittle shear failure or slip (mainly on existing fractures), that is, faulting which commonly generates earthquakes. The study of volcanic earthquakes is therefore one of the most important volcano-monitoring techniques available and plays a large role in our attempts to forecast volcanic eruptions.

4.11 Summary

- Volcanic (volcanotectonic) earthquakes are those that occur inside or close to volcanoes. Most occur at comparatively shallow depths of less than 10 km. Their additional distinctive characteristics are that they commonly occur in swarms of many small and similar-magnitude earthquakes. The earthquakes provide information about the state of stress in the volcano and, following magma-chamber rupture, dike-propagation paths. The study of volcanic earthquakes is commonly referred to as volcano seismology.

- The main types of seismic waves are body waves and surface waves. Body waves travel along paths or rays that may be refracted or reflected at contacts between layers with different Young's moduli. The first body waves to arrive at seismic stations following an earthquake are the P-waves (primary waves), followed by the S-waves (secondary waves). P-waves have common crustal velocities of 2–7 km s^{-1} and travel through solids and fluids. S-waves have common crustal velocities of 2–4 km s^{-1} (50–60% of the corresponding P-wave velocities), but do not travel through fluids (only through solids). S-waves drop out on meeting totally fluid bodies in the crust, forming S-wave shadows that can be used to detect magma chambers. P- and S-waves are used to determine dynamic elastic moduli.

- Surface waves travel along the Earth's surface and generate most of the earthquake damage to human constructions. The main types are Love waves, which have generally a slightly greater velocity of the two types, and Rayleigh waves. But both waves travel at around 90% of the velocity of S-waves.

- The part of the fault where the earthquake begins is referred to as the hypocenter or focus. Its depth is the focal depth. The point at the Earth's surface right above the hypocentre is the epicentre. The focal mechanism or fault-plane solution gives the fault attitude (orientation) and slip direction at the hypocentre. More specifically, the focal mechanism provides information on the type of fault (normal, reverse, strike-slip) slipping during the earthquake.

- Earthquakes resulting from shear movement on a single fault plane are known as double-couple earthquakes; earthquakes not generated in this way are known as non-double-couple earthquakes. The axis of tension (T) gives the direction of the minimum compressive (maximum tensile) principal stress whereas the axis of compression (P) gives the direction of maximum compressive principal stress. For a two-dimensional stress field, T corresponds to σ_3 and P to σ_1. For each focal mechanism there are two possible fault planes (nodal planes): the real one (where the earthquake occurred) and the auxiliary plane at right angles to the real plane. The real fault plane must be determined from geological data (such as observed surface rupture, InSAR data, and GPS data).

- Of the two main earthquake size scales in use, one relates primarily to the damage of human-made constructions, the other to the energy released or transformed during the earthquake. The former is the Mercalli scale, with the range I–XII, where the intensity I is hardly felt by anyone and XII is total destruction. The energy scale began with the Richter local-magnitude scale (M_L) and subsequently the surface-wave (M_s) and the body-wave (m_b) scales. Today, the most commonly used scale is the moment-magnitude scale (M_w), which is directly related to the moment of an earthquake. The earthquake moment (M_0) may be presented in various ways but is commonly given as the product of the average slip on the fault plane (Δu_a), the total slip or rupture area (A), and the shear modulus

(G). There is, theoretically, no clear upper limit to the moment magnitude, but the largest earthquakes recorded so far (in Chile 1960) had a moment magnitude of 9.5.

- Earthquake faults can be modelled as mode II or mode III cracks, or a combination of various modes (mixed-mode cracks). These models can be used to calculate the elastic energy released or transformed during the earthquake. In any one seismogenic zone or area (over a sufficiently long period), as well as for the Earth as a whole, small earthquakes are much more common than large ones. This relationship between size and number is known as the Gutenberg–Richter frequency–magnitude relation and is one of the best known of all power laws in physical science. The fault sizes (rupture areas and fault lengths) also follow power-law size distributions, which is as expected since the earthquake magnitudes are positively related to the fault-rupture areas.

- Volcanic earthquakes belong to four main classes: (1) high-frequency or A-type earthquakes, (2) low-frequency or B-type earthquakes, (3) explosion earthquakes, and (4) volcanic tremors. A-type earthquakes relate to shear failure (faulting) but differ from ordinary tectonic earthquakes by the A-type sources commonly being related to magmatic pressure changes, the tendency of A-type earthquakes to occur in swarms, and the common high b-values (well above the value of 1.0, which is characteristic of tectonic earthquakes). An earthquake swarm is composed of earthquakes all occurring in a given location and all with a very similar magnitude – thereby lacking the foreshock–mainshock–aftershock sequences of typical tectonic earthquakes.

- The B-type earthquakes are characterised by long periods, that is, low frequency (commonly 2–3 Hz) and the lack of S-waves. Many are presumably directly related to fluid (magma and gas) pressure changes in fractures and conduits. Explosion earthquakes are primarily associated with volcanic explosions, occurring mostly in the shallow parts of the main conduit of the eruption.

- Volcanic tremors are continuous (or semi-continuous) ground vibrations or shaking with a duration from minutes to days (or longer). They are related to the migration of fluids (gas and magma) or, more specifically, to magma transport in fractures and boiling of fluids at shallow depths. Although many tremors have occurred without being followed by an eruption, the standard interpretation of a volcanic tremor is that it is an indication of an imminent eruption. In particular, tremors are commonly associated with magma movement at shallow depths in propagating potential feeder-dikes.

- Most eruptions are preceded by increased seismic activity. It follows that seismic monitoring is one of the main methods for assessing the likelihood of an eruption. Earthquake monitoring is always measured against an established background seismicity, namely the average number of earthquakes in the volcano when it is not subject to an unrest period. Often the earthquakes generated in a volcano can be used to monitor the propagation path of a dike. When the potential feeder-dike is very close to the surface, a volcanic tremor may begin, suggesting an imminent eruption.

4.12 Main Symbols Used

A total slip or rupture area associated with seismogenic faulting

a half strike-dimension (rupture length) of an interior crack; total strike dimension for an edge crack (e.g. a fault that reaches the surface and modelled as a mode II crack)

a parameter for earthquake-magnitude distribution; intersection of the straight line with the horizontal axis on the Gutenberg–Richter log–log plot

b parameter for earthquake-magnitude distribution; slope of the straight line on the Gutenberg–Richter log–log plot

E Young's modulus (modulus of elasticity)

G shear modulus (modulus of rigidity)

K bulk modulus (incompressibility)

k constant related to the geometry of a fault plane

L characteristic dimension of the earthquake slip surface

M earthquake magnitude

M_s surface-wave magnitude

M_w moment magnitude

M_0 earthquake moment

m_b body-wave magnitude

N cumulative number of earthquakes

q constant depending on fault geometry

U_t total radiated seismic energy

U_{II} energy released or transformed during faulting modelled as mode II crack slip

U_{III} energy released or transformed during faulting modelled as mode III crack slip

Δu_a average slip on a fault plane

V_p P-wave velocity

V_s S-wave velocity

α parameter for the number of earthquakes per year with a moment larger than 1 N m

β parameter for the seismic moment distribution, a scaling exponent

λ Lamé's constant

v Poisson's ratio

π 3.1415 . . . .

ρ material (solid or fluid) density

τ_d driving shear stress on a fault plane (for fault slip)

τ_f final shear stress (after earthquake) on a fault plane

τ_i initial shear stress (before earthquake) on a fault plane

4.13 Worked Examples

Example 4.1

Problem

The P-wave velocity in a 300-m-thick Holocene pahoehoe lava flow is 2.5 km s^{-1}, its density is 2400 kg m^{-3}, and its Poisson's ratio is 0.25. Calculate the dynamic Young's modulus of the lava flow. Based on the dynamic modulus, what values might the static modulus have and for which volcanotectonic processes should the static modulus be used?

Solution

Using Eq. (4.4) somewhat rewritten, we have:

$$E = \frac{\rho(1 + v)(1 - 2v)V_p^2}{(1 - v)}$$

The dynamic Young's modulus is then:

$$E = \frac{2400 \text{ kg m}^{-3} \times (1 + 0.25) \times (1 - 0.5) \times (2500 \text{ m s}^{-1})^2}{(1 - 0.25)} = 1.25 \times 10^{10} \text{ Pa}$$
$$= 12.5 \text{ GPa}$$

The dynamic Young's modulus can be anywhere between 1 and 13 times the static modulus (Gudmundsson, 2011). Commonly, however, the dynamic modulus is taken as being twice the static modulus, in which case the static modulus here would be around 6.3 GPa. This is a reasonable value and similar to that obtained from studies of Holocene basaltic lava flows in Iceland.

For essentially all volcanotectonic processes, except seismogenic faulting, the static Young's modulus should be used. As indicated by Eq. (4.4) the dynamic Young's modulus relates to the velocities of the P-waves, which are of the order of 2–7 km s^{-1} in the crust. Seismogenic ruptures propagate at velocities similar to those of S-waves in the crust, or 2–4 km s^{-1}, so for modelling these the dynamic Young's modulus should be used. Nearly all other volcanotectonic processes are slower by many orders of magnitude, and for these the static modulus should be used. These latter include dike propagation and inflation–deflation episodes during volcanic unrest.

Example 4.2

Problem

A strike-slip seismogenic fault slip has the surface rupture or strike-dimension of 16 km and a height (width in seismology) or dip-dimension of 10 km. The measured average fault slip is 1.2 m. If the average dynamic Young's modulus of the uppermost 10 km of the crust is 80 GPa and the Poisson's ratio is 0.25, calculate:

(a) The seismic moment.
(b) The moment magnitude.

Solution

(a) From Eq. (4.7) the seismic moment M_0 is given by:

$$M_0 = \Delta u_a A G$$

The seismogenic slip and the rupture area are both given, but in order to use the formula we must first find the shear modulus G. From Eq. (3.19), using the above values for Young's modulus and Poisson's ratio, the shear modulus is:

$$G = \frac{E}{2(1+v)} = \frac{8 \times 10^{10} \text{ Pa}}{2(1+0.25)} = 3.2 \times 10^{10} \text{ Pa} = 32 \text{ GPa}$$

For a seismogenic rupture with a strike-dimension of 16 km and a dip-dimension of 10 km, the rupture or slip-surface area, assuming a rectangular fault geometry, is $A = 10 \text{ km} \times 16 \text{ km} = 160 \text{ km}^2$, or $1.6 \times 10^8 \text{ m}^2$. Since the average slip $\Delta u_a = 1.2$ m, we obtain the seismic moment from Eq. (4.7) as:

$$M_0 = \Delta u_a A G = 1.2 \text{ m} \times 1.6 \times 10^8 \text{m}^2 \times 3.2 \times 10^{10} \text{ Pa} = 6.1 \times 10^{18} \text{ N m}$$

(b) The moment magnitude of the earthquake can be obtained from Eq. (4.15), thus:

$$M_w = \frac{2}{3} \log M_0 - 6.0 = \frac{2}{3} \log(6.1 \times 10^{18} \text{ N m}) - 6.0 = 6.5$$

Thus, the moment magnitude is about 6.5.

All these results, for the dimensions, slips, seismic moment, and the moment magnitude are similar to those obtained for the two June 2000 strike-slip earthquakes in the South Iceland Seismic Zone as calculated by the United States Geological Survey.

Example 4.3

Problem

The largest-magnitude earthquake ever recorded by modern instruments is the 1960 Chile earthquake, commonly referred to as the Valdivia earthquake. The surface rupture or strike-dimension is generally estimated at around 800 km, although some estimates are larger. The dip-dimension (width in seismology) is estimated at about 200 km. The maximum slip is as much as 30 m, but the average is thought to be about 24 m (Kanamori and Cipar, 1974; Barrientos and Ward, 1990). Use a typical dynamic Young's modulus of 100 GPa and a Poisson's ratio of 0.25 for the host rock and calculate for this earthquake:

(a) The seismic moment.
(b) The moment magnitude.

Solution

(a) We first have to calculate the shear modulus based on the available information about the Young's modulus, 100 GPa, and Poisson's ratio, 0.25. From Eq. (3.19) the shear modulus is:

$$G = \frac{E}{2(1+v)} = \frac{1 \times 10^{11} \text{ Pa}}{2(1+0.25)} = 4 \times 10^{10} \text{ Pa} = 40 \text{ GPa}$$

Assuming a rectangular fault, the rupture area is 800 km $\times$ 200 km = 1.6×10^5 km^2 = 1.6×10^{11} m^2. Using this and the shear modulus and the average slip given above, from Eq. (4.7) the seismic moment is:

$$M_0 = \Delta u_a A G = 24 \text{ m} \times 1.6 \times 10^{11} \text{m}^2 \times 4 \times 10^{10} \text{ Pa} = 1.5 \times 10^{23} \text{ N m}$$

(b) The moment magnitude of the earthquake follows from Eq. (4.15), namely:

$$M_w = \frac{2}{3} \log M_0 - 6.0 = \frac{2}{3} \log(1.5 \times 10^{23} \text{ N m}) - 6.0 = 9.5$$

The results in (a) and (b) are both in good agreement with other estimates for this earthquake. Thus, the seismic moment is estimated as being in the range 0.7–2×10^{23} N m, while the moment magnitude is generally considered as about 9.5 (Kanamori and Cipar, 1974; Kanamori and Anderson, 1975; Barrientos and Ward, 1990; Fujii and Satake, 2013).

Example 4.4

Problem

Calculate the energy released or transformed during the 1960 Chile earthquake in Example 4.3 using (a) the maximum estimated fault slip of 30 m, and (b) an assumed typical stress drop of 5 MPa. In both cases, use a mode II crack model and use the same elastic properties of the crust as before, namely a Young's modulus of 100 GPa and Poisson's ratio of 0.25.

Solution

(a) Using the above values, the released or transformed elastic energy U_{II} in terms of maximum slip Δu is obtained from the plane-stress version of Eq. (4.12), namely:

$$U_{II} = \frac{E \Delta u_{II}^2 \pi A}{16a} \tag{4.21}$$

Since we are modelling a dip-slip fault, the parameter a here denotes the total dip-dimension (width) of the fault, that is, 200 km. Using these values, and those above, we obtain the elastic energy transformed or released during the earthquake as:

$$U_{II} = \frac{E \Delta u_{II}^2 \pi A}{16a} = \frac{1 \times 10^{11} \text{ Pa} \times (30 \text{ m})^2 \times 3.1416 \times 1.6 \times 10^{11} \text{ m}^2}{16 \times 2 \times 10^5 \text{ m}} = 1.4 \times 10^{19} \text{ J}$$

(b) For a typical stress drop of 5 MPa, we obtain the elastic energy transformed or released through the plane-stress version of Eq. (4.11), namely:

$$U_{II} = \frac{\tau_d^2 \pi a A}{E} \tag{4.22}$$

from which the energy released during the earthquake becomes:

$$U_{II} = \frac{\tau_d^2 \pi a A}{E} = \frac{(5 \times 10^6 \text{ Pa})^2 \times 3.1416 \times 2 \times 10^5 \text{ m} \times 1.6 \times 10^{11} \text{ m}^2}{1 \times 10^{11} \text{ Pa}}$$
$$= 2.5 \times 10^{19} \text{ J}$$

We have here used the plane-stress equations. However, it could also be argued that the plain-strain equations should be used. This follows because the dip-dimension is used for a so that it is the vertical dimension that is important in these calculations (Gudmundsson, 2011). If the plane-strain equations (Eqs. 4.11 and 4.12) were used, the results in (a) would be by a factor of $1/(1 - v^2) = 1.07$ higher, whereas those in (b) would be by a factor of 1.07 lower. The difference between the plane-strain and plane-stress results in these calculations is thus so small as to be of little concern.

The results show that the elastic energy that was transformed or released in this large earthquake was of the order of 10^{19} J, which is of the same order of magnitude as the elastic energy transformed or released in the largest effusive and explosive eruptions (cf. Gudmundsson, 2014).

Example 4.5

Problem

Repeat Example 4.4 but now using a mode III crack model for the earthquake rupture rather than mode II. Discuss briefly which dimension of the earthquake rupture should be regarded as the controlling dimension, the dip-dimension or the strike-dimension.

Solution

Using first the maximum displacement, we obtain the elastic energy released during the earthquake from Eq. (4.14) as follows:

$$U_{III} = \frac{E \Delta u_{III}^2 \pi A}{16(1 + v)a} = \frac{1 \times 10^{11} \text{ Pa} \times (30 \text{ m})^2 \times 3.1416 \times 1.6 \times 10^{11} \text{ m}^2}{16 \times (1 + 0.25) \times 2 \times 10^5 \text{ m}} = 1.1 \times 10^{19} \text{ J}$$

if we use the dip-dimension as the controlling dimension. For a dip-slip fault modelled as a mode III crack, however, the strike-dimension may sometimes be regarded as the controlling dimension, in which case a would be half the strike-dimension of the fault, or 400 km instead of the dip-dimension (200 km). In the latter case, we obtain the released energy as $U_{III} = 5.5 \times 10^{18}$ J.

When half the strike-dimension is used, it is normally assumed, however, that the bottom of the seismogenic layer is fluid – usually a magma reservoir (Gudmundsson, 2011). For the location of the 1960 Chile earthquake that condition is probably not satisfied. Even if the mantle below the lithosphere was partially molten, it would be most unlikely to be totally molten. Thus, in this case, either dimension may be regarded as the controlling one.

Then, using the driving shear stress or, roughly, the stress drop of 5 MPa, and for the same values as in Example 4.4, from Eq. (4.13) we obtain the elastic energy transformed as:

$$U_{III} = \frac{\tau_d^2(1+v)\pi aA}{E}$$
$$= \frac{(5 \times 10^6 \text{ Pa})^2 \times (1+0.25) \times 3.1416 \times 2 \times 10^5 \text{ m} \times 1.6 \times 10^{11} \text{ m}^2}{1 \times 10^{11} \text{ Pa}}$$
$$= 3.1 \times 10^{19} \text{ J}$$

Here, again, we have used the dip-dimension for a. If half the strike-dimension were used for a, we would obtain the released energy as $U_{III} = 6.3 \times 10^{19}$ J.

In all cases, the elastic energy transformed during the earthquake is of the order of 10^{19} J, or close to that value, and thus similar to the elastic energy released in the largest known eruptions on Earth (Gudmundsson, 2014).

Example 4.6

Problem

Consider a thrust-fault earthquake with a rupture or slip surface strike-dimension of 120 km and dip-dimension of 80 km and an average slip of 2 m. If the average shear modulus of the host rock is 30 GPa, calculate the moment magnitude of the earthquake.

Solution

To find the moment magnitude, we first have to calculate the seismic moment. From the given strike- and dip-dimensions, and assuming a rectangular fault, the rupture area is $A = 9.6 \times 10^9$ m^2. Using the given shear modulus and average slip, from Eq. (4.7) the moment M_0 is:

$$M_0 = \Delta u_a AG = 2 \text{ m} \times 9.6 \times 10^9 \text{ m}^2 \times 3 \times 10^{10} \text{ Pa} = 5.8 \times 10^{20} \text{ N m}$$

The moment magnitude of the earthquake can then be calculated from Eq. (4.15) as:

$$M_w = \frac{2}{3}\log M_0 - 6.0 = \frac{2}{3}\log(5.8 \times 10^{20} \text{N m}) - 6.0 = 7.8$$

All the values given here, as well as the calculated moment magnitude, are very similar to those obtained by the US Geological Survey for the April 2015 Gorkha earthquake in Nepal.

Example 4.7

Problem

For a given seismic zone, the b-value is 1.0 and the zone has, on average, one M6 earthquake every 20 years. Based on this information and the Gutenberg–Richter relation, how common would M7 earthquakes be in the zone?

Solution

We use Eq. (4.16), which has the form:

$$\log N(\geq M) = a - bM$$

It is given that the slope of the line, that is, b on the log–log Gutenberg–Richter plot is 1.0 and that there is one M6 earthquake every 20 years. Since there is one M6 or larger earthquake every 20 years, it means that the cumulative number of earthquakes N larger or equal to M6 per year, on average, is $1/20 = 0.05$. Thus $N(\geq M) = 0.05$. Putting these values for N and b in Eq. (4.16) we get:

$$\log(0.05) = a - (1 \times 6) = -1.3$$

It follows that:

$$a - 6 = -1.3$$

So that:

$$a = 4.7$$

Using these values of a and b in Eq. (4.16), we get the cumulative number of M7 or larger earthquakes each year in this seismic zone as:

$$\log N(\geq 7) = 4.7 - (1 \times 7) = -2.3$$

Taking the antilog (here raising the number -2.3 to a power of 10) we get the number of M7 or larger earthquakes per year in this seismic zone as:

$$N(\geq 7) = 10^{-2.3} = 0.005$$

Thus, on average, there will be 0.005 M7 or larger earthquakes per year in this zone. Translated into a more understandable time framework, this means that there will be, on average, one M7 or larger earthquake in the zone every $1/0.005 = 200$ years.

Example 4.8

Problem

In a given volcanic zone, the b-value is 2.0 and the zone has, on average, 20 earthquakes of M4 or larger each year. Calculate (a) the number of earthquakes of M5 and larger and (b) of M8 and larger per year in the zone. How realistic is the estimate of the number of M8 or larger earthquakes?

Solution

(a) We follow the same procedure as in Example 4.7. Using the values above, Eq. (4.16) gives:

$$\log(20) = a - (2 \times 4) = 1.3$$

It follows that $a = 9.3$

The cumulative number of M5 or larger earthquakes each year in the volcanic zone is therefore:

$$\log N(\geq 5) = 9.3 - (2 \times 5) = -0.7$$

The number of M5 or larger earthquakes per year in the volcanic zone is thus:

$$N(\geq 5) = 10^{-0.7} = 0.2$$

So about 0.2 earthquakes of M5 or larger would occur in the volcanic zone, on average, each year. This means that there will, on average, be one M5 or larger earthquake in the zone every $1/0.2 = 5$ years.

(b) Using the values above, Eq. (4.16) gives the cumulative number of M8 or larger earthquakes each year as follows. We have:

$$\log N(\geq 8) = 9.3 - (2 \times 8) = -6.7$$

So that the number of M8 or larger earthquakes per year in the volcanic zone is:

$$N(\geq 8) = 10^{-6.7} = 2 \times 10^{-7}$$

So about one earthquake of M8 or larger would occur in the volcanic zone every 5 million years. Since a volcanic zone would commonly be active for only 1–2 million years, or even less, the results essentially mean that there is hardly ever going to be a M8 or larger earthquake in the volcanic zone. Indeed, volcanoes and volcanic zones very rarely yield earthquakes larger than M7. This is partly because the zones are simply not large enough to generate great (>M8) earthquakes. The seismic moment, and thus the moment magnitude, depends on the size of the rupture surface, and for a very large rupture surface to occur a very large active zone or area is needed.

This result also shows that the Gutenberg–Richter relation, like other power laws, is valid only for a certain range for a given area or zone. More specifically, the power-law relation, the straight-line slope on the log–log plot, holds over a certain range of values but not outside that range. In the extreme values of both the upper and lower range most power laws no longer hold. Thus, extrapolation of the Gutenberg–Richter relation to very high or very low magnitudes for given seismic zones or areas is normally not warranted.

4.14 Exercises

4.1 What is an earthquake? Provide a definition and explain what is meant by a volcanic earthquake.

4.2 Describe and define primary and secondary waves. What are their typical velocities in the crust?

4.3 Which seismic waves can be used to detect magma chambers and how?

4.4 Describe and define Love waves and Rayleigh waves. What are their typical velocities in the crust and basic difference as regards paths from primary and secondary waves?

4.5 Define and describe the concepts epicentre, hypocentre, and focal depth.

4.6 Define and explain the concepts focal mechanism, double couple, and non-double couple.

4.7 Define and illustrate a beach-ball diagram or symbol and a focal sphere.

4.8 Define the concepts axis of tension and axis of compression and explain how these are used in seismogenic fault analysis and general stress analysis.

4.9 What are nodal planes? How can we determine the location of the real fault plane associated with an earthquake?

4.10 What are the two main scales used to determine earthquake sizes?

4.11 Describe the Mercalli intensity scale, its range, and its use and limitations.

4.12 Name and describe briefly the main earthquake-magnitude (energy-release) scales used in seismology. Which one is the most commonly used today?

4.13 What is seismic moment? Provide an equation, explain all the symbols, and provide their units.

4.14 Define driving shear stress for seismogenic faulting. Indicate the typical magnitude range of this stress.

4.15 Define and explain the characteristic dimensions of an earthquake rupture.

4.16 Provide an equation for the moment magnitude and explain all the symbols and their units.

4.17 Define the Gutenberg–Richter frequency–magnitude relation. What kind of probability or frequency distribution does the relation represent? Provide the appropriate formula for the relation and explain all the symbols and give their units.

4.18 Which are the four main groups of volcanic earthquakes?

4.19 Describe A-type earthquakes and compare and contrast with ordinary tectonic earthquakes.

4.20 What are earthquake swarms? How do they differ from typical tectonic earthquakes with foreshock–mainshock–aftershock sequences.

4.21 Explain why b-values in volcanotectonic earthquakes are commonly significantly higher than those of ordinary tectonic earthquakes.

4.22 Describe B-type earthquakes in terms of stress sources and in relation to other types of earthquakes.

4.23 What are explosion earthquakes?

4.24 Define and explain volcanic tremors and their interpretations.

4.25 Discuss how seismic monitoring improves our understanding of volcanotectonic processes and can contribute to the forecasting of volcanic eruptions.

4.26 A crustal layer has the average density of 2600 kg m^{-3}, Poisson's ratio of 0.25, and a P-wave velocity of 3 km s^{-1}. Calculate (a) the dynamic Young's modulus of the layer, and (b) the S-wave velocity of the layer.

4.27 The average fault slip of 3 m on a strike-slip fault produces a surface rupture length (strike-dimension) of 30 km and a rupture height (width) or dip-dimension of 20 km. The dynamic Young's modulus of the seismogenic layer is 90 GPa and its average Poisson's ratio is 0.25. Calculate (a) the seismic moment, and (b) the moment magnitude.

4.28 Consider an average slip of 20 m on a thrust fault. The rupture length or strike-dimension is 500 km and the rupture width or dip-dimension is 200 km. Using a Poisson's ratio of 0.25 and a Young's modulus of 100 GPa, calculate (a) the seismic moment, and (b) the moment magnitude of the earthquake. How large would this earthquake be in comparison with other instrumentally determined earthquakes?

4.29 Use the mode III crack model to calculate the elastic energy transformed or released during the fault slip in Exercise 4.27.

4.30 A given seismic zone produces, on average, one M5 earthquake every year. Based on this information and the Gutenberg–Richter relation, calculate the average number of M7 earthquakes in the zone.

References and Suggested Reading

Advani, S. H., Lee, T. S., Dean, R. H., Pak, C. K., Avasthi, J. M., 1997. Consequences of fluid lag in three-dimensional hydraulic fractures. *International Journal for Numerical and Analytical Methods in Geomechanics*, **21**, 229–240.

Agustsdottir, T., Woods, J., Greenfield, T., et al., 2016. Strike-slip faulting during the 2014 Bardarbunga–Holuhraun dike intrusion, central Iceland. *Geophysical Research Letters*, **43**, 1495–1503.

Aki, K., Richards, P.G., 2009. *Quantitative Seismology*, 2nd edn. Herndon, VA: University Science Books.

Al Amri, A., Fnais, M., Abdel-Rahma, K., Mogren, S., Al-Dabbagh, M., 2012. Geochronological dating and stratigraphic sequences of Harrat Lunayyir, NW Saudi Arabia. *International Journal of Physical Sciences*, **7**, 2791–2805.

Al Shehri, A., Gudmundsson, A., 2018. Modelling of surface stresses and fracturing during dyke emplacement: application to the 2009 episode at Harrat Lunayyir, Saudi Arabia. *Journal of Volcanology and Geothermal Research*, **356**, 278–303.

Andrew, R. E. B., Gudmundsson, A., 2008. Volcanoes as elastic inclusions: their effects on the propagation of dykes, volcanic fissures, and volcanic zones in Iceland. *Journal of Volcanology and Geothermal Research*, **177**, 1045–1054.

Barrientos, S. E., Ward, S. N., 1990. The 1960 Chile earthquake: inversion for slip distribution from surface deformation. *Geophysical Journal International*, **103**, 589–598.

Becerril, L., Galindo, I., Gudmundsson, A., Morales, J. M., 2013. Depth of origin of magma in eruptions. *Scientific Reports* **3**, 2762, doi:10.1038/srep02762.

Biggs, J., Ebmeier, S. K., Aspinall, W. P., et al., 2014. Global link between deformation and volcanic eruption quantified by satellite imagery. *Nature Communications*, **5**, doi:10.1038/ncomms4471.

Biggs, J., Robertson, E., Cashman, K., 2016. The lateral extent of volcanic interactions during unrest and eruption. *Nature Geoscience*, **9**, 308–311.

Bonaccorso, A., Aoki, Y., Rivalta, E., 2017. Dike propagation energy balance from deformation modeling and seismic release. *Geophysical Research Letters*, **44**, 5486–5494.

Brandsdottir, B., Einarsson, P., 1979. Seismic activity associated with the September 1977 deflation of the Krafla central volcano in NE Iceland. *Journal of Volcanology and Geothermal Research*, **6**, 197–212.

Bunger, A., 2009. *Near-Surface Hydraulic Fracture: Laboratory Experimentation and Modeling of Shallow Hydraulic Fracture Growth*. Saarbrucken: Lambert Academic Publishing.

Carracedo, J. C., Troll, V. R., 2006. Seismicity and gas emissions on Tenerife: a real cause for alarm? *Geology Today*, **22**, 138–141.

Davis, R. J., Mathias, S. A., Moss, J., Hustoft, S., Newport, L. 2012. Hydraulic fractures: how far will they go? *Marine and Petroleum Geology*, **37**, 1–6.

Elshaafi, A., Gudmundsson, A., 2018. Mechanical interaction between volcanic systems in Libya. *Tectonophysics*, **722**, 549–565.

Feigl, K. L., Gasperi, J., Sigmundsson, F., Rigo, A. 2000. Crustal deformation near Hengill Volcano, Iceland 1993–1998: coupling between magmatic activity and faulting inferred from elastic modeling of satellite radar interferograms. *Journal of Geophysical Research* **105**, 25 655–25 670.

Fisher, K., 2014. Hydraulic fracture growth: real data. Presentation given at GTW-AAPG/STGS Eagle Ford plus Adjacent Plays and Extensions Workshop, San Antonio, Texas, February 24–26, 2014.

Flewelling, S. A., Tymchak, M. P., Warpinski, N., 2013. Hydraulic fracture height limits and fault interactions in tight oil and gas formations. *Geophysical Research Letters*, **40**, 3602–3606.

Fujii, Y., Satake, K., 2013. Slip distribution and seismic moment of the 2010 and 1960 Chilean earthquakes inferred from tsunami waveforms and coastal geodetic data. *Pure and Applied Geophysics*, **170**, 1493–1500.

Garagash, D., Detournay, E., 2000. The tip region of a fluid-driven fracture in an elastic medium. *Journal of Applied Mechanics*, **67**, 183–192.

Garcia, A., Ortiz, R., Marrero, J. M., et al., 2006. Monitoring the reawakening of Canary Islands' Teide Volcano. *Eos*, **87**, 61–72.

Gottsmann, J., Wooller, L., Marti, J., et al., 2006. New evidence for the reawakening of Teide Volcano. *Geophysical Research Letters*, **33**, L20311, doi:10.1029/2006GL027523.

Gudmundsson, A. 1983. Stress estimates from the length/width ratios of fractures. *Journal of Structural Geology*, **5**, 623–626.

Gudmundsson, A. 1986. Formation of dykes, feeder-dykes and the intrusion of dykes from magma chambers. *Bulletin of Volcanology*, **47**, 537–550.

Gudmundsson, A., 2011. *Rock Fractures in Geological Processes*. Cambridge: Cambridge University Press.

Gudmundsson, A., 2014. Energy release in great earthquakes and eruptions. *Frontiers in Earth Science*, **2**, doi:10.3389/feart.2014.00010.

Gudmundsson, A., 2017. *The Glorious Geology of Iceland's Golden Circle*. Berlin: Springer Verlag.

Gudmundsson, A., Andrew, R. E. B., 2007. Mechanical interaction between active volcanoes in Iceland. *Geophysical Research Letters*, **34**, doi:10.1029/2007GL029873.

Gudmundsson, A., Brenner, S. L., 2003. Loading of a seismic zone to failure deforms nearby volcanoes: a new earthquake precursor. *Terra Nova*, **15**, 187–193.

Gudmundsson, A., Homberg, C., 1999. Evolution of stress fields and faulting in seismic zones. *Pure and Applied Geophysics*, **154**, 257–280.

Gudmundsson, A., Oskarsson, N., Gronvold, K., et al., 1992. The 1991 eruption of Hekla, Iceland. *Bulletin of Volcanology*, **54**, 238–246.

Gudmundsson, M.T., Thordarson, T., Hoskuldsson, A., et al., 2012. Ash generation and distribution from the April-May 2010 eruption of Eyjafjallajökull, Iceland. *Scientific Reports*, **2**, doi:10.1038/srep00572.

Gudmundsson, A., De Guidi, G., Scudero, S., 2013. Length–displacement scaling and fault growth. *Tectonophysics*, **608**, 1298–1309.

Gudmundsson, A., Lecoeur, N., Mohajeri, N., Thordarson, T., 2014. Dike emplacement at Bardarbunga, Iceland, induces unusual stress changes, caldera deformation, and earthquakes. *Bulletin of Volcanology*, **76**, 869, doi:10.1007/s00445-014-0869-8.

Hill, D. P., Pollitz, F., Newhall, C., 2002. Earthquake-volcano interactions. *Physics Today*, **55**, 41–47.

Jakobsdottir, S., 2008. Seismicity in Iceland: 1994–2007. *Jokull, Icelandic Journal of Earth Sciences*, **58**, 75–100.

Kagan, Y. Y., 2014. *Earthquakes: Models, Statistics, Testable Forecasts*. Oxford: Wiley-Blackwell.

Kanamori, H., 1977. The energy release in great earthquakes. *Journal of Geophysical Research*, **82**, 2981–2987.

Kanamori, H., Anderson, D. L., 1975. Theoretical basis of some empirical relations in seismology. *Bulletin of Seismological Society of America*, **65**, 1074–1095.

Kanamori, H., Brodsky, E. E., 2004. The physics of earthquakes. *Reports on Progress in Physics*, **67**, 1429–1496.

Kanamori, H., Cipar, J., 1974. Focal process of the great Chilean earthquake May 22, 1960. *Physics of the Earth and Planetary Interiors*, **9**, 128–136.

Lopez-Comino, J. A., Cesca, S., Heimann, S., et al., 2017. Characterization of hydraulic fractures growth during the Äspö Hard Rock Laboratory experiment (Sweden). *Journal of Rock Mechanics and Rock Engineering*, **50**, 2985–3001.

Madariaga, R., 1979. On the relation between seismic moment and stress drop in the presence of stress and strength heterogeneity. *Journal of Geophysical Research*, **84**, 2243–2250.

Manga, M., Brodsky, E., 2006. Seismic triggering of eruptions in the far field: volcanoes and geysers. *Annual Review of Earth and Planetary Sciences*, **34**, 263–291.

Marti, J., Geyer, A., Folch, A., Gottsmann, J., 2008. A review of collapse caldera modelling. In Gottsmann, J. and Marti, J. (eds), *Caldera Volcanism: Analysis, Modelling and Response*. Amsterdam: Elsevier, pp. 233–283.

Massa, B., D'Auria, L., Cristiano, E., De Matteo, A., 2016. Determining the stress field in active volcanoes using focal mechanisms. *Frontiers of Earth Science*, **4**, doi:10.3389/feart.2016.00103.

Mathieu, L., De Vries, B. W., Pilato, M., Troll, V. R., 2011. The interaction between volcanoes and strike-slip, transtensional and transpressional fault zones: analogue models and natural examples. *Journal of Structural Geology*, **33**, 898–906.

Mavko, G., Mukerji, T., Dvorkin, J., 2009. *The Rock Physics Handbook: Tools for Seismic Analysis of Porous Media*. Cambridge: Cambridge University Press.

McNutt, S. R., 2000. Volcanic seismicity. In Sigurdsson, H., Houghton, B. F., McNutt, S. R., Rymer, H., Stix, J. (eds.), *Encylopedia of Volcanoes*. New York, NY: Academic Press, pp. 1015–1033.

McNutt, S. R., 2005. A review of volcanic seismology. *Annual Reviews of Earth and Planetary Sciences*, **33**, 461–491.

Murase, T., McBirney, A. R., 1973. Properties of some common igneous rocks and their melts at high temperatures. *Geological Society of America Bulletin*, **84**, 3563–3592.

Namiki, A., Rivalta, E., Woith, H., Walter, T. R., 2016. Sloshing of a bubbly magma reservoir as a mechanism of triggered eruptions. *Journal of Volcanology and Geothermal Research*, **320**, 156–171.

Ohnaka, M., 2013. *The Physics of Rock Failure and Earthquakes*. Cambridge: Cambridge University Press.

Pallister, J., McCausland, W., Jónsson, S., et al., 2010. Broad accommodation of rift-related extension recorded by dyke intrusion in Saudi Arabia. *Nature Geoscience*, **3**, 705–712.

Passarelli, L., Rivalta, E., Cesca, S., Aoki, Y., 2015. Stress changes, focal mechanisms, and earthquake scaling laws for the 2000 dike at Miyakejima (Japan). *Journal of Geophysical Research*, **120**, 4130–4145.

Peng, Z., Gomberg, J., 2010. An integrated perspective of the continuum between earthquakes and slow-slip phenomena. *Nature Geoscience*, **3**, 599–607.

Reiter, L., 1990. *Earthquake Hazard Analysis: Issues and Insight*. New York, NY: Columbia University Press.

Ritchie, D., Gates, A. E., 2001. *Encyclopedia of Earthquakes and Volcanoes*. New York, NY: Facts on File.

Roman, D. C., Cashman, K. V., 2018. Top-down precursory volcanic seismicity: implications for 'stealth' magma ascent and long-term eruption forecasting. *Frontiers in Earth Science*, **6**, doi:10.3389/feart.2018.00124.

Scholz, C. H., 1990. *The Mechanics of Earthquakes and Faulting*. Cambridge: Cambridge University Press.

Secor, D. T., 1965. Role of fluid pressure in jointing. *American Journal of Science*, **263**, 633–646.

Segall, P., 2010. *Earthquake and Volcano Deformation*. Princeton: Princeton University Press.

Sigmundsson, F., Hreinsdottir, S., Hooper, A., et al., 2010. Intrusion triggering of the 2010 Eyjafjallajökull explosive eruption. *Nature*, **468**, 426–430.

Sigmundsson, F., Hooper, A., Hreinsdottir, S., et al., 2015. Segmented lateral dyke growth in a rifting event at Bardarbunga Volcanic System, Iceland. *Nature*, **517**, 191–195.

Sigurdsson, H., Houghton, B. F., McNutt, S. R., Rymer, H., Stix, J. (eds.), 2000. *Encyclopedia of Volcanoes*. New York, NY: Academic Press.

Stacey, F. D., Davis, P. M., 2008. *Physics of the Earth*, 4th edn. Cambridge: Cambridge University Press.

Stein, S., Wysession, M., 2002. *An Introduction to Seismology, Earthquakes and Earth Structure*. Oxford: Wiley-Blackwell.

Tarasewicz, J., White, R. S., Woods, A. W., Brandsdottir, B., Gudmundsson, M. T., 2012. Magma mobilization by downward-propagating decompression of the Eyjafjallajökull volcanic plumbing system. *Geophysical Research Letters*, **39**, doi:10.1029/2012GL053518.

Udias, A., Madariaga, R., Buforn, E., 2014. *Source Mechanisms of Earthquakes: Theory and Practice*. Cambridge: Cambridge University Press.

Vogfjord, K., Hjaltadottir, S., Slunga, R., 2005. Volcano-tectonic interaction in the Hengill region, Iceland, during 1993–1998. *Geophysical Research Abstracts*, **7**, 09947.

Warpinski, H. R. 1985. Measurement of width and pressure in a propagating hydraulic fracture. *Journal of the Society of Petroleum Engineers*, February, 46–54.

Watt, S. F. L., Pyle, D. M., Mather, T. A., 2009. The influence of great earthquakes on volcanic eruption rate along the Chilean subduction zone. *Earth and Planetary Science Letters*, **277**, 399–407.

Xu, W., Jonsson, S., Corbi, F., Rivalta, E., 2016. Graben formation and dike arrest during the 2009 Harrat Lunayyir dike intrusion in Saudi Arabia: insights from InSAR, stress calculations and analog experiments. *Journal of Geophysical Research*, **121**, doi:10.1002/2015JB012505.

Yew, C. H., 2013. *Mechanics of Hydraulic Fracturing*. Houston, TX: Gulf Professional Publishing.

Zobin, V. M., 2003. *Introduction to Volcanic Seismology*. London: Elsevier.

5 Volcanotectonic Processes

5.1 Aims

One principal aim of volcanotectonic studies is to provide a theoretical framework that makes it possible to make reliable deterministic or probabilistic forecasts of volcanotectonic events. These events, in turn, depend on volcanotectonic processes inside the volcanoes. In the previous chapters we have discussed some of the main observational aspects of volcanotectonics, both geological and geophysical, and defined several of the basic concepts. In order to bring into focus those field observations that are useful for understanding the main processes leading to eruptions, vertical or lateral collapses, and related events, we need to know the basic physics that controls the processes. Here we provide an overview of some principal processes that control volcanotectonic events, emphasising elementary physics, particularly mechanics, and the quantitative aspects of volcanotectonics. Many of these processes are elaborated in later chapters. The primary aims of this chapter are to outline and illustrate, by examples, the basic principles of:

- Magma-chamber initiation.
- Magma-chamber rupture and dike initiation.
- Dike propagation, deflection, and arrest.
- Emplacement of sills.
- Emplacement of laccoliths and large plutons.
- Vertical collapses (collapse-caldera formation).
- Lateral/sector collapses (landslide formation).

5.2 Magma-Chamber Initiation

We recall that a magma chamber is the principal reason for the formation of a polygenetic volcano (cf. Chapter 9). But how do the magma chambers themselves form? What are their geometries or shapes, and how do they develop those geometries? These and related questions are of fundamental importance for understanding polygenetic volcanoes and their behaviour. In order to address them, let us first define and clarify some of the important concepts.

By magma chamber we mean a **crustal magma body**, capable of supplying magma to **intrusions** (dikes, sills, and inclined sheets) and **eruptions**. We distinguish between shallow crustal **magma chambers**, as are discussed here, and deep-seated **magma reservoirs** (Chapter 6). The latter are normally much larger and located at much greater depths than the shallow magma chambers. The shallow chambers are commonly at 1–5 km depth below the surface. These depths refer to that of the uppermost contact between the fluid (magma) and the host rock, the roof, below the **average regional elevation** of the area or crustal segment within which the chamber is located. Thus, the chamber depths below the tops of the associated polygenetic volcanoes themselves would normally be greater, particularly when the volcanoes form large edifices that rise one or more kilometres above their surroundings. Thus, shallow chambers may occasionally be as deep as 7–8 km below the tops of tall volcanic edifices, but in this book we normally regard the maximum depth of the roof of a shallow chamber as **5–6 km** (cf. Fig. 6.17; Chapter 6) By contrast, deep-seated reservoirs are generally at depths in excess of 10 km (except at mid-ocean ridges where they may be as shallow as 6–7 km) and commonly at **15–30 km** depth or deeper. The chambers are located in the **upper crust**, whereas the reservoirs are located in **the lower crust or upper mantle**. Here we shall review the main aspects of magma-chamber initiation while the formation of reservoirs, and the chamber-reservoir interactions, are discussed in Chapters 6 and 8.

Many shallow magma chambers are located in areas of extension, such as divergent plate boundaries and various types of rift zones. The most common type of intrusion in areas undergoing extension is the vertical dike (Figs. 2.1 and 2.2). This follows because in areas of extension one of the horizontal principal stresses is the minimum compressive principal (maximum tensile) stress σ_3 while the vertical stress is the maximum compressive principal stress σ_1. Such a stress field favours the formation of extension fractures and, in the vicinity of magma, the emplacement of vertical magma-driven fractures, namely dikes. The dike intensity can be very high in rift zones; witness the ophiolites and the ocean ridges where dikes are close to **100% of the rock** in the so-called **sheeted dike complexes**. Similar swarms of **inclined sheets** occur close to many shallow magma chambers (Fig. 5.1), the dip being shallower than in ophiolites because the local stress field has σ_1 inclined rather than vertical (and σ_3 inclined rather than horizontal). Yet, the dike and sheet clusters do not develop shallow magma chambers, or else they would not be seen as clusters or swarms today (Fig. 5.1). Thus, some other conditions than numerous dikes or inclined sheets must be satisfied for a shallow magma chamber to develop. The conditions for magma-chamber formation discussed below, while initially developed for rift zones, apply, in fact, to chamber formation in all tectonic regimes.

The **principal condition** for the formation of a shallow magma chamber is the emplacement of a sill or sills. While a horizontal or gently dipping sill stays liquid it acts as a barrier to vertical dike propagation and tends to absorb part of the magma of the dikes that meet it (Fig. 5.2). The initial sill may be a single one, or a cluster of sills that eventually merge into a single magma body (Fig. 5.3). The main condition for a sill (or sills) to evolve into

Fig. 5.1 Dense swarms of dikes and inclined sheets normally do not evolve into magma chambers. View northwest, a part of the local sheet swarms in Kalfafellsdalur in Southeast Iceland, where the entire cliff is composed of (mostly basaltic) sheets dipping towards a shallow crustal chamber (not exposed). Similar sheet intrusion densities are observed elsewhere in central volcanoes (Iceland, Scotland) and in ophiolites. The floor of the valley is at about 2 km below the original top of the volcano to which the sheets belong. The person standing in the lower right part of the photograph provides a scale.

a shallow chamber is that frequency of dikes meeting the sill must be high enough for it to **remain liquid** during the elastic-plastic expansion necessary for the chamber formation (Fig. 5.2). If the initial sill is thick enough so as to stay liquid for a while and thus be able to absorb the magma of subsequent dikes that meet it, the sill has a chance of developing into a shallow magma chamber (Fig. 5.3). The minimum thickness necessary for the initial sill to develop into a magma chamber depends on the **rate of injection** of dikes. Thus, if the dike-injection rate is high – which, for a rift zone, implies that the spreading rate is high – then even a comparatively thin initial sill, say with a thickness of 10–20 m, has a chance of developing into a chamber. If, however, the dike-injection rate is low, such as at slow-spreading ridges, only thick sills, of the order of many tens of metres or more, have a chance of developing into magma chambers (Fig. 5.4).

The main limiting factor is the time it takes the sill to **solidify**, which depends on the thickness of the sill to the **second power** (power two). More specifically, the time to complete solidification t_s of a basaltic sill can be obtained from the standard theory of heat conduction in solids (Carslaw and Jaeger, 1959; Jaeger, 1964). For a sill emplaced in a basaltic pile such as at divergent plate boundaries the time in years is $t_s \cong 0.0825w^2$ (cf. Chapter 7; Gudmundsson, 1990), where w is half the thickness of the sill, or approximately $2w = b$, where b is the sill aperture or opening displacement. This implies that very thin sills, of the order of metres or less, have little chance of developing into magma chambers. Thus, some thick multiple sills composed of numerous thin sills or

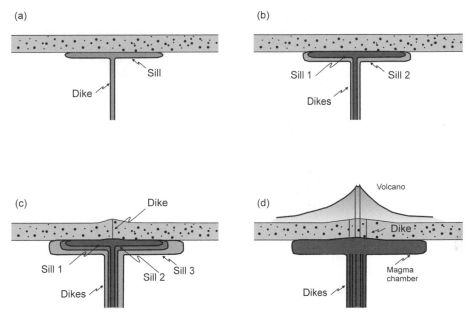

Fig. 5.2   Formation of a shallow magma chamber through the injection of sills. (a) A sill forms at the contact between mechanically dissimilar rock layers (cf. Gudmundsson, 2012). (b) Subsequent dike injections become arrested and their magmas are partly absorbed by the original sill. (c) The sill cluster expands and (d) forms a magma chamber that supplies magma to a central volcano. This schematic illustration shows the earlier sills as cooling somewhat, but they still stay liquid, before a new sill emplacement occurs (sills number 1–3 in (c)), as is sometimes observed (Fig. 5.6). When the rate of dike injection is very low, earlier sills solidify before the new injections occur, resulting in a multiple sill that does not develop into a magma chamber (Fig. 5.5).

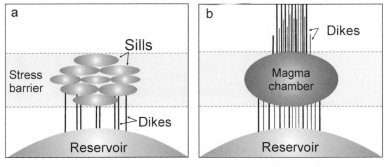

Fig. 5.3   Formation of a magma chamber within a stress barrier (also, commonly, at the lower contact of the barrier, Fig. 5.2). The origin of the chamber may be (a) a cluster of sills that merge into a single large one while liquid (Fig. 5.2), or (b) a single thick sill. Depending on the stress and mechanical conditions in the layer or unit that acts as a barrier, the sills may form beneath the barrier (Fig. 5.8) or within it. Modified from Gudmundsson (1990).

columnar rows do not evolve into magma chambers, presumably because the rate of dike injections is so low that individual columnar rows became solidified before the subsequent ones were injected (Fig. 5.5). By contrast, when **multiple sills** are composed of many thick

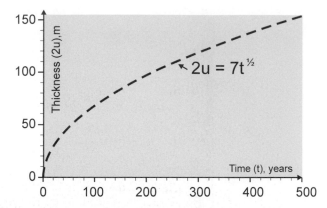

Fig. 5.4   Average thickness of the solidified part of a sill *2u* as a function of time *t* since sill emplacement. The thickness is given in metres and the time in years. Modified from Gudmundsson (1990).

Fig. 5.5   Part of a multiple basaltic 120-m-thick sill in East Iceland. The sill is composed of at least 16 columnar rows or sheets (some indicated). The overall shape of the sill is concave (Fig. 5.7b), as is seen from a greater distance (Fig. 2.5). At the time of its emplacement the sill was about 800 m below the top of the active volcanic zone to which it belongs. The exposed lateral dimension of the sill is about 3 km, the remainder being eroded away so that its initial dimension is unknown. The sill did not develop into a magma chamber. The sill and the nearby felsic dike (the top part of a composite dike) are seen in Fig. 2.5.

sills they have a chance of functioning as magma chambers (Fig. 5.6), even in areas of comparatively low rate of dike injection, such as in slow-spreading rift zones. Thus, the conditions for a sill cluster or a multiple sill to evolve into a magma chamber (Fig. 5.3) depends not only on the dike-injection rate but also on the initial thicknesses of the sills or

Fig. 5.6   The extinct Slaufrudalur magma chamber, now a pluton, was generated through the emplacement of many granophyre sills. Some of the sills can be seen as crude layering, most of the layers (sills) being 15–50 m thick. The deepest exposed part of the magma chamber is about 2 km below the original top of the central volcano to which it acted as a magma source (cf. Figs. 1.17 and 3.15).

rows that constitute the multiple sill. If the initial sills are very thin, of the order of metres or less, their chance of evolving into magma chambers is slim (Fig. 5.5), but they are much greater when the initial sills are of the order of tens of metres in thickness (Fig. 5.6).

More specifically, since the time for solidification depends on the thickness of the sill to the second power, thick sills take a much longer time to solidify than thin sills. This implies that, even if the sill receives magma (through dikes) only infrequently, it may still have a chance of evolving into a shallow magma chamber if it is initially very thick (many tens of metres or more). By contrast, if the rate of dike injection is high, such as at fast-spreading ridges where dikes are injected once **every 2–5 years**, a thin sill, say 10–20 m thick, would still be partly liquid after 2–5 years and thus absorb a large fraction of the magma in a dike that would meet it. When absorbing new magma, heat is added to the sill so that it may partly re-melt and thus stay liquid until the next dike meets it, and so on. Gradually, such a sill may absorb more and more magma and evolve into a magma chamber (Figs. 5.1 and 5.2; Gudmundsson, 2012).

But why do sills form in rift zones in the first place? Why are all the intrusions not subvertical dikes, as are favoured by the most common state of stress in a rift zone? The principal reason for sill formation in rift zones (and, in fact, everywhere) is the **deflection of dikes** into sills. Dike deflection is most common at boundaries, that is, contacts between rock layers or units with different mechanical properties. There are **three principal mechanisms** by which dikes become deflected into sills (or, alternatively, halt, that is, become arrested, at the contact). These mechanisms are:

- The Cook–Gordon delamination or debonding mechanism.
- The generation of a stress barrier.
- An unfavourable elastic mismatch between the layers above and below the contact.

These mechanisms are discussed in detail in Chapter 7. Briefly, however, they operate as follows.

Cook–Gordon Delamination. The essence of this mechanism is that when a vertical dike propagates towards a contact, the tensile stress ahead of the dike tip tends to **open up** a mechanically **weak contact** (Fig. 3.39). When, subsequently, the dike tip hits the open contact, the magma supplied by the dike tends to fill the open cavity or fracture generated by opening the contact and it thus spreads laterally to form a sill. In a vertical section, the sill may propagate primarily in one direction, and thus become asymmetric (Fig. 5.7c). Alternatively, the sill may propagate in both directions, and thus become symmetric.

This mechanism was initially proposed and tested within the field of materials science (Gordon, 1976). It applies to all extension fractures, that is, all fractures that develop perpendicular to the minimum compressive principal (maximum tensile) stress σ_3 and thus parallel with maximum (σ_1) and intermediate (σ_2) stresses (cf. Chapter 3). Since dikes are extension fractures it follows that the same mechanism applies to them. Presumably, this mechanism is most effective at shallow depths in the crust where it is comparatively easy to open up weak contacts in active volcanic areas (Fig. 3.39). By weak, here I mean weak in tension, that is, the tensile strength of the contact should be close to zero.

Stress Barrier. A stress barrier is simply a rock layer or rock unit that has a local **stress field** which is **unfavourable** to the propagation of a particular type of fracture. Thus, a stress barrier to sill propagation could be a rock unit where σ_1 is vertical whereas outside that unit,

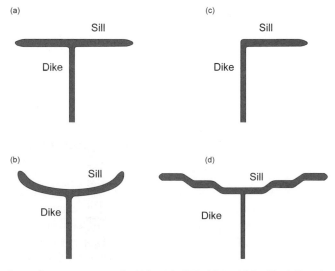

Fig. 5.7 Sills have various forms, the most common of which are indicated here. (a) Double-deflected straight sill. (b) Concave or upward-bending sill. (c) Single-deflected straight sill. (d) Staircase-shaped double-deflected sill, a subgroup of which is the saucer-shaped (disc-shaped) sill. Modified from Gudmundsson (2012).

where the sill happens to be propagating, σ_1 would be horizontal (as it normally has to be for the sill to propagate). Similarly, a rock layer with σ_1 horizontal and σ_3 vertical is a barrier to the propagation of vertical dikes, but favours the propagation of horizontal sills (Fig. 5.8). The stress barrier was the first mechanism identified as favouring dike deflection into sills at contacts (Anderson, 1942; Gretener, 1969; Gudmundsson, 1986) and is presumably very common.

Stress barriers to dike propagation, so as to favour either dike arrest or dike deflection into a sill at a contact, may form in various ways (Fig. 5.8). One way is through the stress effects of an older dike. When a dike is injected, its overpressure, that is, its magma pressure over or above the stress acting perpendicular on the dike (which is normally σ_3), can reach tens of mega-pascals (Chapter 8). Even if the dike contracts or shrinks (normally by about 10%) as its magma cools down, much of the overpressure remains in the crust as compressive stress perpendicular to the dike, for years or decades following the dike emplacement, until eventually it is relaxed (such as through plate movements). In fact, regional dike emplacement in rift zones commonly results in horizontal compressive stresses that are only relaxed through **excess spreading**. Such excess spreading, which means a measured spreading rate in excess of the long-term normal spreading rates, has, for example, been measured following the 1975–1984 rifting episode with dike injections in North Iceland and the 2005–2014 rifting episode (with dike injections) in the African Rift (Wright et al., 2012).

The magmatic overpressure in a dike can easily change the regional σ_3, which is normally perpendicular to the dike strike, temporarily to σ_2 or σ_1. Consider, for example, the state of stress at 1.5 km depth in a rift zone, the depth at which many sills form, some of which develop into magma chambers. Before dike injection, σ_1 would be vertical and given by $\sigma_v = \rho_r g z$ (Chapter 3), where σ_v is the vertical stress, ρ_r is the average crustal density (here 2500 kg m^{-3}), g is the acceleration due to gravity (9.81 m s^{-2}), and z is the crustal depth

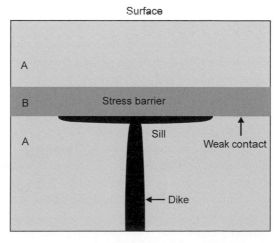

Fig. 5.8 A dike meeting a stress barrier may either become arrested or deflected into a sill. This schematic illustration shows the beginning of the sill formation; as the sill propagates further it normally becomes much thicker than its feeder-dike. A thick sill then has a chance of developing into a magma chamber (Fig. 5.2).

(1500 m). Using these values, the vertical stress $\sigma_v = \sigma_1$ is about 37 MPa. The minimum principal stress σ_3 must be equal to or less than σ_1 and is likely to be somewhere between 30 and 37 MPa, and the intermediate stress σ_2 has a magnitude somewhere between those of σ_1 and σ_3. Let us take σ_3 as 35 MPa before the dike injection. The magma overpressure in a basaltic dike at the depth of 1.5 km is commonly 5–20 MPa (Chapter 8). Here, we take the overpressure as 10 MPa. Once the dike is injected, the horizontal stress perpendicular to the dike strike becomes 35 + 10 MPa = 45 MPa, and thus is higher than the vertical stress (37 MPa). This means that the magmatic overpressure associated with the dike injection flips the principal stresses by 90°. Thus, following the dike emplacement, the **horizontal stress** perpendicular to the dike will, temporarily, **be σ_1** and with a magnitude 45 MPa, whereas the vertical stress will be σ_2 or σ_3 (here they are of essentially the same), with a magnitude 37 MPa. Thus, a **stress barrier** has formed that encourages subsequent dike(s) to become arrested (Fig. 4.11), change into sills (Figs. 5.3 and 5.8) or, occasionally, propagate laterally beneath the barrier until a path for vertical magma propagation is found again (Fig. 2.23).

If the initial state of stress before dike emplacement was such that the magnitudes of σ_1 and σ_2 were quite different – a situation not likely in active rift zones – then subsequent to the dike emplacement the vertical stress could be σ_2 rather than σ_3. Then the stress field would, theoretically, encourage a 90° horizontal flip in the strike of the dike rather than its change into a sill. Such abrupt changes in the strikes of individual dikes are very rarely seen in nature, suggesting that this mechanism of stress change does not often happen during the propagation of individual dikes. One reason why a 90° change in the dike path in a vertical section (a dike deflection into a sill) is more common than a similar change in a horizontal section (an abrupt change in the dike strike) is presumably the Cook–Gordon delamination mechanism. Accordingly, when the dike meets with the horizontal contact marking the stress barrier, the contact is already open – and thus has zero tensile strength – ahead of the dike tip (Fig. 3.39). It is normally easier for the dike to deflect into the open contact and form a sill than to overcome the tensile strength of the rock and form a new path at 90° to its previous strike.

There are other mechanisms for the formation of stress barriers, but here we mention only one additional mechanism, namely **graben formation** or subsidence (Fig. 5.9). The state of stress that encourages graben formation or subsidence along existing graben boundary faults is basically the same as that which encourages dike emplacement: σ_1 is vertical and σ_3 is horizontal and perpendicular to the strike of the graben faults, whereas σ_2 is parallel with the strike of the faults. When the graben forms, or an existing graben subsides, the vertical displacement on the boundary faults may 'overshoot', that is, the subsidence may become more than needed to bring the state of stress to lithostatic (where all the principal stresses are the same and equal to the vertical stress). Because graben subsidence is a dynamic process—commonly associated with earthquakes—the vertical displacement on the normal graben faults may press the graben floor, the 'wedge', to such a depth that it generates a **horizontal compression** in the footwalls of the graben faults. The horizontal stress, thus generated, may result in a 90° flip of the principal stress, just like in the case of the dike emplacement and thus to the formation of a stress barrier. It follows that grabens commonly capture dikes and either arrest them (Fig. 2.31) or deflect them into sills.

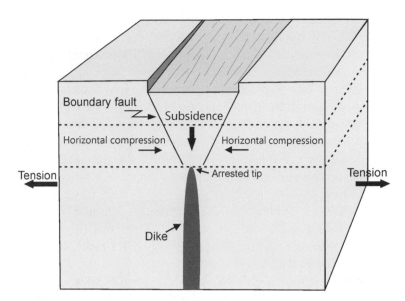

Fig. 5.9 The slip or subsidence of a graben in a rift zone may temporarily alter the stress field so that an injected dike becomes arrested, as seen here and in Fig. 2.31, or, alternatively, it is deflected into a sill.

Both these types of 90° stress flips, through dike emplacement and through graben subsidence, are encouraged when the layers that constitute the crustal segment have contrasting mechanical properties (Figs. 1.16 and 2.11). When there are compliant layers, such as soft tuff, scoria, or soil layers that alternate with stiff lava flows, most of the extra compressive loading is taken up by the stiff layers, which thereby become highly stressed in compression. Abrupt variations in horizontal stresses are well documented in crustal segments composed of widely varying mechanical properties, particularly with different stiffness (Young's moduli), such as in sedimentary basins and in active volcanic rift zones.

Elastic Mismatch. This mechanism is basically the result of the layers having a different Young's modulus or stiffness on either side of a contact. Results from materials science show that as the stiffness contrast across the contact increases – that is, as the elastic mismatch increases – the likelihood that an extension fracture does not penetrate the contact increases (Section 7.4). The extension fracture can either become arrested, that is, stop or halt altogether at the contact or, alternatively, become deflected along the contact. When the extension fracture is a dike, then the fracture that becomes deflected along the contact is a sill (Fig. 5.10).

This mechanism is very common in materials science, particularly in artificial materials such as composite materials. In fact, **composite materials** are made strong – they are made so as to maximise the likelihood of fracture arrest at contacts, that is, made resilient to fracture propagation – by their being composed of alternating layers of different stiffness. But the same principles apply equally well to layered rock bodies. These are commonly composed of alternating compliant or soft layers, such as of shale, silt, scoria, and soil, and stiff layers, such as lava flows of various types, horizontal intrusions (such as sills), and welded rocks. The basic results are the same. When a dike meets a contact across which

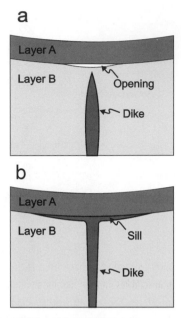

Elastic mismatch between layer A (stiff) and layer B (soft) commonly results in dike arrest or, alternatively, deflection into a sill, as indicated here. In this numerical model, the contact between layers A and B is, in addition, weak in tension (with a low tensile strength) so that it (a) opens up ahead of the dike which, on meeting the contact (b) changes into a sill. This is one example where two of the main mechanisms for dike arrest/deflection operated together: here the process of elastic mismatch and the Cook–Gordon mechanism of delamination.

there is significant to large elastic mismatch, there is a strong tendency for the dike to become deflected along the contact to form a sill (Fig. 5.10).

All the three mechanisms – Cook–Gordon delamination, stress-barrier formation, and elastic mismatch – can operate individually or together. The Cook–Gordon mechanism is likely to be most effective at comparatively **shallow depths** in the crust, say in the uppermost few kilometres, but the other two can operate at **any depth**. These three mechanisms are the most likely reason for the deflection of dikes into sills, and for the formation of sills in general. Since a sill is the type of intrusion that is most likely to develop into a shallow magma chamber, these mechanisms are also the primary means of initiating shallow magma chambers.

5.3 Magma-Chamber Rupture and Dike Injection

Once a sill has formed it has a chance of developing into a shallow magma chamber if the conditions discussed above are met. The main conditions for this to happen depend on the **dike-injection rate** (Figs. 5.2 and 5.3), which, in turn, is a function of the spreading rate (Fig. 5.4). For a slow spreading rate, the thickness of the initial sill must be many tens of metres or more. For a fast spreading rate, the initial sill need not be more than a few tens of

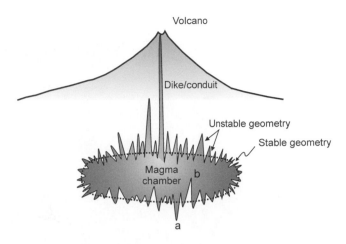

Fig. 5.11 Magma chambers are unlikely to have very irregular boundaries because such chambers are mechanically and thermally unstable. Mechanical instability is because the notches that project into the host rock raise or concentrate stress and tend to inject dikes into the roof for any non-zero chamber excess pressure. The thermal instability is because the jogs that project into the chamber tend to melt rapidly whereas the notches that project into the host rock tend to solidify rapidly. An active chamber thus normally assumes a more stable, smoother geometry, as indicated here and seen in extinct chambers (Figs. 1.17, 5.12, 5.18 and 5.19a). Modified from Gudmundsson (2012).

metres. Both conditions reflect the main requirement that the initial **sill must stay liquid** if it is to have a chance of evolving into a shallow magma chamber. Thus, the main limitation on the chances of the sill to evolve into a chamber is the cooling of the sill and, if a new magma input is not received frequently enough to keep it liquid, its solidification.

The shallow magma chamber that evolves from the initial sill or sill cluster can take on many forms. A magma chamber is a **dynamic system** so that its shape is changing throughout its lifetime. But a mechanically and thermally stable magma chamber – one that is able to function as such over tens or hundreds of thousands of years or more – cannot be very irregular in shape (Fig. 5.11). Accordingly, long-lived magma chambers tend to have comparatively smooth boundaries with the host rock, as are commonly seen in fossil chambers (plutons) whose contacts with the host rock are exposed at shallow crustal depths (Figs. 1.17, 5.12, and 5.18; Chapter 7).

Many magma chambers, particularly at fast-spreading ridges, maintain their sill-like geometries throughout most of their lifetime. Other chambers, however, evolve into oblate ellipsoids (Fig. 5.3), spheres, or, occasionally, prolate ellipsoids (Fig. 5.13). No real magma chambers, however, have exactly these ideal geometries – there are always some irregularities along their boundaries or contacts with the host rock (Figs. 1.17 and 5.12). But based on the arguments above, many chambers approximate these geometries to a greater or lesser degree, and we will often refer to these geometries when describing magma chambers. It is to be understood that, while very useful for descriptions, these geometries are **idealisations**.

Now the question arises: under what conditions will a shallow magma-chamber rupture and inject a dike into the host rock? The answer is that there are three main processes that may result in magma-chamber rupture and dike injection, namely the following:

Fig 5.12   Contacts between extinct magma chambers (plutons) and their host rocks tend to be smooth, as indicated by the roof of a part of the Austurhorn pluton, a fossil magma chamber in Southeast Iceland (Furman et al., 1992). This part is exposed in the mountain Krossanesfjall (its height is 716 m). Also indicated are the primarily felsic and primarily mafic parts of the pluton. On the sandy coast, a part of a net-veined complex, of cross-cutting small-scale felsic and mafic intrusions, is seen (Blake, 1966).

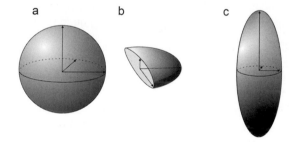

Fig. 5.13   Three main ideal geometries of magma chambers are (a) a sphere, (b) an oblate ellipsoid (a sill-like chamber, only half of which is shown here), and (c) a prolate ellipsoid. Modified from Gudmundsson and Nilsen (2006).

1. **Magma is added to the chamber** from a deep-seated reservoir below, usually through a dike. As the volume of magma in the chamber increases, local tensile stresses develop at the boundary of the chamber with its host rock. When the tensile stress reaches roughly the tensile strength of the host rock, the host rock ruptures at one (sometimes more than one) locality and a magma-filled fracture (a dike or an inclined sheet) forms.

2. A **gradual extension** of the crustal segment hosting the chamber, such as in continental rift zones or at divergent plate boundaries in general, results in the concentration (raising) of tensile stress at the boundary between the chamber and the host rock.

When this stress reaches the tensile strength of the host rock, the host rock ruptures at one (sometimes more than one) locality and a magma-filled fracture (a dike or an inclined sheet) forms.

3. Processes (1) and (2) commonly **operate together**. Thus, at divergent plate boundaries, the gradual extension of the crustal segment hosting the chamber is a continuous process. The magma flow from the deeper source reservoir up into the chamber, however, is normally a discontinuous process. But when the latter occurs, both processes (1) and (2) operate simultaneously. Then it depends on the volumetric rate of flow of magma into the shallow chamber, in relation to the chamber size, as to which of the two processes dominates.

When the conditions listed in items 1–3 above are satisfied, the magma chamber will rupture. Mathematically, the **condition for rupture** and dike injection is given by:

$$p_l + p_e = \sigma_3 + T_0 \tag{5.1}$$

Here, p_l denotes the lithostatic stress at the rupture site at the boundary (the roof or the walls, but rarely the floor) of the magma chamber, p_e is the excess magmatic pressure in the chamber – in excess of σ_3, the minimum compressive principal (maximum tensile) stress – and T_0 is the local *in situ* tensile strength at the rupture site. Equation (5.1) can also be written in a somewhat different form as:

$$p_t = \sigma_3 + T_0 \tag{5.2}$$

where $p_t = p_l + p_e$ is the total fluid pressure in the chamber at the time of rupture. What Eqs. (5.1) and (5.2) mean is that when the total fluid pressure in the chamber reaches the combined value of the minimum compressive principal (maximum tensile) stress and the *in situ* tensile strength, the chamber will rupture and inject a magma-filled fracture, an inclined sheet, or a dike (Fig. 1.17).

There are several points that need to be clarified regarding Eqs. (5.1) and (5.2), namely the following:

- The magma-filled (sometimes, initially, gas-filled) fracture that forms is an **extension fracture**. Accordingly, the conditions do not apply to the formation of shear fractures, that is, faults of any kind. The formation of faults is discussed in Section 5.7 (cf. Gudmundsson, 2011a). Field observations show that, as a rule, dikes, inclined sheets, and sills are extension fractures and thus not shear fractures. Over the years, many (including the author) have suggested that some magma-driven fractures are shear fractures, particularly inclined (cone) sheets. Such models were potentially plausible decades ago before systematic studies of cross-cutting relations between sheets and dikes among themselves, and with other structures such as lava flows, were undertaken. Thousands of measurements in the past decades, however, show clearly that the great majority – commonly 90% of all sheet-like intrusions – are extension fractures (Chapter 2). And those that occupy shear fractures (faults) do so normally along short parts of their paths, and then mostly those parts that were steeply dipping and with close to zero tensile strengths at the time of dike or inclined-sheet emplacement (cf. Chapter 10).
- The extension fracture is formed by the magmatic pressure. It is the **magma itself** that breaks or **ruptures the rock**, in a manner analogous to artificial hydraulic fracturing used

to increase the permeability in reservoirs of various types. There are no wide-open extension fractures at many kilometres depth that are waiting to be filled with magma, neither in rift zones nor anywhere else in the Earth's crust. Griffith's theory of fracture provides a theoretical reason why large tension fractures cannot form at greater depths than about 1 km, and usually do not extend from the surface to depths exceeding a few hundred metres (Gudmundsson, 2011a). And field observations in caldera walls, erosional cliffs, and other sections into active and inactive volcanoes and rift zones show that **large tension fractures** only exist at very **shallow depths.** Thus, dikes, inclined sheets, and sills essentially all form their own fractures.

- The host rock behaves as if it is brittle and fails in a **brittle manner,** that is, through fracture. This is what field observations show. Even close to or at the contacts with the magma chambers, rock failure during magma-chamber rupture and dike or sheet injection is predominantly brittle. Many have suggested viscoelastic, plastic, and viscoplastic behaviour of the host rocks of shallow magma chambers. But where the ruptured margins between the chambers and their host rocks can be observed in detail, the results indicate that the failure occurred, as a rule, in a brittle rather than ductile manner.

- It follows that the strength that needs to be reached for the magma to form its own fracture is the **tensile strength** of the crust. This is the strength appropriate for brittle failure during extension-fracture formation. Rocks have very low *in situ* tensile strengths or between 0.5 and 9 MPa. The laboratory tensile strengths are commonly much higher, the maximum values for many rocks being 20–30 MPa. However, laboratory measurements are made on centimetre-scale samples that normally do not contain any significant weaknesses or 'flaws' (large vesicles, pores, cracks, or other inhomogeneities). All natural rocks at the scale of dike-fractures contain numerous flaws and fractures (Figs. 1.7, 1.21, 4.11 and 4.12) that lower the *in situ* tensile strength to the *in situ* values of 0.5–9 MPa. In fact, most commonly the *in situ* tensile strength is measured using small hydraulic fractures in drill holes or wells (Gudmundsson, 2011a), so that the analogy with magma-chamber rupture and dike injection is clear. While the *in situ* tensile strength has the range above, most values are between 2 MPa and 4 MPa, so that the *in situ* tensile strength of solid rocks is close to being a constant.

- The **total fluid pressure** $p_t = p_l + p_e$ includes that of the **magma** as well as that of any **gas** in the chamber. In chambers containing primarily mafic magma at the depth of several kilometres, water (H_2O) is contained within the magma – the H_2O does not form a separate phase, is not exsolved – while some other gases, particularly carbon dioxide (CO_2), may already be exsolved to form a **separate phase**. A chamber containing felsic magma has many more volatiles, and the gases form separate phases (bubbles) at much greater depths than in mafic magma. Thus, much of the CO_2 and some H_2O form bubbles in felsic magma in chambers at depths of 1–4 km. In basaltic magma, such bubble formation occurs normally at depths of less than about 1 km, and thus rarely in magma chambers but rather in feeder-dikes or shallow conduits. Irrespective of how much of the gas is exsolved, the fluid pressures, both the total pressure p_t as well as the excess pressure p_e, result from and include the **combined pressure** effects (also any buoyancy effect) of all the fluids (gases and liquids) in the chamber.

- Stress concentrations around ellipsoids of revolution (Fig. 5.13) subject to external tensile stress and internal fluid pressure, or both, suggest that the tensile stress around magma chambers during unrest periods varies with location at the chamber boundary (Chapter 3). This might suggest that Eqs. (5.1) and (5.2) are not appropriate. In particular, a spherical chamber gives rise to stress concentrations that differ from those either of a prolate ellipsoid (rare shape for chambers) or an oblate ellipsoid (common shape of chambers). However, rupture and dike injection occurs always at some **irregularities at the boundary of the chamber** (Figs. 1.17, 5.12, 6.18). And at such irregularities the stress concentration is significantly higher than that around the magma chamber as a whole. It is therefore the **local stress concentration** at an irregularity in the roof or the walls of the magma chamber (rarely the floor of the chamber) that results in rupture rather than the concentration around the entire chamber of a given crude general geometric shape. Hence, Eqs. (5.1) and (5.2) are generally appropriate irrespective of the overall approximate shape of the chamber.

5.4 Dike Emplacement

When Eqs. (5.1) and (5.2) are satisfied, the chamber ruptures and a dike (or an inclined sheet) is injected into its roof. For a dike that propagates up into the crustal layers of the chamber roof, the magmatic driving pressure or **overpressure** p_o is given by:

$$p_o = p_e + (\rho_r - \rho_m)gh + \sigma_d \tag{5.3}$$

Here, ρ_r is the average host-rock density, ρ_m is the average magma density, g is acceleration due to gravity, h is the dip-dimension (height) of the dike above its rupture site at a particular time during its propagation, as measured from the point of rupture and dike initiation at the chamber boundary, and σ_d is the differential stress ($\sigma_d = \sigma_1 - \sigma_3$) in the crustal layer (at the crustal level) which the dike has reached during its propagation when examined (Fig. 5.14). For a feeder-dike, the dip-dimension h is the vertical distance between the point of initiation at the boundary of the chamber and the Earth's surface where the volcanic fissure fed by the dike forms. For a non-feeder-dike, h may refer either to a propagating (liquid) dike (in real-time monitoring) or to a solidified (old) dike exposed, through erosion or collapse, at some depth in the crust. For a propagating dike, h is the vertical distance from the chamber to the tip of the propagating dike at a particular time. For an old dike in an eroded volcano or the walls of a collapse caldera, h refers to the vertical distance from the boundary of the source (commonly fossil) chamber to the layer where the measured dike is exposed.

Equation (5.3) can be used to estimate the magmatic overpressure of a dike, while keeping in mind the following points:

- At the time of dike initiation from a reasonably large magma chamber, the excess pressure p_e is positive, as it must be to rupture the chamber roof or walls. Normally, when the excess pressure reaches roughly the *in situ* tensile strength of the host rock at the

Fig. 5.14 Schematic illustration of the state of stress in the host rock close to a dike, with reference to Eq. (5.3). The unit cross-sectional area of a rock column used is indicated, as well as the differential stress σ_d at the crustal level where the dike is measured. The depth of the unit area below the Earth's surface at the time of dike emplacement is z in Eqs. (3.5) and (3.6), whereas the height of the dike exposure above the magma source, the dike dip-dimension, is h in Eq. (5.3). Modified from Gudmundsson (2012).

chamber boundary, in which case $p_e = T_0$, the reservoir ruptures in tension and a dike (or an inclined sheet or a sill) is initiated. The overpressure available for driving the dike propagation the first hundreds of metres is primarily the **excess pressure**. This follows because so long as the dip-dimension or height of the dike h is only a few hundred metres or less the second term on the right-hand side of Eq. (5.3), the **buoyancy term**, does not contribute much to the overpressure. In fact, for high-density basaltic magma injected through dikes from shallow magma chambers, this term (in Eq. (5.3)) may be **zero** (magma and host-rock density equal) or even **negative** (magma denser than the host rock), in which case the only overpressure available to drive the dike propagation is the excess pressure in the chamber.

- The **differential stress** $\sigma_d = \sigma_1 - \sigma_3$ must be either zero or positive (Fig. 5.14). The differential stress σ_d cannot be negative because $\sigma_1 \geq \sigma_2 \geq \sigma_3$ (Chapter 3) so that σ_1 cannot be less than σ_3. When $\sigma_d = 0$, then $\sigma_1 = \sigma_3$ so that the state of stress is isotropic (in two dimensions at least) and if applicable to three dimensions (in which case we also have $\sigma_1 = \sigma_2$), then the state of stress is lithostatic.
- The **density difference** $\rho_r - \rho_m$ can be as follows: (1) negative, when the magma is denser than rock; (2) zero, when the density of the magma is equal to that of the rock; or (3) positive, when the rock is denser than the magma. The density of most crustal rocks is in the range of $2000 \leq \rho_r \leq 3000$ kg m^{-3} while the density of typical magmas is primarily in the range $2250 \leq \rho_m \leq 2750$ kg m^{-3}.
- For stresses around magma chambers and dikes and other sheet intrusions, it is the excess pressure (for the chamber) and the overpressure (for the dike) that drive the fracture propagation and associated brittle deformation. The total pressure is rarely used. Thus, in the analysis and equations above, the **effect of gravity** is automatically taken into account in such an analysis. In this book, the magmatic excess pressure/driving pressure is used rather than the total pressure.

Once the dike is initiated and starts to propagate, the local stress field will control its propagation path. The details are given in Chapters 7 and 10, but here I shall indicate briefly how the likely paths can be forecasted. Because dikes are extension fractures, they must, by definition, follow paths that are perpendicular to the trajectories or ticks of the minimum compressive principal stress σ_3 and thus parallel with the trajectories or ticks of the intermediate σ_2 and maximum compressive principal stress σ_1. For dikes propagating in a homogeneous, isotropic crustal segment, plotting the likely paths of dikes is thus easy. For a magma chamber with a circular cross-section and excess pressure as the only loading, the dikes are injected radially from the chamber boundary, that is, vertical above the centre of the roof of the magma chamber, and inclined (forming inclined sheets) from the sides or the walls of the chamber (Fig. 3.18). The stress trajectories and the theoretical dike paths are different, however, if the chamber is close (in relation to its size) to the free surface (surface of a volcano or volcanic zone), as seen in Fig. 3.19. Also, in the case of the magma chamber subject to external tension rather than excess pressure as the only loading, the trajectories and the dike paths are different (Figs. 3.19 and 3.20). Notice that a magma chamber with a circular cross-section may be spherical, but does not have to be; the chamber can be a prolate ellipsoid of revolution (Fig. 5.13) with the major axis horizontal – for example along the axis of a volcanic rift zone.

No crustal segments are homogeneous and isotropic but rather heterogeneous and anisotropic. The latter applies particularly to active volcanic areas where the mechanical properties of the layers commonly vary abruptly across contacts. This has important implications for dike-propagation paths. The simple models in Chapter 3 show the main difference between the trajectories and likely paths of dikes, injected from a chamber of a circular cross-section and with an excess pressure as the only loading, for isotropic and anisotropic crustal segments/volcanoes. The orientation of the ticks or trajectories that represent the local directions of σ_1 and likely dike paths in the isotropic models (Figs. 3.18 and 3.19) and the anisotropic or layered models (Figs. 3.25 and 3.26) are very different. This difference is particularly clear in the central upper parts above the tops of the magma chambers. At some contacts in these parts in the anisotropic models the ticks for σ_1 change from vertical to horizontal. No such change occurs in the isotropic models. We know that a vertical orientation of σ_1 favours the propagation of vertical dikes, whereas a horizontal direction of σ_1, particularly when σ_3 is vertical, favours the propagation of horizontal sills. Consequently, when the direction of σ_1 changes abruptly from vertical to horizontal across a contact – and at the same time, but not shown here, the direction of σ_3 changes from horizontal to vertical – then a vertically propagating dike is likely to become either arrested at the contact or, alternatively, deflected into a sill. And it is to sill emplacement that I shall now turn briefly.

5.5 Sill Emplacement

I have already discussed the principal mechanisms by which dikes may change into sills. These are: (1) the Cook–Gordon delamination or debonding mechanism, (2) the generation of a stress barrier, and (3) an unfavourable elastic mismatch between the layers above and below the contact. I also indicated that the formation of a sill or a cluster of sills is the main mechanism by which shallow crustal magma chambers are initiated (Figs. 5.2, 5.3 and 5.6). Here I explain briefly the mechanism of sill emplacement itself (cf. Gudmundsson, 2011b).

Sills are normally **fed by dikes** (Figs. 5.2, 5.3, 5.8 and 5.15). Even when sills at a higher stratigraphic level are fed by sills at lower levels, as is common, the connection between the sills is normally a dike or an inclined sheet. As a rule, the **feeder-dike** of a sill is much **thinner** than the sill (Fig. 5.15). This follows because of two main reasons:

Fig. 5.15　Deflection of a dike into a sill in Southwest Iceland. Here the dike is much thinner than the sill, as is normal. More specifically, the dike is about 20 cm thick at the contact with the 60-cm-thick asymmetric sill (the pencil is 19 cm long).

1. The tip-part of the dike, just before it becomes deflected into a sill, is gradually entering a local stress field that is **unfavourable** for dike propagation, and favourable for sill propagation.
2. The **effective stiffness** (Young's modulus) of a horizontally layered crustal segment is normally lower in response to vertical loading (sill emplacement) than to horizontal loading (dike emplacement). This difference is because it is generally easier to bend a pile of horizontal layers, with numerous mechanically soft or weak contacts, through vertical loading than through horizontal loading. Thus, for the same magmatic overpressure and sheet dimensions, a sill is normally thicker in relation to its other dimensions than a dike.

Once a dike has become deflected into a sill, normally along a contact between mechanically dissimilar layers, the sill may propagate asymmetrically or symmetrically, as seen in the vertical section (Figs. 5.7 and 5.15). Whether the sill propagates in one or both directions in a vertical section depends on several factors, including:

- The **attitude**, particularly the dip, of the feeder-dike – if the dike is inclined, an inclined sheet, it opens up a contact in an asymmetric way, encouraging asymmetric sill emplacement (Fig. 5.15).
- The **dip** of the contact – a non-horizontal contact may encourage asymmetric sill emplacement, with the sill tending to propagate up-dip.
- The local **stress field** – the variation in vertical stress, such as topographical effects (higher vertical stress under topographic highs), may encourage asymmetric sill emplacement.
- The rock **strength** – the variation in rock strength and material **toughness**, particularly along the contact along which the sill is emplaced, may encourage asymmetric sill emplacement.

All hydrofractures propagate in the same way. The rate of propagation is controlled by the rate of fluid (here magma or gas) flow to the tip of the fracture. The fluid pressure generates tensile stress at the tip of the fracture. When the concentrated tensile stress reaches the tensile strength of the rock, the fracture (here the sill-fracture) propagates. More specifically, when the conditions of Eq. (5.2) are satisfied, the tip of the liquid sill will propagate. Since the tensile strength T_0 for rocks is basically a constant, the conditions of Eq. (5.2) can be reached in two principal ways. First, by increasing the total pressure p_t in the sill and, second, by lowering the (normally) compressive value of σ_3 to close to zero or, at the tip, to absolute tension (which means a negative value of σ_3). The value of p_t can be increased by magma flowing into the tip region of the sill-fracture; the compressive value of σ_3 can be reduced to zero and then made tensile through tensile-stress concentration at the fracture tip.

As the sill-fracture propagates, it expands laterally and also vertically. This follows because, for a given driving pressure (overpressure) and mechanical properties of the host rock, there is a constant ratio between the strike- or dip-dimensions of an extension fracture and its opening. More specifically, the smaller of these two dimensions has the greater effect on the opening of the extension fracture, the sill, and is the controlling dimension. For a tunnel-shaped sill, that is, a sill with a dimension into the wall of a vertical cross-section (say, a vertical cliff) that is much larger than that of the lateral dimension seen in the section (Figs. 5.16 and 5.17), the relation between the maximum aperture (roughly the thickness of the solidified sill) and the lateral dimension in the vertical section is given by (cf. Eq. (7.13)):

$$\Delta u_I = \frac{2p_o(1 - v^2)L}{E} \tag{5.4}$$

whereas if the sill is circular, the same relation is given by:

$$\Delta u_I = \frac{8p_o(1 - v^2)a}{\pi E} \tag{5.5}$$

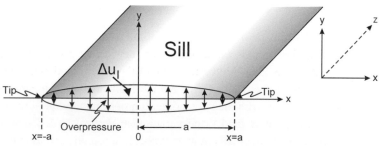

Fig. 5.16   Tunnel-shaped sill. This shape means that one dimension, here along the z-axis, is much larger than the other two dimensions (along the x- and y-axes). For constant overpressure, the opening or thickness Δu_I of the sill is that of a flat ellipse, with zero opening at the lateral tips ($x = a$ and $x = -a$). The strike-dimension of the sill is $2a = L$.

Fig. 5.17   Crudely tunnel-shaped basaltic sill in the lava pile of East Iceland. At the time of its emplacement the sill was about 1200 m below the top of the active volcanic zone to which it belongs. The farmhouse provides a scale.

In Eqs. (5.4) and (5.5) the symbols have the following meaning: Δu_I is the maximum aperture or opening of the sill-fracture, p_o is the magmatic overpressure (driving pressure), v is the Poisson's ratio of the host rock, $L = 2a$ is the lateral dimension (length) of the sill as seen in a vertical cross-section, a is the half length of a tunnel-shaped sill and the radius of a circular sill, π is the constant pi (3.1416), and E is the Young's modulus (stiffness) of the host rock. It should be noted that Eq. (5.4) can also be applied to dikes, particularly feeder-dikes, where L is the strike-dimension (length) of the dike and Δu_I its maximum opening or thickness.

At any time during the sill propagation, the theoretical or potential **stress concentration** at its tip may be estimated crudely from the equation for an elliptical hole. Modern fracture mechanics uses different estimates, as discussed in Chapter 7. However, for a rough estimate of the potential maximum tensile stress σ_{max}, the following formula may be used:

$$\sigma_{max} = p_o \left(\frac{L}{\Delta u_I} - 1 \right) \tag{5.6}$$

where all the symbols are as defined above. The magnitude of σ_{max} for a sill (or a dike) is normally many times higher than that of the actual maximum tensile principal stress σ_3 at rock rupture and the advance of the sill tip. As indicated in Chapter 4, the propagation of a sill and other hydrofractures occurs in **steps,** with the fluid front commonly lagging behind the fracture front at any particular time (Fig. 4.10). When the tensile stress in Eq. (5.6) reaches the tensile strength of the rock, T_0, the fracture tip will propagate, that is, advance laterally by a certain distance. When this happens, the liquid sill does not only expand laterally but also increases its aperture or opening and thus its **volume**. The lateral propagation of the tips is very rapid; its velocity is similar to the rate of propagation of earthquake fractures – of the order of kilometres per second. It follows that the propagation of the sill tip (or a dike tip) by, say, some tens of metres takes only a fraction of a second. The rate of transport of the magma front (or the gas fluid front) is much slower – of the order of $0.1–1 \, \mathrm{m \, s^{-1}}$ – so that the magma (or gas) front cannot keep up with the rate of the fracture propagation. This means that there is no fluid for a while in the newly expanded tip-part of the fracture (Fig. 4.10). Consequently, in this part, p_o temporarily falls to zero, and so does the stress concentration in Eq. (5.6), and the fracture propagation stops, that is, the sill tip becomes temporarily **arrested**.

The tip may become permanently arrested, particularly if it stops at a contact. But so long as the propagation continues, all the arrests are temporary. The temporary arrest of the tip continues until the magma (and the associated gas if any) has reached the tip so as to build up overpressure in the tip-part and thereby raise the tensile stress at the tip. When the conditions of Eq. (5.2) are reached again, the fracture tip will propagate a certain distance and then become arrested again. The tip remains arrested until the magma (or gas) has again reached the tip so as to satisfy Eq. (5.2), which results in a new sill-tip propagation. This process repeats itself until the tip reaches the location of its **permanent arrest** (Fig. 5.17). This arrest location and thus the size of the resulting sill depends on many factors, including the mechanical properties of the contacts and rock units along the potential path of the sill-fracture as well as the supply of magma from the feeder-dike to the sill. Cooling of the magma at or close to the tip may also contribute to the halting of its propagation. However, since the tensile strength (or toughness, as discussed in Chapter 7) of the solidified magma is commonly very similar to that of the host rock – particularly for basaltic dikes and sills in a basaltic lava pile – it is not clear that a solidified part of a dike or sill would offer any more resistance to the advancement of the still fluid part of the intrusion than the rest of the host rock.

The propagation of all fluid-driven fractures follows basically the principles outlined above. The fluid must flow into the tip region so as to build up pressure there and thereby raise the tensile stress at the tip until the conditions of Eq. (5.2) are satisfied, whereby the fracture advances a certain distance. The process of rapid fracture propagation, arrest, build-up of fluid pressure, and further propagation is very clearly seen on the surfaces of many **joints** (Chapters 4 and 9; Secor, 1965). Many, presumably most (excluding columnar joints), joints are **hydrofractures** and propagate as such. For many joints, the fluid comes from a source deeper in the sedimentary pile, say a layer of sand, silt, or mud, in which case the propagation is entirely analogous to that of dikes and sills. For other joints, however, the fluid source is local. Then the flow is into the joint-cavity from the surrounding rock and is driven in the direction of the minimum potential energy – for water, in the direction of the negative hydraulic gradient. In the latter case, the development of a joint can take, geologically, a long time, particularly in rocks with very low permeability. Then, following joint propagation, the joint volume increases (the joint dimensions increase) so that the fluid pressure in the joint decreases. This is because during the rapid expansion or propagation of the joint tips (edges) the amount of fluid in the joint remains constant. So when the same amount of fluid must occupy a larger joint volume, the fluid pressure must decrease. The next propagation of the joint then occurs only after the slow migration of fluid into the joint from the surrounding rocks has increased the fluid pressure sufficiently so that the conditions of Eq. (5.2) are satisfied again. And the process can be repeated many times. Each arrest of the joint, while waiting for the fluid pressure to build up again, is marked by so-called ribs on the joint surface.

The volume of a sill while propagating, as well as its volume once its emplacement is completed but the sill is still fluid, can be estimated. By analogy with Eq. (5.5), the volume V_s of a circular sill is given by:

$$V_s = \frac{16(1 - v^2)p_o a^3}{3E} \tag{5.7}$$

where all the symbols are as defined above. Equation (5.7) applies when the sill is forming, or just after its formation, when the magmatic overpressure driving the sill emplacement is known or can be inferred. This equation can also be used to forecast the thickness of a sill the lateral dimensions of which are known, for example from seismic images from sedimentary basins. Then the magmatic overpressure must be estimated using general values based on the calculated overpressures associated with sill emplacement in other areas, such as can be inferred from Eqs. (5.4) and (5.5).

Sills directly observed in the field or on seismic images are long solidified. If their thicknesses and lateral dimensions can be estimated (Figs. 5.16 and 5.17), the volume follows from the general equation for the volume of an ellipsoid (Fig. 5.13), namely:

$$V = \frac{4}{3}\pi abc \tag{5.8}$$

where a, b, c are the semi-axes of the ellipsoid. For a circular sill, $a = b$, in which case Eq. (5.8) reduces to the formula:

$$V = \frac{4}{3}\pi a^2 c \qquad (5.9)$$

Some magma chambers, even those that start as sills, eventually become spherical. For a spherical magma chamber, $a = b = c$ and Eq. (5.8) reduces to the well-known formula for the volume of a sphere, namely:

$$V = \frac{4}{3}\pi a^3 \qquad (5.10)$$

While some sills may gradually develop into spherical intrusions or magma chambers, they normally do so in several steps. And these steps commonly include a laccolith as an intermediate geometric form.

5.6 Laccolith Emplacement

A **laccolith** is a dome or mushroom-shaped body. Ideally, the lower margin is straight, whereas the upper margin is curved and convex. The diameter of the intrusion can reach many kilometres and the thickness hundreds of metres and, occasionally, one or two kilometres. The best-known laccoliths presumably are those of the Henry Mountains in Utah (United States), the first laccoliths to be described, and the Torres del Paine laccolith in Chile (Fig. 5.18). The Torres del Paine laccolith, primarily of granite, is unusually thick; its maximum thickness is about 2 km. These laccoliths are hosted by sedimentary rocks, as is common. There are, however, also several well-exposed laccoliths in basaltic lava piles, such as in Iceland (Fig. 5.19).

Many laccoliths are made of one type of magma – some felsic (Figs. 5.18 and 5.19), others are mafic. Many laccoliths are formed in several intrusions, that is, are multiple. Some multiple laccoliths form tree-like structures – commonly referred to as cedar-tree laccoliths. Multiple laccoliths are made of the same, or very similar, magma. Other laccoliths, however, are composite, that is, made of more than one type of magma, such as felsic and mafic. Laccoliths generate space primarily by lifting or up-doming of their roof layers, lopoliths primarily by down-bending of the floor layers, and some plutons by up-doming of the roof as well as down-bending of the floor, resulting in 'hybrid' structures (Fig. 5.20). In particular, **lopoliths** are commonly large plutons, similar to large sills, but while sills have essentially straight upper and lower margins, lopoliths down-bend the host-rock layers in the central part, while having a straight upper margin. The overall shape of an ideal lopolith is thus lenticular or like that of a planoconvex lens, while some show more funnel-shaped geometry at the lower central boundary. Lopoliths are commonly layered with composition ranging from ultramafic to felsic. Examples include the Skaergaard Complex in Greenland, the Bushveldt Complex in South Africa, and the Stillwater Complex in Montana (United States).

Fig. 5.18 The Torres del Paine laccolith in Chile, a fossil shallow magma chamber. The laccolith/chamber is mostly of granite but, underlain by a thick mafic pluton, it was emplaced about 12 million years ago during many thick sill injections over a period of some 90 000 years (Michel et al., 2008; Lauthold et al., 2014). The depth of emplacement was 2–3 km below the top of the volcanic area to which the laccolith belongs, the area of the laccolith is about 80 km^2, and the maximum thickness of the felsic part is about 2 km. Numerous felsic dikes and inclined sheets dissect the roof of the extinct chamber. View north, showing the main part of the laccolith. Photo: Evelyn Proimos/Flickr. A black and white version of this figure will appear in some formats. For the colour version, please refer to the plate section.

Here the focus, however, is on laccoliths, many of which function as magma chambers (Figs. 5.18 and 5.19). Ideally, a laccolith has a straight lower margin and curved and convex upper margin (Figs. 5.18 and 5.20). Laccoliths are thus like sills inflated in the middle and are generally thought to develop from sills. If the sill radius is less than its depth below the Earth's surface, Eqs. (5.4) and (5.5) can be used to estimate the sill thickness. Sills are normally not inflated in the centre. Although they have many geometric forms (Fig. 5.7), their thickness tends to vary comparatively little from the centre until close to the tips. By contrast, the thicknesses of laccoliths vary much from the centre, where they reach their greatest thickness, to their lateral tips (Figs. 5.18 and 5.20).

The evolution of a **sill into a laccolith** requires the magmatic pressure to be high enough to bend the layers above the sill so as to inflate its centre. The larger the lateral dimensions (diameter) of the sill in relation to its depth below the Earth's surface, then, for given mechanical properties of the host rocks, the easier it is for the sill to expand into a laccolith. When the diameter of the sill is similar to or larger than the depth of the sill below the Earth's surface, the model of a large sill is used. Then deflection or doming of the over-burden may happen. For some laccoliths, depending on the stiffness of the layers above and below the sill, there may also be significant **downward bending** of the host rock (Fig. 5.20). Here, however, we focus on the **upward bending** of the host-rock layers, which is generally the main mechanism for generating the space for the laccolith. For a circular,

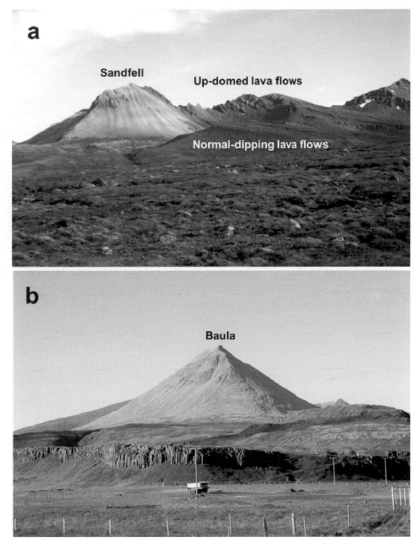

Fig. 5.19 Laccoliths may be felsic or mafic. Here are two felsic laccoliths in the Tertiary lava pile of Iceland. Both rise some 500 m above their surroundings. (a) The laccolith Sandfell in East Iceland, with part of the roof (up-domed basaltic lava flows) well preserved. (b) The laccolith Baula in West Iceland, with most of the roof having been eroded away (cf. Gudmundsson et al., 2018). A black and white version of this figure will appear in some formats. For the colour version, please refer to the plate section.

large sill, the maximum deflection u_{max} of the upper boundary (roof) of the sill is given by the following equation:

$$u_{max} = \frac{p_o a^4}{64 D_f} \left(\frac{5+v}{1+v} + \frac{4d^2}{(1-v^2)a^2} \right) \qquad (5.11)$$

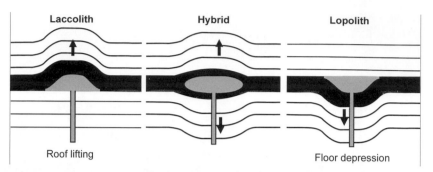

Fig. 5.20 Schematic illustration of some typical plutons/magma chambers. Laccoliths primarily generate space by up-doming or bending of the roof layers. Lopoliths primarily generate space by down-bending of the floor layers. Hybrid plutons generate space partly by up-bending of the roof layers and down-bending of the floor layers. Each pluton is shown in grey, whereas the main intruded layer/unit is in black (cf. Cruden, 1998).

where p_o is the overpressure in the sill, a is the radius of the circular sill, v is the Poisson's ratio of the host rock, and d is the depth to the top of the sill below the Earth's surface. The flexural rigidity D_f of the bent part of the crustal segment above the sill is given by:

$$D_f = \frac{Ed^3}{12(1 - v^2)} \tag{5.12}$$

where E is the Young's modulus of the bent crustal segment above the sill. For a layered crustal segment, the **effective flexural rigidity** D_e is commonly used. It is then assumed that the layers can slip along their contacts, in which case the rigidity decreases (Reddy, 2003). The effective flexural rigidity is given by the equation:

$$D_e = \frac{ET_e^{\,3}}{12(1 - v^2)} \tag{5.13}$$

Here, T_e is the effective thickness of the layered crustal segment above the sill. The effective thickness depends on the number of layers that constitute the bent crustal segment, and also on how the layers slip during bending. The layers are normally assumed to undergo **flexural slip** during the sill emplacement. There have been several assessments of T_e for various crustal segments (Pollard and Johnson, 1973; Gudmundsson, 1990). In the absence of detailed knowledge of the reaction of the layers during sill emplacement and bending, these can only be regarded as crude estimates.

From Eqs. (5.11)–(5.13) it follows that the thickness of a sill/laccolith intrusion can be estimated crudely in relation to its lateral dimension, magmatic overpressure, and the elastic constants. Such estimates have been made by many (Pollard and Johnson, 1973; Bunger and Cruden, 2011; Galland and Scheibert, 2013). All the results indicate that elasticity theory, as presented here, is a useful first approximation for the mechanical understanding of laccolith formation. It is clear, however, that during laccolith development the roof normally becomes fractured. This is clearly seen in the exceptionally well-exposed laccoliths of Torres del Paine (Fig. 5.18; Michel et al., 2008) and in Sandfell (Fig. 5.19; Gudmundsson et al., 2018) in Iceland. Thus, while part of the deformation in the roof during laccolith development can be described as elastic bending, part is related to fracture formation and is thus permanent. Permanent deformation is often best described as plastic, so that perhaps the best description of the host-rock behaviour during laccolith development is elastic-plastic.

The brittle deformation of the laccolith roof is partly through dike and sheet injections, and partly through faulting. Most or all the faults were presumably formed during the growth of the laccolith. Faults, primarily normal faults, are common in the roofs of many shallow intrusions, showing that part of the space generated for the intrusion is though roof-extension that exceeded the elastic limit. While some of the dikes may also have been injected during laccolith development, many dikes were presumably injected after the laccolith formed, in which case the laccolith acted as a shallow crustal magma chamber. And shallow magma chambers are responsible for the local stresses that control the formation of collapse calderas – a topic to which I turn now.

5.7 Vertical Collapses (Caldera Collapses)

Vertical collapses occur along dip-slip faults that constitute the ring-faults of collapse calderas. The ring-faults are primarily **shear fractures,** although some are injected by magma to form ring-dikes. Thus, the ring-faults of collapse calderas are either pure dip-slip faults, that is, shear fractures, or **mixed-mode fractures**, that is, partly **ring-faults** (shear fractures) and partly **ring-dikes** (extension fractures). All fractures depend on a **stress concentration** for their formation. For ring-faults to form there must be a suitable stress concentration above the margins of the associated shallow magma chamber. That is, there must be a concentration of shear stress that favours the development of a ring-fault, or a slip on an existing one (Chapters 3 and 9). This concentration happens only rarely, as is evident from the fact that caldera collapses in a given volcano – even one that already has developed a collapse caldera – are very infrequent in comparison with the number of dike injections and eruptions in the volcano. Thus, although many collapse calderas erupt frequently, there is seldom any movement on the existing ring-fault during eruptions. Evidently, therefore, very special conditions must be satisfied for the stresses to concentrate at the right location so as to cause a ring-fault to form or slip.

Since ring-faults are shear fractures, that is, faults, their condition of formation or slip follows the **Coulomb criterion**, a two-dimensional criterion for shear failure (assuming $\sigma_2 = 0$), namely:

$$\tau = \tau_0 + \mu\sigma_n \tag{5.14}$$

Here τ is the shear stress on the slip plane – for seismogenic fault slip it is referred to as the **driving shear stress** – τ_0 is the cohesion, cohesive strength, or inherent **shear strength** of the rock (or granular material in general), μ is the coefficient of **internal friction**, and σ_n is the **normal stress** on the fault plane (Fig. 3.3). The (inherent) shear strength of the material is its shear strength when no normal stress is applied. The Coulomb criterion was originally developed for granular materials (such as sand) and might therefore be thought to be most applicable to sediments and soft sedimentary rocks. It has, indeed, been much used to explain shear failure or faulting in soft soils, but it has also been used successfully to explain faulting of brittle rocks, such as igneous rocks (including lava flows and intrusions).

When applied to solid rocks, the failure criterion is also referred to as the **Navier–Coulomb** criterion and sometimes as the Mohr–Coulomb criterion. It is then commonly rewritten in a form named the **Modified Griffith criterion**, whereby Eq. (5.14) becomes:

$$\tau = 2T_0 + \mu\sigma_n \tag{5.15}$$

in which the cohesion or shear strength is substituted with twice the tensile strength, that is, $\tau_0 = 2T_0$. The Modified Griffith criterion follows, as the name implies, from the **Griffith theory** of brittle failure (fracture), which is discussed in detail with application to rock fractures in Gudmundsson (2011a). Equation (5.15) implies that the inherent shear strength is about twice as large as the tensile strength, which is in good agreement with measurements.

Equations (5.14) and (5.15) are for dry rocks. All rocks, however, below a certain crustal depth contain fluids, mostly water. On dry land, this depth is normally a few metres – several tens of metres in very dry areas – and the boundary below which the pore space (composed of cavities, vesicles, contacts, and fractures) of the rock is filled with fluids is, for groundwater, the water table. Increasing pore-fluid pressure increases the likelihood of fault initiation or slip, for all types of faults, including ring-faults.

Presumably, all tectonic faults, including ring-faults, form – or at least are initiated – within zones or areas of **high pore-fluid pressure.** Volcanic areas commonly contain geothermal fields and these are subject to high pore-fluid pressure during volcano deformation. For example, in Iceland there were over 100 000 earthquakes in Hengill, a central volcano with a major geothermal field over a period of only 6 years, from 1994 to 2000 (Chapter 4; Jakobsdottir, 2008). In addition to the pore-fluid pressure, there can be **pressurised fluids** on the potential or actual fault plane itself. For ring-faults, these fluids may be groundwater, volcanic gas, geothermal water, and magma. When any such fluids are present on the fault plane, Eqs. (5.14) and (5.15) become modified to the following equations:

$$\tau = \tau_0 + \mu(\sigma_n - p_t) \tag{5.16}$$

$$\tau = 2T_0 + \mu(\sigma_n - p_t) \tag{5.17}$$

where p_t is total fluid pressure in the rock, on and in the vicinity of the fault plane, at the time of fault initiation or slip.

Eqs. (5.16) and (5.17) have important implications for fault development in general, and for ring-faults in particular, namely:

- When the total fluid pressure p_t approaches or equals the normal stress σ_n, the term $\mu(\sigma_n - p_t)$ approaches or equals zero and the driving shear stress for slip becomes τ_0, which is equal to $2T_0$.
- Since the *in situ* tensile strength of rocks is typically in the range of 0.5–6 MPa, it follows that, for high-fluid-pressure fault zones, the typical driving shear stress for slip should be 1–12 MPa. These values are in agreement with common estimated **stress drops**.

- Fault slip is greatly facilitated when $\mu(\sigma_n - p_t)$ approaches or equals zero. Not only because the normal stress σ_n on the fault plane becomes zero or negative (negative, for example, if a ring-dike forms), but also because the friction along the fault plane is very much reduced. That is, the fluid (water, magma, gas) **lubricates the fault plane** and makes displacement easier.

Notice that a ring-dike is not a necessary condition for a ring-fault to form or slip. However, a ring-dike makes slip much easier and it is less likely that the propagating ring-fault will become arrested (Fig. 5.21). That is, a ring-dike emplacement makes it more likely that, for otherwise suitable mechanical conditions, a comparatively large caldera collapse will occur.

Ring-faults are **dip-slip faults**. Some appear to be normal faults, whereas others appear to be reverse faults. One characteristic of most ring-faults, however, is that, whether they are normal faults or reverse faults, they are normally very **steeply dipping**. In fact, many ring-faults are close to vertical (Fig. 5.22). So why do so many ring-faults have close-to-vertical dips instead of the typical dips of 50–70° for normal faults and reverse faults?

The answer is that the stress concentration around the shallow chamber generating the ring-fault modifies the stress field so as to favour close-to-vertical dip-slip faults. This is seen in Fig. 5.23. Here, the maximum compressive principal stress σ_1 dips in such a way that if the fault makes the angle of, say, 30° to σ_1, it is clear that the fault (here the ring-fault) would be somewhat inward-dipping or very close to **vertical**.

The vertical displacement on the ring-fault during collapse may vary from a few metres or less to several kilometres. In fact, during some caldera collapses part of the ring-fault

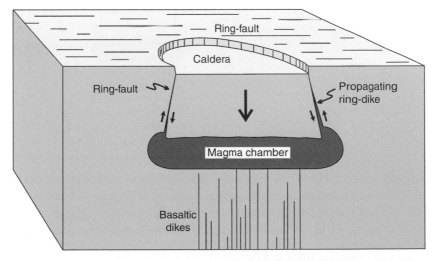

Fig. 5.21 Schematic illustration of the beginning stages of ring-dike emplacement along an outward-dipping (reverse) ring-fault. As the caldera block subsides, the opening or aperture of the ring-fault increases so that the ring-dike gradually occupies much of the, or the entire, fault, and becomes thicker until, eventually, the dike may erupt at the surface (Fig. 5.24). For an outward-dipping fault, the slip is unstable (cf. Gudmundsson, 2015).

Part of the ring-fault of the Hafnarfjall caldera, an extinct Tertiary central volcano in West Iceland. View east, the ring-fault is subvertical, with an estimated displacement of about 200 m. A multiple ring-dike, a few metres thick, occupies (the lower) part of the exposed fault. The lava flows inside the caldera (right of the fault) are more steeply dipping than those outside the caldera (cf. Gautneb et al., 1989; Browning and Gudmundsson, 2015b). A black and white version of this figure will appear in some formats. For the colour version, please refer to the plate section.

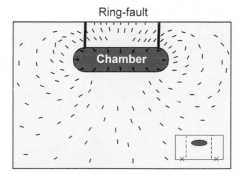

Numerical model of a sill-like magma chamber subject to an internal excess pressure of 5 MPa as the only loading. The ticks represent the stress trajectories or directions of the maximum compressive principal stress σ_1. A dip-slip fault, such as a ring-fault, commonly makes an angle of about 30° to the direction of σ_1, in which case the fault would be close to vertical as indicated schematically by the thick, vertical lines. The mechanics of ring-fault formation is also discussed in Chapter 3.

does not slip at all. Examples include **trapdoor collapses**, such as have been observed in the calderas of the Galapagos Islands (Amelung et al., 2000). Normally, however, most of the, or the entire, ring-fault is involved in the collapse.

The vertical displacement on ring-faults usually occurs in many slips. That is, the entire displacement of, commonly, hundreds of metres, is normally not reached in a single slip. There are many examples of this process. For example, the Lake Öskjuvatn collapse in the Askja central volcano in Iceland, which started in 1874, apparently took many years to complete (Hartley and Thordarson, 2013). Also, the well-documented collapse of Miyakejima in Japan in 2000, of as much as 450 m, took many weeks to complete (Geshi et al., 2002). Similarly, the 2018 summit collapse of Kilauea, Hawaii, reaching a maximum displacement of about 500 m, occurred in 62 slips over a period of more than 3 months (Neal et al., 2019). While the maximum displacement of 350 m during the collapse of Fernandina in Galapagos in 1968 – the first instrumentally monitored collapse – happened quickly, the displacement did not occur as a single slip. Earthquake studies (Filson et al., 1973) suggest that the displacement of 350 m occurred in some 75 slips, each slip being, on average, 4–5 m.

As mentioned above, a ring-dike is not needed for a caldera collapse to occur. In fact, of the six instrumentally documented caldera collapses during the past century or so, no ring-dike eruption occurred in at least five of them. These are the collapses of Katmai, Alaska (1912), Fernandina, Galapagos (1968), Miyakejima, Japan (2000), Piton de la Fournaise, Reunion (2007), and Kilauea (2018). Although part of the eruption during the caldera collapse in Pinatubo, Philippines (1991) may have been fed by a ring-dike, most of the eruption was apparently along the central conduit.

The 60-m maximum, largely elastic subsidence in the centre of Bardarbunga, Iceland, in 2014–2015, cannot be regarded as an ordinary caldera collapse – and was not associated with any ring-dike eruption (Browning and Gudmundsson, 2015a). The 1975–1976 collapse of the Plosky Tolbachik pit crater (Fedotov et al., 1980) may, however, be regarded as the seventh instrumentally recorded caldera collapse. The Plosky Tolbachik Volcano is a basaltic edifice and part of its summit collapsed during the 1975–1976 fissure eruption (Fedotov et al., 1980). The collapse was associated with an existing pit crater. The diameter of the floor of the collapse was only about 700 m and thus was less than the minimum diameter of 1 km of a typical caldera. The maximum diameter at the rims (upper edge of the caldera), however, is about 1300 m, in which case the collapse may be regarded as that of a caldera with a rather gently inward-dipping (normal) ring-fault. There was no ring-dike eruption during the collapse. The 2018 Kilauea collapse was also associated with a pit crater (and a caldera, of course) and occurred on inward-dipping (normal) faults. The maximum diameter of the collapse, however, is much larger than that of Plosky Tolbachik, 2.5–3 km (Neal et al., 2019).

While a ring-dike is thus not needed for caldera collapses to occur, such a dike will almost always be injected during collapse along **outward-dipping faults**, that is, reverse faults. This follows because the geometric consequence of the outward-dipping ring-fault is a cavity, an **open ring-fault**, at the contact with the magma chamber itself. As the floor of the caldera subsides, the opening of the ring-fault increases (Fig. 5.24). Magma in the chamber tends to flow to regions of such open cavities or fractures which are, technically, regions of lowered potential energy. Thus, magma would tend to flow into the gradually opened fault to form a ring-dike.

The evidence for outward-dipping versus inward-dipping ring-faults is discussed in greater detail in Chapter 8. The ring-fault dips, however, have implications for the collapse

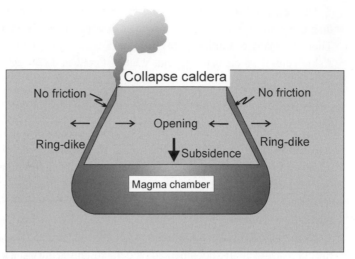

Fig. 5.24 The aperture (opening) of an outward-dipping ring-fault increases (indicated by horizontal arrows) as the caldera subsides, and the fault will normally be occupied by a ring-dike. It follows that there is essentially no friction along the ring-fault and the caldera block may subside to the bottom of the chamber and, practically, squeeze out all the magma in the chamber, thereby giving rise to (commonly) a large eruption. Also, the volumetric flow or effusion rate should increase greatly with increasing subsidence of the caldera block. Modified from Gudmundsson (2015).

mechanisms and the associated eruptions that will now be summarised briefly. In addition to the formation of a ring-dike, which will necessarily be injected from the magma chamber into an outward-dipping, open ring-fault, there are two importance consequences of an outward dip of a ring-fault (Gudmundsson, 2015).

The first consequence is that the volume of erupted magma per unit time supplied by the ring-dike feeding a surface eruption should **increase** with increasing subsidence of the caldera floor. As the piston-like caldera block subsides into the magma chamber, the **opening** or aperture of the fracture between the host rock and the block gradually **increases**. The **volumetric flow rate Q**, the volume per unit time that flows up through the ring-dike, is related to the ring-dike aperture or opening through the equation (Fig. 5.25):

$$Q = \frac{\Delta u_I^3 W}{12\mu_m} \left[(\rho_r - \rho_m)g \sin \alpha - \frac{\partial p_e}{\partial L} \right] \tag{5.18}$$

Here, Δu_I is the opening or aperture of the ring-dike (or any dike or volcanic fissure, for that matter), W is its length or strike-dimension of the erupting fissure at the surface, μ_m is the dynamic viscosity of the magma in Pa s, ρ_m is the density of the magma in kg m^{-3} (assumed to be constant), and ρ_r is the average density of the crustal segment (which includes the volcano) through which the ring-dike (or any feeder-dike) propagates to the surface. Also, g is the acceleration due to gravity in m s^{-2}, α is the dip of the ring-dike in degrees (for a vertical ring-dike, sin 90° = 1), and $\partial p_e/\partial L$ is the vertical magmatic excess-pressure gradient in the direction of the magma flow, that is, in the direction of the dip-dimension of the dike L in metres. Since the volumetric flow rate through the ring-dike depends on the aperture to the third power, that is, Δu^3, the aperture increase with caldera subsidence along

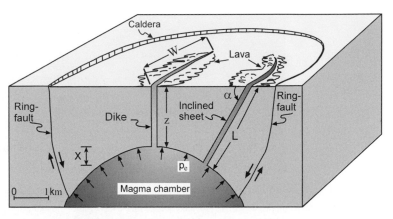

Fig. 5.25 Schematic illustration of magma transport to the surface of a volcano, here a collapse caldera, through a vertical dike and an inclined sheet. The dip-dimension of the dike is denoted by z, that of the inclined sheet (with a dip a) is denoted by L. The strike-dimension of the dike, half of which is shown here, is denoted by W, the excess pressure in the chamber at the time of rupture by p_e. The vertical difference in the point of initiation of the sheet and the dike is denoted by X; for a sill-like chamber that difference is normally close to zero. Modified from Gudmundsson (2015).

an outward-dipping ring-fault should result in a very great increase in the volumetric flow rate Q during the subsidence. This dependence of Q on the cube of the aperture, Δu^3, applies to any flow of any fluid (groundwater, geothermal water, gas, oil, magma) through rock fractures and is known as the **cubic law.**

The second consequence is that because of the injected fluid ring-dike there will be essentially no friction along the ring-fault or, more specifically, only the 'frictional effects' related to the viscosity of the magma that forms the ring-dike (Fig. 5.24). As the ring-dike opening increases with increasing caldera-block subsidence on the outward-dipping fault, there is theoretically nothing to stop the piston-like block from **sinking to the bottom** of the chamber. The consequence, for an outward-dipping ring-fault, is that a ring-dike would form, and essentially the **entire magma chamber would erupt**. More specifically, very little if any magma would be left in the chamber following collapse on an outward-dipping ring-fault. By analogy with fracture propagation, such a caldera collapse can be referred to as **unstable.**

The other basic model of ring-fault attitude is the **inward-dipping** or normal fault (Fig. 5.26). Inward-dipping ring-faults tend to form when the crustal segment, including the volcano itself, above the shallow magma chamber is subject to extension. Such a stress field is very common. Not only does it occur in all volcanoes related to rift zones, such as at divergent plate boundaries and continental grabens, but it is also the stress field generated by doming of the volcanic field or zone to which the volcano belongs. More specifically, extension sufficient for inward-dipping ring-fault formation can be generated through centi-metre-scale doming induced by a deeper reservoir (Gudmundsson, 2007), rift-related extension, gravity sliding resulting in volcano spreading, or any other similar mechanism.

Partly because the shallow magma chamber acts as a cavity that raises stress, and partly because of the developing or slipping (existing) ring-fault itself, tensile stresses concentrate

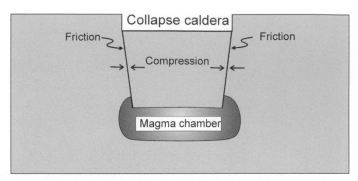

Fig. 5.26   As caldera block subsides along an inward-dipping normal ring-fault it becomes subject to higher horizontal compression (compressive stress). Provided that no ring-dike occupies the fault, the friction and normal stress on the ring-fault ensure that the fault remains closed and the displacement stable. Thus, the collapse on an inward-dipping normal ring-fault is normally stable and is less likely than that on an outward-dipping ring-fault to result in an increasing volumetric flow rate during the subsidence. Modified from Gudmundsson (2015).

in and around the piston-like caldera block. This tensile stress reduces friction along the ring-fault and promotes caldera subsidence. If the inward-dipping ring-fault is close to vertical, injected by a ring-dike, or both, then a **total collapse** is possible. This implies, as in the case of the outward-dipping ring-fault, that the magma chamber may be, to a large degree, **emptied during the eruption**. For other conditions, the friction on inward-dipping ring-faults (Fig. 5.26) limits the subsidence to a maximum of, in most cases, a few hundred metres and, occasionally, 1–2 km. Also, on parts of a ring-fault where there is high friction, such as parts where no ring-dike forms, or parts where the dip of the ring-fault is unusually gentle and the normal stress high, there may be little or no fault slip during the collapse. These are presumably the main reasons for the formation of **trapdoor calderas**. Thus, for inward-dipping ring-faults, the slips are more controlled than for outward-dipping faults. By analogy with fracture propagation, we refer to caldera collapse on inward-dipping faults as **stable.**

As discussed above, and in more detail in Chapter 9, it is not known if the majority of ring-faults are outward dipping or inward dipping. What is clear, though, is that many, perhaps most, ring-faults are close to vertical (Figs. 5.22 and 5.23). As for ring-dikes, many calderas, for example in Iceland, have very thin or non-existent ring-dikes (Browning and Gudmundsson, 2015b). The calderas where the **ring-dikes are missing** tend to be steeply inward-dipping (Fig. 5.22). This is in agreement with the points raised above, namely that an outward-dipping ring-fault would normally be expected for be occupied with a ring-dike.

As for the volumetric (effusion) rate (Eq. (5.18)), it is not clear if this increases as the caldera subsidence continues. We also do not know whether or not caldera eruptions are primarily related to the ring-dikes or central eruptions. Some caldera-forming eruptions are partly along a ring-dike, while others appear to be primarily in the central part of the caldera. The central part may then either be the main conduit of the stratovolcano that subsequently collapses or form the central part of an already existing caldera. And, of course, many collapses, particularly in basaltic edifices, occur with little or no eruptions.

The outward-dipping ring-fault model implies that magma chambers are largely or entirely destroyed during the collapse – that effectively all the magma is squeezed out of the chamber. Observations suggest that some magma chambers are destroyed by caldera collapses. In some calderas there may be little, if any, volcanic activity following a major collapse (e.g. Marti and Gudmundsson, 2000). Most calderas, however, appear to be highly **active after collapse** (Newhall and Dzurisin, 1988), suggesting that the magma chamber is not destroyed during the collapse. Of the well-documented collapses in the past century and discussed above, there have been eruptions in Fernandina (following the 1968 collapse), in Pinatubo (following the 1991 collapse), in Miyakejima (following the 2000 collapse), and in Piton de la Fournaise (following the 2007 collapse). These observations suggest that many, perhaps most, caldera collapses do not destroy the associated magma chambers and that the chambers remain active, even soon after the collapse.

5.8 Lateral/Sector Collapses (Landslides)

Lateral/sector collapses are common in volcanic edifices. All major lateral collapses occur, as is the case for vertical collapses, along shear fractures, that is, faults. These types of collapses, however, do not cover all types of landslides, which are many and include debris flows, for example. Here I provide a brief discussion of the typical mechanical conditions for landslides in volcanic edifices (Fig. 5.27). While landslides of various types are common in all volcanoes, the frequency and style is somewhat different between different

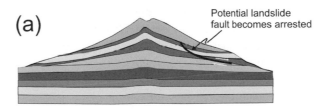

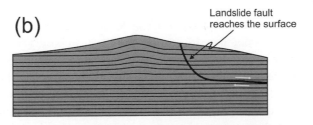

Fig. 5.27   For a large landslide to occur in a volcanic edifice, the shear fracture (the landslide fault or slip surface) must reach the surface along the slopes of the volcano. No large landslide occurs if the propagating fracture becomes arrested at contacts between layers, which is more likely to happen in (a) a stratovolcano than in (b) a basaltic edifice.

types. There are several factors that affect how frequent major landslides are in different volcano types. The main points are the following:

1. The **slope** of the volcano. Other things being equal, major landslides are expected to be more common in volcanoes with steep slopes than in volcanoes with gentle slopes.
2. The **strength** of the volcano. By strength I mean here primarily how easily a fracture (for a lateral collapse, a shear fracture) propagates through a part of the volcanic edifice and to its surface. Thus, the main strengths under consideration are the tensile and shear strengths. Other things being equal, major landslides are expected to be more common in low-strength edifices than in high-strength ones.
3. For most volcanic edifices there is a positive relationship between slope and strength. This means that edifices with **steep slopes** are normally mechanically **stronger** than those with gentle slopes. More specifically, stratovolcanoes (steep slopes) are generally mechanically stronger than basaltic edifices (gentle slopes).

As indicated in the third point, stratovolcanoes tend to have steeper slopes and are mechanically stronger than basaltic edifices. More specifically, stratovolcanoes tend to be long-lived, tall, and strong structures (Figs. 1.2 and 1.3). The tallest **stratovolcanoes** reach 6–6.9 km above sea level or 4–5.8 km above their surroundings. The slopes vary from very gentle at the margins of the volcanoes to as steep as 35–42° (Gudmundsson, 2009). While small landslides or rock-falls are common, which is understandable given the steep slopes, large landslides, which take 20–30% of the cone material of the volcano appear to be rare. Such landslides or lateral collapses in stratovolcanoes seem to require specific external loading, such as shallow magma intrusion or earthquakes (Tibaldi et al., 2006).

Yet in some stratovolcanoes, over certain time periods, there may be many significant landslides. Examples include stratovolcanoes such as Stromboli in Italy, where four large landslides have occurred in the past 15 thousand years (Tibaldi, 2001), and Augustine in Alaska where 11 lateral collapses have occurred in the past 2000 years (Beget and Kienle, 1992). Both these stratovolcanoes are very active. For example, Augustine is the most active volcano in the eastern part of the Aleutian Arc. Similarly, Stromboli is one of the most active volcanoes in the world. The high activity of these two volcanoes may be one reason for their relative instability and frequent landslides over these time periods. In general, large landslides appear to be **less frequent** in typical stratovolcanoes than in typical basaltic edifices.

The largest **basaltic edifices** (shield volcanoes) on Earth rise as much as 9 km above the sea floor; in fact they rise 15–17 km if the depression of the sea floor is taken into account. Most basaltic edifices, however, are much smaller. In comparison with the stratovolcanoes, the slopes of the basaltic edifices are generally gentle. For example, the slopes of the subaerial parts of Kilauea and Mauna Loa are mostly 4–8°. Large landslides are common in basaltic edifices. Well-known large landslides occur, for example, around the Big Island of Hawaii, the Canary Islands (some of which are primarily basaltic), and Reunion. Some large landslides are submarine, others are subaerial. The largest landslides around Hawaii are among the largest on Earth, some of them reaching lengths of about 200 km and volumes of as much as 5000 km^3 (Moore et al., 1994).

Because the rock layers are very similar in basaltic edifices, but dissimilar in stratovolcanoes, the chances that a propagating fracture becomes arrested at contacts between layers is generally much greater in stratovolcanoes than in basaltic edifices (Fig. 5.27). This follows because all the processes of fracture arrest discussed earlier, that is, the Cook–Gordon delamination, stress barriers, and elastic mismatch, operate more efficiently in volcanoes composed of layers with contrasting mechanical properties. Thus, stratovolcanoes show more **resistance** to failure through fracture propagation, which includes the ring-faults of collapse calderas and the shear fractures or faults of major landslides. Other things being equal, it should therefore normally be **easier** to form a landslide in a basaltic edifice than in a stratovolcano. No systematic studies of landslide or lateral-collapse frequencies have been undertaken, however. And such a study is urgently needed to assess the hazards and risks from large landslides in volcanic edifices, and the associated risk of tsunamis in volcanic islands and other edifices that are close to large water bodies such as large lakes and the sea.

5.9 Summary

- Most magma chambers evolve from sills. While a horizontal sill stays liquid it acts as a barrier to vertical dike propagation and tends to absorb part of the magma of the dikes that meet it. The initial sill may be a single one, or a cluster that eventually merges into a single magma body. The main condition for a sill (or sills) to evolve into a shallow chamber is that the frequency of dikes meeting the sill is high enough for it to remain liquid during the elastic-plastic expansion necessary for the chamber to form.
- Three mechanisms encourage the emplacement of sills: the Cook–Gordon delamination, stress barriers, and elastic mismatch. The essence of the Cook–Gordon mechanism for sills is that when a vertical dike approaches a contact, the tensile stresses ahead of the dike tip tend to open the contact so that, once the contact is reached, the dike may deflect into the contact to form a sill. A stress barrier is simply a rock layer or unit with a local state of stress that is unfavourable to the propagation of a particular type of fracture. For a dike to deflect into a sill, such a barrier would have σ_1 horizontal and σ_3 vertical. The essence of elastic mismatch for sills is difference in the stiffness or Young's modulus on either side of a contact or discontinuity in relation to the contact properties. The stiffer the layer above (for a subhorizontal contact) the contact in relation to the layer below the contact (and hosting the dike), the greater is the tendency for the dike to become deflected along the contact to form a sill.
- A magma chamber ruptures and injects a dike (or an inclined sheet) when the excess pressure p_e in the chamber reaches the tensile strength T_0 of the host rock (the chamber roof), as specified in Eq. (5.1). Alternatively, a magma chamber ruptures when its total fluid pressure reaches the combined value of the minimum compressive principal σ_3 stress and the *in situ* tensile strength (Eq. (5.2)). The conditions of Eqs. (5.1) and (5.2) can be reached by (1) adding magma to the chamber, (2) reducing σ_3 through extension (local or regional), and (3) a combination of added magma and reduced σ_3.

- Once a dike (or an inclined sheet) starts to propagate to shallower crustal layers above the magma-chamber roof, buoyancy becomes added to the excess pressure that ruptured the chamber. The total driving pressure or overpressure p_o of the dike at any crustal level is then given by Eq. (5.3). For some high-density mafic magmas, the buoyancy effect may be negative along the entire dike path from the shallow chamber to the surface. For many mafic magmas, and all intermediate and felsic magmas, however, the buoyancy effect is positive (it adds to the driving pressure) on the path close to the shallow magma chamber and may be so all the way to the surface.

- The shape, in particular the thickness (aperture), of a dike, an inclined sheet, or a sill is determined by Eq. (5.4), which applies to a general sheet-like intrusion. This equation assumes that the strike- and dip-dimensions are very unequal in size; either the strike-dimension is much larger than the dip-dimension, or vice versa. Fractures where these dimensions are very different in size are known as tunnel cracks and intrusions modelled as such are referred to as tunnel-shaped. Many sills, however, are close to circular in plan view, in which case Eq. (5.5) is appropriate for describing their shape and Eq. (5.7) for calculating their volume. Crudely, the condition for sheet propagation may be inferred from Eq. (5.6). However, fracture mechanics, as discussed in Chapter 7, offers a more accurate description of the conditions for their propagation.

- Many sills develop into laccoliths, that is, mushroom-shaped intrusions, many of which function as shallow magma chambers. This applies particularly to sills with lateral dimensions that are large in relation to their depth below the Earth's surface. The expansion of a sill into a laccolith is complex, and involves fracture development in the roof (dike injection and faulting) and general elastic-plastic deformation of the host rock. To a first approximation, however, the maximum uplift of the roof, to generate space for the magma constituting the laccolith, can be estimated from Eqs. (5.11)–(5.13).

- Vertical and lateral collapses are common in volcanic edifices. Both relate to failure of the volcanic edifice through the development of shear fractures, that is, faults (Eqs. (5.14)–(5.17)). Vertical collapses occur along ring-faults and generate collapse calderas. A necessary condition for a ring-fault to form is a shear-stress concentration in zones above the lateral edges of the associated shallow magma chamber. Most collapses appear to occur in steps, sometimes rapidly (over days, weeks, or months), and sometimes slowly (over years or decades). Ring-fault formation or slip, particularly in basaltic edifices, does not require any major eruptions. However, all large explosive eruptions are associated with caldera collapses.

- All ring-faults are dip-slip faults and most have dips close to vertical. The steep dip of the ring-fault, in contrast to the more gentle dips typical of normal and reverse faults, is due to the inclined trajectories of the principal stresses around the associated shallow magma chamber. For slightly outward-dipping (reverse) ring-faults, a ring-dike is expected to form and the collapse is likely to be unstable. This means that most or all of the magma in the chamber is erupted during the collapse, and the chamber itself is effectively destroyed. For slightly inward-dipping (normal) ring-faults, a ring-dike may inject but it is not a necessary result of the collapse, which is likely to be stable. This means that the chamber is normally not destroyed and much magma remains in it following the collapse. The different types of ring-faults (outward- or inward-dipping) have different implications for volcanic hazards. There are, however, few accurate measurements of ring-fault and ring-dike attitudes so that, at this stage, it is not known whether the outward- or the inward-dipping ring-fault is the more common.

• Lateral collapses occur along gently dipping faults and generate large landslides. These are mainly rotational or translational landslides. The probability of large landslides in volcanoes depends on several factors, primarily the slope and the mechanical strength of the volcanic edifice. Stratovolcanoes, particularly their upper parts, have steeper slopes than basaltic edifices. But stratovolcanoes are also mechanically stronger – more resistant to large-scale fracture propagation – than basaltic edifices. The frequency of large land-slides appears to be greater in basaltic edifices than in stratovolcanoes. However, no detailed systematic studies have been made of landslide frequencies in relation to slopes, strengths, and other factors. Since these frequencies, and their causes, have great hazard implications, such studies are urgently needed.

5.10 Main Symbols Used

a	radius of a circular (penny-shaped) interior crack, here a sill
a	major semi-axis of an ellipsoid, or in two-dimensions, an ellipse, here a sill
b	intermediate semi-axis of an ellipsoid, here a sill
C	constant
c	minor semi-axis of an ellipsoid, here a sill
D_e	effective flexural rigidity
D_f	flexural rigidity
d	depth to the top of a sill (below the Earth's surface)
E	Young's modulus (modulus of elasticity)
g	acceleration due to gravity
h	height or dip-dimension of a dike (above its source magma chamber)
L	the lateral dimension (length) of a tunnel-shaped fracture
L	length of fluid transport along a dike or an inclined sheet
p_e	excess magmatic pressure (in a chamber)
p_l	lithostatic stress or pressure
p_t	total fluid pressure
p_o	magmatic overpressure
Q	volumetric flow rate (of a fluid, here magma)
T_e	effective thickness of a layered crustal segment (above a sill)
T_0	tensile strength
u	half thickness of an intrusive sheet
Δu_I	maximum aperture (total opening displacement) of a mode I crack, here a sill
V	volume of an ellipsoid, here a sill

V_s volume of a circular (penny-shaped) sill

W strike-dimension of a dike or volcanic fissure

α dip of hydrofracture, here a ring-dike

μ coefficient of internal friction

μ_m dynamic viscosity of magma

v Poisson's ratio

π 3.1416

ρ_m magma density

ρ_r rock or crustal density

σ_{max} maximum tensile stress

σ_d differential stress $(\sigma_1 - \sigma_3)$

σ_n normal stress (usually compressive) on a fault plane

σ_3 the minimum compressive principal (maximum tensile) stress

τ shear stress for faulting

τ_0 inherent shear strength (or cohesion or cohesive strength)

5.11 Worked Examples

Example 5.1

Problem

Consider a 1.5-km-long feeder-dike, supplying magma to a volcanic fissure. The host rock at the surface has a Young's modulus of 10 GPa and a Poisson's ratio of 0.25. Calculate the opening or aperture of the volcanic fissure for a magmatic overpressure of 10 MPa.

Solution

As always, we use standard SI units, so that giga-pascals (GPa) and mega-pascals, given in the example, must be changed to pascals (Pa). Similarly, lengths given in kilometres must be changed into metres. Because it is a feeder-dike, the controlling dimension is the strike-dimension or length L of the volcanic fissure at the surface. From Eq. (5.4) we then have:

$$\Delta u_I = \frac{2p_o(1-v^2)L}{E} = \frac{2 \times 1 \times 10^7 \text{ Pa} \times (1 - 0.25^2) \times 1500 \text{ m}}{1 \times 10^{10} \text{ Pa}} = 2.8 \text{ m}$$

This result is a very reasonable one for a feeder-dike. Many basaltic feeder-dikes worldwide have thicknesses from a fraction of a metre to many metres at the surface (Fig. 1.21; Geshi et al., 2010; Galindo and Gudmundsson, 2012; Geshi and Neri, 2014).

Example 5.2

Problem

During caldera collapse on an outward-dipping ring-fault, the opening or aperture Δu_I of a ring-dike feeding the eruption increases abruptly from its initial value of 1 m to 10 m (Fig. 5.24). If all the other parameters affecting the volumetric flow rate Q remain unchanged (constant) during this aperture increase, roughly by how much would Q increase?

Solution

From Eq. (5.18) we have:

$$Q = \frac{\Delta u_I^3 W}{12\mu_m} \left[(\rho_r - \rho_m)g \sin \alpha - \frac{\partial p_e}{\partial L} \right]$$

It is stated, however, in the example that the only thing that is supposed to change during the collapse is the aperture Δu_I of the ring-dike. It follows that the remainder of the right-hand side of the equation may here be regarded as a constant, C, in which case Eq. (5.18) becomes:

$$Q = C\Delta u_I^3 \qquad (5.19)$$

It follows from Eq. (5.19) that when Δu_I is increased by 10 times, then Q will increase by $10^3 = 1000$ times. Thus, if the aperture of the ring-dike increased abruptly from 1 m to 10 m, and all other parameters remained the same, the volumetric flow rate up through the ring-dike would increase by a factor of 1000.

We have here explained the ring-dike aperture increase as a consequence of subsidence along an outward-dipping ring-fault. As discussed in the chapter, however, so long as a ring-dike forms and supplies magma to the caldera eruption – whether along an outward-dipping, an inward-dipping, or a vertical ring-fault – increase in the dike aperture will have large effects on the volumetric flow rate, that is, the power of the eruption. Large-aperture ring-dikes are presumably one reason for the enormous power at the peak of many caldera-forming eruptions.

Example 5.3

Problem

A circular sill is emplaced at 5 km depth in a lava pile in a rift zone. The sill radius is 2 km and the estimated overpressure at its time of formation is 15 MPa. The Young's modulus of the host rock is 30 GPa and its Poisson's ratio 0.25.

(a) Calculate the maximum opening or thickness of the sill.
(b) Estimate crudely the time of solidification for the sill, using Eq. (5.20) below.

Solution

(a) Since the sill is circular and its diameter is less than the sill depth below the surface, it is appropriate to use Eq. (5.5) to calculate the sill opening or thickness. Using the values given for the various parameters, we have:

$$\Delta u_I = \frac{8p_o(1-v^2)a}{\pi E} = \frac{8 \times 1.5 \times 10^7 \mathrm{Pa}(1-0.25^2) \times 2000 \text{ m}}{3.1416 \times 3 \times 10^{10} \text{ Pa}} = 2.4 \text{ m}$$

(b) All the basic equations for the solidification of a sill are discussed in Chapter 7. In the present chapter the only quantitative assessment provided is the following approximate formula, which follows from Eq. (7.27), namely that the approximate time for solidification in years is:

$$t_y \cong 0.073u^2 \tag{5.20}$$

which we hereby give an equation number. Here, the implied unit of the number (the factor 0.073) is m^{-2} yr. Since u is half the thickness of the sill at the time of emplacement, we have from (a) that $u = 1.2$ m. It follows from Eq. (5.20) that the time of complete solidification of the sill is 0.10 years or about 5.5 weeks. This thin sill has thus essentially no chance of developing into a magma chamber.

Example 5.4

Problem

A circular sill is emplaced at the depth of 2 km below the surface of an active rift zone. The sill radius is 6 km and the estimated overpressure at its time of emplacement is 10 MPa. The Young's modulus of the host rock is 15 GPa and its Poisson's ratio is 0.25. Assume no slip along the contacts of the host-rock layers.

(a) Calculate the maximum opening or thickness of the sill.
(b) Estimate crudely the time of solidification of the sill.

Solution

(a) The sill has a diameter of 12 km but is located at 2 km below the Earth's surface. Thus, its lateral dimension is much larger than its depth below the surface. It follows that the appropriate formula to use to calculate its thickness is Eq. (5.11), suggesting that the sill may develop into a laccolith. To use Eq. (5.11) we need to know the flexural rigidity. It is assumed that there is no flexural slip along the host-rock layers, so that the appropriate formula for the flexural rigidity D_f is Eq. (5.12). Thus, we have:

$$D_f = \frac{Ed^3}{12(1-v^2)} = \frac{1.5 \times 10^{10} \text{ Pa} \times 2000^3}{12(1-0.25^2)} = 1 \times 10^{19} \text{ Nm}$$

Substituting this result in Eq. (5.11) yields:

$$u_{max} = \frac{p_o a^4}{64D_f}\left(\frac{5+v}{1+v} + \frac{4d^2}{(1-v^2)a^2}\right)$$

$$= \frac{1 \times 10^7 \text{ Pa} \times (6000 \text{ m})^4}{64 \times 1 \times 10^{19} \text{ Nm}}\left(\frac{5+0.25}{1+0.25} + \frac{4 \times (2000 \text{ m})^2}{(1-0.25^2) \times (6000 \text{ m})^2}\right)$$

$$u_{max} = 94.6 \text{ m}$$

Since the half thickness of the sill is 94.6 m, the total thickness is twice this value, or 189.2 m.

(b) To calculate the time for a complete solidification of the sill in years, we again use the approximate formula Eq. (5.20), which yields:

$$t_s \cong 0.073u^2 = 0.073 \times (94.6)^2 = 653 \text{ years}$$

A sill of this thickness would therefore remain partially fluid for hundreds of years. Thus, this sill would have a great chance of developing into a shallow magma chamber, particularly in rift zones with spreading rates similar to those at intermediate- and fast-spreading ridges (4–17 cm per year). At these spreading rates, whether in continental grabens, at mid-ocean ridges, or in general volcanic rift zones such as on volcanic edifices (volcano spreading) or ocean islands, the dike-injection frequency is commonly one dike reaching the shallow crustal level of a sill of this type once every 5–50 years. Such an injection rate would commonly be sufficient to maintain a shallow sill-like magma chamber. At slow spreading rates (0.5–4 cm per year), however, as in many continental volcanic areas and at slow-spreading ridges and rift zones, the rate may be too low to develop even such a thick sill into a shallow magma chamber. This is one reason why many thick sills never develop into magma chambers. The other reason is that many thick sills are multiple (Figs. 5.3, 5.5), that is, formed in many magma injections where each injection solidifies (forms a columnar sheet or row) before the next injection (Figs. 2.5 and Fig. 5.5)

5.12 Exercises

5.1 A circular (penny-shaped) basaltic sill is emplaced at 3 km depth in a rift zone. The Young's modulus of the host rock is 20 GPa and its Poisson's ratio is 0.25. Find the maximum thickness of the sill if the magmatic overpressure at the time of its emplacement is 10 MPa.

5.2 Measurements in a drill hole show that a sill cut by the hole at 5.5 km depth is 3.8 m thick. On seismic images, the lateral dimension of the sill is estimated at 4 km. Assume the sill to be circular and with a diameter equal to the detected lateral dimension, 4 km. If the Young's modulus and Poisson's ratio of the host rock at the time of sill emplacement were 25 GPa and 0.25, respectively, what was the magmatic overpressure driving the formation of the sill?

5.3 Calculate the volume of the sill in Exercise 5.2. If this volume had erupted as lava flow at the surface, instead of forming a sill, would it be regarded as a small, medium (moderate), or large effusive eruption?

5.4 Estimate crudely the time for complete solidification of the sill in Exercise 5.2 and assess the chances of the sill developing into a shallow magma chamber.

5.5 A circular sill intrusion is emplaced at the depth of 3 km below the surface. The sill radius is 5 km and the estimated overpressure at its time of emplacement is 15 MPa.

         The Young's modulus of the host rock is 20 GPa and its Poisson's ratio is 0.25. There is no evidence of slip along the contacts of the host-rock layers. Calculate the maximum total opening displacement or thickness of the sill.

5.6   Estimate crudely the time for complete solidification of the sill in Exercise 5.5 and assess the chances of the intrusion developing into a shallow magma chamber. Compare with the results in Exercise 5.4 and discuss the implication for shallow-magma-chamber formation in general.

5.7   Consider a 5-km-long (strike-dimension) feeder-dike, supplying magma to a volcanic fissure. The host rock at the surface has a Young's modulus of 10 GPa and a Poisson's ratio of 0.25. Calculate the opening or aperture of the volcanic fissure for a magmatic overpressure of 5 MPa.

5.8   When a horizontal contact between a compliant pyroclastic layer and a basaltic lava flow is approached by a vertically propagating dike, the contact opens up. What is this mechanism of dike-induced contact opening called and what is likely to happen to the dike once it meets the contact?

5.9   What is the general stress and pressure condition for magma-chamber rupture and dike injection? Provide the equation and explain all the symbols.

5.10 Explain the concepts excess pressure and overpressure, as used in volcanotectonics.

5.11 How does the magma overpressure (driving pressure) change with height above the source magma chamber or reservoir? Provide a suitable equation and explain all the symbols.

5.12 Why is buoyancy of little importance for driving dike propagation close to the source magma chamber or reservoir?

5.13 What is a stress barrier? How do stress barriers form and affect dike and inclined-sheet propagation in volcanoes?

5.14 Define and describe a laccolith.

5.15 A vertical feeder-dike is injected from a magma chamber located at 3 km depth below the surface. Provide the typical range of the excess-magmatic-pressure gradient associated with that feeder-dike. Explain under what conditions that gradient would be the only pressure source available to sustain the eruption.

5.16 Explain why an outward-dipping ring-dike feeder would normally increase its volumetric flow rate during a (comparatively abrupt) caldera collapse.

5.17 Explain why caldera collapse along an inward-dipping ring-fault is referred to as stable but that along an outward-dipping ring-fault as unstable.

5.18 Discuss the theoretical reasons for the stress drop in earthquakes, including those associated with caldera collapses, being commonly in the range 1–12 MPa. Provide an appropriate equation and explain all the symbols (cf. Chapter 4).

5.19 Provide theoretical reasons as to why stratovolcanoes may be regarded as mechanically stronger than basaltic edifices and the implications this has for vertical and lateral collapses.

5.20 Relate the concepts of edifice strength to those of the Cook–Gordon delamination, stress barriers, and elastic mismatch. In view of these concepts, why would you expect fracture arrest to be more common in stratovolcanoes than in basaltic edifices and relate your conclusions to eruption frequencies of these types of volcanoes.

References and Suggested Reading

Amelung, F., Jonsson, S., Zebker, H., Segall, P., 2000. Widespread uplift and 'trapdoor' faulting on Galapagos volcanoes observed with radar interferometry. *Nature*, **407**, 993–996.

Anderson, E. M., 1942. *The Dynamics of Faulting and Dyke Formation with Application to Britain*. Edinburgh: Oliver and Boyd.

Beget, J. E., Kienle, J., 1992. Cyclic formation of debris avalanches at Mount St Augustine Volcano. *Nature*, **356**, 701–704.

Blake, D. H., 1966. The net-veined complex of the Austurhorn intrusion, southeastern Iceland. *Journal of Geology*, **74**, 891–907.

Browning, J., Gudmundsson, A., 2015a. Surface displacements resulting from magma-chamber roof subsidence, with application to the 2014–2015 Bardarbunga-Holuhraun volcanotectonic episode in Iceland. *Journal of Volcanology and Geothermal Research*, **308**, 82–98.

Browning, J., Gudmundsson, A., 2015b. Caldera faults capture and deflect inclined sheets: an alternative mechanism of ring-dike formation. *Bulletin of Volcanology*, **77**, 889, doi:10.1007/s00445-014-0889-4.

Bunger, A. P., Cruden, A. R., 2011. Modeling the growth of laccoliths and large mafic sills: role of magma body forces. *Journal of Geophysical Research*, **116**, B02203, doi:10.1029/2010JB007648.

Carslaw, H., Jaeger, J.C., 1959. *Conduction of Heat in Solids*. Oxford: Oxford University Press.

Cruden, A. R., 1998. On the emplacement of tabular granites. *Journal of the Geological Society of London*, **155**, 853–862.

Fagents, S. A., Gregg, T. K. P., Lopes, R. M. C. (eds.), 2013. *Modeling Volcanic Processes: The Physics and Mathematics of Volcanism*. Cambridge: Cambridge University Press.

Fedotov, S. A., Chirkov, A. M., Gusev, N. A., Kovalev, G. N., Slezin, Yu. B., 1980. The large fissure eruption in the region of Plosky Tolbachik Volcano in Kamchatka, 1975–1976. *Bulletin of Volcanology*, **43**, 47–60.

Filson, J., Simkin, T., Leu, L. 1973. Seismicity of a caldera collapse: Galapagos Islands 1968. *Journal of Geophysical Research*, **78**, 8591–8622.

Furman, T., Meyer, P. S., Frey, F., 1992. Evolution of Icelandic central volcanoes: evidence from the Austurhorn intrusion, southeastern Iceland. *Bulletin of Volcanology*, **55**, 45–62.

Galindo, I., Gudmundsson, A., 2012. Basaltic feeder dykes in rift zones: geometry, emplacement, and effusion rates. *Natural Hazards and Earth System Sciences*, **12**, 3683–3700.

Galland, O., Scheibert, J., 2013. Analytical model of surface uplift above axisymmetric flat-lying magma intrusions: implications for sill emplacement and geodesy. *Journal of Volcanology and Geothermal Research*, **253**, 114–130.

Gautneb, H., Gudmundsson, A., Oskarsson, N., 1989. Structure, petrochemistry, and evolution of a sheet swarm in an Icelandic central volcano. *Geological Magazine*, **126**, 659–673.

Geshi, N., Neri, M., 2014. Dynamic feeder dyke systems in basaltic volcanoes: the exceptional example of the 1809 Etna eruption (Italy). *Frontiers in Earth Science*, **2**, doi:10.3389/feart.2014.00013.

Geshi, N., Shimano, T., Chiba, T., Nakada S., 2002. Caldera collapse during the 2000 eruption of Miyakejima volcano, Japan. *Bulletin of Volcanology*, **64**, 55–68.

Geshi, N., Kusumoto, S., Gudmundsson, A., 2010. The geometric difference between non-feeders and feeder dikes. *Geology*, **38**, 195–198.

Gordon, J. E., 1976. *The New Science of Strong Materials*. London: Penguin.

Gretener, P. E. 1969. On the mechanics of the intrusion of sills. *Canadian Journal of Earth Sciences*, **6**, 1415–1419.

Gudmundsson, A., 1986. Formation of crustal magma chambers in Iceland. *Geology*, **14**, 164–166.

Gudmundsson, A., 1990. Emplacement of dikes, sills and crustal magma chambers at divergent plate boundaries. *Tectonophysics*, **176**, 257–275.

Gudmundsson, A., 2007. Conceptual and numerical models of ring-fault formation. *Journal of Volcanology and Geothermal Research*, **164**, 142–160.

Gudmundsson, A., 2009. Toughness and failure of volcanic edifices. *Tectonophysics*, **471**, 27–35.

Gudmundsson, A., 2011a. *Rock Fractures in Geological Processes*. Cambridge: Cambridge University Press.

Gudmundsson, A., 2011b. Deflection of dykes into sills at discontinuities and magma-chamber formation. *Tectonophysics*, **500**, 50–64.

Gudmundsson, A., 2012. Magma chambers: formation, local stresses, excess pressures, and compartments. *Journal of Volcanology and Geothermal Research*, **237–238**, 19–41 .

Gudmundsson, A., 2015. Collapse-driven large eruptions. *Journal of Volcanology and Geothermal Research*, **304**, 1–10.

Gudmundsson, A., Nilsen, K., 2006. Ring-faults in composite volcanoes: structures, models and stress fields associated with their formation. In Troise, C., De Natle, G., Kilburn, C. R. J. (eds.), *Mechanism of Activity and Unrest at Large Calderas. Geological Society of London Special Publications, 269*. London: Geological Society of London, pp. 83–108.

Gudmundsson, A., Pasquare, F.A., Tibaldi, A., 2018. Dykes, sills, laccoliths, and inclined sheets in Iceland. In Breitkreuz, C., Rocchi, S. (eds), *Physical Geology of Shallow Magmatic Systems: Dykes, Sills and Laccoliths*. Berlin: Springer, pp. 363–376.

Hartley, M. E., Thordarson, T., 2013. Formation of Öskjuvatn caldera at Askja, North Iceland: mechanism of caldera collapse and implications for the lateral flow hypothesis. *Journal of Volcanology and Geothermal Research*, **227–228**, 85–101.

Hartley, M. E., Thordarson, T., de Joux, A., 2016. Postglacial eruptive history of the Askja region, North Iceland. *Bulletin of Volcanology*, **78**, doi:10.1007/s00445-016-1022-7.

Jaeger, J. C., 1964. Thermal effects of intrusions. *Reviews of Geophysics*, **2**, 443–466.

Jakobsdottir, S., 2008. Seismicity in Iceland 1994–2007. *Jokull*, **58**, 75–100.

Lauthold, J., Muntener, O., Baumgartener, L. P., et al., 2014. A detailed geochemical study of a shallow arc-related laccolith: the Torres del Paine Mafic Complex (Patagonia). *Journal of Petrology*, **54**, 273–303.

Marti, J., Gudmundsson, A., 2000. The Las Canadas caldera (Tenerife, Canary Islands): an overlapping collapse caldera generated by magma-chamber migration. *Journal of Volcanology und Geothermal Research*, **103**, 161–173.

Michel, J., Baumgartner, L., Putlitz, B., Schaltegger, U., Ovtcharova, M., 2008. Incremental growth of the Patagonian Torres del Paine laccolith over 90 k.y. *Geology*, **36**, 459–462, doi:10.1130/G24546A.1.

Moore, J.G., Normark, W. R., Holcomb, R.T., 1994. Giant Hawaiian landslides. *Annual Review of Earth and Planetary Sciences*, **22**, 119–144.

Neal, C. A., Brantley, S. R., Antolik, J. L., et al., 2019. The 2018 rift eruption and summit collapse of Kilauea Volcano. *Science*, **363**, 367–374.

Newhall, C. G., Dzurisin, D., 1988. *Historical Unrest of Large Calderas of the World*. Reston, VA: US Geological Survey.

Pollard, D. D., Johnson, A. M., 1973. Mechanics of growth of some laccolithic intrusions in the Henry mountains, Utah, II. Bending and failure of overburden layers and sill formation. *Tectonophysics*, **18**, 311–354.

Reddy, J. N., 2002. *Energy Principles and Variational Methods in Applied Mechanics*, 2nd edn. Hoboken, NJ: Wiley.

Reddy, J. N., 2003. *Mechanics of Laminated Composite Plates and Shells: Theory and Analysis*, 2nd edn. Boca Raton, FL: CRC Press.

Ritchie, D., Gates, A. E., 2001. *Encyclopedia of Earthquakes and Volcanoes*. New York, NY: Facts on File.

Secor, D. T., 1965. Role of fluid pressure in jointing. *American Journal of Science*, **263**, 633–646.

Sigurdsson, H., Houghton, B. F., McNutt, S. R., Rymer, H., Stix, J. (eds.), 2000. *Encylopedia of Volcanoes*. New York, NY: Academic Press.

Tibaldi, A., 2001. Multiple sector collapses at Stromboli volcano, Italy: how they work. *Bulletin of Volcanology*, **63**, 112–125.

Tibaldi, A., Corazzato, C., Apuani, T., Cancelli, A., 2003. Deformation at Stromboli volcano (Italy) revealed by rock mechanics and structural geology. *Tectonophysics*, **361**, 187–204.

Tibaldi, A., Bistacchi, A., Pasquare, F. A., Vezzoli, L., 2006. Extensional tectonics and volcano lateral collapses: insights from Ollague volcano (Chile–Bolivia) and analogue modelling. *Terra Nova*, **18**, 282–289.

Williams, H., McBirney, A. R., 1979. *Volcanology*. San Francisco, CA: Freeman.

Wright, T. J., Sigmundsson, F., Pagli, C., et al., 2012. Geophysical constraints on the dynamics of spreading centres from rifting episodes on land. *Nature Geoscience*, **5**, 250.

6 Formation and Dynamics of Magma Chambers and Reservoirs

6.1 Aims

A magma chamber is the heart of every polygenetic volcano. Many, presumably most, polygenetic volcanoes have two magma chambers: one shallow crustal chamber and another deep-seated chamber, which we here refer to as a reservoir. Together, the reservoir and the shallow chamber constitute a double magma chamber. The complex interaction between the source reservoir and the chamber determines the frequency of injection of inclined sheets and dikes. Together with the mechanical layering and local stresses in the crustal segment, the double chamber also largely controls the frequency and sizes of eruptions in the volcano to which it supplies magma. We have learned that most shallow chambers evolve from sills and are located in the upper crust. The deep-seated reservoirs, by contrast, are normally located in the lower crust or upper mantle. If located in the crust, they may also evolve from sills; if located in the upper mantle, they may evolve as magma accumulations in regions of low potential energy. The accurate determination of the location of active magma chambers is generally difficult. In this chapter I review some of the methods used, primarily geodetic and seismic techniques, and provide typical chamber depths. The main aims of this chapter are to:

- Explain the formation of deep-seated source reservoirs.
- Discuss magma migration in porous deep-seated reservoirs.
- Explain the general formation of double magma chambers.
- Describe magma transport between reservoirs and chambers.
- Discuss the dynamics of deep-seated reservoirs.
- Explain the intrusion and extrusion frequencies of double magma chambers.
- Describe the main methods used to detect active magma chambers.

6.2 Formation of Deep-Seated Reservoirs

For a double magma chamber, the deep-seated reservoir is normally located either in the lower part of the crust or in the upper mantle. The focus here is on polygenetic volcanoes fed by double magma chambers. A **double magma chamber** is composed of a shallow magma chamber, which, in turn, is supplied with magma from a deep-seated reservoir. It should be noted, however, that there are many volcanoes that are supplied with magma by more than two chambers. Such volcanoes are most common where the lithosphere is thick, such as at some convergent plate boundaries. There are indications that many parts of the lower crust, especially at convergent plate boundaries, are partially molten (e.g. Cavalcante et al., 2014) in which case deep-seated reservoirs may form there through porous flow, in a similar way as they do in the upper mantle. Other deep-seated reservoirs/chambers in the thick parts of the lithosphere, however, may develop from sills, as shallow chambers do in most tectonic regimes (Chapter 5). Thus, while a double chamber, one shallow and one deeper, is by no means the only possible configuration of a volcano plumbing system, the double chamber can serve to illustrate all the salient points of such a system. Here the focus is thus on a double magma chamber, with the lower one – the reservoir – being located in a partially molten lower crust or upper mantle (Fig. 6.1).

A **deep-seated reservoir** is initiated when magma or melt accumulates in the lower crust or at the boundary between the lower crust and the upper mantle. At divergent plate boundaries, these are commonly regions of faulting, graben formation, and crustal thinning (Fig. 6.2). The magma is driven to regions of low potential energy and tends to accumulate

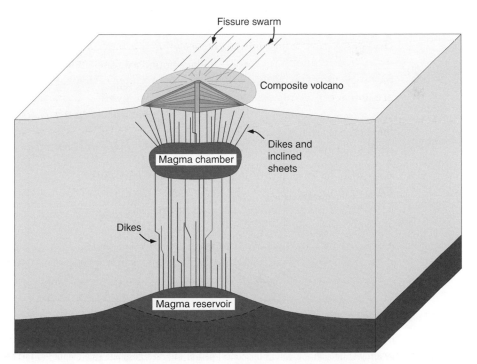

Fig. 6.1 Schematic illustration of a double magma chamber, composed of a shallow magma chamber and a deep-seated source reservoir. The volume of the shallow chamber is normally a small fraction of that of the source reservoir. The shallow chamber channels magma to the polygenetic (composite, central) volcano and this is the main reason for its formation. Modified from Gudmundsson (2006).

where the potential energy reaches a local minimum. For quasi-static conditions, the regions of local minimum potential energy normally coincide with those where the depth to the partially molten lower crust or upper mantle reaches a local minimum (Fig. 6.2). At divergent plate boundaries in general, as exemplified by mid-ocean ridges, the minimum depth to the reservoirs is at around 6–8 km (Fig. 6.3). For comparison, the minimum depth

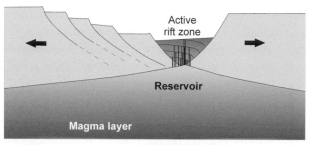

Fig. 6.2   A deep-seated reservoir forms when magma or melt accumulates in the lower crust or at the boundary between the lower crust and the upper mantle in regions of reduced (local minimum) potential energy. In areas of extension, such as divergent plate boundaries, these are commonly the areas of faulting, graben formation, and crustal thinning. A 'magma layer' is any extensive partially molten layer; it commonly coincides with the uppermost part of the asthenosphere beneath volcanic zones or fields. Modified from Gudmundsson (2017).

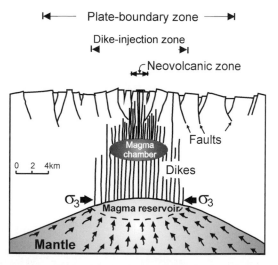

Fig. 6.3   Schematic illustration of a double magma chamber at an intermediate-spreading ridge. A sill-like shallow chamber is at 3 km depth, whereas the top of the deep-seated reservoir is at 7 km depth. Melt or magma migrates (indicated by arrows) through the porous uppermost part of the asthenosphere (the mantle) to the site of the local minimum potential energy below the dike-injection zone. The width of the neovolcanic zone is 1 km, that of the dike-injection zone is 6 km, and that of the plate-boundary zone is 12 km. The minimum compressive (maximum tensile) principal stress, σ_3, is perpendicular to the dikes. The location of the faults is loosely based on a profile from the intermediate-spreading East Pacific Rise at 21°N. Modified from Gudmundsson (1990).

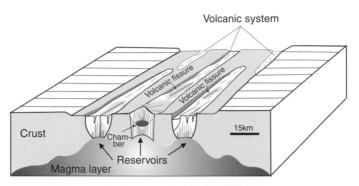

Fig. 6.4   Volcanic systems and associated reservoirs in a rift zone. The cross-sectional shapes of the reservoirs reflect the geometries of the volcanic systems to which they supply magma, either directly or through a shallow magma chamber. More specifically, the reservoirs are semi-ellipsoidal, dome-shaped regions of low potential energy to which the magma and melt are driven (Fig. 6.3). Here, the three parallel reservoirs are separated by regions of thicker crust, so that the magma in each reservoir can develop independently of the magmas in the neighbouring reservoirs. The internal structure of the central volcanic system is characterised by vertical dikes and inclined sheets. The section shows the shallow chamber and the associated central volcano of the system in the centre, but for the other two systems the central volcanoes and associated chambers are not seen (they are outside the parts seen in the illustration). The scale is approximate. Modified from Gudmundsson (1987).

to the partially molten upper mantle beneath the active volcanic zones of Iceland is from about 10 km to 20–25 km or more (Fig. 6.4). This is based on the currently widely accepted 'thick-crust-model' interpretation of the lithosphere of Iceland (Bjarnason, 2008; Brandsdottir and Menke, 2008). The basic difference in interpretation between the thick-crust model and the earlier thin-crust model, however, is that an upper mantle layer in the thin-crust model is simply interpreted as the lower crustal layer in the thick-crust model. Since the layer is the same in both models, and it is likely to be partially molten, the deep reservoirs below the volcanic zones in Iceland may have their tops at depths anywhere between 10 and 20 km, as is supported by recent studies in Iceland (de Zeeuw-van Dalfsen et al., 2004, 2012; Sturkell et al., 2006; Reverso et al., 2014; Browning and Gudmundsson, 2015a). Thus, deep-seated reservoirs beneath the active volcanic zones of Iceland may be either located in the partially molten lower crust or the partially molten upper mantle, and the same applies to many other volcanic areas.

   Gradually, a quasi-static magma reservoir becomes **density stratified**. This means that the low-density magma accumulates at the top of the reservoir whereas the high-density magma accumulates at its bottom or floor. Under these conditions, the contacts between the magmas of different densities are generally horizontal. Because the thickness of the lithosphere has normally a local minimum at the site of a reservoir (Figs. 6.2 and 6.3), and also because of uplift and bending due to magma pressure and plate pull, reservoirs tend to be dome-shaped (Figs. 6.1 and 6.4). Normally, the horizontal cross-sectional area of a reservoir is similar to that of the volcanic field, zone, or system to which it supplies magma at the surface (Fig. 6.4). Reservoirs supplying magma to volcanic systems of rift zones, such as in Iceland and elsewhere at divergent plate boundaries, thus have elongated,

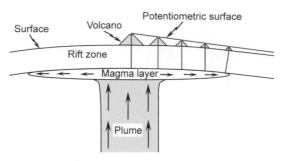

Fig. 6.5   The flow of magma and melt is generally from the centre of a mantle plume towards and along adjacent rift zones. The potentiometric surface – the surface to which magma could rise on its own in a vertical 'pipe' – reaches its highest elevation above the centre of the plume and falls with distance from the centre and along the rift zones. This schematic illustration applies particularly to the plume and rift zones in Iceland and the decrease in the elevation of the potentiometric surface is supported by the gradual decrease in the heights of monogenetic table mountains along the rift zones.

elliptical, cross-sectional areas in plan view. The long axes of the reservoirs normally follow roughly the axes of the associated volcanic systems. At convergent plate boundaries, or in general in areas outside divergent plate boundaries, the reservoirs may be of any shape. For mechanical and thermal reasons, many such reservoirs may have **circular** to slightly elliptical horizontal cross-sections.

The **quasi-static conditions**, described above, are probably a common situation for reservoirs. However, under certain dynamic conditions, the contacts between magma compartments or layers of different density need no longer be horizontal but rather are inclined or tilted. This applies in particular in volcanic zones that connect to sources such as mantle plumes. Then, the flow of the magma is from the centre of the plume and along the rift zones (Fig. 6.5). Hydrodynamic effects for trapping oil and gas with inclined contacts are well known from petroleum reservoirs, although they are not as common as once thought. In volcanic areas with magma migration along rift volcanic zones, magma-dynamic environments favouring **tilted contacts** may be common. In such reservoirs, low-density magmas and, depending on the depth of the reservoir, gas could have tilted (dipping) contacts with high-density magmas (Fig. 6.6). The dip of the contact between the low- and high-density magmas depends on the intensity of the flow of the high-density magma. However, the dip is always **downstream**, that is, in the direction of the flow beneath the volcanic zone.

The greatest **magma accumulation** does not have to be in the centre of the reservoir, but it can be either in the upstream or the downstream part. For example, many central volcanoes in areas such as Iceland are not located above the middle part of the associated reservoir but rather above either of the lateral end parts of the reservoir – as inferred from the geometry of the associated volcanic system. The greatest magma accumulation in a reservoir is normally beneath the main polygenetic volcano to which the reservoir supplies magma, either directly or through one (or more) shallow magma chambers (Figs. 6.1 and 6.4). Thus, if the main volcano, the central volcano (a stratovolcano, basaltic

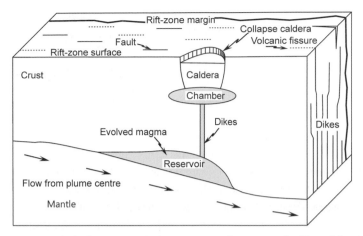

Fig. 6.6 Formation of a reservoir in a hydrodynamic trap. Such reservoirs may form as a result of flow of the magma from the centre of the plume and along the rift zones (Fig. 6.5). Low-density magmas and, depending on the depth of the reservoir, gas could have tilted (dipping) contacts with high-density magmas. In hydrodynamic reservoirs the dip of the contacts between magmas is always downstream (Fig. 6.5).

edifice, or a collapse caldera) is located at one end of the volcanic system or field to which it belongs, then most likely the **maximum thickness** of the reservoir, its maximum magma accumulation, occurs beneath that end.

6.3 Magma Migration in Deep-Seated Reservoirs

Let us now consider magma or melt accumulation at the top of a partially molten porous layer, the '**magma layer**', so as to form a reservoir (Figs. 6.1–6.5). Fluid flow in porous media is controlled by **Darcy's law.** This law can be expressed in many ways, one of which is (Bear, 1972):

$$\vec{q} = \frac{Q}{A} = -\frac{k\rho_m g}{\mu_m}\nabla\varphi \tag{6.1}$$

Here $\vec{q}$ is the **discharge velocity** (a vector, hence the arrow above q), also known as specific discharge or Darcy velocity. More specifically, $\vec{q}$ is the volumetric flow rate Q per unit surface area A normal to the direction of flow and has the unit of velocity, m s^{-1}. The symbol k denotes the intrinsic permeability, often referred to simply as permeability, which has the dimensions of area and thus the unit of m^2. The symbol ρ_m denotes the density of the magma or melt (unit kg m^{-3}), g is the acceleration due to gravity (unit m s^{-2}), and μ_m is the dynamic (absolute) viscosity of the magma (unit kg m^{-1} s^{-1}, which has the name pascal second, Pa s). The symbol ∇ is the mathematical operator del (nabla), and φ is the potential energy or total head of the magma. When del is followed by a scalar field, such as that of the potential energy φ, the result is known as the gradient of the field (see below). The negative sign is to indicate that magma migration is **driven** from regions of higher potential energy – higher head – to regions of **lower potential energy**. Recall that potential energy may be defined as the ability or capacity to do work by virtue of the configuration or location of the system under consideration. Once it is known and understood that the flow is driven in the direction of decreasing total head or potential energy, the minus sign is commonly omitted – as will be done here where no confusion can arise.

The **physical meaning** of Eq. (6.1) is that the magma flows from regions of comparatively high potential energy towards regions of comparatively low potential energy. It is only in the special case of purely horizontal migration that the magma is driven from higher to lower pressure. A well-known example from hydrogeology is rain that falls on a mountain top and migrates as groundwater down into the interior of the mountain. This water normally migrates down towards a higher rock pressure until it comes again to the surface in a valley. Part of the groundwater may reach the surface in the slopes of the valley, forming springs. Another part goes deeper into the mountain and comes to the surface only at the bottom of the valley, commonly a river bed, where the local potential energy reaches its minimum.

Similarly, in a partially molten magma layer within which a reservoir commonly forms, magma is driven to the region of local **minimum potential energy** (Fig. 6.3). This region is the top region of the partially molten layer. For rift zones, or divergent plate boundaries in general, the formation of regions of minimum potential energy can be attributed to three related factors: (a) divergent plate movements, (b) reduced lithosphere/crustal thickness, and (c) relative tensile-stress concentration and fracturing of the lithosphere/crust (Figs. 6.3 and 6.4).

Coming back to Eq. (6.1), it can be expressed in various different forms, some of which may be easier to grasp intuitively. Let us first clarify the mathematical operator del. It represents the gradient of the scalar field, the potential energy φ, and it may be written as:

$$\nabla \varphi = grad\varphi = \bar{i}\frac{\partial \varphi}{\partial x} + \bar{j}\frac{\partial \varphi}{\partial y} + \bar{k}\frac{\partial \varphi}{\partial z} \qquad (6.2)$$

where $\bar{i}$, $\bar{j}$ and $\bar{k}$ are unit vectors in the directions of the coordinate axes x, y, and z, respectively. The **gradient** $\nabla \varphi$ is a vector that represents the rate of change of the potential φ and the direction of $\nabla \varphi$ coincides with that in which φ changes fastest. More specifically, at a given point, the gradient is the vector that points in the direction of the steepest slope at that point. If the gradient is positive, then it points in the direction of fastest increase of the potential energy φ – or other function of interest such as temperature. By contrast, when the gradient is negative, as in Eq. (6.1), then it points in the direction of the fastest decrease of the potential energy φ, namely in the direction of the magma flow towards and inside the reservoir. In analogy with a measured structure, such as an inclined sheet or a lava flow, the gradient corresponds to the dip of the structure.

Again, the landscape (mountains and valleys) may be of help to visualise, in this case Eq. (6.2). Supposed you were standing on the top of a mountain. Then, for a negative gradient, the direction of the **steepest slope** towards an adjacent valley floor, that is, the direction of maximum descent, is the direction of the gradient. Similarly, the maximum ascent is the steepest slope up the mountain. The measured slope along the direction of the gradient, that is, the change in elevation over horizontal distance, is the magnitude of the gradient. The gradient is always **perpendicular** to the contour lines that mark the elevation and shape of the mountain on a map (cf. Fleisch, 2012; Griffith, 2014).

More specifically, the steepness of the slope at a point is indicated by the magnitude $|\nabla\varphi|$, normally the length of the arrow representing the gradient vector. If the coordinate axes are oriented such that x and y are horizontal and z is vertical and positive up (as is appropriate for magma migration in a reservoir), then it follows that if the slope of the gradient is up, the gradient is positive, if down, the gradient is negative, as was in the case of the mountain slope discussed above. (Notice, however, that in this book the coordinate systems sometimes have z considered positive down vertically.)

The potential energy can be expressed as follows (Bear, 1972):

$$\varphi = z + \int_{p_i}^{p} \frac{dp}{g\rho_m(p)} \tag{6.3}$$

where z is the elevation head, that is, the energy of the magma due to its position or location and refers to the elevation of a magma element above or below a horizontal reference plane (here the floor or bottom of the magma layer, Figs. 6.4 and 6.5), p_i and p are the limiting values of the magmatic pressure over the interval considered, that is, the distance of magma migration within the magma layer or reservoir, g is the acceleration due to gravity, and ρ_m is density of the magma, which is a function of pressure (p). Since the integral in Eq. (6.3) is a measure of the pressure head and z a measure of the elevation head, both representing energy per unit weight ($\gamma = \rho_m g$) and with the **unit of length** (m), it follows that φ, the **potential energy**, is also the **total head** of the magma (the velocity head for magma migration can be neglected, cf. Example 6.6).

The floor or lower margin (bottom) of the magma layer (Figs. 6.4 and 6.5) is defined as $z = 0$, namely as the reference plane, and z is taken as vertical and positive upwards. On substituting Eq. (6.3) for φ in Eq. (6.1) and rewriting, the average or Darcy velocity of magma migration from the magma layer and into the reservoir is obtained:

$$\vec{q} = -\frac{k}{\mu_m}(\nabla p + \rho_m g \nabla z) \tag{6.4}$$

where all the symbols are as defined above (cf. Example 6.6).

The 'force' driving the flow of the magma or melt into the reservoir and towards its top is the gradient of the potential energy φ or $\nabla\varphi$. The steeper the gradient, the greater is the force. This follows from the well-known principle in physics that whenever the potential energy varies within a given region, the variation gives rise to a force. This applies to all **conservative forces**; they are derivable from the gradient of the potential energy (cf. Chapter 1). This implies that the force conserves the mechanical energy so that there is, for example, no friction or otherwise dissipation of energy. More specifically, a conservative force is one where the work done, say in moving a particle from one point in space to another, is **independent of the path** taken between these points.

In the present context, the important thing is that a conservative force is the negative derivative of the potential energy. Well-known examples are the force of gravity and the spring force in one form of Hooke's law – Eq. (3.10). Consider the gravitational force. For an object of mass m in Earth's gravitational field at a vertical distance z above the reference level (say, the Earth's surface) the potential energy U of that mass is given by:

$$U = mgz \tag{6.5}$$

Differentiating Eq. (6.5) we get:

$$-\frac{dU}{dz} = -mg = F \tag{6.6}$$

Here mg, namely mass $\times$ acceleration due to gravity, is the gravitational force close to the Earth's surface. This is the force that we read as our 'weight' when we step on a scale. Because the force is acting 'downwards' towards the Earth's surface, whereas here z is positive upwards, it follows we use minus signs for the change in potential energy.

Equation (6.3) gives the value of the potential energy or total head of the magma at any point within the magma layer and the associated reservoir. If a reasonably wide borehole were drilled from the surface and down to the magma layer, the magma would rise in that hole to a certain level. The level to which the magma rises in many such drill holes defines a surface. By analogy with hydrogeology, it is named the **potentiometric surface** (Fig. 6.5), which in hydrogeology is also called the piezometric surface. Thus, for the magma layer and the associated magma reservoir, the potentiometric surface is the level to which magma could rise if there were no unfavourable layers or contacts or cooling to stop it on its path.

If the potentiometric surface is below the Earth's surface, then **no magma** coming directly from the magma layer/reservoir would normally be able to reach the surface. If, however, the potentiometric surface is above the Earth's surface, then magma from the magma layer/reservoir can theoretically rise above the Earth's surface. What this means is that the magma is able to form volcanoes that rise at least to the height of the potentiometric surface above the surroundings. The **height of monogenetic volcanoes**, supposed to be fed directly from deep magma reservoirs, is thus an indication of the height of the potentiometric surface above the surroundings (Fig. 6.5).

This idea is particularly applicable to volcanic areas associated with **mantle plumes**. The plume then provides the 'magma layer', and the top parts of the magma layer constitute the reservoirs (Figs. 6.4 and 6.5). The centre of the mantle plume is where the magma layer reaches its greatest thickness and the magma has the greatest buoyancy. Hence, we might expect the monogenetic volcanoes to reach the greatest heights above their surroundings close to the centre of the mantle plume. By contrast, as the magma migrates and cools somewhat along the lower contact of the volcanic zones, the magma layer becomes thinner, the magma less buoyant, and the heights of the monogenetic volcanoes above their surroundings might be expected to decrease. The idea is schematic and does not consider some factors that might affect the heights of the monogenetic volcanoes; for example, the potential effects of ice thickness during eruption of mostly (supposed-to-be) monogenetic table mountains. Nevertheless, there is an approximate agreement between the heights of the table mountains in the rift zone of North Iceland – where this relation has been studied – and their distances from the centre of the mantle plume (Walker, 1965). The elevation of the mountains above their surroundings gradually decreased from the centre of the mantle plume towards the coast, that is, with increasing distance from the plume centre.

6.4 Formation of Double Magma Chambers

A double magma chamber is composed of a shallow magma chamber and a deep-seated magma reservoir (Figs. 6.1, 6.3, and 6.4). There is plenty of evidence for the existence of double magma chambers in many volcanic areas. For example, in Iceland, there is clear evidence of **double magma chambers** beneath the volcanoes of Grimsvötn (Reverso et al., 2014), Katla (Oladottir et al., 2008), Askja (Sturkell et al., 2006; Soosalu et al., 2010), and Krafla (Sturkell et al., 2006). The recent volcanotectonic episode in the Bardarbunga Volcanic System may also be interpreted as being related to a double magma chamber (Gudmundsson et al., 2014; Browning and Gudmundsson, 2015a). The deep-seated reservoirs detected so far in Iceland are mostly at depths of 15–20 km, as was predicted based on theoretical considerations (Gudmundsson, 1987). Studies of magma chambers and reservoirs in other volcanic areas also suggest some double or even triple chambers, while the interpretations are still being worked out (Chaussard and Amelung, 2014). Some of the reservoirs, however, may be still deeper, perhaps at 25 km depth. I have already discussed how the deep-seated reservoirs in the lower crust or uppermost mantle form (Section 6.2). And in Chapter 5 I discussed how shallow magma chambers form. Briefly, the main points in the formation of chambers and reservoirs may be summarised as follows.

- A deep-seated reservoir forms in a region of local **minimum potential energy** in a partially molten layer. Such a region must occur at the top of the partially molten layer (Figs. 6.3 and 6.4). The reservoir becomes density stratified with the most evolved (lightest) magma and, depending on the depth of the reservoir, some gas at the top, and more primitive and denser (heavier) magma at greater depths (Fig. 6.7).

- In most reservoirs, presumably, the contacts between the lighter and heavier fluids are close to **horizontal** (Fig. 6.7). However, in some reservoirs where there is continuous magma migration in a certain direction, the contacts may be **tilted** or dipping in that (downstream) direction (Fig. 6.6).

- The greatest **magma accumulation** is normally in the centre of the reservoir (Figs. 6.1, 6.3, and 6.7). However, the reservoir itself need not be in the centre of the volcanic field, zone, or system to which it supplies magma. Many reservoirs may be at or close to one end of the zone or system. As a rule, the main polygenetic volcano (stratovolcano, basaltic edifice, caldera, central volcano) in the zone or system is located above the thickest part of magma reservoirs (Figs. 6.1, 6.4, and 6.7). Thus, the location of the main volcano is an indication of the location of the main magma accumulation in the reservoir.

- Most shallow magma **chambers** are thought to develop from **sills**. They stay liquid so long as the sill receives magma frequently enough to hinder complete solidification. A chamber may develop from one, reasonably large, and often multiple, sill or, perhaps more commonly, from many sills – a cluster (Chapter 5).

- The chamber generates space for itself by **elastic-plastic expansion** or bending of the crustal layers above (and below) the developing chamber (Figs. 5.18–5.20), as well as through melting (anatexis) and stoping (blocks falling into the chamber) of the host rock.

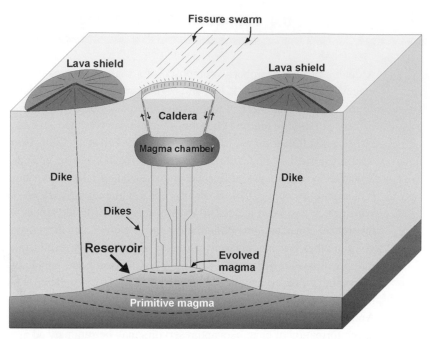

Fig. 6.7   The reservoir normally becomes density stratified with the most evolved (lightest) magma and, depending on the depth of the reservoir, possibly some gas at the top, and the more primitive and denser (heavier) magma at greater depths. Lava shields (shield volcanoes), of basalt, are supplied with magma (through regional dikes) directly from the reservoir, whereas a shallow chamber supplies magma (through radial dikes and inclined sheets) to most of the eruptions of the central volcano, here a collapse caldera (cf. Gudmundsson, 2016).

- A double magma chamber of this sort remains **active**, or 'lives', so long as there is enough **magma supply** from the source reservoir to keep the shallow chamber at least partially fluid. Typical lifetimes are of the order of 10^5–10^6 years, during which time the double chamber injects regional dikes (primarily from the reservoir) and radial dikes and inclined sheets (primarily from the chamber), forming different types of dike swarms (Fig. 6.8). Occasionally, there are caldera collapses associated with the shallow chamber (Fig. 6.7), some of which generate ring-dikes (Chapter 5).
- A reservoir becomes **inactive** if it ceases to occupy a region of minimum potential energy. This may happen through thickening of the crust/lithosphere (for example because of magma solidification and/or underplating) at the location of the reservoir, when the reservoir becomes shifted out of the location of minimum potential energy (through plate movements), or both.
- A shallow magma chamber becomes inactive and solidifies as a **pluton** (Figs. 1.17, 5.12, 5.18, 5.19, 7.5) once the magma supply from the source reservoir diminishes below a certain threshold and, eventually, is cut off entirely. This may happen because the reservoir itself has become inactive or, alternatively, because the crustal segment hosting the chamber has shifted (because of plate movements) out of the magma source region and a new chamber has formed.

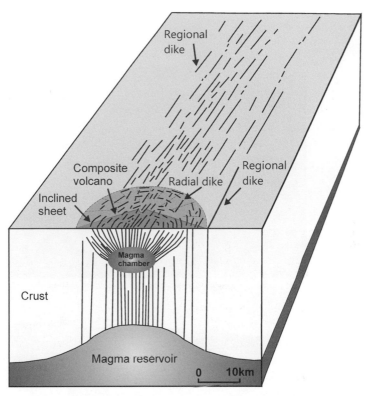

Fig. 6.8 A double magma chamber may remain active for 10^5–10^6 years. During this time two distinct dike/sheet swarms develop within the volcanic system or field: a local swarm of comparatively thin radial dikes and inclined sheets (Figs. 2.34, 5.1, and 6.13), primarily injected from the chamber, and a regional swarm of comparatively thick subvertical dikes (Figs. 2.1, 2.4, and 6.9), which is largely injected from the reservoir. Modified from Gudmundsson (2017).

While a shallow magma chamber is normally located vertically above its source reservoir there are occasions when the arrangement may be somewhat different. If the general shape of the reservoir is not perfectly symmetric, the local stress field that the reservoir induces in the crustal segment above may favour the formation and maintenance of a shallow chamber that is somewhat **shifted** to one side of the reservoir. This would be likely to happen, for example, when the reservoir location and shape is related to magma dynamic forces (Fig. 6.6). Another effect in shifting the location of a magma chamber in relation to its source reservoir is the general tilting, that is, general increase in dip of the strata (lava flows, sedimentary layers, intrusions) with depth, as is common in sedimentary basins and in volcanic zones (Fig. 6.9). Depending on the tilting, the polygenetic volcano supplied with magma from the shallow chamber may then also be **shifted in the down-dip** direction of the pile and thus not be located right above the centre of the source magma chamber (Fig. 6.10). This volcanotectonic situation is quite different from the common one (Figs. 6.1, 6.4, 6.7, and 6.8) and needs to be considered when inferring the sources of eruptive materials.

Fig. 6.9 The dip of a lava pile increases with depth, and the same applies to beds in sedimentary basins. Depending on the rate of increase of dip with depth, the location of a polygenetic volcano supplied with magma from a shallow chamber may also be shifted in the down-dip direction of the pile and thus may not be located right above the centre of the source magma chamber. (a) View north, the average dip of the lava flows in East Iceland increases by about 1° for every 150 m depth, as indicated here where the sea level is at about 1200–1300 m below the original top of the pile. The pyramid-shaped peak (indicated) has an elevation of 1069 m and thus is only 100–200 m below the original top of the lava pile. The rectangular scree in the central part of the photograph is a felsic intrusion. (b) View northeast, the tilting of the lava pile continues at greater depths in the crust, as seen here in a 400-m-tall mountain slope in Southeast Iceland. The valley floor is at nearly 2 km beneath the original top of the pile, that is, cut some 700 m deeper into the crust than in (a). A regional swarm of subvertical, basaltic dikes cuts through the pile. The dikes are up to 13 m thick; and many are offset, while some are arrested at layer contacts. A black and white version of this figure will appear in some formats. For the colour version, please refer to the plate section.

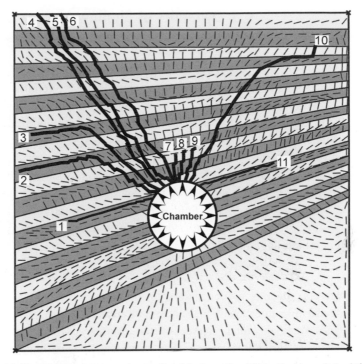

Fig. 6.10 Numerical model showing how an increase in the dip of the lava pile with depth can result in shift of the location of the polygenetic or central volcano, in the down-dip direction, in relation to the shallow source chamber (cf. Fig. 6.9). Thus, because the dike paths meet the surface (here paths 4, 5, and 6) somewhat down-dip (less likely up-dip) from the point at the surface right above the top of the chamber, the volcano need not be right above the centre of the source magma chamber (cf. Gudmundsson, 2006).

Part of the magmatism associated with reservoirs is independent of the shallow chamber. Many monogenetic volcanoes (mostly monogenetic basaltic volcanoes, Chapter 2) derive their magma **directly** from the deep-seated **reservoirs**, in which case the dynamics of the reservoir alone determines the style, volume, and duration of the eruption. Similarly, most eruptions from shallow chambers occur at a much faster volumetric flow rate or effusion rate than the rate of replenishment (inflow of magma) from the reservoir, so that the eruptive volume is largely determined by the constitution and magma volume of the chamber itself. The long-term eruption frequency of a chamber, however, is much affected by its being a part of a double magma chamber. In particular, the relative sizes (magma volumes) of the shallow chamber and the source reservoir are an important factor in eruption/intrusion statistics. In the analysis of the dynamics and normal eruptive volumes of chambers/ reservoirs, I start by exploring the simplest case, namely that of a single shallow magma chamber. Then I move on to the more complex case of a single magma reservoir. And finally, the effects of a double chamber on eruption frequencies of volcanoes are explored. The focus here is on 'normal' eruptions; unusually large eruptions from chambers and reservoirs are treated in Chapter 8.

6.5 Dynamics of Shallow Magma Chambers

Many crustal chambers at fast-spreading ocean ridges are likely to be **totally molten**. Most other chambers, however, are presumably **partially molten**. Here we want to estimate the maximum volume that can be erupted from a shallow chamber during an ordinary eruption. By 'ordinary eruptions' I mean eruptions without any large-scale volcanotectonic events such as a caldera collapse, a lateral collapse, or a large graben subsidence (the effects of these on eruptions are discussed in Chapter 8). Let us focus first on totally molten chambers; the extension to partially molten ones is trivial, as is indicated below. The total volume of magma V_{er} 'erupted', that is, transported or flowing out of a totally fluid magma chamber through a feeder-dike (including the volume of the feeder itself) during an eruption, may be estimated as follows (e.g. Gudmundsson, 1987):

$$V_{er} = p_e(\beta_r + \beta_m)V_c \tag{6.7}$$

Here p_e denotes the magma excess pressure in the chamber at the time of rupture and feeder-dike injection, β_r is the host-rock compressibility, β_m is the magma compressibility, and V_c is the total volume of the chamber. The transport of magma out of the chamber through the feeder-dike stops and the eruption comes to an end when the excess pressure becomes too small ($p_e \rightarrow 0$) to be able to keep the dike-fracture open (Chapter 8).

The **fraction of magma** that is transported out of the chamber through the feeder-dike before the feeder closes at its contact with the chamber can be estimated as follows, considering first basaltic magma. The static compressibility β_m for tholeiite (basaltic) magma at temperatures of 1100–1300 °C is about 1.3×10^{-10} Pa^{-1}, whereas the static host-rock compressibility β_r is estimated at about 3×10^{-11} Pa^{-1} (Gudmundsson, 1987). The latter value is actually for the crust of Iceland, but would be similar for many chambers at depths of 1–2 km. This estimate of the static β_r uses the dynamic/static ratio of Young's modulus of 2. For a shallow magma chamber located in a highly fractured crust, as is common in active volcanic zones, a more reasonable dynamic/static ratio might be 10 (Gudmundsson, 2011), in which case the compressibility β_r for a chamber at 1–2 km depth would be about 1.5×10^{-10} Pa^{-1}. Using the total range in the measured *in situ* tensile strength T_0, 0.5–9 MPa, and remembering that the tensile strength T_0 is roughly equal to magmatic excess pressure p_e, Eq. (6.7) can be used to calculate the ratio of the eruptive volume V_{er} to the total volume of the chamber V_c. For the above excess-pressure (tensile-strength) range, the V_{er}/V_c ratio ranges from about 2.5×10^{-3} ($p_e = 9$ MPa) to 1.4×10^{-4} ($p_e = 0.5$ MPa). For a typical or average $p_e = 3$ MPa, the approximate general ratio becomes:

$$V_{er} \approx 7 \times 10^{-4} V_c \tag{6.8}$$

Thus, typically, about 0.07% or, more generally, less than 0.1% (0.1% if $p_e = 4$ MPa) of the volume of a totally fluid basaltic magma chamber would be transported out of the chamber, that is, would be erupted/injected, during an ordinary eruption.

Volcanic eruptions may be classified in various ways. One is to base the classification on **eruptive volumes**, calculated as solid-rock or magma equivalents. Using this criterion, we divide eruptions into three classes:

(1) Small, for eruptive volumes from ≤ 0.001 km^3 to < 0.1 km^3.
(2) Moderate, for eruptive volumes from 0.1 km^3 to < 10 km^3.
(3) Large, for eruptive volumes from 10 km^3 to ≥ 1000 km^3.

Some eruptions may be very small; in fact, as small as we like to recognise them as such. For example, a very small lava eruption may be 0.0001 km^3 or less. Part of the magma that is transported out of the chamber during an eruption, however, forms the feeder-dike, the volume of which is rarely less than 0.001 km^3. Very small eruptions thus normally imply **at least 0.001 km^3** of magma being transported out of the chamber during the eruption. In all the calculations in this chapter of the volume leaving the chamber/reservoir during an eruption, the (estimated) dike volume is included in the calculated volume.

In all calculations in this book the estimated erupted material **includes** the estimated volume of the **feeder-dike** (or feeder sheet, if the eruption is fed by an inclined sheet). Also, it is assumed, in agreement with observations, that the volumetric rate of flow of magma from the deep-seated source reservoir into the shallow chamber is so slow during most chamber eruptions that the **inflow** during the eruption can normally be **neglected**. On these assumptions, Eq. (6.8) can be used to estimate the magma-chamber volume needed to supply magma to basaltic eruptions. If the volume of magma transported out of the chamber during an eruption is 0.001 km^3, then the chamber volume must, from Eq. (6.8), be at least about 1.4 km^3. Similarly, if the volume transported out of the chamber during an eruption is 0.1 km^3, Eq (6.8) yields a minimum magma-chamber volume of about 140 km^3.

Caldera areas provide a rough indication of the **cross-sectional areas** of the associated magma chambers in plan view (Chapters 5 and 8). Calderas on Earth range from less than 2 km in diameter to as much as 80 km, and are commonly 5–20 km in diameter (Chapters 5 and 9). Normally, the area of the caldera is somewhat smaller than the cross-sectional area of the associated magma chamber, but it is nevertheless a good proxy to the chamber size (Chapter 5; Gudmundsson, 2007). A circular caldera of 5 km in diameter has an area of close to 20 km^2. A corresponding totally molten sill-like, circular (penny-shaped) magma chamber with a thickness of 75 m has a volume of about 1.4 km^3 and could thus yield an eruptive volume of 0.001 km^3. Such a chamber, however, would be less likely to give rise to an eruptive volume of 0.1 km^3 in an ordinary eruption, because for a chamber of 5 km in diameter the thickness would have to exceed 7 km to reach the necessary volume of about 140 km^3. While there are chambers/plutons that have greater heights than their lateral dimensions – being prolate ellipsoids (Fig. 5.13) or cylinders with a vertical long axis – they are presumably not very common.

A penny-shaped **sill-like chamber** of diameter 10 km has an area of about 79 km^2, however, and need only be 180 m thick to reach a total volume of 140 km^3 and yield an eruptive volume of 0.1 km^3. Similarly, a penny-shaped chamber of diameter 20 km has an area of about 314 km^2 and need only be 45 m thick to produce an eruptive volume of 0.1 km^3. Even a moderate eruption of 1 km^3, such as occurred in St Helens in the United States in 1980, does not require an exceptionally large chamber. More specifically, the chamber volume needed for such an eruption is about 1430 km^3. For a penny-shaped (circular) chamber with a diameter of 20 km, the thickness for a chamber volume of 1430 km^3 is around 4.5 km. Thus, sill-like magma chambers of lateral dimensions similar to those of typical calderas, as are presumably common in many central volcanoes, and with plausible thicknesses, are large enough to produce typical small to moderate eruptions through an ordinary eruption mechanism.

The overall **compressibility** of intermediate magma (Murase and McBirney, 1973) and acid magma (Kress and Carmichael, 1991; Dobran, 2001) are similar to those of typical basaltic magmas. The gas or volatile content, however, can have significant effects on the magma compressibility (Woods and Huppert, 2003; Malfait et al., 2011; Guo, 2013; Seifert, 2013) because gas has much higher compressibility than either liquid or solid. Gas bubbles in magma have much higher compressibility than either the magmatic liquid itself or the solid host rock. The main volatiles are water (H_2O,) and carbon dioxide (CO_2). Sulphur (as hydrogen sulphide (H_2S) and sulphur dioxide (SO_2)) is also common but less so than H_2O and CO_2, on which I focus here.

One main difference in the behaviour of **carbon dioxide** and **water** in shallow magma chambers is that CO_2 exsolves to form bubbles normally at much **greater depths** (at higher total pressures) than H_2O (Gonnermann and Manga, 2013). More specifically, in **acid magma**, CO_2 may exsolve and form bubbles at pressures of 100 MPa or higher and in basaltic magma at pressures of 25 MPa or higher. By contrast, in acid magma H_2O exsolves at pressures less than 100 MPa, and in basaltic magma at pressures less than 25 MPa. In volcanic zones worldwide, the average crustal density of the uppermost 4 km is generally 2500–2600 kg m^{-3}, so that 25 MPa corresponds to about 1 km depth and 100 MPa to 4 km depth. Most shallow magma chambers are in roughly this depth range, that is, normally with roofs at depths between 1 km and 5 km, although some reach depths of 6–8 km. It follows that for shallow chambers much CO_2 (and some H_2O) in acid magmas is readily exsolved to form bubbles. Since bubbles in acid magma have negligible mobility (Gonnermann and Manga, 2013), they mostly **remain within the magma**. The one exception is when vigorous convection occurs in the chamber. Then, the gas could accumulate at the top of the acid magma and form a separate phase.

In **basaltic magma**, by contrast, exsolution and bubble formation is unlikely to occur at the depth of most shallow magma chambers. Much of the gas exsolution in basaltic magma takes place at very shallow depths, particularly in the feeder-dikes on their paths to the surface. Field studies in Hawaii, for example, suggest that most of the gas exsolution in basaltic magma occurs in the uppermost **few hundred metres** of the feeder-dike/conduit (Greenland et al., 1985, 1988). Similar results have been obtained through direct observations of dikes, sills, and inclined sheets in eroded volcanic areas. In most sheet-like intrusions, vesicles (formed by expanding gas bubbles) are small and few at depths exceeding several hundred metres below the surface at the time of sheet emplacement. By contrast, some feeder-dikes contain abundant large vesicles in their parts that are close to the surface (Galindo and Gudmundsson, 2012).

The effects of gas exsolution and bubble formation in magma chambers on the compressibility of the chamber magma depends on the fraction of gas in the magma. The **compressibility of the gas** can be 100–1000 times greater than the compressibility of the liquid magma (Woods and Huppert, 2003; Malfait et al., 2011; Guo, 2013; Seifert, 2013). The **gas bubbles** are therefore much more compressible than the liquid magma and associated crystals. The overall compressibility of the magma plus gas, however, depends on the gas fraction. For a small volume fraction of gas in the chamber, the compressibility of the liquid magma (β_m) dominates over the compressibility of gas (β_g), in which case the effects of the bubbles on the overall compressibility of the magma (gas plus liquid) is small. Conversely, when the volume fraction of gas in the magma becomes larger than the β_g/β_m ratio then the high compressibility of the gas bubbles can increase significantly the overall compressibility of the liquid magma plus gas in the chamber (Woods and Huppert, 2003). The bubble-rich and highly compressible parts of the magma chamber are unlikely to be uniformly distributed but rather confined to certain layers or compartments (Gudmundsson, 2012). The formation of bubble-rich compartments implies that the compressibility of the magma chamber may be highly heterogeneous, with compartments of high compressibility alternating with compartments or layers of much lower compressibility, resulting in uneven volume changes and overall response of different magma-chamber compartments to volume decrease during eruptions (Gudmundsson, 2012).

Here, the gas-rich compartments will not be treated as such since the focus is on the overall compressibility of the magma in the chamber. From the above considerations it follows that a very gas-rich acid magma has much **higher compressibility** than a basaltic magma, perhaps by several orders of magnitude. Depending on the amount of acid magma in the chamber, the compressibility of gas-rich acid magma (gas plus liquid) could be at least 10 and possibly 100 times that of a chamber composed entirely of basaltic magma (Woods and Huppert, 2003). For such high magma compressibility, $\beta_m \gg \beta_r$, in which case Eq. (6.7) reduces to:

$$V_{er} = p_e \beta_m V_c \qquad (6.9)$$

which implies that the ratio of the volume of erupted/injected magma transported out of the chamber during an eruption and/or intrusion to the total magma volume depends almost entirely on the magma compressibility β_m. That is, the host-rock compressibility may be ignored in most such calculations. For a purely **acid** bubble-rich magma chamber, using the typical excess pressure p_e (= T_0) of 4 MPa, the ratio in Eq. (6.8) could then reach the maximum value of:

$$V_{er} \approx 4 \times 10^{-2} V_c \qquad (6.10)$$

Equation (6.10) indicates that as much as 4% (3% if p_e = 3 MPa) of the magma in an acid chamber – or the bubble-rich acid compartment of a larger chamber – could leave the chamber (as intrusive and extrusive material) during chamber rupture and eruption.

So far, only totally molten magma chambers have been considered. The analysis can, however, easily be extended to **partially molten chambers**. If the melt is located within a solid crystal mush, then the pore compressibility of the crystal mush needs to be considered in addition to the compressibility of the rock hosting the chamber. The appropriate theory is then that of a poroelastic material, which is the same as that developed for partially molten reservoirs in the next section. If, however, the outer parts of the initial magma chamber have solidified so that only a small central part is still liquid, the molten fraction f becomes a multiplying factor and Eq. (6.7) should be rewritten as:

$$V_{er} = fp_e(\beta_r + \beta_m)V_c \qquad (6.11)$$

6.6 Dynamics of Deep-Seated Magma Reservoirs

Deep-seated magma reservoirs are normally partially molten and behave as **poroelastic** media. Except at fast-spreading ridges (where the reservoirs may be as shallow as 6–7 km), most reservoirs are at depths exceeding 10 km, and many are much deeper. For reservoirs at depths greater than 10 km, the vertical stresses or total pressures at the location of the reservoirs exceed 250 MPa, and are commonly 300–600 MPa. This follows because the vertical stress σ_v in the crust is given by (Chapter 3):

$$\sigma_v = \rho_r g d \qquad (6.12)$$

where ρ_r is the average density of the crustal layers above the reservoir, g is the acceleration due to gravity (9.81 m s^{-2} close to the Earth's surface), and d is the depth to the top of the reservoir. Thus, if the average crustal density is 2700 kg m^{-3} for a continental crust, the vertical stress would be about 270 MPa at 10 km depth and 540 MPa at 20 km depth. Similarly, for an oceanic crust and upper lithosphere of density 2900 kg m^{-3}, the vertical stress would be about 290 MPa at 10 km depth. The lowest vertical stress would be for a shallow chamber in a sedimentary basin, with an average density of around 2500 kg m^{-3} (e.g. Tenzer and Gladkikh, 2014), for which the vertical stress at 10 km depth would be about 250 MPa. From these vertical stress considerations it follows that neither acid magma, which might be produced in the top parts of the reservoirs, nor the andesitic and basaltic magmas, which constitute their bulk magmas, would normally be subject to exsolution of either CO_2 or H_2O. Consequently, the compressibility of the magma or melt is unlikely to be affected by bubble formation.

For a poroelastic magma reservoir filled with magma or melt, four compressibilities need to be considered (Bear, 1972; cf. Wang, 2000). In the discussion of these compressibilities isothermal (constant-temperature) conditions are assumed. The first one is the **bulk compressibility** β_b. It is a measure of the fractional change in the bulk volume ΔV_b of the entire poroelastic reservoir due to a change in magmatic excess pressure Δp_e and is given by:

$$\beta_b = \frac{\Delta V_b}{\Delta p_e V_b} \qquad (6.13)$$

where V_b is the original bulk volume of the poroelastic reservoir. When Δp_e is positive, so that the excess pressure in the reservoir increases, then the reservoir expands and its bulk volume change ΔV_b is positive.

The second one is the **solid-matrix compressibility** β_s. It is a measure of the fractional change in volume of the solid matrix ΔV_s of the reservoir (commonly the crystal mush) when the excess pressure changes. The solid-matrix compressibility is given by:

$$\beta_s = -\frac{\Delta V_s}{\Delta p_e V_s} \tag{6.14}$$

where V_s is the original volume of the matrix of the reservoir. The negative sign is because when Δp_e is positive the excess pressure in the reservoir increases and the matrix volume decreases or shrinks.

The third one is the **pore compressibility** β_p. It is a measure of the fractional change in the pore volume ΔV_p, that is, the volume occupied by the magma, when the excess pressure changes. The pore compressibility is given by:

$$\beta_p = \frac{\Delta V_p}{\Delta p_e V_p} \tag{6.15}$$

where V_p is the original pore volume (the magma volume) in the reservoir. When Δp_e is positive, the excess pressure in the reservoir increases. The pores holding the magma then expand and the pore volume increases.

The fourth one is **magma** (melt or fluid) **compressibility** β_m. It is a measure of the fractional change in magma volume ΔV_m when the excess pressure changes. The magma compressibility is given by:

$$\beta_m = -\frac{\Delta V_m}{\Delta p_e V_m} \tag{6.16}$$

where V_m is the original magma volume in the reservoir. The negative sign can be understood as follows. When new magma is added to the reservoirs then Δp_e increases. Part of the volume needed for the newly added magma, however, is obtained by compressing (and increasing the density of) the existing (original) magma in the pores of the reservoir.

Let us now combine the compressibilities into a single equation giving the ratio between the reservoir volume and the eruptive volume during '**ordinary eruptions**', such as those giving rise to lava shields erupted directly from deep-seated reservoirs in Iceland (Fig. 6.7). Here, as in the discussion of eruptions from magma chambers, the 'eruptive volume' is not just the volume of materials or magma transported onto the Earth's surface but includes also the volume of the intrusive material (dikes, inclined sheets, and sills). That is, the eruptive volume is the total volume of magma transported out of the reservoir during the event. We first relate the bulk compressibility β_b and the solid-matrix compressibility β_s through the equation (Bear, 1972):

$$\beta_b = (1 - \eta)\beta_s + \eta\beta_p \tag{6.17}$$

where η is the porosity (magma or melt fraction) of the reservoir. The solid grains of the matrix are much stiffer, that is, have a higher Young's modulus, than the pores. It follows that the first term on the right-hand side of Eq. (6.17) makes a negligible contribution to β_b, so that $(1 - \eta)\beta_s << \beta_b$. It follows that the equation reduces to (Bear, 1972):

$$\beta_b = \eta\beta_p \tag{6.18}$$

When the reservoir receives new magma or melt from the mantle, part of the volume needed for the new magma is made by **compressing** the existing magma and part by **expanding** the pores containing the magma. Whether pore expansion or magma compression dominates depends on the ratio of the respective compressibilites, that is, on β_m/β_p. Thus, when magma or melt of volume ΔV_m is added to the reservoir of original total pore or magma volume $V_m(=V_p)$, the increase in excess pressure Δp_e is given by:

$$\Delta p_e = \frac{\Delta V_m}{V_m \beta_m} \left(1 - \frac{\beta_p}{\beta_m + \beta_p} \right) \tag{6.19}$$

Equation (6.19) can also be rewritten in the more suitable form:

$$\Delta p_e = \frac{\Delta V_m}{V_m (\beta_m + \beta_p)} \tag{6.20}$$

It is assumed that the reservoir is initially in mechanical or **lithostatic equilibrium** with its surroundings, so that its excess pressure is zero, $p_e = 0$, before magma of volume ΔV_m is added to the reservoir. Following the notation used in this chapter, the eruptive volume V_{er} is the total volume transported out of the reservoir during the eruption or event and includes intrusive material (dikes, inclined sheets, sills) as well as the eruptive material reaching the surface. The term ΔV_m is therefore identified with the eruptive volume so that $\Delta V_m = V_{er}$. Using η again for reservoir porosity:

$$V_m = V_p = \eta V_b \tag{6.21}$$

From the considerations above and Eqs. (6.20) and (6.21), the **bulk volume** of the reservoir V_b is obtained as follows:

$$V_b = \frac{V_{er}}{p_e \eta (\beta_m + \beta_p)} \tag{6.22}$$

Equation (6.22) can now be used to obtain a typical ratio between the bulk volume of a reservoir and its eruptive volume during an ordinary eruption. Recall that the condition for reservoir rupture and dike (the potential feeder) injection is (Chapter 5):

$$p_l + p_e = \sigma_3 + T_0 \tag{6.23}$$

Here, p_l is the lithostatic pressure (equal to overburden pressure or vertical stress (Eq. (6.12)) when the reservoir is in mechanical equilibrium with its surroundings), σ_3 is the minimum compressive principal (maximum tensile) stress, and T_0 is the *in situ* tensile strength of the host rock (the roof) of the reservoir. When the reservoir is in lithostatic equilibrium and before it receives new magma, the lithostatic fluid pressure in the reservoir is equal to the minimum principal stress, that is, $p_l = \sigma_3$. Recall that for lithostatic equilibrium all the principal stresses are equal, so that $\sigma_3 = \sigma_2 = \sigma_1$ (Chapter 3). When magma of volume ΔV_m is added to the reservoir, rupture and dike injection occurs when the excess pressure is equal to the *in situ* tensile strength of the host rock (the roof), that is, when $p_e = T_0$.

The *in situ* tensile strength can vary from 0.5 to 9 MPa (Appendix E.2), but here the same typical or average value of 4 MPa is used, as for the shallow chambers above in deriving the ratio in Eq. (6.10). We use the static compressibility β_m for tholeiite (basaltic) magma at temperatures of 1100–1300 °C estimated in Section 6.5 as 1×10^{-10} Pa^{-1}. Consider, as an example, a deep-seated reservoir beneath volcanic zones of Iceland. For these, the pore compressibility is estimated at about $9 \times 10^{-11} \, Pa^{-1}$ (Gudmundsson, 1987) and the average porosity at 0.25 (25%). Both these values are poorly constrained and could easily vary by a factor of two. For example, it is possible that the topmost parts of typical reservoirs beneath the volcanic zone of Iceland are totally molten. It is trivial, however, to modify the formulas so as to be suitable for totally molten reservoirs (or compartments of reservoirs), and here let us consider the general case of a partially molten reservoir.

Solving Eq. (6.22) for the eruptive volume V_{er}:

$$V_{er} = \eta p_e (\beta_p + \beta_m) V_b \qquad (6.24)$$

Inserting the values obtained above in Eq. (6.24) gives the following ratio between the bulk volume of the reservoir V_b and the **eruptive volume** in an 'ordinary' eruption V_{er}:

$$V_{er} \approx 2 \times 10^{-4} V_b \qquad (6.25)$$

As an example, let us now apply these results, again, to the well-constrained reservoirs of Iceland, although the equations are, of course, applicable to any volcanic area. The cross-sectional areas of reservoirs beneath the volcanic zones of Iceland are generally similar to the cross-sectional areas of the volcanic systems to which the reservoirs supply magma. Most of the systems, hence the reservoirs, have (lateral) **cross-sectional areas** between about 100 km^2 and 2500 km^2, with typical values between about 300 km^2 and 1600 km^2 (Thordarson and Larsen, 2007). For a typical small fissure eruption of 0.2 km^3 (a lava flow of 0.1 km^3 and a feeder-dike of 0.1 km^3), the bulk volume of the reservoir would, from Eq. (6.25), have to be 1000 km^3. For lateral cross-sectional areas of 300 and 1600 km^2, the **thicknesses** of the reservoirs would have to be about 3.3 km and 0.6 km, respectively. Given the estimated lateral dimensions of the reservoirs, such thicknesses are very reasonable, so that typical small fissure eruptions do not pose any volume problems for associated reservoirs.

Moderately large Holocene lava shields in Iceland (Fig. 6.7) are also commonly around 1 km^3 or less, including the estimated feeder-dike volumes (e.g. Rossi, 1996; Andrew and Gudmundsson, 2007; Thordarson and Larsen, 2007). From Eq. (6.25) the bulk volume of the corresponding reservoir would have to be about 5000 km^3. For a reservoir of a lateral cross-sectional area between 300 km^2 and 1600 km^2, as used above, to produce 1 km^3, the thicknesses of the reservoir would have to be 16.6 km and 3.1 km, respectively. The former value is not impossible, but somewhat thick, but the latter is very reasonable. It can thus be concluded that the reservoirs in Iceland can, as regards their typical volumes, produce basaltic eruptions that are **small to moderate** in size, that is, up to at least 1 km^3 through the ordinary mechanism discussed here. Much larger Holocene eruptions, however, have occurred in Iceland – at least up to 25 km^3, and even the largest reservoirs could not produce such volumes in a single eruption through the ordinary poroelastic mechanism presented above. And worldwide, there are many effusive eruptions with volumes reaching tens or hundreds, and, occasionally, thousands of cubic kilometres that were apparently produced in single, presumably long-lasting, eruptions. The general dynamics of large eruptions, effusive and explosive, is treated in Chapter 8. Here, however, I continue with the mechanism of ordinary eruptions, and turn now to the mechanical interaction between the shallow chamber and its deep-seated source reservoir.

6.7 Dynamics of Double Magma Chambers

Since the shallow chamber is normally much smaller than the source reservoir (Figs. 6.1, 6.4, 6.7 and 6.8) it follows that a **single magma flow** from the reservoir, the volume of which is partly or almost entirely received by the chamber, may trigger **tens** of magma-chamber ruptures and dike/sheet injections (Chapter 9). Many of these dikes may reach the surface and feed eruptions (Fig. 6.11). The difference in volume between the chamber and the source reservoir is also one reason why the **eruption frequency** of a volcanic system/field is normally much higher in the main polygenetic volcano (stratovolcano, caldera, basaltic edifice, central volcano) than in the rest of that system/field, that is, outside its central volcano.

Another reason for the high eruption frequency of the central volcano in comparison with the other parts of the associated volcanic system/field is **tensile-stress concentration** around the shallow chamber (Chapter 3). The resulting local stress field commonly favours the formation of a swarm of inclined sheets (Fig. 6.12), with various directions and dips (Figs. 6.1, 6.8, and 6.13). All sheet intrusions, including dikes, sills, and inclined sheets, are primarily extension fractures. It follows that the host-rock displacement is perpendicular to a sheet. For a shallow-dipping sheet, the host-rock displacement across the sheet close to the surface is mainly crustal uplift (Fig. 6.14). Because there are many shallow-dipping sheets in a typical sheet swarm (Fig. 6.13), the crustal dilation associated with such a swarm is partly

Fig. 6.11   The 1991 eruption of the Hekla Volcano in South Iceland was partly fed by a vertical dike (the main fissure at the top of the volcano) and partly by inclined sheets (the main crater and the tiny lava flow, both on the left, that is, the southeast part of the volcano). After the first days of the eruption, only one of the sheets continued to supply magma to the eruption at the location of the main crater cone for some 50 days (cf. Gudmundsson, 2016).

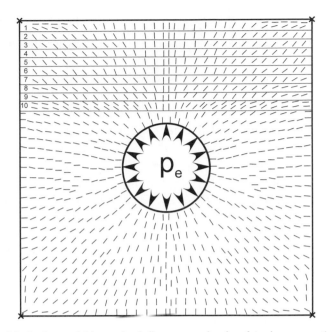

Fig. 6.12 Numerical model of the local stress field around a shallow magma chamber of circular cross-section. The trajectories or ticks of the maximum compressive principal stress σ_1 are indicated. The ticks map out the likely paths of dikes and inclined sheets (Chapter 10), primarily radial dikes in the centre of the chamber and inclined sheets above its marginal parts. The only loading is a magmatic excess pressure p_e of 10 MPa. There are 10 layers above the magma chamber. The surface layer has a stiffness of 10 GPa, a stiff basaltic lava flow, and then the layers below gradually increase in stiffness by 2 GPa for each layer. The layer hosting the chamber has a stiffness of 30 GPa. Modified from Gudmundsson and Brenner (2004).

accommodated by uplift rather than by horizontal crustal extension – the latter dominating in regional dike swarms (Fig. 2.1; Chapter 2). Since the intensity of a sheet swarm is normally much greater than that of a regional dike swarm (Figs. 2.1, 5.1 and 6.13), the sheet-injection and eruption frequencies and the associated crustal dilation are normally much higher inside the polygenetic (central) volcanoes than in the surrounding volcanic field/system.

Mechanically, the **intrusion and eruption statistics** of central volcanoes can partly be explained using the double-magma-chamber model (Figs. 6.1, 6.4, and 6.8; Chapter 9). The total volume erupted, that is, the total volume that is transported out of a deep-seated reservoir during rupture is given by Eq. (6.24). Similarly, the total volume erupted or transported out of a shallow magma chamber during rupture is given by Eq. (6.11). When the shallow chamber is totally molten, the total volume transported out of it during rupture is given by Eq. (6.7).

As indicated, the shallow-chamber volume is normally less than the reservoir volume (Fig. 6.7). We therefore have:

$$V_c = \lambda V_r \tag{6.26}$$

where $0.0 \leq \lambda \leq 1.0$. Normally, the shallow magma chamber would not receive all the magma that is transported out of the source reservoir during rupture and magma flow out of

Fig. 6.13 Part of a swarm of inclined sheets in the extinct Geitafell Volcano in Southeast Iceland. View north, most of the sheets are basaltic and with thickness less than 1 m. Many are cross-cutting (some cross-cuttings are marked by the letter c), indicating changes in the local stress field around the nearby shallow magma chamber over time.

Fig. 6.14 Inclined basaltic sheet that abruptly changes its path on meeting a stiff lava flow. View northeast, the sheet is hosted by soft lake sediments and changes its dip and becomes thinner under the basaltic lava flow on the top of the sediments. The average sheet thickness is about 0.5 m. Since sheets are extension fractures, the opening is normally perpendicular to the intrusion.

the latter. Part of the magma solidifies in the feeder-dike connecting the two chambers (Figs. 6.1, 6.7, and 6.8), and part may be erupted directly from the reservoir and outside the chamber (Figs. 6.7 and 6.8). Consider a shallow magma chamber which **receives** a fraction $q(0.0 \leq q \leq 1.0)$ of the total magma that flows out of a reservoir during its rupture and dike injection, and denote the **fraction of magma** in the shallow chamber by f. Measurements in drill holes worldwide, using hydraulic fracturing, suggest that the *in situ* tensile strength T_0 of rocks varies little, and does not show any systematic change in magnitude with depth (Appendix E). We therefore assume that the tensile strength of the roof of the source reservoir is the same as that of the roof of the shallow chamber. Since the excess pressure at rupture equals roughly the *in situ* tensile strength, this implies that the excess magmatic pressure p_e in the reservoir before rupture is the same as that in the shallow chamber. The magma compressibility β_m is also roughly the same for the reservoir and the chamber. By contrast, the host-rock compressibility of the chamber β_r is likely to be different from the pore compressibility of the reservoir β_p, partly because of the much higher temperature of the porous matrix of the reservoir. Based on these considerations and Eqs. (6.11), (6.24), and (6.26), the **number of dikes and inclined sheets** N_s from the chamber triggered by a single magma flow from the source reservoir is:

$$N_s = \frac{\eta p_e q(\beta_p + \beta_m)}{f p_e \lambda(\beta_r + \beta_m)} = \frac{\eta q(\beta_p + \beta_m)}{f \lambda(\beta_r + \beta_m)} \tag{6.27}$$

As an example of applying these results, let us consider the double magma chamber associated with the Krafla Volcano in North Iceland (Fig. 6.15). The porosity of the deep-seated reservoir may be taken as $\eta = 0.25$ (Gudmundsson, 1987). The fraction of the magma volume received by the shallow chamber during flow from the reservoir is $q = 0.5$, using rough estimates from Tryggvason (1984, 1986). If the volume of the shallow chamber is 0.5% of the volume of the reservoir, that is, $\lambda = 0.005$, and the compressibilities of the magma and the host rock are $\beta_m - 1.3 \times 10^{-10}$ Pa^{-1}, $\beta_p = 9 \times 10^{-11}$ Pa^{-1}, and $\beta_r = 3 \times 10^{-11}$ Pa^{-1} (Gudmundsson, 1987), and the shallow chamber is totally molten ($f = 1$) then Eq. (6.27) yields the number of dikes and inclined sheets injected by the shallow chamber during a single magma flow from the source reservoir as $N_s = 34$.

The estimated number of 34 is clearly only approximate since many of the parameters involved are poorly known and constrained. Nevertheless, the results show that a single magma flow from the magma reservoir, which lasted for 9 years during the Krafla rifting episode 1975–1984 (Björnsson, 1985), may have triggered **some 30** dike and inclined-sheet intrusions, some of which feed eruptions, from the shallow crustal chamber of the Krafla Volcano. This conclusion agrees well with geodetic and seismic data, which indicate that during the 1975–1984 rifting episode the shallow chamber injected dikes and inclined sheets at least 20–30 times, of which nine were feeder-dikes that supplied magma to eruptions (Björnsson, 1985; Tryggvason 1984, 1986; Harris et al., 2000).

Commonly, the **interaction** between shallow magma chambers and their source reservoirs may be somewhat more complex than indicated above. This follows because the composition and volatile content of the magma in the shallow chamber evolves with time, and so does the excess pressure, partly independently of the magma injections or flows from the source reservoir. Also, the local stress concentration around the shallow magma

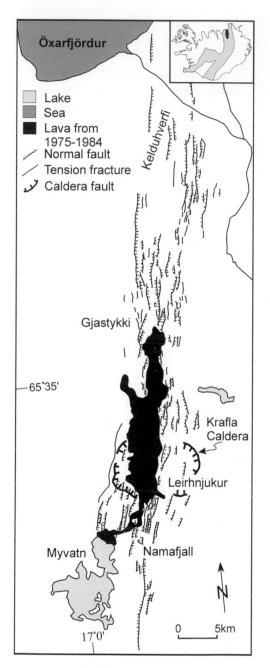

Fig. 6.15 Krafla Volcanic System in the North Volcanic Zone of Iceland. The system is close to 100 km long and from 4 km to 10 km wide, the main surface structures being tension fractures, normal faults, and volcanic fissures. The polygenetic volcano is a collapse caldera (the Krafla caldera) with a centre close to the hill Leirhnjukur, which was also the location of the main measured inflation and deflation during the 1975–1984 Krafla Fires. The presently exposed fissure swarms extends from Gjastykki to Kelduhverfi. During the early rifting events in the Krafla Fires, extension or dilation across the swarm was measured from Lake Myvatn, in the south, north to the coast of the fjord Öxarfjördur. Data from Opheim and Gudmundsson (1989) and Saemundsson and Sigmundsson (2013).

chamber, hence its rupture tendency, depends on the chamber shape which, again, may change with time, partly independently of the magma supply from the source reservoir. Furthermore, the proportion of sheets/dikes injected from the shallow chamber that reach the surface to feed eruptions is a function of several factors, including the mechanical properties of the layers that constitute the crustal segment hosting the chamber as well as the local stresses. Nevertheless, the analysis above, and in particular Eq. (6.27), suggest that one reason for the normally much higher dike/sheet injection and **eruption frequencies** of polygenetic volcanoes (stratovolcanoes, central volcanoes, and basaltic edifices), but smaller volumes on average in each eruption, than in the remaining part of the associated volcanic field/system is the difference in volume between the source reservoir and the shallow chamber. The shallow chamber acts as a sink for magma from the reservoir, and as a source for the polygenetic volcano – and, indeed, as the reason for the comparatively frequent eruptions and, hence, the formation of that volcano (cf. Chapter 9).

6.8 Detecting Magma Chambers and Reservoirs

Active magma chambers can be detected through geophysical measurements. The main techniques used are **geodetic** and **seismic**, both of which are described in Chapter 1 and applied in Chapters 3 (geodetic) and 4 (seismic) for volcano deformation and volcanic earthquakes. In addition, there are several **geochemical** and **petrological** indicators as to the depths of the sources for erupted materials. These latter focus primarily on the pressure–temperature conditions at the location where the magma resided before it was erupted and are not very accurate as to the location of the main magma chambers and reservoirs. Here, the focus is on the geophysical methods for detecting active magma chambers, and on field methods for identifying fossil chambers. Unless stated otherwise, when the word 'chamber' is used without further qualification, it refers to a shallow chamber, a deep-seated reservoir, or both, as the case may be.

6.8.1 Detecting Active Magma Chambers and Reservoirs

Seismic studies for detecting magma chambers are of two main types: (1) those that focus on the host rock of the chamber, and (2) those that focus on the magma chamber itself. Studies on earthquakes in the host rock are mainly '**passive**' in the sense that they focus on recording earthquakes in the rock hosting the magma chamber during unrest periods. The foci or focal points of the earthquakes form a '**cloud**', the density of which is normally highest close to the associated magma chamber (Fig. 1.12). From the distribution of earthquake foci it is often possible not only to infer the depth of the magma chamber, but also, crudely, its dimensions (Fig. 6.16). Independent studies, such as of volcanic fissures and dikes supposed to be injected from the same magma chamber, can then be used to test and constrain further the location and, possibly, the dimensions of the chamber (Becerril et al., 2013; Browning et al., 2015). If and when the magma chamber ruptures, the earthquake

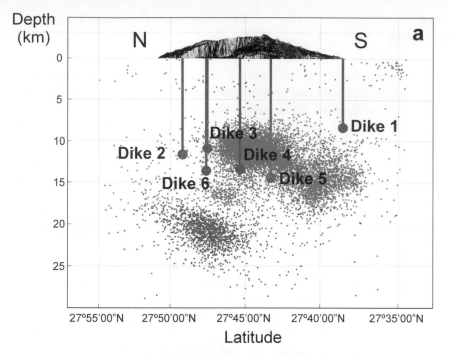

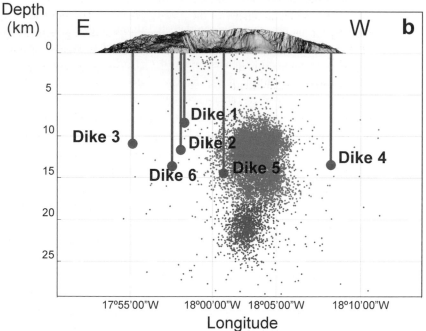

Fig. 6.16 Magma-chamber detection. Here the chamber beneath the volcanic island of El Hierro (Canary Islands) is detected by two independent methods. The first is based on the depths of the hypocentres (Fig. 4.2) of the earthquakes that occurred during the unrest period prior to and during the 2011–2012 eruption offshore El Hierro. The main earthquake cloud at 11–15 km depth is assumed to occur primarily inside and close to the magma chamber. The second method is based on measured aspect ratios (Chapter 5) of several feeder-dikes (dikes 1–6) measured at the surface of El Hierro. From the aspect ratio of the dikes, the depths to their sources (magma chambers) are calculated. The source depths of the dikes coincide with the depth of the main earthquake cloud, suggesting that the feeder-dikes may have come from the presently active magma chamber (as determined by the earthquake hypocentres). (a) Cross-section from north to south parallel to the main earthquake swarm of the 2011–2012 eruption. (b) Cross-section from east to west perpendicular to the main earthquake swarm (Becerril et al., 2013).

foci become concentrated at and around the resulting dike (Fig. 1.13), the propagation of which can then be monitored through real-time earthquake data.

The other main seismic method is **tomography**, that is, imaging of the magma chamber (or other crustal bodies) using seismic data. The velocities of the seismic waves indicate the mechanical properties and densities of the rocks or other materials through which the waves pass (Eqs. (4.1)–(4.5), Chapter 4). From these data, dense parts of partly or totally solidified magma chambers, such as gabbro or granitic bodies, can be inferred. The partially molten parts of the chambers can also be inferred, particularly from the **attenuation** (reduction in the energy or decay) of the seismic waves as they pass through these parts. Furthermore, for magma chambers, or parts of chambers, that are totally molten, the S-waves do not penetrate, but rather drop out, resulting in **S-wave shadows** (Fig. 4.1; Chapter 4). Seismic attenuation and, in particular, S-wave shadows have been used to identify magma chambers at many locations, such as at fast-spreading mid-ocean ridges.

While some tomography studies are passive in the sense that waves from natural earthquakes are used, there are also many **active** studies. In the latter, the seismic waves are generated through human-made explosions. A dense network of seismometers is then used to detect the earthquakes, and attenuation and S-wave shadows can be used to map the three-dimensional geometry of the magma chamber. The same technique is used in the oil industry for detecting likely oil and gas reservoirs and for general imaging of the upper part of the crust, particularly in sedimentary basins. Tomography of this kind is also used to detect much larger structures, such as deep-seated reservoirs, plutons, and mantle plumes.

Geodetic measurements for detecting magma chambers have historically primarily relied on the use of the Mogi model. The model has many limitations (Chapter 3) and is basically only useful to indicate, crudely, the **likely depth** of a magma chamber or reservoir during eruptions and/or unrest periods. The model cannot really be used to infer the size or shape of the chamber/reservoir, whereas numerical models are more appropriate for such tasks. When geodetic and seismic results are combined with detailed modelling, preferably with numerical models and layered crustal segments and volcanoes (Chapter 3), reasonably accurate information about the likely dimensions and depth of a magma chamber may be obtained.

In particular, when using numerical models, all the available data on crustal layering, including the effects of ice sheets and caldera lakes, can be used to estimate the magma **chamber depths and dimensions** (Browning and Gudmundsson, 2015a). The more information there is on the mechanical layering and general anisotropy in the inflating or deflating volcanoes, the more accurate will be the estimated sizes, shapes, and depths of the associated chambers. When the crustal layering can only be inferred, for example by analogy with other similar but better known crustal segments, the measured geodetic and seismic data can still be used through numerical models to obtain reasonably reliable models of the chamber associated with an unrest period and, eventually, an eruption.

Another way to estimate **magma-chamber volume** is to combine geodetic (and seismic) data with data on eruptive and intrusive volumes. The seismic data together with the geodetic data then indicate the depth and, to a degree, the dimensions of the chamber and the dimensions of the feeder-dike during one or more eruptions. Combining the estimated volume of typical feeder-dikes with typical volumes of eruptive materials (such as lava flows) can then be used as input for V_{er} in the poroelastic models in the present chapter to calculate the bulk or total volume of a chamber (Eqs. (6.7) and (6.11)) and a reservoir (Eqs. (6.22) and (6.24)). Similar calculations are made in Examples 6.1 and 6.3.

6.8.2 Depths of Active Magma Chambers and Reservoirs

As indicated earlier, shallow magma chambers are normally in the upper part of the crust whereas deep-seated reservoirs are in the lower part of the crust, or in the upper mantle. A compilation of magma-chamber depths from 15 plate-boundary regions (Chaussard and Amelung, 2014) indicates that most of the shallow chambers are with roofs at **depths of 0.1–5 km** and mostly 1–5 km (Fig. 6.17). By contrast, the roofs of the deeper chambers or reservoirs at these boundaries extend to depths of at least **18 km**. These estimates refer to the depth below the **average regional elevation** of the area or crustal segment within which the chamber is located. These are thus not depths below the tops of the associated stratovolcanoes (or calderas); with reference to volcano tops the given depths would normally be greater. The depths are based on geodetic, seismological, and petrological/

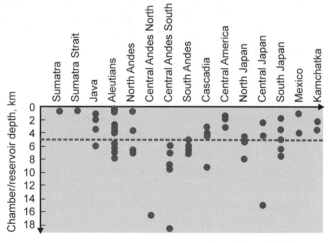

Fig. 6.17   Common depths of magma chambers and reservoirs. Shallow magma chambers are mostly located in the upper part of the crust whereas deep-seated reservoirs are located in the lower part of the crust, or in the upper mantle. This compilation of magma-chamber/reservoir depths from 15 plate-boundary regions (Chaussard and Amelung, 2014) suggests that shallow chambers generally have roofs at depths of 0.1–5 km and mostly 1–5 km below the surface. The roofs of the deeper chambers or reservoirs, however, extend to depths of at least 18 km. These estimates refer to the depths of the roofs below the average regional elevation of crustal segment within which the chamber/reservoir is located. It follows that when referred to the depths below the tops of the associated polygenetic volcanoes the depths would normally be greater, so that some reservoir roofs reach depths in excess of 20 km, and the reservoirs themselves extend to greater depths. The depths are based on geodetic, seismological, and petrological/geochemical data and refer to chambers/reservoirs of 70 polygenetic volcanoes from subduction zones (arcs) in North, Central, and South America, as well as in Sumatra, Java, and Japan in Asia.

geochemical data of the type discussed above and refer to chambers/reservoirs of 70 polygenetic volcanoes.

While there is some spread in the depth estimates, the distinction between the roof depths of the shallow chambers and the deeper reservoirs is generally clear, at **about 5 km**. Because many of the chamber depths are based on Mogi models and seismicity, which tend to emphasise the top parts and the roofs of the chambers, we refer to these as depths to the contacts between the roofs and the chambers rather than the depths of the centres of the chambers. This is in agreement with the method used by Chaussard and Amelung (2014). The shallowest chambers have roofs at only **0.1 km depth**, and several chambers are in the range of 0.4–0.9 km, but most are deeper than about 1 km.

Similar results on the depths of active magma chambers have been obtained from studies in different volcanotectonic regimes. Thus, at **fast-spreading ridges**, that is, divergent plate boundaries, shallow chambers commonly have roofs at depths of **1–2 km** below the sea floor and continuous for tens of kilometres. For example, below the East Pacific Rise shallow chambers have been detected at depths of 1.2–2.4 km below the sea floor, and in places shallower than 1 km (Orcutt, 1987; Kent et al., 1990; Searle, 2013). The chambers are very thin – the thickness is less than 100 m and perhaps only several tens of metres – and have widths (perpendicular to ridge axis) of only 1–2 km, but can be detected along the axis for tens of kilometres. In fact, the shallow magma chambers appear almost **continuous** along the entire ridge segments (Searle, 2013). Similar magma chambers, but generally having roofs at greater depths (Searle, 2013), have been detected at intermediate-spreading ridges such as the Southeast Indian Ridge (offshore Australia; Baran et al., 2005), the Juan de Fuca Ridge (offshore western United States; Carbotte et al., 2006), and the Galapagos Spreading Centre (offshore Equador; Detrick et al., 2002).

By contrast, very few shallow magma chambers appear to be located below **slow-spreading ridges** (Pagli et al., 2012; Searle, 2013). Perhaps the best known are those detected at one location on the Mid-Atlantic Ridge (Singh et al., 2006) and another at the south end of the Reykjanes Ridge (south of Iceland; Sinha et al., 1998). The tops of these magma chambers are at **2.5–3 km**, and thus considerably deeper than those below fast-spreading ridges. Generally, the depth to the tops of shallow magma chambers at mid-ocean ridges increases with decreasing spreading rate, from occasionally less than 1 km for the fastest-spreading ridges to about 3 km for the slow-spreading ridges (Chen, 2004; Baran et al., 2005; Searle, 2013).

Deeper **reservoirs** have also been suggested for the ocean ridges, partly based on field observations of **ophiolites** (primarily the Oman or Semail ophiolites). Sill-like reservoirs are proposed at the base of the lower crust, and these supply magma to the shallow chambers through dikes (Crawford and Webb, 2002; Canales et al., 2009). Such reservoirs were proposed on theoretical grounds decades ago (Gudmundsson, 1990), and are now beginning to be detected through seismic studies (Searle, 2013).

Deep and shallow magma chambers/reservoirs have also been detected in volcanic areas outside the subduction zones (Fig. 6.17) and mid-ocean ridges discussed above. For example, in Italy, sill-like magma chambers at 7–8 km depth below the nearby Vesuvius Volcano and the Campi Flegrei caldera have been detected (Auger et al., 2001; Zollo et al., 2008). Also, a magma chamber has been detected at 6–7 km depth below the Soufriere Hills

Volcano in Montserrat, a Caribbean island in the Lesser Antilles of the British West Indies (Beachly et al., 2012; Paulatto et al., 2012).

The **Kluchevskoy Volcano**, part of a group of volcanoes in Kamchatka, Russia, is located in one of the volcanic areas sampled in Fig. 6.17. However, it receives an extra discussion here as one of the volcanoes that is apparently fed by a **triple magma chamber** (Koulakov et al., 2011). A **shallow chamber** with a roof at 1–2 km depth is directly beneath the volcano. Its shape is unclear, but the diameter is of the order of 3–4 km. With a roof at a depth of about 8 km beneath the volcano is another chamber. This **middle chamber** is much larger, with a diameter of 10–12 km, and it apparently supplies magma to the shallower chamber. The roof of the deepest chamber, a **reservoir** with a lateral diameter of at least 10–15 km and unknown thickness, is at a depth of about 25 km, at the boundary between the lower crust and upper mantle. The deep-seated reservoir supplies magma to the middle chamber, but it may also supply magma to other volcanoes and eruptions outside the Kluchevskoy Volcano.

Magma chambers have been detected below Mauna Loa and Kilauea in Hawaii (Poland et al., 2014a,b). **Mauna Loa** seems to have a shallow chamber at a depth of 3–4 km and possibly a deeper reservoir. Due to its continuous activity in recent decades, **Kilauea's magma chambers** are better constrained than those beneath Mauna Loa. There appear to be two sill-like chambers beneath the summit of the volcano (Poland et al., 2014a,b). One of the shallow chambers has a roof at about 1–1.5 km below the surface, the other at 2.5–3 km. These sill-like chambers appear to extend into the rift zones: the shallower chamber primarily into the Southwest Rift Zone, and the deeper chamber primarily into the East Rift Zone. Many studies also indicate that, in addition to these shallow chambers, there are deeper reservoirs beneath the rift zones, as are observed in Iceland and, increasingly, at mid-ocean ridges. Recent seismic studies add further support to the existence of a **deep, partially molten reservoir** at a depth of 8–11 km beneath the East Rift Zone of Kilauea (Lin et al., 2014). Most likely, this reservoir supplies magma to many dike injections in the East Rift Zone, in addition to those injected from the shallow magma chambers.

In the Erte Ale Volcanic System or segment in the Afar Region of Ethiopia, the location of the **Erte Ale Volcano**, an elongated, sill-like magma chamber has been detected at a depth of about 1 km (Pagli et al., 2012). The chamber is apparently divided into two parts, but both parts supplied magma to an eruption in 2008. The rate of spreading in the Erte Ale System is low, about 1.2 cm yr^{-1}. The Erte Ale chamber is shallow in comparison with slow-spreading mid-ocean ridges, as discussed above. However, a comparison with Iceland, also a slow-spreading area, suggests that in some slow-spreading regimes the shallow magma chambers occur at depths of about 1 km.

In **Iceland**, many shallow and deep-seated reservoirs have been detected (Section 6.4). Examples of shallow chambers include the one in the **Katla Volcano** in South Iceland where a sill-like shallow chamber occurs with a roof at 1.5–2 km below the bottom of the associated collapse caldera (Sturkell et al., 2010). The chamber is about 1 km thick and has a lateral diameter of about 5 km – and thus is strikingly similar in depth and thickness to some well-exposed fossil chambers in Iceland (e.g. Fig. 7.5). While no deep-seated reservoir has been detected, petrological arguments indicate that the large eruptions in the Katla Volcanic System do not derive from the shallow chamber but rather from a much deeper and larger reservoir (Fig. 6.7; Oladottir et al., 2008). Similarly, the **Grimsvötn Volcano** in the Vatnajökull ice sheet has a shallow sill-like chamber with a roof at a depth of about 1.7 km (Hreinsdottir et al., 2014). The chamber is presumably less than 1 km thick and 4–5 km in diameter (Alfaro et al., 2007). A deeper magma reservoir was inferred to exist below the shallow chamber (Reverso et al., 2014). The deeper reservoir is also likely to be sill-like and with a lateral diameter of about 10 km. Its depth, however, remains poorly constrained and could be anywhere between 10 km and 30 km. In the nearby **Bardarbunga Volcano,** recent studies suggest a shallow sill-like magma chamber, located at a depth of a few kilometres, supplied with magma from a deep-seated reservoir, with a roof at a depth of at least 12–15 km, and perhaps deeper – similar to the situation indicated in Fig. 6.8 (cf. Gudmundsson et al., 2014; Browning and Gudmundsson, 2015a). The **Askja** and **Krafla** calderas both have deep-seated reservoirs in addition to shallow chambers. In Krafla, the shallow chamber has a roof at about 2–3 km depth, but its shape is poorly constrained. The roof of the deep-seated reservoir is at a depth of about 20 km, at the contact between the upper mantle and the lower crust (Sturkell et al., 2008). In Askja, the shallow sill-like chamber has a roof at 2–3 km depth, perhaps shallower, and the deep-seated reservoir has a roof at about 16 km depth (Sturkell et al., 2006; Soosalu et al., 2010).

The location and shapes of double magma chambers detected so far in Iceland and elsewhere are in good agreement with the **theory** developed in this chapter. This theory was initially proposed for divergent plate boundaries (Gudmundsson, 1990), and for reservoirs and chambers in Iceland in particular (Figs. 6.3 and 6.4; Gudmundsson, 1987). The double chambers detected in recent decades are as predicted by the general poroelastic models presented in these early papers and developed in the present chapter (cf. Gudmundsson, 2016). There is little doubt that deep-seated reservoirs will eventually be found under most, perhaps all, volcanic systems in Iceland, and at intermediate- and fast-spreading ridges. While the physical principles of poroelasticity and fluid transport are the same at convergent boundaries, their different stress regimes require further development of the models presented here, a task for the near future.

6.8.3 Depths of Fossil Magma Chambers

Fossil magma chambers and reservoirs are exposed as plutons. Many plutons in Iceland have roofs at depths similar to the currently active shallow magma chambers. Most of the roofs of the exposed fossil magma chambers in Iceland are at depths of 0.5–2 km below the surface of the volcanic system within which they formed. Elsewhere, there are numerous plutons found at depths of erosion of **one to several kilometres** below the top of the volcanic area to which they belonged at the time of emplacement of the plutons (e.g. Sheth, 2018).

Fossil magma chambers include the laccoliths Sandfell in East Iceland and Baula (Fig. 5.19) and Stardalshnjukar, all in West Iceland, as well as Geitafellsbjorg (Fig. 7.5) in Southeast Iceland. Sandfell and Baula are both acid and have roofs at very shallow depths – presumably about **0.5 km** below the surface at their time of emplacement. These are small plutons and while they must have fed dikes that very likely reached the surface, thereby acting as shallow chambers supplying magma to eruptions, they were presumably active for a short time. Stardalshnjukar is another very shallow fossil chamber

(Gudmundsson, 2017). Its roof was at about **0.7 km** depth below the Stardalur Volcano, a collapse caldera, to which it supplied magma. In contrast to Sandfell and Baula, both acid, Stardalshnjukar is a mafic pluton (of microgabbro) and sill-like (oblate ellipsoidal) in shape. The pluton is small, however, and presumably acted as a magma chamber only for a short time. Similarly, Geitafellsbjorg (Fig. 7.5) is a mafic pluton (of gabbro) and is also sill-like in shape and located within a collapse caldera (the Geitafell Volcano). But its roof is at much greater depth than that of Stardalshnjukar, or about **1.8 km** below the surface at the time of its emplacement. This pluton functioned as a magma chamber for the Geitafell Volcano for many hundreds of thousands of years, and is thus a typical long-lived chamber. While the **roofs** are not very well exposed in these plutons (except, partly, for Sandfell), their depths below the surface at the time of emplacement follow from various considera-tions, such as heights of nearby mountains and secondary mineralisation.

There are, however, several particularly well-exposed plutons where not only the **roof,** but also parts of the **walls**, are exposed. The granophyre pluton Slaufrudalur in Southeast Iceland is an excellent example (Figs. 1.17 and 5.6). Here, the upper parts of the walls as well as a large part of the roof are easily seen (Fig. 1.17). Furthermore, because of the high viscosity of its magma, the way that the pluton formed is still clear – it was generated through multiple sill injections, the contacts of which are still seen (Fig. 5.6). There is little doubt that this pluton acted as a magma chamber. Many **acid dikes** are seen cutting through its roof (Fig. 1.17), some of which almost certainly reached the surface to supply magma to eruptions (Gudmundsson, 2012).

The exposed part of the Slaufrudalur pluton/magma chamber is located at 1.5–2 km below the original surface, the contact between the basaltic roof and the granophyre pluton (Fig. 1.17) being at about 1.5 km below the surface when the chamber was active. There are several other fossil magma chambers in Southeast Iceland where part of the roof is seen at a similar depth below the surface. One is Austurhorn, a pluton composed partly of acid (granophyre) and partly of mafic (gabbro) rocks (Fig. 6.18). As is also seen for Slaufrudalur (Fig. 1.17), the contact between pluton and the host rock (a basaltic lava pile) is smooth. This follows from thermal considerations which indicate that large-scale contacts between fossil magma cham-bers/plutons and the host rock tend to be so. The smooth contact seen here is exposed in the mountain Krossanesfjall, again at about 1.5 km below the surface at the time of magma-chamber emplacement. Also indicated is the contact between the primarily felsic and primarily mafic intrusions in part of the mountain. On the sandy coast, a part of the net-veined complex in Krossanesfjall is seen (Blake, 1966; Furman et al., 1992; Thorarinsson and Tegner, 2009; Gudmundsson, 2012). The other fossil magma chamber discussed here and also composed partly of acid (granophyre) and partly of mafic (gabbro) rocks, is Vesturhorn (Fig. 6.19). This is the largest exposed pluton/ fossil chamber in Iceland (with an area of about 20 km^2), is felsic (granophyre) and mafic (gabbro), with part of its roof (a basaltic lava pile) seen at about 1.5 km below the surface at the time of emplacement. The pluton is composed of more than 70 individual intrusive bodies of various sizes (Roobol, 1974), which in the felsic part are commonly sills similar to those that constitute the felsic part of the Slaufrudalur pluton.

Fig. 6.18 Contact between the roof and the host rock (basaltic lava flows) of a part of the Austurhorn pluton, a fossil shallow magma chamber in Southeast Iceland seen here in the mountain Krossanesfjall (its height is 716 m). View northwest, the large-scale roof contact, which occurs at about 1.5 km below the original surface of the lava pile, is comparatively smooth, and the same applies to the contact between the primarily felsic (granophyre) and primarily mafic (gabbro) intrusions in part of the mountain (cf. Gudmundsson, 2012).

Fig. 6.19 Part of the Vesturhorn pluton, a fossil shallow magma chamber in Southeast Iceland (the maximum mountain height is 889 m). View northwest, the pluton is composed partly of felsic (granophyre) and partly of mafic (gabbro) bodies (some are indicated). Also indicated are some dikes, inclined sheets, and sills that cut through the roof, a basaltic lava pile. The pluton has an exposed area of some 20 km^2, is composed of more than 70 individual intrusive bodies of various sizes (Roobol, 1974), and is the largest exposed pluton in Iceland. A black and white version of this figure will appear in some formats. For the colour version, please refer to the plate section.

From this overview it is clear that there are many fossil magma chambers in Iceland exposed with the **roof** (the contact between the top boundary of the pluton and the host rock) at depths between **0.5 and 2 km below the surface** at the time when the chambers were active. Similar fossil magma chambers (plutons), although normally less well exposed than in Iceland, are found in many other places, including some ophiolites (particularly the Semail or Oman ophiolites). In particular, there are many thick sills and laccoliths exposed in Canada, the United States, the United Kingdom, South Africa and, in particular, Antarctica (e.g. Sheth, 2018) some of whom presumably acted as shallow magma chambers. Perhaps the best known, and certainly the best exposed, is the Torres del Paine laccolith in Chile (Figs. 5.18, 10.1). The fossil chamber is mostly of granite and was formed, like many of the shallow felsic chambers in Iceland, through many thick sill injections at a depth of 2–3 km below the surface at the time (Michel et al., 2008; Lauthold et al., 2014). With an area of some 80 km^2, the **floor** pluton is exposed so that the maximum thickness of the felsic part can be determined at about 2 km. Just like in the Icelandic fossil chambers, many felsic dikes and inclined sheets dissect the roof of the fossil chamber.

6.9 Summary

- Many volcanic systems, zones, and fields are supplied with magma through double magma chambers. There is a deep-seated large reservoir that supplies magma to a much smaller, shallow magma chamber. The shallow chambers have roofs most commonly at depths of 1–5 km, although some are deeper. The deeper reservoirs have roofs at depths of 10–25 km or deeper.
- Double magma chambers are presumably the most common configuration of volcano plumbing systems, particularly at divergent plate boundaries. However, there exist more complex configurations, particularly at convergent plate boundaries where triple magma chambers may be common. In this chapter the focus is on double magma chambers.
- Deep-seated reservoirs form when magma or melt accumulates in regions of low potential energy, usually in the lower crust or at the boundary between the crust and upper mantle. The regions of low potential energy tend to coincide with the thinnest parts of the crust/lithosphere, that is, regions where the depth to a partially molten lower crust or upper mantle reaches a local minimum.
- Magma or melt flows from regions of comparatively high potential energy to regions of comparatively low potential energy, in accordance with Darcy's law of fluid transport in porous media. The force driving the magma to the low-potential regions (where the reservoirs form) is the gradient of the potential energy. The steeper the gradient, the greater is the force driving the flow into the reservoir. This follows because the variation in potential energy gives rise to a force. More specifically, all conservative forces (those

forces that conserve their mechanical energy) are derivable from the gradient of the potential energy.

- The level to which magma in a reservoir (or water in a closed aquifer) could rise in vertical 'pipes' connected to the reservoir is referred to as the potentiometric surface. If the potentiometric surface is below the Earth's surface, no magma injected directly from the reservoir would normally be able to reach the surface, but could form non-feeder-dikes and other intrusions. If the potentiometric surface is above the Earth's surface, eruptions are theoretically possible. The heights of monogenetic volcanoes fed directly by reservoirs above their surroundings indicate the elevation of the potentiometric surface above Earth's surface.

- Many and perhaps most shallow magma chambers initiate from sills. Many stay as sill-like through their lifetimes. The shallow chamber is supplied with magma from its deep-seated source reservoir. The resulting double magma chamber remains active so long as there is enough magma supply from the reservoir to keep the chamber at least partially molten. When this magma supply is cut off, the shallow chamber becomes inactive and solidifies as a pluton.

- For a totally molten basaltic shallow chamber, less than 0.1% of its magma volume is erupted/injected, that is, flows out of the chamber during an ordinary eruption. For an acid magma chamber, as much as 4% of the magma may flow out of the chamber during an ordinary eruption. 'Ordinary eruptions' are those that occur without any large associated volcanotectonic events such as vertical (caldera) or lateral (landslide) collapses.

- For a partially molten (poroelastic) basaltic reservoir, the ratio between the eruptive/intrusive volume in a single eruption and the total or bulk volume (magma and solid matrix) of the reservoir is commonly 0.0002. For a reservoir with 25% porosity, this means that about 0.08% of the magma in the reservoir flows out (as eruptive and/or intrusive materials) during an ordinary eruption. All small (less than 0.1 km^3) and many medium-sized (less than 10 km^3) eruptions can normally be fed directly by poroelastic reservoirs during ordinary eruptions.

- The volumes of shallow chambers are commonly 1–10% of the volumes of their source reservoirs. It follows that a single magma flow from the reservoir can trigger tens of magma flows (eruptions and/or dike intrusions) from the associated chamber. This is one reason why the eruption frequencies of central or polygenetic volcanoes (fed by chambers) are normally much higher – but with much smaller average volumes per eruption – than in the rest of the volcanic system/field outside the central volcano (and mostly fed by the reservoir).

- Using geophysical methods, primarily volcano earthquakes and geodetic results, shallow magma chambers have been detected in many volcanic areas. The chamber roofs (the uppermost contact between the chamber and the host rock) are mostly at depths of 1 km to 5 km below the general elevation of the surface above them. Some chambers, however, are shallower, at about 0.5 km or even less, others are deeper, perhaps at 6–8 km. There are not clear natural boundaries as regards depth between what is regarded as a shallow chamber and what as a deep-seated reservoir, but most of the shallow chambers have roofs or ceilings (upper margins) at depths of less than 5–6 km, and 5 km is the general reference in this book.

• Shallow magma chambers have, in particular, been detected at 1–2 km depth below many fast-spreading ridges, and at somewhat greater depths below intermediate-spreading ridges. Few shallow chambers, however, have been detected below slow-spreading ridges, and seem to be generally rare there. Overall, the results indicate that shallow chambers are much more common – almost continuous beneath many ridge segments – and shallower at fast-spreading ridges than at slow-spreading ridges.

• Deep-seated reservoirs were proposed on theoretical grounds decades ago, and have been increasingly confirmed through geophysical studies in recent years. Deep-seated reservoirs supply magma either directly to the surface – and then mostly primitive (basaltic) magma – or to shallow chambers, in which case the chamber and the reservoir act as a double chamber. There is increasing evidence for double magma chambers in Iceland, Hawaii, some mid-ocean ridges, and elsewhere. There is little doubt that deep-seated reservoirs, located in the lower crust or in the upper mantle, will be found under many volcanoes in the coming years and decades.

• Direct observations of fossil shallow magma chambers are possible in many deeply eroded volcanic areas. In particular, in Iceland the tops, including parts of the roofs and walls, of many fossil chambers, that is, felsic and/or mafic plutons, are exposed at depths of 0.5–2 km below the surface at the time when the chambers were active. Other areas of fossil shallow chambers include Canada, the United States, the United Kingdom, South Africa, and Antarctica. Perhaps the best-exposed fossil shallow magma chamber, however, is the Torres del Paine laccolith in Chile where, in addition to the roof, part of the floor of the 2-km-thick felsic part of the magma chamber is exposed.

6.10 Main Symbols Used

A	area
d	depth to the top of a reservoir, distance
F	force
f	fraction of magma in a shallow chamber
g	acceleration due to gravity
$\bar{i}, \bar{j}, \bar{k}$	unit vectors
k	intrinsic permeability
N_s	number of dikes/inclined sheets injected from a shallow chamber
p	final magmatic pressure (e.g. in a reservoir)
p_e	excess magmatic pressure

Δp_e	change in excess magmatic pressure
p_i	initial magmatic pressure (e.g. in a reservoir)
p_l	lithostatic pressure
Q	volumetric flow rate (of a fluid, here magma)
q	the fraction of magma volume 'erupted' by a reservoir that is received by its chamber
$\vec{q}$	discharge or Darcy velocity
T_0	tensile strength
U	potential energy
V_b	bulk volume of a reservoir
V_c	total volume of a shallow chamber
V_{er}	'eruptive' volume of magma, including both intrusive and eruptive volumes
V_m	magma volume of a reservoir
V_p	pore volume of a reservoir
V_s	volume of the solid matrix of a reservoir
ΔV_b	fractional change in the bulk volume of a reservoir
ΔV_m	fractional change in the magma volume of a reservoir
ΔV_p	fractional change in the pore volume of a reservoir
ΔV_s	fractional change in the volume of the solid matrix of a reservoir
x	horizontal coordinate, horizontal distance
y	horizontal coordinate, horizontal distance
z	vertical coordinate, vertical distance
z	elevation head
β_b	bulk compressibility of a reservoir
β_m	magma compressibility
β_p	pore compressibility
β_r	host-rock compressibility
β_s	solid-matrix compressibility
γ	specific weight
η	porosity (magma or melt fraction) of a reservoir
λ	ratio of the volume of a shallow chamber and that of its source reservoir
μ_m	dynamic viscosity of magma
ρ_m	magma density
ρ_r	host-rock or crustal density
σ_v	vertical stress
σ_3	the minimum compressive (maximum tensile) principal stress
φ	potential energy or total head of the magma
∇	mathematical operator del (nabla)
$\nabla\varphi$	gradient of the potential energy of the magma
∂	partial derivative

6.11 Worked Examples

Example 6.1

Problem

Field measurements show that basaltic fissure eruptions in a certain central volcano (a collapse caldera) over the past several hundred years produce lava flows with an average volume of 0.027 km^3. Exposed basaltic feeder-dikes in the walls of the caldera are, on average, 1.7 m thick and the associated volcanic fissures are on average 900 m long. Geodetic measurements suggest that the shallow chamber is at the depth of 2 km below the surface of the caldera. Make a crude estimate of the volume of the shallow magma chamber, assumed to be totally molten, feeding these eruptions.

Solution

If we had detailed information about the compressibility of the host rock of the magma chamber, and the magma itself, we would use Eq. (6.7). In the present case, however, we do not have such information, so we use Eq. (6.8) to get a crude estimate of the magma-chamber volume V_c. To do so, we must first find the total volume of injected and erupted magma during the typical fissure eruptions. The average volume of the lava flows is given as 0.027 km^3. Average feeder-dikes have a thickness of 1.7 m, a strike dimension of 900 m (the length of a typical volcanic fissure), and a depth of 2000 m (the depth to the source magma chamber, based on geodetic data).

    If we assume that the feeder-dike has the same strike-dimension from the surface to the depth of the source chamber (this assumption may hold reasonably well for such a small dip-dimension), then the volume of the feeder-dike is 900 m × 2000 m × 1.7 m = $3 \times 10^6 \, m^3$ = 0.003 km^3. Adding this to the lava volume, we get $V_{er} = 0.027 + 0.003 = 0.03 \, km^3$. From Eq. (6.8) we then have:

$$V_c \approx \frac{V_{er}}{7 \times 10^{-4}} = \frac{0.03 \, km^3}{7 \times 10^{-4}} = 43 \, km^3$$

If, for example, the magma chamber was sill-like and circular in plan view with a diameter of 6 km, then a thickness of about 1.5 km, for a totally molten chamber, would correspond to the calculated volume of 43 km^3. This would be a rather small magma chamber.

Example 6.2

Problem

Geodetic measurements suggest that the shallow magma chamber of an active caldera has been receiving new magma from its source reservoir over an unrest period of 5 years. The new magma volume is estimated as $2 \times 10^7 \, m^3$ and the depth to the chamber – from geodetic and seismic data – as 5 km. Dike thicknesses in the caldera walls are typically about 2 m.

Recent basaltic lava flows in the caldera have an average volume of 1.5×10^7 m³. Only two volcanic fissures are well exposed, both forming crater rows each about 1.5 km long. Use this information, together with that of typical length–thickness ratios of basaltic dikes, to assess the likelihood of chamber rupture and eruption in view of the new magma being added to the chamber in the past 5 years.

Solution

Here the main question is whether the added volume of magma to the chamber is sufficiently large to rupture the chamber roof and inject a dike that is able to reach the surface to feed an eruption. On the assumption that the chamber is totally molten and was in lithostatic equilibrium with its host rock before it received the new magma, then V_{er} in Eqs. (6.7) and (6.8) is an indication of the magma received by the chamber before it reaches the condition of rupture and possible eruption. If the new magma received so far by the chamber is similar to or larger than V_{er}, then rupture and dike injection is likely. If, however, the volume of the new magma received by the chamber is significantly less than V_{er}, dike injection (and therefore eruption) is unlikely.

Here, the typical volume of erupted lava flows is known, so we need to find the volume of the feeder-dike. The dip-dimension of the dike is 5000 m, that is, the depth to the shallow chamber, and its thickness may be taken as 2 m. It is true that the thickness commonly changes with depth, partly because of changes in the Young's moduli or stiffness of the crustal layers and partly because of changes in magmatic overpressure or driving pressure (cf. Eq. (5.4); Fig. 2.22). In the present case, the overpressure presumably increases with depth because the buoyancy effect (Eq. (5.3)) is either zero or negative. This follows because the crustal layers above shallow chambers have densities that are normally similar to, or less than, those of typical basaltic magmas. Thus, the strike-dimension and thickness of this dike may not change much with depth. Since the volcanic fissures at the surface are 1500 m long, and the fissures are normally shorter than the entire strike-dimension of the feeder-dike, we may take the dike strike-dimension as 2000 m. This is also in agreement with the typical length–thickness ratios of basaltic dikes at shallow crustal depths, about 1000.

From these considerations we estimate the volume of the feeder-dike as 2000 m × 5000 m × 2 m = 2×10^7 m³. It then follows that:

$$V_{er} = 1.5 \times 10^7 \text{m}^3 + 2 \times 10^7 \text{m}^3 = 3.5 \times 10^7 \text{m}^3 = 0.035 \text{ km}^3$$

From these calculations it appears that the magma that flows out of the chamber during typical eruptions is considerably greater in volume than that of the new magma that has been added to the chamber in the past 5 years. The present difference between these volumes is a factor of 3.5/2 = 1.75. We therefore conclude that, at this stage, the excess pressure and associated strain energy in the chamber (Eq. (6.7)) is unlikely to have reached the level necessary for an eruption. As we discuss in detail in Chapters 7 and 10, even if the chamber did rupture and form a dike, that in itself is not a sufficient condition for an eruption since many dikes become arrested. Here, however, the focus is on simple necessary (but commonly not sufficient) conditions for dike-fed eruptions based on simple volume considerations.

These estimates are crude. The estimated depth of the magma chamber may be reasonably accurate, when geodetic and seismic data both indicate a similar depth. The estimated new volume of magma, however, is normally not highly accurate; in most cases it would be the result of using a rather unreliable Mogi model for an elastic half-space (Chapter 3). Also, although the lava-flow volume should be reasonably accurate (at least for eruptions during the past decades), the feeder-dike volume is less so, but still reasonably reliable. Furthermore, as indicated above, since many injected dikes become arrested (Chapters 7 and 10), an injected dike need not result in an eruption. Nevertheless, this new method may be useful as complementary to other methods for assessing the likelihood of eruptions during unrest periods in well-monitored volcanoes (cf. Browning et al., 2015).

Example 6.3

Problem

The largest Holocene volcanic system in Iceland is the Bardarbunga–Veidivötn System, about 190 km long and as wide as 28 km (Thordarson and Larsen, 2007). This volcanic system gave rise to the largest Holocene eruption in Iceland, the 25 km^3 basaltic Thjorsarhraun lava flow (erupted about 8600 years ago). Assume that the lava was issued from a reservoir with a cross-sectional area similar to that of the volcanic system and that the feeder-dike to the lava was 10 m thick and with an average length (strike-dimension) of 20 km. The reservoir is supposed to be at 20 km depth. How thick would the reservoir have to be to be able to erupt the Thjorsarhraun lava?

Solution

Since we have no specific information about the reservoir, we use the standard formula (Eq. (6.25)) to estimate the reservoir volume. The lava-flow volume is known, and so to have a correct estimate of V_{er} we must also estimate the feeder-dike volume. For a dike with a strike-dimension of 20 km, a dip-dimension of 20 km, and a thickness of 10 m, the total volume is 4 km^3. Thus, in our estimate, we have:

$$V_{er} = 25 \, km^3 + 4 \, km^3 = 29 \, km^3$$

Rewriting Eq. (6.25), we then get:

$$V_b \approx \frac{V_{er}}{2 \times 10^{-4}} = \frac{29 \, km^3}{2 \times 10^{-4}} = 1.4 \times 10^5 \, km^3$$

The volcanic system of Bardarbunga–Veidivötn has roughly the geometry of a flat ellipse. With a semi-major axis a = 95 km ($2a$ = 190 km) and semi-minor axis b = 14 km ($2b$ = 28 km), the lateral cross-sectional area of the reservoir becomes $A = \pi ab = 4178 \, km^2$. It then follows from the estimated volume of the reservoir that in order for magma of volume 29 km^3 to be able to flow out of the reservoir in a single, ordinary eruption, the crude reservoir thickness $2c$ would have to be:

$$2c = \frac{V_b}{A} = \frac{1.4 \times 10^5 \, km^3}{4178 \, km^2} = 33.5 \, km$$

This assumes that the reservoir is 'dome-shaped' (Fig. 6.4). Alternatively, we may assume that the reservoir is a prolate ellipsoid (Fig. 5.13) with a horizontal long axis parallel with the axis of the rift zone. The volume of the reservoir is then given by:

$$V_b = \tfrac{4}{3}\pi abc = 1.4 \times 10^5 \text{ km}^3$$

from which it follows that the maximum semi-thickness c is about 25 km and the maximum total thickness $2c$ is thus 50 km. Both thickness estimates are very large. The dome shape is presumably more likely, in which case the estimated thickness is somewhat less. Nevertheless, even if 33.5 km is only about 18% of the length of the reservoir, it exceeds the maximum width of the reservoir by about 20%. Such geometry is not impossible, but perhaps not very plausible. It is more likely that collapses, in particular graben subsidence, took place during this large eruption, which would have reduced the effective thickness of the reservoir needed to produce the eruptive volume (Chapter 8).

Example 6.4

Problem

Radial or lateral dike propagation (Fig. 6.8) is common in many volcanoes: well-known examples include Etna in Italy and Kilauea in Hawaii. Normally, the propagation distance from the source is, at most, several kilometres, and, rarely, tens of kilometres. One reason for the limited propagation length for lateral dikes is a lack of driving pressure; the buoyancy term in Eq. (5.3) is normally close to zero for lateral or radial dikes so that the main overpressure or driving pressure available to drive the propagation is the excess pressure in the shallow chamber before dike injection. That excess pressure is similar to the *in situ* tensile strength and thus normally small, that is, a few mega-pascals. In the absence of caldera collapse, the pressure available to drive the dike laterally and radially out from the chamber is thus limited.

Earthquake swarms (Chapter 4) can also be used to estimate the possible dimensions of laterally propagating dikes. Some interpretations of earthquake swarms (Fig. 4.9) suggest lateral dike propagation for much longer distances, many tens of kilometres or more, than the theoretical considerations above would indicate as being likely. Such a scenario has been proposed for the first major rifting event in the Krafla Fires (1975–1984) in North Iceland (Fig. 6.15; cf. Saemundsson and Sigmundsson, 2013): a lateral dike is supposed to have propagated from a shallow magma chamber at around 3 km depth below the Krafla caldera (with an area of about 60 km^2). The first rifting, with earthquakes and associated dike injection, occurred to a distance of about 60 km from the centre of the magma chamber. The maximum subsidence in the centre of the Krafla caldera measured during that dike injection was about 2.3 m.

Based on these assumptions and for the case of no eruption (there was a tiny eruption with a lava volume less than 0.0001 km^3, which can be omitted here), calculate the maximum volume of magma flowing out of the chamber during the first dike-injection event in the Krafla Fires.

Solution

Laterally injected dikes are often thought to have comparatively small dip-dimensions. A likely dip-dimension of a 60-km-long dike would be at least 10 km, but let us assume as a minimum a dip-dimension of 5 km. Field observations show that dikes of these dimensions are never less than 4–5 m thick, commonly much more (Chapter 2). Here, we take the dike thickness as 4 m (measured surface dilation is, for a non-feeder, much smaller than the dike thickness). Since the dike is not a feeder, the volume of magma flowing out of the chamber is only the dike volume and given by:

$$V_{er} = 60\,\text{km} \times 5\,\text{km} \times 0.004\,\text{km} = 1.2\,\text{km}^3$$

When a dike is injected laterally from a shallow chamber, the subsidence of the surface, the deflation (Chapter 3), should agree roughly with the volume of the dike (plus eruptive materials, if an eruption takes place).

If the average subsidence was about 1 m, and distributed evenly within the caldera (certainly an overestimate because the chamber itself is today of much smaller area than the caldera – see below), then the maximum volume of magma flowing out of the chamber would be:

$$V_{er} = 60\,\text{km}^2 \times 0.001\,\text{km} = 0.06\,\text{km}^3$$

Even if there was inflow into the chamber from the deeper reservoir, the volumetric rate of inflow would have been far too low to affect individual dike-injection events. That is, the rate of outflow from the chamber during dike injection is much higher than the rate of inflow from the source reservoir. Clearly, therefore, the measured deflation does not match with the volume needed to form the large dike – the deflation is about 20 times too small. It is thus likely that the large regional dike formed during the Krafla Fires was only partly derived from the shallow chamber, and largely from a deep-seated reservoir located beneath the entire volcanic system (Gudmundsson, 1995; Hollingsworth et al., 2012).

Example 6.5

Problem

The volume of the shallow magma chamber of the Krafla Volcano (cf. Example 6.4; Fig. 6.15), located at the depth of roughly 3 km, is estimated as somewhere between 12 km^3 and 54 km^3 (Saemundsson and Sigmundsson, 2013). It is unclear how much of this volume is really molten magma, however. Using the maximum volume and assuming that the entire chamber is molten, make a crude estimate of (a) the maximum volume of basaltic magma that can flow out of the Krafla chamber during a single chamber rupture, and (b) the maximum strike-dimension of a laterally injected dike. Assume that the outflow is so rapid that the inflow of magma from the deeper reservoir during the outflow event can be neglected.

Solution

(a) From Eq. (6.8) we have:

$$V_{er} \approx 7 \times 10^{-4} V_c = 7 \times 10^{-7} \times 54 \, \text{km}^3 = 0.04 \, \text{km}^3$$

Thus, the absolute maximum volume of magma (forming a lava and/or dike) that can flow out of the Krafla chamber in a single event is about 0.04 km³. We see that this is very similar to the magma-flow value estimated from the maximum subsidence or deflation of the Krafla caldera during the Krafla Fires (Example 6.4).

(b) Since the top of the chamber is at 3 km depth, it is clear that a feeder-dike from the Krafla chamber cannot have a dip-dimension that is less than 3 km. For the dike to be 'laterally emplaced' it must somehow come from the lateral margin of the chamber and thus be somewhat deeper than the top of the chamber. As a minimum, we use 4 km for the dike dip-dimension, that is, slightly less than in Example 6.4. Based on measured aspect ratios of dikes and theoretical considerations (Chapters 2 and 5), such a dike is likely to be 2–4 m thick on average, but possibly thinner towards the surface because of negative buoyancy (Chapter 5). However, let us assume that the dike is as thin as 1 m. The vertical cross-sectional area of the dike is then $A = 0.004 \, \text{km}^2$ and the maximum strike-dimension or length L of the dike:

$$L = \frac{V_{er}}{A} = \frac{0.04 \, \text{km}^3}{0.004 \, \text{km}^2} = 10 \, \text{km}$$

If the dike was 2 m thick, as is certainly more likely, or if we used the average estimated chamber volume of 33 km³ rather than the maximum, the length would be reduced to about 5 km. This is similar to the measured maximum distances of radial dikes and inclined sheets from some well-exposed, eroded central volcanoes in Iceland (Figs. 3.10 and 3.11).

The results of Example 6.5 thus agree with those of Example 6.4, suggesting that dikes with lengths or strike-dimensions of tens of kilometres must have derived part of their magma from sources other than the shallow chamber associated with the Krafla Volcano. The most likely source for part of the magma in the dike is a large reservoir at about 15–20 km depth. Such a reservoir, suggested long ago (Gudmundsson, 1995), is supported by recent geodetic measurements which indicate extensive magma accumulation in a large deep reservoir following the Krafla Fires (Hollingsworth et al., 2012; Saemundsson and Sigmundsson, 2013).

That the magma in the dike was partly derived from such a deep-seated reservoir is also supported by the difference in chemistry of the erupted lavas in the Krafla Fires. Lavas inside the Krafla caldera, and thus clearly derived from the shallow chamber, are of tholeiite and thus are rather evolved, whereas those that erupted outside the caldera are of olivine tholeiite and are much less evolved (Saemundsson and Sigmundsson, 2013). Thus, most likely, within the caldera the dikes were injected from the small shallow chamber and erupted evolved basalts, whereas the dikes and dike segments outside the caldera were injected from the much larger deep-seated reservoir and erupted more primitive basalts.

Example 6.6

Problem

Use the following equation (the Bernoulli equation) as a basis to:

(a) Show that all the 'heads' in magma transport have the unit of length (metre).
(b) Explain why the velocity head can be ignored in calculations.
(c) Provide an equation for the total head and use the result to derive Eq. (6.4).

For a laminar or streamline steady flow (flow where the velocity at a particular point does not change with time), the general equation for a compressible fluid is (Bear, 1972):

$$\int \frac{dp}{g\rho_m(p)} + \frac{v^2}{2g} + z = constant \tag{6.28}$$

For a steady incompressible flow (where the fluid density does not change with pressure), Eq. (6.28) reduces to the well-known Bernoulli equation, given by:

$$\frac{p}{\rho_m g} + \frac{v^2}{2g} + z = constant \tag{6.29}$$

In Eqs. (6.28) and (6.29) p is the fluid (here magma) pressure, g is the acceleration due to gravity, v is the velocity (of magma flow), z is the elevation (above some reference level, here the bottom of the magma layer or reservoir, Fig. 6.4), and ρ_m is the magma density, where $\rho_m(p)$ means that the density is a function of pressure.

Solution

(a) It is clear that z, the elevation (the elevation head), must be in metres. Using Eq. (6.29), the unit of the pressure head is obtained as follows:

$$\frac{p}{\rho_m g} = \frac{\text{kg m s}^{-2} \times \text{m}^{-2}}{\text{kg m}^{-3} \times \text{m s}^{-2}} = \text{m} \tag{6.30}$$

so that the unit is length (m). Similarly, for the velocity head we get:

$$\frac{v^2}{2g} = \frac{(\text{m/s})^2}{\text{m s}^{-2}} = \frac{\text{m}^2 \text{ s}^{-2}}{\text{m s}^{-2}} = \text{m} \tag{6.31}$$

so again the unit is length (m). Notice that since we are concerned only with the unit, the factor 2 for g is dropped. Thus, all the terms in Eq. (6.29) have the unit of length (metre). More specifically, the first term, the pressure head, is energy due to magma (or other fluid) pressure, the second term, the velocity head, is energy due to magma movement, and the third term, the elevation head, is energy due to the position or dip-dimension of the theoretical magma column. All the terms represent energy per unit weight (or specific weight). Denoting energy by U and unit weight of magma by $\gamma_m = \rho_m g$, and remembering that energy is force times distance, $F \times d$, we have:

$$\frac{U}{\gamma_m} = \frac{Fd}{\rho g} = \frac{\text{kg m s}^{-2}\text{m}}{\text{kg m}^{-3}\text{m s}^{-2}} = \text{m} \tag{6.32}$$

showing, again, that the unit for all these heads is length (metre).

(b) During fluid migration (groundwater, magma, oil, gas) in porous crustal materials (rocks or sediments), the velocity is normally very low. The migration rate of magma or melt in a porous reservoir or magma layer is not well constrained but is presumably lower than typical rates of groundwater flow, 10^{-6} m s^{-1}, which we take as an upper limit for the velocity of porous magma flow. For the pressure head in Eq. (6.29), using a pressure of 600 MPa (corresponding to crustal depth of about 20 km), magma density of 2800 kg m^{-3}, and acceleration due to gravity of 9.81 m s^{-2}, the head becomes about 22 km. Similarly, the elevation head, that is, the location above the bottom of the reservoir, may be taken as several kilometres, say 5 km. For comparison, for a velocity of 10^{-6} m s^{-1} the velocity head would be about 5×10^{-14} m or about 5×10^{-17} km, and thus clearly negligible. Even if the magma velocity was, under special conditions, several orders of magnitude greater, the velocity head would still be negligible for porous magma flow.

(c) Ignoring the velocity head in Eq. (6.29), the total head reduces to:

$$\varphi = z + \frac{p}{\rho_m g} \tag{6.33}$$

Substituting Eq. (6.33) for φ, Eq. (6.1) becomes:

$$\vec{q} = -\frac{k\rho_m g}{\mu_m} \nabla \left(z + \frac{p}{\rho_m g} \right) \tag{6.34}$$

which can be simplified to Eq. (6.4) thus:

$$\vec{q} = -\frac{k}{\mu_m} \left(\nabla p + \rho_m g \nabla z \right)$$

6.12 Exercises

6.1 What is a double magma chamber?

6.2 How do deep-seated reservoirs form?

6.3 What is a typical size (volume) difference between a deep-seated reservoir and an associated shallow magma chamber?

6.4 Where are deep-seated reservoirs thought to be located?

6.5 What is Darcy's law?

6.6 What are the main physical parameters that control fluid flow into and within magma reservoirs?

6.7 What is a shallow crustal magma chamber and how is it thought to form?

6.8 What is the typical crustal depth range of shallow magma chambers?

6.9 What is the relationship between a shallow magma chamber and a polygenetic volcano?

6.10 How do shallow magma chambers become inactive – extinct?

6.11 How are extinct shallow chambers recognised in the field? What structures do they form?

6.12 What are the volumes in terms of intrusive and extrusive (eruptive) materials for: small, moderate, and large eruptions?

6.13 What indications do we have as to the sizes of shallow magma chambers? What would be typical volumes of such chambers?

6.14 In a typical basaltic eruption from a shallow magma chamber, what percentage of the total magma volume flows out of the chamber?

6.15 In an acid eruption from a shallow chamber, what percentage of the total magma volume can be expected to flow out of the chamber?

6.16 For an eruption from a basaltic deep-seated reservoir, what percentage of the total magma volume would normally flow out of the reservoir?

6.17 In a typical double magma chamber, which one erupts more frequently: the shallow magma chamber or the deep-seated reservoir?

6.18 Explain how a single magma flow from a deep-seated reservoir could trigger tens of eruptions and/or dike injections from the associated shallow chamber.

6.19 A typical basaltic shallow chamber located at 3 km depth produces a fissure eruption (the fissure being 1500 m long and the feeder-dike 2 m thick) and a lava flow of estimated volume 0.05 km^3. Estimate the size (the magma volume) of the magma chamber.

6.20 A typical acid shallow chamber located at 4 km depth produces a fissure eruption (the fissure being 3 km long and the feeder-dike 5 m thick) and a lava flow of estimated volume 0.8 km^3. Estimate the size (the acid magma volume) of the magma chamber.

References and Suggested Reading

Alfaro, R., Brandsdottir, B., Rowlands, D. P., et al., 2007. Structure of the Grimsvötn central volcano under the Vatnajökull icecap. *Geophysical Journal International*, **168**, 863–876.

Andrew, R. E. B., Gudmundsson, A., 2007. Distribution, structure, and formation of Holocene lava shields in Iceland. *Journal of Volcanology and Geothermal Research*, **168**, 137–154.

Auger, E., Gasparini, P., Virieax, J., Zollo, A., 2001. Seismic evidence of an extended magmatic sill under Mt. Vesuvius. *Science*, **294**, 1510–1512.

Baran, J. M., Cochran, J. R., Carbotte, S. M., Nedimovic, M. R., 2005. Variations in upper crustal structure due to variable mantle temperature along the Southeast Indian Ridge. *Geochemistry, Geophysics, Geosystems*, **6**, doi:10.1029/2005GC000943.

Beachly, M. W., Hooft, E. E. E., Toomey, D. R., Waite, G. P., 2012. Upper crustal structure of Newberry Volcano from P-wave tomography and finite difference waveform modeling. *Journal of Geophysical Research*, **117**, doi:10.1029/2012JB009458.

Bear, J., 1972. *Dynamics of Fluids in Porous Media*. Amsterdam: Elsevier.

Becerril, L., Galindo, I., Gudmundsson, A., Morales, J. M., 2013. Depth of origin of magma in eruptions. *Scientific Reports*, **3**, 2762, doi:10.1038/srep02762.

Bjarnason, I. Th., 2008. An Iceland hotspot saga. *Jokull*, **58**, 3–16.

Björnsson, A., 1985. Dynamics of crustal rifting in NE Iceland. *Journal of Geophysical Research*, **90**, 10 151–10 162.

Blake, D. H., 1966. The net-veined complex of the Austurhorn intrusion, southeastern Iceland. *Journal of Geology*, **74**, 891–907.

Brandsdottir, B., Menke, W. H., 2008. The seismic structure of Iceland. *Jokull*, **58**, 17–34.

Browning, J., Gudmundsson, A., 2015a. Surface displacements resulting from magma-chamber roof subsidence, with application to the 2014–2015 Bardarbunga–Holuhraun volcanotectonic episode in Iceland. *Journal of Volcanology and Geothermal Research*, **308**, 82–98.

Browning, J., Gudmundsson, A., 2015b. Caldera faults capture and deflect inclined sheets: An alternative mechanism of ring-dike formation. *Bulletin of Volcanology*, **77**, 889, doi:10.1007/s00445-014-0889-4.

Browning, J., Drymoni, K., Gudmundsson, A., 2015. Forecasting magma-chamber rupture at Santorini volcano, Greece. *Scientific Reports*, **5**, doi:10.1038/srep15785.

Canales, J. P., Nedimovic, M. R., Kent, G. M., Carbotte, S. M., Detrick, R. S., 2009. Seismic reflection images of a near-axis melt sill within the lower crust at the Juan de Fuca ridge. *Nature*, **460**, 89–93.

Carbotte, S. M, Detrick, R. S., Harding, A., et al., 2006. Rift topography linked to magmatism at the intermediate spreading Juan de Fuca Ridge. *Geology*, **34**, 209–212.

Cavalcante, G. C. G., Vauchez, A., Merlet, C., et al., 2014. Thermal conditions during deformation of partially molten crust from Titani geothermometry: rheological implications for the anatectic domain of the Araçuaí belt, Eastern Brazil. *Solid Earth Discussions*, **6**, 1299–1333.

Chaussard, E., Amelung, F., 2014. Regional controls on magma ascent and storage in volcanic arcs. *Geochemistry, Geophysics, Geosystems*, **15**, doi:10.1002/2013GC005216.

Chen, Y., 2004. Modelling the thermal structure of the oceanic crust. In German, C., Lin, J., Parson, L. M. (eds.), *Mid-Ocean Ridges: Hydrothermal Interactions between the Lithosphere and Oceans. Geophysical Monograph 148*. Washington, DC.: American Geophysical Union.

Crawford, W. C., Webb, S. C., 2002. Variations in the distribution of magma in the lower crust and at the Moho beneath the East Pacific Rise at 9 degrees–10 degrees N. *Earth and Planetary Science Letters*, **203**, 117–130.

de Zeeuw-van Dalfsen, E., Pedersen, R., Sigmundsson, F., Pagli, C., 2004. Satellite radar interferometry 1993–1999 suggests deep accumulation of magma near the crust-mantle boundary at the Krafla volcanic system, Iceland. *Geophysical Research Letters*, **31**, doi:10.1029/2004GL020059.

de Zeeuw-van Dalfsen, E., Pedersen, R., Hooper, A., Sigmundsson, F., 2012. Subsidence of Askja caldera 2000–2009: Modelling of deformation processes at an extensional plate boundary, constrained by time series InSAR analysis. *Journal of Volcanology and Geothermal Research*, **213–214**, 72–82.

Detrick, R. S., Sinton, J. M., Ito, G., et al., 2002. Correlated geophysical, geochemical, and volcanological manifestation of plume-ridge interaction along the Galapagos Spreading Center. *Geochemistry, Geophysics, Geosystems*, **3**, doi:1029/2002gc000350.

Dobran, F., 2001. *Volcanic Processes. Mechanisms in Material Transport*. New York, NY: Kluwer.

Fagents, S. A., Gregg, T. K. P., Lopes, R. M. C. (eds.), 2013. *Modeling Volcanic Processes: The Physics and Mathematics of Volcanism*. Cambridge: Cambridge University Press.

Fleisch, D., 2012. *A Students Guide to Vectors and Tensors*. Cambridge: Cambridge University Press.

Furman, T., Meyer, P. S., Frey, F., 1992. Evolution of Icelandic central volcanoes: evidence from the Austurhorn intrusion, southeastern Iceland. *Bulletin of Volcanology*, **55**, 45–62.

Galindo, I., Gudmundsson, A., 2012. Basaltic feeder dykes in rift zones: geometry, emplacement, and effusion rates. *Natural Hazards and Earth System Sciences*, **12**, 3683–3700.

Gonnermann, H. M., Manga, M., 2013. Dynamics of magma ascent in the volcanic conduit. In Fagents, S. A., Gregg, T. K. P., Lopes, R. M. C. (eds.), *Modeling Volcanic Processes*. Cambridge University Press, Cambridge, pp. 55–84.

Greenland, L. P., Rose, W. I., Stokes, J. B., 1985. An estimate of gas emissions and magmatic gas content from Kilauea volcano. *Geochimica et Cosmochimica Acta*, **49**, 125–129.

Greenland, L. P., Okamura, A. T., Stokes, J. B., 1988. Constraints on the mechanics of the eruption. In Wolfe, E. W (ed.), *The Puu Oo Eruption of Kilauea Volcano, Hawaii: Episodes Through 20, January 3, 1983 Through June 8, 1984. US Geological Survey Professional Paper, 1463*. Denver, CO: US Geological Survey, pp. 155–164.

Griffith, D. J., 2014. *Introduction to Electrodynamics*. Cambridge: Pearson.

Gudmundsson, A., 1987. Formation and mechanics of magma reservoirs in Iceland. *Geophysical Journal of the Royal Astronomical Society*, **91**, 27–41.

Gudmundsson, A., 1990. Emplacement of dikes, sills and crustal magma chambers at divergent plate boundaries. *Tectonophysics*, **176**, 257–275.

Gudmundsson, A., 1995. The geometry and growth of dykes. In Baer, G., Heimann, A. (eds.), *Physics and Chemistry of Dykes*. Rotterdam: Balkema, pp. 23–34.

Gudmundsson, A., 2006. How local stresses control magma-chamber ruptures, dyke injections, and eruptions in composite volcanoes. *Earth-Science Reviews*, **79**, 1–31.

Gudmundsson, A., 2007. Conceptual and numerical models of ring-fault formation. *Journal of Volcanology and Geothermal Research*, **164**, 142–160.

Gudmundsson, A., 2011. *Rock Fractures in Geological Processes*. Cambridge: Cambridge University Press.

Gudmundsson, A., 2012. Magma chambers: formation, local stresses, excess pressures, and compartments. *Journal of Volcanology and Geothermal Research*, **237–238**, 19–41.

Gudmundsson, A., 2016. The mechanics of large volcanic eruptions. *Earth-Science Reviews*, **163**, 72–93.

Gudmundsson, A., 2017. *The Glorious Geology of Iceland's Golden Circle*. Berlin: Springer Verlag.

Gudmundsson, A., Brenner, S. L., 2004. How mechanical layering affects local stresses, unrests, and eruptions of volcanoes. *Geophysical Research Letters*, **31**, doi.org/10.1029/2004GL020083.

Gudmundsson, A., Oskarsson, N., Gronvold, K., et al., 1992. The 1991 eruption of Hekla, Iceland. *Bulletin of Volcanology*, **54**, 238–246.

Gudmundsson, A., Lecoeur, N., Mohajeri, N., Thordarson, T., 2014. Dike emplacement at Bardarbunga, Iceland, induces unusual stress changes, caldera deformation, and earthquakes. *Bulletin of Volcanology*, **76**, 869, doi:10.1007/s00445-014-0869-8.

Guo, X., 2013. *Density and Compressibility of FeO-Bearing Silicate Melt: Relevance to Magma Behavior in the Earth*. PhD Thesis, University of Michigan, Ann Arbor, MI.

Harris, A. J. L., Murray, J. B., Aries, S. E., et al., 2000. Effusion rate trends at Etna and Krafla and their implications for eruptive mechanisms. *Journal of Volcanology and Geothermal Research*, **102**, 237–270.

Hollingsworth, J., Leprince, S., Ayoub, F., Avouac, J. P., 2012. Deformation during the 1975–1984 Krafla rifting crisis, NE Iceland, measured from historical optical imagery. *Journal of Geophysical Research*, **117**, doi:10.1029/2012JB009140.

Hreinsdottir, S., Sigmundsson, F., Roberts, M. J., et al., 2014. Volcanic plume height correlated with magma-pressure change at Grimsvötn Volcano, Iceland. *Nature Geoscience*, **7**, 214–218, https://doi.org/10.1038/ngeo2044.

Kent, G. M., Harding, A. J., Orcutt, J. A. 1990. Evidence for a smaller magma chamber beneath the East Pacific Rise at 9°30' N. *Nature* **344**, 650–653.

Koulakov, I., Gordeev, E. I., Dobretsov, N. L., et al., 2011. Feeding volcanoes of the Kluchevskoy group from the results of local earthquake tomography. *Geophysical Research Letters*, **38**, L09305, doi:10.1029/2011GL046957.

Fig. 1.3 Stratovolcanoes are formed by repeated eruptions within a limited surface area. They commonly rise high above their surroundings and have a cone shape, as exemplified here by the Augustine Volcano, forming one of the islands offshore Alaska (United States), with an elevation (height above sea level) of 1260 m and a maximum diameter at sea level of about 12 km. A highly active volcano with frequent eruptions. This photograph shows gas rising during the 2005–2006 eruption. Photo: USGS/Cyrus Read.

Fig. 1.7 Dike thickness is measured as indicated by the black horizontal line. This dike, in the caldera wall of the island of Santorini, Greece, is about 1.5 m thick (cf. Browning et al., 2015).

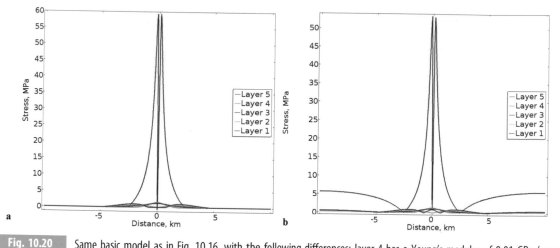

Fig. 10.20 Same basic model as in Fig. 10.16, with the following differences: layer 4 has a Young's modulus of 0.01 GPa (very soft/compliant), the surface layer (layer 1) has a very high stiffness of 20 GPa, and the magmatic overpressure in the dike is 15 MPa (rather than 5 MPa). (a) Plots of the variation in the magnitude of the maximum tensile principal stress σ_3 at the contacts between the layers. (b) Plots of the variation in the magnitude of the von Mises shear stress τ at the contacts between the layers. Layer 5 denotes the contact between layer 5 and 4; layer 4, that between layer 4 and 3; layer 3, that between layer 3 and 2; layer 2, that between layer 3 and 2; and layer 1, the contact between layer 1 and the atmosphere, that is, the free surface of the volcanic zone/volcano (Bazargan and Gudmundsson, 2019).

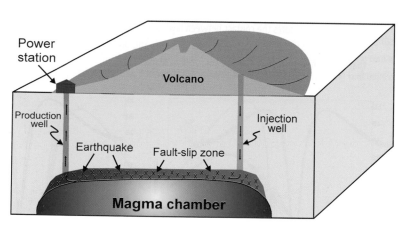

Fig. 10.24 Schematic illustration of hydro-shearing around a magma chamber so as to try to reduce the stress difference in the roof of the chamber (or in a different layer at a shallower depth) and thus to prevent dike propagation and a large eruption. Cold water is pumped into the roof of the chamber, above one of the edges of the chamber, through an injection well. Through increased pore-fluid pressure and slip on existing fractures (Figs. 10.23 and 10.25), the permeability of rock layers that constitute the roof increases and allows the water to circulate through the layers. As it does so, the water increases the pore-fluid pressure, thereby generating a fault slip and earthquakes in gradually the larger part of the roof and, eventually, the entire roof . The fault slip reduces the stress difference, as presented for example by Mohr circles (Fig. 10.25), and brings the state of stress closer to a lithostatic state, which is unfavourable (acts as a barrier) to dike propagation. Eventually, above the other edge of the chamber, a production well may be drilled for pumping up the hot water and using it in a power plant (power station) for space heating and the production of electricity.

Kress, V. C., Carmichael, I. S. E., 1991. The compressibility of silicate liquids containing Fe_2O_3 and the effect of composition, temperature, oxygen fugacity and pressure on their redox states. *Contributions to Mineralogy and Petrology*, **108**, 82–92.

Lauthold, J., Muntener, O., Baumgartener, L. P., et al., 2014. A detailed geochemical study of a shallow arc-related laccolith; the Torres del Paine Mafic Complex (Patagonia). *Journal of Petrology*, **54**, 273–303.

Lin, G., Amelung, F. Y., Lavallée, Y., Okubo, P. G., 2014. Seismic evidence for a crustal magma reservoir beneath the upper east rift zone of Kilauea volcano, Hawaii. *Geology*, **42**, 187–190.

Macdonald, K. C., 1982. Mid-ocean ridges: fine scale tectonic, volcanic and hydrothermal processes within a plate boundary. *Annual Review of Earth and Planetary Sciences*, **10**, 155–190.

Malfait, W. J., Sanchez-Valle, C., Ardia, P., Médard, E., Lerch, P., 2011. Compositional dependent compressibility of dissolved water in silicate glasses. *American Mineralogist*, **96**, 1402–1409.

Michel, J., Baumgartner, L., Putlitz, B., Schaltegger, U., Ovtcharova, M., 2008. Incremental growth of the Patagonian Torres del Paine laccolith over 90 k.y. *Geology*, **36**, 459–462, doi:10.1130/G24546A.1.

Murase, T., McBirney, A. R., 1973. Properties of some common igneous rocks and their melts at high temperatures. *Geological Society of America Bulletin*, **84**, 3563–3592.

Oladottir, B., Sigmarsson, O., Larsen, G., Thordarson, T., 2008. Katla volcano, Iceland: magma composition, dynamics and eruption frequency as recorded by Holocene tephra layers. *Bulletin of Volcanology*, **70**, 475–493.

Opheim, J. A., Gudmundsson, A., 1989. Formation and geometry of fractures, and related volcanism, of the Krafla Fissure Swarm, Northeast Iceland. *Geological Society of America Bulletin*, **101**, 1608–1622.

Orcutt, J. A. 1987. Structure of the earth: oceanic crust and uppermost mantle. *Reviews of Geophysics*, **25**, 1177–1196.

Pagli, C., Wright, T. J., Ebinger, C. J., et al., 2012. Shallow axial chamber at the slow-spreading Erta Ale Ridge. *Nature Geoscience*, doi:10.1038/NGEO1414.

Paulatto, M., Annen, C., Henstock, T. J., et al., 2012. Magma chamber properties from integrated seismic tomography and thermal modeling at Montserrat. *Geochemistry, Geophysics, Geosystems*, **13**, Q01014, doi:10.1029/2011GC003892.

Poland, M. P., Takahashi, T. J., Landowski, C. M. (eds.), 2014a. *Characteristics of Hawaiian Volcanoes. US Geological Survey Professional Paper, 1801*. Denver, CO: US Geological Survey.

Poland, M. P., Miklius, A., Montgomery-Brown, E. K., 2014b. Magma supply, storage, and transport at shield-stage Hawaiian volcanoes. In Poland, M. P., Takahashi, T. J., Landowski, C. M. (eds.), *Characteristics of Hawaiian Volcanoes. US Geological Survey Professional Paper, 1801*. Denver, CO: US Geological Survey, pp. 179–234.

Reverso, T., Vandemeulebrouck, J., Jouanne, F., et al., 2014. A two-magma chamber model as a source of deformation at Grimsvötn Volcano, Iceland. *Journal of Geophysical Research*, **119**, 4666–4683.

Ritchie, D., Gates, A. E., 2001. *Encyclopedia of Earthquakes and Volcanoes*. New York, NY: Facts on File.

Roobol, M. J., 1974. Geology of the Vesturhorn intrusion, SE Iceland. *Geological Magazine*, **111**, 273–285.

Rossi, M. J., 1996. Morphology and mechanism of eruption of postglacial lava shields in Iceland. *Bulletin of Volcanology*, **57**, 530–540.

Saemundsson, K., Sigmundsson, F., 2013. The North Volcanic Zone, Krafla. In Solnes, J., Sigmundsson, F., Bessason, B. (eds.), *Natural Hazards in Iceland: Volcanic Eruptions and Earthquakes*. Reykjavik: Vidlagatrygging/Haskolautgafan, pp. 324–337 (in Icelandic).

Searle, R., 2013. *Mid-Ocean Ridges*. Cambridge: Cambridge University Press.

Seifert, R., 2013 *Compressibility of Volatile-Bearing Magmatic Liquids*. PhD Thesis, ETH, Zurich.

Sheth, H., 2018. *A Photographic Atlas of Flood Basalt Volcanism*. Berlin: Springer.

Singh, S. C., Crawford, W. C., Carton, H., et al., 2006. Discovery of a magma chamber and faults beneath a Mid-Atlantic Ridge hydrothermal field. *Nature*, **442**, 1029–1032.

Sinha, M. C., Constable, S. C., Peirce, C., et al., 1998. Magmatic processes at slow-spreading ridges: implications of the RMESSES experiment at 57°45' N on the Mid-Atlantic Ridge. *Geophysical Journal International*, **135**, 731–745.

Soosalu, H., Key, J., White, R.S., et al. 2010. Lower-crustal earthquakes caused by magma movement beneath Askja volcano on the north Iceland rift. *Bulletin of Volcanology*, **72**, 55–62.

Sturkell, E., Einarsson, P., Sigmundsson, F., et al., 2006. Volcano geodesy and magma dynamics in Iceland. *Journal of Volcanology and Geothermal Research*, **150**, 14–34.

Sturkell, E., Sigmundsson, F., Geirsson, H., et al., 2008. Multiple volcano deformation sources in a post-rifting period: 1989–2005 behaviour of Krafla, Iceland constrained by levelling, tilt and GPS observations. *Journal of Volcanology and Geothermal Research*, **177**, 405–417.

Sturkell, E., Einarsson, P., Sigmundsson, F. et al., 2010. Katla and Eyjafjallajökull volcanoes. *Developments in Quaternary Sciences*, **13**, 5–21.

Tenzer, R., Gladkikh, V., 2014. Assessment of density variations of marine sediments with ocean and sediment depths. *The Scientific World Journal*, **2**, doi:10.1155/2014/823296

Thorarinsson, S. B., Tegner, C., 2009. Magma chamber processes in central volcanic systems of Iceland: constraints from layered gabbro of the Austurhorn intrusive complex. *Contributions to Mineralogy and Petrology*, **158**, 223–244.

Thordarson, T., Hoskuldsson, A., 2008. Postglacial volcanism in Iceland. *Jokull*, **58**, 197–228.

Thordarson, T., Larsen, G., 2007. Volcanism in Iceland in historical time: volcano types, eruption styles and eruptive history. *Journal of Geodynamics*, **43**, 118–152.

Tryggvason, E. 1984. Widening of the Krafla Fissure Swarm during the 1975–1981 volcano-tectonic episode. *Bulletin of Volcanology*, **47**, 47–69.

Tryggvason, E. 1986. Multiple magma reservoirs in a rift zone volcano: ground deformation and magma transport during the September 1984 eruption of Krafla, Iceland. *Journal of Volcanology and Geothermal Research*, **28**, 1–44.

Walker, G. P. L., 1965. Some aspects of Quaternary volcanism in Iceland. *Quaternary Journal of the Geological Society*, **49**, 25–40.

Wang, H. F., 2000. *Theory of Linear Poroelasticity*. Princeton, NJ: Princeton University Press.

Williams, H., McBirney, A. R., 1979. *Volcanology*. San Francisco, CA: Freeman.

Woods, A. W., Huppert, H. E., 2003. On magma chamber evolution during slow effusive eruptions. *Journal of Geophysical Research*, **108**, 2403, doi:10.1029/2002JB002019.

Zollo, A., Maercklin, N., Vassallo, M., et al., 2008. Seismic reflections reveal a massive melt layer feeding Campi Flegrei caldera. *Geophysical Research Letters*, **35**, L12306, doi:10.1029/2008GL034242.

7 Magma Movement through the Crust: Dike Paths

7.1 Aims

How does magma move or rise from its source chamber to the surface? More specifically, how does magma generate a path to the surface so as to supply magma to an eruption? Or, in general, under what conditions do dike-fed eruptions occur? While these questions have been briefly mentioned in some of the earlier chapters, they and the answers have not been discussed in detail. That I shall do in the present chapter. While magma moves through the crust by different mechanisms (e.g. as diapirs), the main mechanism is magma-driven fractures. The general name for all magma-driven fractures, once solidified, is sheet intrusions or **sheets**, which include dikes, inclined sheets, and sills. Unless stated otherwise, the theoretical discussion in this chapter applies equally to all these three types of sheets. Here, the focus is on mostly dikes, partly for the simple reason that dikes supply magma to most eruptions. For general theoretical considerations, **dike** denotes both subvertical dikes, regional and local, and commonly also inclined sheets, although in some instances a distinction will be made between these structures. In Chapter 5, some aspects of dike initiation, propagation, and arrest were discussed. In this chapter, these topics are developed further and many new topics, such as magma transport, rock toughness, and thermal aspects of intrusions – their cooling and solidification – are considered. The energy and 'action' aspects of dike propagation, however, are discussed in Chapter 10. The main aims of this chapter are to:

- Explain the conditions for dike injection from a magma chamber.
- Quantify and explain how dikes select their propagation paths.
- Describe the condition for dike arrest and associated surface effects.
- Explain the conditions for a dike-fed eruption.
- Explore the relation between magma flow rates and dike aperture.
- Discuss the rate of cooling, solidification, and thermal effects of dikes and sills.

7.2 Dike Initiation

Dikes are injected from a magma chamber when the stress conditions at the chamber margin, usually in the chamber roof, are such that the roof ruptures to form a magma-filled **extension**

fracture (Fig. 1.17). Recall that the great majority of ruptures of magma-chamber margins are extension fractures, not shear fractures. Shear failure, or faulting, is very common during volcanic unrest periods, particularly during magma-chamber inflation, and is monitored as such through seismic activity (Chapter 4). But the faulting is then primarily not related to the chamber rupture but rather to shear stresses generated in the crustal segment above the chamber, through chamber expansion or contraction, that is, during inflation and deflation (Fig. 1.12). Once the rupture has occurred and a dike or dikes begin to propagate away from the chamber, **earthquake swarms** (Figs. 1.13, 4.9, and 4.13) are commonly associated with the dike propagation. Indeed, even if part of dike propagation may be aseismic, earthquake swarms are commonly used to **monitor dike propagation** (Chapter 4). The earthquakes of the swarm occur mostly on existing fractures in the host rock of the dike – to either side of the dike – but some are associated with faulting ahead of the upper dike tip. The tip itself, however, is primarily generated through extension-fracture failure and does therefore not produce ordinary double-couple earthquakes (Chapter 4).

The conclusion that the **rupture** of the magma chamber to form a dike is mostly through the formation of **extension fractures** rests on extensive field evidence (e.g. Figs. 1.17, 1.21, 2.1, 3.3, 4.11, and 4.12). Additional evidence on a regional scale is provided by dikes dissecting lava piles where the cross-cutting relations between the regional dikes and lava flows demonstrate that the dikes are extension fractures (Fig. 7.1). On a local scale, cross-cutting relations between sheets and dikes in local swarms of inclined sheets also show that the intrusions are primarily pure extension fractures (Fig. 7.2).

Many dikes and inclined sheets use existing **cooling joints** in the host rock as parts of their paths (Fig. 7.3), particularly where the joint attitude coincides with that of the local

Fig. 7.1   Regional dikes (numbered 1–10) dissecting a basaltic lava pile in Southeast Iceland (arrows indicate the dip). The dikes dissect the lava flows at right angles. The absence of vertical displacements parallel to the dikes suggests that the dikes are extension fractures (cf. Figs. 2.1 and 6.9b).

Fig.7.2 Cross-cutting dikes and inclined sheets show no sheet and dike parallel displacements, indicating that they are all extension fractures. The basaltic dikes and inclined sheets dissect lake sediments in South Iceland. The length of the hammer is about 30 cm.

Fig. 7.3 An inclined sheet, about 2 m thick, cutting through a gabbro body, the outermost part of the magma chamber seen in Fig. 7.5. At this stage the outermost part of the chamber, located at about 2 km depth below the surface of the associated polygenetic Geitafell Volcano (a collapse caldera) in Southeast Iceland, had already solidified. The inclined sheet from the still-liquid inner part of the chamber was injected partly along existing cooling joints in the gabbro.

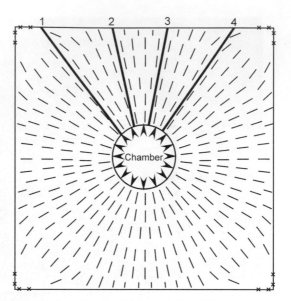

Fig. 7.4 Potential paths (four indicated) of dikes and inclined sheets and sills, all following the σ_1-trajectories, injected from a magma chamber in a homogeneous isotropic crustal segment. In this numerical model (corners fastened) the only loading is an internal excess magmatic pressure p_e of 10 MPa. So long as the magma has an overpressure (Eq. (7.5)) all the dikes and inclined sheets should reach the surface. The primary reason that most do not reach the surface is the tendency to sheet and dike arrest at layer contacts of real crustal segments (cf. Fig. 3.18).

attitude of the maximum compressive principal stress (Fig. 7.4). In Fig. 7.3 a dike is seen dissecting a gabbro body, which is the outermost part or zone, the solidified envelope, of a magma chamber that was active 5–6 million years ago (Fig. 7.5). The field criteria show that the dike paths are partly along cooling joints in the solidified gabbro envelope. The joints are extension fractures, and so are the dikes.

Generally, matching **jogs and notches** at the contacts and marker layers can be used to determine if dikes occupy extension fractures or shear fractures. The results in dike and sheet swarms worldwide show that the great majority of dikes, inclined sheets, and sills are extension fractures. The same applies to other fluid-driven fractures, such as mineral veins (Gudmundsson et al., 2012; Philipp, 2008, 2012), and human-made hydraulic fractures (Valko and Economides, 1995; Bunger, 2009; Yew and Weng, 2014). This conclusion is of great importance for **forecasting** likely dike paths during unrest periods and magma-chamber rupture. This is because, as extension fractures, dikes follow the local trajectories of the maximum compressive principal stress σ_1 (and the intermediate principal stress σ_2 as well) and are perpendicular to the minimum compressive principal stress σ_3. The potential dike paths can therefore be forecasted crudely, based on stress trajectories derived from numerical models (Section 7.3). Before a dike is initiated and begins to propagate along its path, the source magma chamber must rupture, to which I turn now.

The simplest general condition for magma-chamber rupture and dike (dike/inclined-sheet) injection is, from Eq. (5.1), given by:

$$p_l + p_e = \sigma_3 + T_0 \qquad (7.1)$$

Fig. 7.5 The exposed shallow magma chamber, now a gabbro pluton, and the associated local sheet swarm of the Geitafell Volcano, Southeast Iceland. When active, the top (roof) of the chamber was at about 2 km below the surface of the associated volcano. (a) View north, the main contact between the magma chamber/pluton and the sheet swarm is indicated. Only the uppermost part of the chamber is seen here. To the west (left) is the Hoffellsjökull glacier. (b) View south, part of the inclined sheets and radial dikes of the local swarm constitute 80–100% of the rock close to the magma chamber. The main contact between the chamber and the sheet swarm is indicated (as in (a)). A black and white version of this figure will appear in some formats. For the colour version, please refer to the plate section.

where p_l is the lithostatic stress at the location of the rupture and dike initiation at the magma-chamber boundary, p_e is the excess magmatic pressure in the chamber, and T_0 is the local *in situ* tensile strength at the rupture site. As is discussed below, the flow of magma out of the chamber through the dike normally stops once the excess pressure has fallen very close to zero. In Fig. 1.17, however, the dike contacts with the magma chamber remained open until the dike (and the chamber) had solidified, suggesting that the excess pressure was not zero when the dike propagation stopped. This commonly happens when the dike is a **non-feeder**, and particularly when the dike is composed of magma of **high viscosity** (here felsic magma). The open contact is thus an indication that the dike became arrested while there was still excess pressure in the chamber. By contrast, at the end of an eruption (which implies feeder-dike formation), particularly of low-viscosity basaltic magma, nearly all the excess pressure is normally relieved so that $p_e \rightarrow 0$ and the connection between the dike and the magma chamber **closes**.

In fracture mechanics, the criterion normally used for conditions for propagation of a fluid-driven fracture is toughness. The toughness of a rock and other solid materials is a measure of the material resistance to fracture. The two closely related toughness measures used are material toughness and fracture toughness. The relevant definitions are as follows (Gudmundsson, 2011a):

- **Material toughness** is a measure of the energy absorbed in a material per unit area of fracture or crack in that material. It is also known as the critical strain energy release rate and has the unit of J m^{-2}, that is, energy (work) per unit area. The unit J m^{-2} can also be expressed as N m^{-1}, force per unit length of crack. The material toughness is thus sometimes referred to as the **crack extension force.** One measure of toughness is the area under the stress–strain curve, that is, the elastic strain energy (Fig. 3.6). The **energy release rate** of a material is usually denoted by G and its critical value, the material toughness, by G_c. In this book, the energy release rate G and material toughness G_c are used interchangeably, although the measured or estimated material toughness is strictly the critical energy release rate needed for a crack to propagate. The material toughness depends on the mode of the crack surface (or wall) displacement. These are modes I, II, and III, and combinations of these – known as mixed mode. Mode I applies to extension fractures (tension fractures and fluid-driven fractures – hydrofractures – such as dikes, inclined sheets, and sills), whereas modes II and III apply to shear fractures, including all faults. Material toughness for mode I fractures – extension fractures – is denoted by $G_{Ic.}$
- **Fracture toughness**, usually denoted by the symbol K_c, is the critical **stress-intensity factor** for a fracture to propagate. It has the unit of N m$^{-3/2}$ or Pa m$^{1/2}$. Since the pascal (Pa) is a very small measure of stress and pressure, the fracture toughness is normally given in MPa m$^{1/2}$. Fracture toughness varies positively with increasing volume of material at the fracture tip that deforms plastically, or through microcracking, or both. Fracture toughness for mode I fractures, such as dikes, is denoted by K_{Ic}.

The energy release rate and stress intensity, and therefore material toughness and fracture toughness, are related. They do, however, provide different measures of the fracture resistance of the material and have different units. The **energy release rate** G measures the energy per unit area of crack extension; the **stress-intensity factor** K measures the magnitude (intensity) of the stress field close to the crack tip.

The conditions for dike initiation in the roof of a magma chamber (Fig. 1.17) in terms of toughness can be formulated in various ways (Gudmundsson, 2011a). Here we present some of the basic formulations. For a **dike to be initiated** in the roof, the general condition to be satisfied is:

$$p_e = \frac{K_{Ic}}{(\pi a)^{1/2}} \tag{7.2}$$

where p_e is the magmatic excess pressure in the chamber at the location of the rupture and dike initiation, K_{Ic} is the fracture toughness for an extension (mode I) fracture in the roof at that location and a is the dip-dimension (height) of the dike. Equation (7.2) applies equally to plane-strain and plane-stress conditions, discussed in the next paragraph.

When the material toughness G_c is used instead of the fracture toughness K_c, then the condition for **dike initiation** may be written as:

$$p_e = \left(\frac{EG_{Ic}}{\pi a}\right)^{1/2} \tag{7.3}$$

where E is the Young's modulus. Equation (7.3) is for plane-stress conditions, which are most suitable when dealing with fractures that are hosted by crustal segments that are very wide (have large lateral dimensions) in relation to their thickness. For dike injection from a magma chamber, the vertical dimension – the dip-dimension – of the dike is commonly the most important one, in which case the plane-strain formulation is normally more suitable. Generally, plane-strain conditions apply when all the dimensions are similar or when one dimension, often the dip-dimension or thickness, is somewhat larger than the others. The difference in the equation is the factor $1 - v^2$, where v is the Poisson's ratio, that is, the ratio between axial extension and lateral contraction of a rock specimen or a crustal segment (Chapter 3). When using the plane-strain formulation, Eq. (7.3) becomes:

$$p_e = \left(\frac{EG_{Ic}}{\pi(1 - v^2)a}\right)^{1/2} \tag{7.4}$$

Equations (7.2)–(7.4) show that the **excess pressure** in the chamber needed for dike initiation is inversely proportional to the dike dip-dimension at any instant. All extension fractures originate from 'flaws' in rocks. On the scale of dikes, the flaws are mostly joints, and, in igneous rocks, primarily columnar (cooling) joints (Figs. 1.21, 2.8, and 5.5). The dimensions of such joints are from tens of centimetres to several metres, which is thus the minimum size of a in these equations.

When the dike begins to propagate up into the roof and away from the contact with the magma chamber (Figs. 1.17, 7.3, and 7.5), **buoyancy** contributes to the driving pressure, which then becomes the overpressure p_o, defined as (Eq. (5.3)):

$$p_o = p_e + (\rho_r - \rho_m)gh + \sigma_d \tag{7.5}$$

where ρ_r is the average host-rock density, ρ_m is the average magma density, g is acceleration due to gravity, h is the dip-dimension of the dike, and σ_d is the differential stress (the difference between the maximum and the minimum principal stress) in the host rock where the dike is studied. For dike propagation, in contrast to dike initiation, Eqs. (7.2)–(7.4) are thus more appropriately given in terms of over-pressure p_o rather than excess pressure p_e, thus:

$$p_o = \frac{K_{Ic}}{(\pi a)^{1/2}} \tag{7.6}$$

$$p_o = \left(\frac{EG_{Ic}}{\pi a}\right)^{1/2} \tag{7.7}$$

$$p_o = \left(\frac{EG_{Ic}}{\pi(1 - v^2)a}\right)^{1/2} \tag{7.8}$$

where all the symbols are as defined above. Eqs. (7.1)–(7.8) may be regarded as defining the conditions for the initiation and propagation of dikes (and other hydrofractures) in terms of toughness. However, these equations were initially derived for homogeneous and isotropic materials. Even when they can be applied to heterogeneous and anisotropic materials, such as those constituting volcanic edifices, these equations do not allow us to understand, let alone forecast, the commonly complex propagation paths, to which I turn now.

7.3 Dike-Propagation Paths

Once a dike is initiated and begins to propagate, it can theoretically choose infinitely many paths up into (and in the case of eruption, through) the roof. How a particular overall dike path is eventually chosen depends on physical principles that can be presented in terms of the **general principles** of least action or minimum potential energy (Chapter 10). At any particular moment during the propagation, however, the selected path is primarily determined by the **local stresses** in the layers, units, and contacts ahead of the dike tip. The local stresses, in turn, depend on the **mechanical properties** of the layers and their contacts. How far into the roof a dike propagates depends also on the **elastic energy** (the potential mechanical energy) available to form the dike-fracture and squeeze magma out of the source chamber and into the dike. The detailed energy aspects of dike-propagation paths are discussed further in Chapter 10.

Thermal effects also play a role. When a dike is propagating, heat is continuously transferred from its magma to the host rock, so that the dike magma cools down. Eventually, for a very long path, the magma in the dike tip may solidify, suggesting that the rate of cooling can offer a limit to the length of dike propagation. However, as discussed later (Section 7.7), the rate of cooling is **very slow** in comparison with the normal rate of dike propagation. Furthermore, even if the tip became solidified, dike magma with a sufficiently high overpressure – equal to the tensile strength of the solidified tip (Section 5.3) – would be able to rupture the solidified tip or, alternatively, the surrounding host rock. The tensile strength of a solidified tip would normally be very similar to that of the host rock. It therefore does not make much difference whether the magma ruptures a path through the solidified dike tip or through the surrounding host rock.

For exploring the details of potential dike paths, we must use numerical modelling. The common approach is **two-dimensional modelling**, which applies to most of the results shown in this chapter. In two-dimensional modelling it is assumed that one dimension of the modelled dike is **large** in comparison with the two dimensions used and shown in the model. This is often a well-justified assumption when modelling dike-propagation paths. One reason is that the part of the path of the greatest interest is the one seen in a section, such as in a sea cliff, a caldera wall, or other extensive subvertical outcrops. We have normally no way of knowing what the path looks like inside the cliff or caldera wall, only in the **two-dimensional sections**. A second reason for the use of two-dimensional models is that the statistical data that can be applied to test the models are also mostly two-dimensional. Abrupt changes in the paths, deflection along a contact, arrest, or, alternatively, a path reaching the surface, are all primarily observed in sections and are thus two-dimensional.

Consider first the model of a potential path of an injected dike propagating through a crustal segment that is both **homogeneous and isotropic** (Fig. 7.4). For such a model, if the overpressure is sufficiently high and maintained along the entire potential path, the dike should normally reach the surface. The only factors that could hinder the dike reaching the surface would be (1) lack of overpressure and (2) lack of energy for squeezing out the magma in the source chamber. Let us briefly consider these two factors.

From Eq. (7.5) the parameter $\sigma_d = \sigma_1 - \sigma_3$ may contribute to the overpressure and can only be positive or zero. During rifting episodes with inclined-sheet/dike injections there is normally an **extension** operating in the upper part of the crustal segment through which the dike is propagating, so that σ_1 is either close to vertical (for a dike) or inclined and parallel with the dip-dimension (for an inclined sheet). By definition, σ_3 is perpendicular to σ_1 (σ_2 is not considered since the models under discussion are two-dimensional). Thus, σ_d cannot contribute to the dike stopping or being arrested on its path to the surface. Since the excess pressure p_e must also be positive, and roughly equal to the *in situ* tensile strength in order for the chamber to rupture and inject a dike, p_e also cannot contribute to the stopping of the dike. The thermal effects are discussed again in Section 7.7. However, as regards Eq. (7.5), the only factor that can become **negative** on the potential path of the dike to the surface is the **buoyancy term** $(\rho_r - \rho_m)gh$. And since g (acceleration due to gravity) and h (the dip-dimension of the dike) are both positive, it is only the factor $(\rho_r - \rho_m)$, that is, the difference between the density of the host rock ρ_r and the magma ρ_m, that can become negative and contribute to dike arrest.

This **density difference** is, indeed, **often negative** for basaltic dikes that are injected from shallow magma chambers. Many mafic shallow magma chambers are located in rocks with densities similar to those of mafic magmas. Since the density of crustal rocks generally decreases towards the surface, the average density above the shallow magma chamber may be less than that of a typical basaltic magma (2600–2800 kg m^{-3}). Thus, for example, in Iceland, many shallow magma chambers, active and fossil, have their ceilings or tops at about 2 km depth (Fig. 7.5; Chapter 6). The average crustal density of the uppermost 2 km of the crust in the volcanic zones of Iceland is about 2550 kg m^{-3} and thus of considerably less density than that of typical basaltic magma. Gas expansion may decrease the magma density, but it operates primarily at very shallow depths in basaltic dikes (Chapters 6 and 8).

Many other mafic chambers have upper boundaries at similar depths below the surface (Fig. 6.17; Chapter 6). Thus, for many shallow magma chambers and the volcanoes they feed, the effect of buoyancy on the propagation of basaltic dikes is either **zero or negative**. Under these circumstances the only factor contributing to a positive driving pressure is the **excess pressure** in the chamber p_e at magma-chamber rupture and dike injection. It may be concluded that while some basaltic dikes in a homogeneous and isotropic crustal segment may halt because of lack of overpressure most should theoretically **reach the surface.**

But it is well known that most dikes neither reach the surface nor have simple straight paths. Dike arrest is treated in the next section and here the focus is on the geometries of the paths. As soon as we begin to consider the effects of layering, some of the reasons for the typically **complex dike paths** become clear. Consider the seven-layer model in Fig. 7.6. The chamber itself is hosted by a thick layer with a stiffness of 40 GPa. The thin layers above the chamber are very soft, with a stiffness of 1 GPa, and could correspond to piles of soft pyroclastic layers or sediments. The thick layers are very stiff, with a stiffness of 100 GPa, and could correspond to piles of lava flows. The chamber has a circular cross-section subject to a magmatic excess pressure of 5 MPa as the only loading. The trajectories or ticks of σ_1, which indicate the paths of injected dikes, change their orientation abruptly in the central part of the layered segment and become horizontal and then inclined at the second contact between soft and stiff layers. On meeting this contact the dike could (Fig. 7.9b):

- Become arrested (path A).
- Become deflected along the contact and then become arrested (path B).
- Become deflected along the contact for a part of its path and then continue as an inclined sheet to shallower crustal levels (path C). An example of the latter is seen in Figs. 7.7 and 7.8 (as well as in Figs. 1.18 and 2.25).

A sill-like chamber hosted by a crustal segment with the same layering and subject to the same loading as the model in Fig. 7.6 yields **similar results** (Fig. 7.9). A dike injected into the central part of the chamber roof would very likely either become arrested at the contact between the layer hosting the chamber and the lowermost thin layer (Fig. 7.9b). Alternatively, the dike could change into a **sill** at the contact and subsequently into an inclined sheet. If the magma-chamber rupture occurred closer to the lateral ends of the chamber, however, the intrusion would be likely to be initiated as an inclined sheet and then take on different attitudes (strike and/or dip) when propagating through the different layers (Fig. 7.9b) .

A more detailed model, in terms of layering, is shown in Fig. 7.10. Here, the shallow magma chamber is located in a single, thick layer or unit. The only loading is an excess pressure of 10 MPa. Above the chamber there are **30 layers**. The thick layer hosting the chamber has a Young's modulus of 40 GPa, a typical upper crustal value. By contrast, the 30 thin layers alternate between very soft ones, 1 GPa, and very stiff ones, 100 GPa. For a magma chamber located with its top at 1.5–2 km, as is common at divergent plate boundaries, for example, the thickness of each of the 30 layers would be somewhere between **50 m and 67 m**. Many layers of pillow lava, hyaloclastite, and sediments reach these thicknesses, whereas most ordinary lava flows would be thinner (5–20 m for basaltic

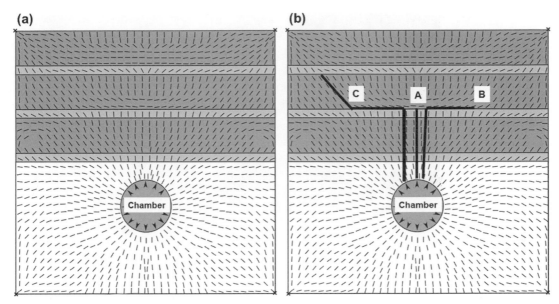

(a) **(b)**

Fig. 7.6 Potential paths of dikes injected from a shallow magma chamber of a circular cross-section subject to an internal excess pressure of 5 MPa as the only loading. (a) Since dikes are extension fractures their paths follow the indicated σ_1-trajectories (ticks). The thin layers are compliant (soft, 1 GPa), the thick layers are stiff (100 GPa). (b) The dike following path A becomes arrested at the contact where the σ_1-trajectories flip by 90° (cf. Gudmundsson, 2006). The dike following path B changes into a sill along that contact. The dike following path C changes into a sill along the same contact, and then, subsequently, into an inclined sheet (cf. Figs. 7.7 and 7.9).

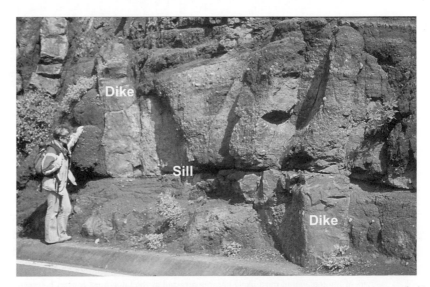

Fig. 7.7 A basaltic dike becomes deflected into a thin sill along a contact between pyroclastic layers in Tenerife (Canary Islands). The lower dike segment is 0.43 m thick, strikes N12°E and dips 87°W, whereas the upper segment is 0.46 m thick, strikes N1°E and dips 89°W. Thus, the dike strike changes by about 11° on crossing the contact, whereas the dike thickness remains similar. At the contact itself, however, the dike changes into a sill with a thickness of as little as 0.02 m (2 cm) for a lateral distance of about 3.4 m (cf. Figs. 7.8 and 7.9).

Fig. 7.8 Zig-zag propagation path of an inclined sheet, 0.5–1 m thick, in Southwest Iceland. The basaltic sheet becomes deflected for a while along a contact between a stiff basaltic lava flow and a compliant or soft scoria layer. Subsequently, the sheet follows an inclined path through the lava flow (cf. Figs. 7.6 and 7.9).

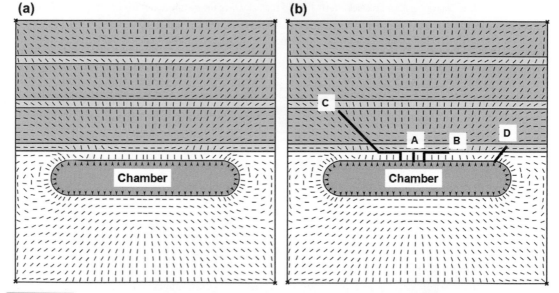

Fig. 7.9 Potential paths of dikes injected from a shallow sill-like magma chamber subject to 5 MPa internal excess pressure as the only loading (cf. Gudmundsson, 2006). (a) The dike paths follow the indicated σ_1-trajectories (ticks). The thin layers are compliant (soft, 1 GPa), the thick layers are stiff (100 GPa). (b) A dike injected into the central part of the roof would most likely become arrested at the lowermost contact between soft and stiff layers (path A), change into a sill at the contact (path B), or continue its propagation as an inclined sheet (path C). Rupture close to the lateral ends of the chamber would normally result in the formation of inclined sheets (path D) (cf. Figs. 7.6, 7.7 and 7.8).

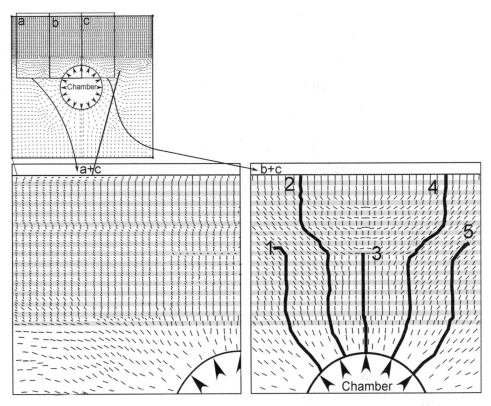

Fig. 7.10 Some potential paths of dikes and sheets injected from a magma chamber of circular cross-section into a roof composed of 30 layers of alternating Young's moduli of 1 GPa and 100 GPa. An internal magmatic excess pressure of 10 MPa is the only loading. The paths follow the indicated σ_1-trajectories (ticks). Only a few potential paths (1–5) are indicated (cf. Figs. 7.6, 7.7, 7.8, and 7.9; Gudmundsson and Philipp, 2006).

lava flows being common). The layering in the model may thus be regarded as an approximation to the layering exposed in Iceland and many divergent plate boundaries and rift zones, and also appropriate for many crustal segments at convergent plate boundaries.

The results show that, for this local stress field, there could be a **variety of dike paths**. Close to the chamber, within the thick unit, the dikes are inclined in a manner similar to that in Fig. 7.4. Within the 30 layers, however, the paths are much more complex. The lower parts of some of the paths are straight in the region just above the magma-chamber top. Other paths are irregular, many with a **zig-zag geometry** (Figs. 7.7 and 7.8) through all the 30 layers (Fig. 7.10). Because the layers in basaltic edifices are more similar in mechanical properties than those that constitute stratovolcanoes, dike paths in basaltic edifices have generally simpler geometries than those in stratovolcanoes. In the latter, dike deflection into sills or inclined sheets are more common, resulting in irregular paths. Many dikes would also be expected to become arrested, particularly in the central part of some of the upper layers. This follows from the abrupt 90° flip in the orientation of σ_1 in the central parts of many of the layers.

In the uppermost five layers or so, however, the orientation of σ_1 again flips by 90°, back to vertical as in the lower layers.

There is, in principle, no specific limit as to how many and how thin the layers in such models may be. Numerical programs today allow for many more layers than are used here. For example, 200 layers, each with a thickness of **10 m**, could constitute the uppermost 2 km above a typical shallow magma chamber (Fig. 7.5). Also, the thickness can be varied and matched with the actual geology. More reliable dike-path numerical models, combined with analytical considerations (Chapter 10), are clearly needed in volcanotectonics, particularly for hazard assessments. Such models should use the actual layers – obtained from drill holes, such as in geothermal wells or from well-exposed sections (Fig. 1.16). Only through models where the real stratigraphy is considered, including the contact properties and accurate elastic properties and thicknesses of the layers, can we hope to understand and forecast dike paths and, in particular, dike arrest.

7.4 Dike Arrest

We have already discussed and described the three main mechanisms of dike arrest in connection with magma-chamber initiation (Section 5.2). The mechanisms are (1) the Cook–Gordon **delamination** or debonding, (2) the formation of a **stress barrier**, and (3) the generation of an unfavourable **elastic mismatch** across a contact. All these mechanisms may lead to direct arrest of the dike tip on meeting a contact or they may result in deflection of the dike into and along the contact, forming a short sill-like path (Figs. 1.18, 6.14, 7.7, and 7.8) or a real sill, which may be large (Figs. 5.5 and 5.6) or small (Figs. 5.15 and 5.17). The sill-like path of the dike normally ends laterally along the contact. Alternatively, it may change into an inclined sheet (Fig. 7.8) or a dike (Figs. 1.18 and 7.7) that propagates to higher stratigraphic levels (Fig. 7.6). In addition to arrest by these three mechanisms, a dike may become arrested because of lack of driving pressure and, possibly, because of cooling of the dike tip, as discussed above. Thus, there are the following general mechanisms for dike arrest:

1. Cook–Gordon debonding or delamination.
2. Development of stress barriers.
3. Development of unfavourable elastic mismatch across contacts.
4. Magma overpressure being reduced to zero.
5. Solidification of the dike tip.

Here the focus is on the **first three** because field observations show these to be the most common. Mechanisms 4 and 5 have already been discussed above, and the solidification is discussed further in Section 7.7. The first three mechanisms relate to dike arrest at **discontinuities**, which include joints and faults, but are most commonly contacts between mechanically dissimilar rocks.

When a dike or any extension fracture meets a discontinuity – here a horizontal contact (Fig. 7.11) – there are **four possible scenarios** (as seen in a vertical, two-dimensional, section):

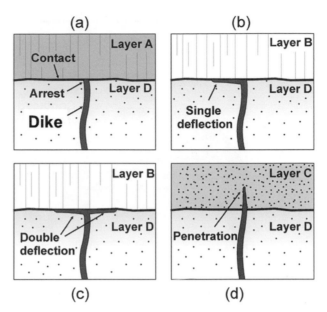

Fig. 7.11 When a dike meets a contact or other discontinuity in a crustal segment or a volcano, a dike may (a) become arrested, (b) become singly deflected or (c) doubly deflected to form a sill, or (d) penetrate the contact/discontinuity and continue its propagation and perhaps, eventually, reach the surface and erupt. Here layers A and B are stiffer than layer D, whereas layer C has the same stiffness as layer D. Modified from Hutchinson (1996) and Gudmundsson (2011b).

(a) The fracture tip becomes arrested at the contact.
(b) The fracture becomes deflected in one direction along the contact, resulting in an asymmetric deflection.
(c) The fracture becomes deflected in two directions along the contact, resulting in a symmetric deflection.
(d) The fracture penetrates the contact and continues its vertical propagation.

What exactly happens when the fracture meets the contact depends primarily on the three mechanisms mentioned above, and which are now discussed further.

7.4.1 Cook–Gordon Delamination

Consider first a vertical dike approaching a contact (Fig. 7.12). As discussed in Section 5.2, a dike approaching a contact induces dike-parallel tensile stresses that may **open** up the contact. This applies particularly to mechanically weak contacts, that is, contacts with a low tensile strength. The dike-parallel **tensile stress** reaches about 20% of the dike-perpendicular tensile stress (Fig. 7.13), the latter being the stress that ruptures the rock and advances the dike-fracture. The *in situ* tensile strength of rocks in general is low, mostly between 0.5 MPa and 6 MPa and only occasionally as high as 9 MPa. If, for example, the dike-fracture itself opens at 2–3 MPa, which would correspond to common dike-

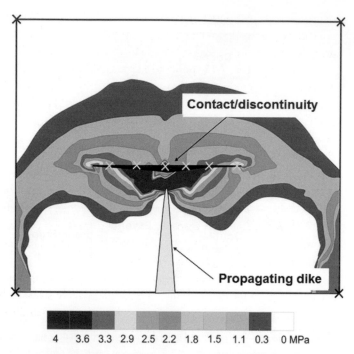

| 4 | 3.6 | 3.3 | 2.9 | 2.5 | 2.2 | 1.8 | 1.5 | 1.1 | 0.3 | 0 MPa |

Fig. 7.12   Dike approaching a weak (low-tensile-strength) horizontal contact or discontinuity tends to open the contact, as shown in this numerical model. This is an example of the Cook–Gordon delamination mechanism.

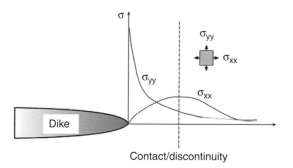

Fig. 7.13   The maximum magnitude of the tensile stress induced ahead of, and parallel with the dip-dimension of, a dike (or mode I cracks in general) is about 20% of the magnitude of the induced tensile stress ahead of and perpendicular to the dike. Comparatively close to a contact or a discontinuity the dike-induced tensile stress may thus open the contact/discontinuity, provided the contact tensile strength (in a direction parallel with the dike dip-dimension) is less than about 20% of tensile strength of the adjacent rock layers in a direction perpendicular to the dike dip-dimension.

perpendicular stress at rock failure, then the corresponding dike-parallel tensile stress, the one that operates on the contact, would be 0.4–0.6 MPa. The latter is the same as the minimum *in situ* tensile strength, and thus similar to the tensile strength of many weak contacts. Many dikes, however, develop higher overpressures than 2–3 MPa; some reach 20–30 MPa along parts of their paths, and the dike-parallel stress ahead of the tip would then be correspondingly higher, perhaps **4–6 MPa**. As a dike approaches such a contact, the contact may open up. When the dike tip hits the open contact, the tip may either become arrested or become deflected along the contact to form a sill.

The model in Fig. 7.12 applies to a vertical dike meeting a **horizontal contact**, so that the angle between the dike and the contact is 90° (Figs. 1.18 and 7.7). If the contact is already open when the dike meets it, the angle between a dike hitting the contact and the contact itself must be 90°. This follows because an open contact is a **free surface** and the principal stress trajectories must be either parallel or perpendicular to a free surface. In rift zones, and most areas of dike-fed volcanism, it is σ_1 that is vertical at and close to the surface. However, most contacts are not open at depth in the crust, but may become opened by an approaching dike through the Cook–Gordon mechanism. It is observed that for many dikes/inclined sheets the angle between the contact and the intrusion is less than 90° (Figs. 4.11 and 7.8). When this is the case, the opening of the contact is no longer symmetric, but rather asymmetric (Fig. 7.14). If the dike becomes deflected into the contact, the result is normally an asymmetric sill (Figs. 5.15 and 7.8).

The Cook–Gordon delamination may theoretically operate at any depth in the crust. However, given the need for the dike-parallel tensile stress to be large enough to open the contact, the mechanism is most likely to be effective at **shallow depths**, particularly in the uppermost several hundred metres of the crust.

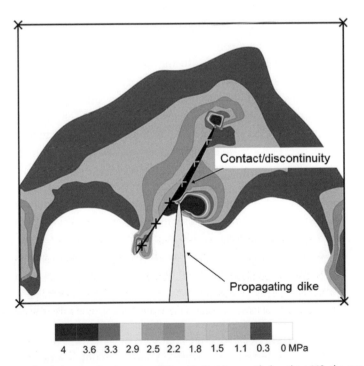

Fig. 7.14 When the dike or inclined sheet meets the contact/discontinuity at an angle less than 90°, the opening of the contact/discontinuity will be asymmetric. In this numerical model we show the contact as inclined, but the same result happens when an inclined sheet meets a horizontal contact (Fig. 7.8).

7.4.2 Stress Barrier

This mechanism is explained in considerable detail in Section 5.2, particularly the formation of the stress barrier itself, so that the present discussion will be brief. Stress barriers are layers or units in the crust that develop local stress fields which are unfavourable to the propagation of a particular type of fracture. For all extension fractures such as dikes, a **stress barrier** is a layer where the orientation of σ_1 changes from being parallel to the fracture below the barrier to being perpendicular to that fracture in the barrier itself. This means that when the extension fracture meets the barrier there are two possible scenarios: the fracture becomes arrested (Figs. 2.11, 2.29, and 2.30) or deflected along one of the contacts of the barrier (Figs. 2.25, 7.7, and 7.8). These scenarios are modelled in Figs. 7.6 and 7.9.

A particular stress barrier **prevents** the propagation of a specific type of fracture. Here the focus is on extension fractures, while barriers to shear fractures, or faults, also exist. A stress barrier to dike propagation is normally not a stress barrier to sill propagation, and vice versa. This is exemplified in Fig. 7.6. Based on the orientation of the trajectories of σ_1, a vertical dike could here propagate from the top part of the shallow magma chamber to the upper contact between the central layers (path A). In the central part of that contact, the orientation of σ_1 flips by 90°. On reaching the upper contact the dike cannot propagate further vertically. This follows because, as an extension fracture, the dike must follow the orientation of σ_1, its strike- and dip-dimensions must be parallel with σ_1, and thus cannot be perpendicular to σ_1. The dike will thus halt its propagation altogether or become deflected into a sill, which would then propagate parallel with the horizontal σ_1 trajectories (and thus perpendicular to σ_3), as is normal for an extension fracture. The sill may be the end stage of the dike propagation (path B). Alternatively, the sill may propagate for a while along the contact and then deflect upwards into an inclined sheet (path C).

Another aspect of the model is that the stress barrier to vertical dike propagation is more extensive, or wider, in the uppermost part of the model, close to the surface (Fig. 7.6). This implies that even if a vertical dike is injected somewhat to the side of the absolute top of the chamber, so as to avoid becoming arrested at the first stress barrier, the dike might still become arrested at the **second barrier**, the one close to the surface.

The trajectories of σ_1 indicate many other possible dike paths (Fig. 7.10). If the dike is not injected from the absolute top part of the chamber but rather from the roof parts to either side of the top part, the dike would be initiated not as a vertical dike but rather as an inclined dike or an inclined sheet. Then the path of the dike/sheet would be somewhat zig-zag, such as in the model in Fig. 7.6. However, based solely on the orientation of the trajectories of σ_1, there is no obvious reason why the inclined sheet should become arrested at any of the contacts.

Stress barriers can operate at **any crustal depth**. They may be generated by the contrast in elastic properties, primarily the Young's modulus between adjacent layers. Alternatively, earlier intrusions or faulting, particularly graben formation or slip, may generate stress barriers for vertical dike propagation (Section 5.2).

7.4.3 Elastic Mismatch

This is a well-known **major mechanism** for extension fracture deflection and/or arrest in fracture mechanics and materials science (He and Hutchinson, 1989; Freund and Suresh 2004; Sun and Jin, 2011). It is only recently, however, that this mechanism has been used to explain dike deflection and dike arrest in volcanic edifices (Gudmundsson, 2011a,b). Elastic mismatch relates to the difference or change in elastic properties across a discontinuity. Here, again, the discontinuities we focus on are contacts. More specifically, the elastic mismatch refers to that between the **contact itself** and the **adjacent layers**, that is, the layers above and below the contact. The chief elastic properties that bear on this mechanism are the material toughness and the Young's modulus (stiffness).

The elastic-mismatch mechanism is illustrated in Figs. 7.11 and 7.15. In relation to Fig. 7.15, the total energy release rate G_{total} of any fracture (assuming plane-strain conditions) subject to mixed-mode (extension and shear) loading is given by (Gudmundsson, 2011a,b):

$$G_{total} = G_I + G_{II} + G_{III} = \frac{(1 - v^2)K_I^2}{E} + \frac{(1 - v^2)K_{II}^2}{E} + \frac{(1 + v)K_{III}^2}{E} \qquad (7.9)$$

where, as before, G_I, G_{II}, and G_{III} and K_I, K_{II}, and K_{III} are the energy release rates and the stress-intensity factors, respectively, for modes I, II, and III cracks, v is the Poisson's ratio, and E is the Young's modulus.

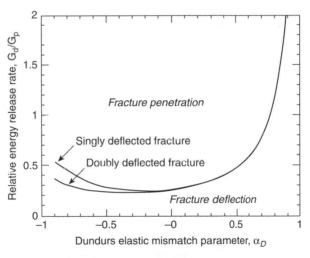

Fig. 7.15   When a dike meets a contact or discontinuity (generally, an interface), the ratio of strain-energy release rate for fracture deflection G_d to that of fracture penetration G_p controls the likelihood of dike deflection into the contact versus dike penetration of the contact (cf. Fig. 7.11). This ratio is here shown as a function of the Dundurs elastic mismatch parameter a_D (Eq. (7.11)). There is little difference in the elastic strain-energy release rate for a single or double deflection. For negative values of a_D, layer 2 (the dike-hosting layer) is stiffer than layer 1 and there is little tendency to dike deflection into the contact. However, as the stiffness of layer 1 increases in relation to that of layer 2, the tendency to dike deflection into a sill along the contact greatly increases (Fig. 7.8). If there is no Young's modulus mismatch across the contact/discontinuity/interface, dike (or other extension-fracture) deflection occurs only if the contact toughness for deflection G_d is about 26% of the toughness of the material on the other side of the contact, G_p. However, when the mismatch increases (to the right), dike deflection will still occur even if the contact toughness G_d becomes equal to or higher than the toughness G_p of layer 1 (cf. Gudmundsson, 2011a). Modified from He et al. (1994).

A dike is normally an extension fracture so that the total energy release rate is G_I and is thus given by the **first term** on the right-hand side of the equality sign in Eq. (7.9). Deflection of propagating extension fractures, including dikes, along a contact normally involves more than one mode of crack displacement (He and Hutchinson, 1989; Hutchinson, 1996). Thus, the deflection generally implies **mixed-mode** fracture propagation. It follows that the total energy release rate is then normally a combination of G_I and either G_{II} or G_{III}. The energy required to deflect a mixed-mode fracture (such as the dike in Fig. 7.7 and the inclined sheet in Fig. 7.8) into and along a contact between layers is therefore **greater** per unit extension than a pure mode I propagation (such as the dikes in Figs. 7.1 and 7.16).

Fig. 7.16   A basaltic dike dissecting a Tertiary lava pile in North Iceland. View south, the dike strike is N37°E, the dip is 83°E, and the thickness about 6 m. The walls of the dike are straight, indicating that the mechanical properties (primarily the Young's modulus) of the lava pile had become essentially uniform (homogenised) at the time of dike emplacement. Such a homogenisation takes time, so that the dike emplacement most likely took place late in the active history of the volcanic zone to which the dike belongs.

The **path length** of a dike deflected into a sill at a contact normally exceeds that of a dike that propagates vertically through the same lava pile (Figs. 2.25, 7.6–7.9). A vertical dike deflected into a sill, even if the dike continues as such when the sill has reached a certain length, therefore needs more energy for its propagation than a purely vertically propagating dike because the deflected dike propagates a longer distance and in a mixed mode. This is a direct consequence of the stress-intensity factor – its critical value the fracture toughness K_c – depending on the path length of the dike-/sill-fracture (Eqs. (7.3) and (7.6)). More specifically, from Eq. (7.3), the general stress-intensity factor K_I for a mode I crack model of a dike is given by (Gudmundsson, 2011a):

$$K_I = p_o(\pi a)^{1/2} \tag{7.10}$$

where p_o is the overpressure in the dike and a is its dip-dimension, if the focus is on the vertical section of an arrested dike, or half the strike-dimension, if the focus is on the lateral dimension of the dike.

The likelihood that a **dike** becomes **deflected into a sill** at a contact is indicated by the Dundurs elastic extensional mismatch parameter α_D (He and Hutchinson, 1989; Hutchinson, 1996):

$$\alpha_D = \frac{E_1 - E_2}{E_1 + E_2} \tag{7.11}$$

where E is the plane-strain extensional Young's modulus (stiffness). An arrest or deflection of a fracture along the contact is encouraged when the Young's modulus of the layer hosting the fracture (E_2) and below the contact is lower than that of the layer above the contact (E_1). The likelihood of a dike arrest or deflection into a sill at the contact varies positively with α_D (Eq. (7.11); Fig. 7. 11) or, more specifically, with **increasing elastic mismatch** or difference between E_1 and E_2. These conclusions are supported by many experimental results in fracture mechanics and materials science (Kim et al., 2006). They are also supported by analogue experiments on sill emplacement (Kavanagh et al., 2006; Menand, 2008) and by direct field observations (e.g. Figs. 2.25, 7.7, and 7.8).

In the elastic-mismatch mechanism, a dike arrest or deflection into a sill is encouraged when there is a **large elastic mismatch** across the contact. By contrast, dike penetration of the contact is encouraged when the elastic mismatch is small or zero (Figs. 7.11 and 7.15). An arrest or deflection into a sill is therefore more likely when the layer above the contact in a vertical section is stiffer than the layer below the contact.

7.4.4 Examples of Arrested Dikes

Many arrested dikes have been studied in various volcanic edifices and zones, both active and extinct. Several examples of **arrested dikes** are given in Figs. 2.11, 2.29, and 2.30. Here, these and other examples are discussed in more detail.

The dike in Fig. 7.17 forms a part of a dike swarm in the caldera wall of the island of Santorini, Greece. The dike propagates through several layers of different mechanical properties, including a stiff lava flow at the foot of the exposure as well as several pyroclastic layers. It is clear that the **attitude** of the dike **changes** when

Fig. 7.17 An arrested dike in the caldera wall of Santorini, Greece. View north, at the bottom of the outcrop, the dike is about 0.4 m thick but becomes gradually thinner upwards. The dike dissects lava flows and pyroclastic layers and ends vertically at the contact between a stiff lava flow and a comparatively soft scoria layer.

passing through the pyroclastic layers; in particular, the dike dip within these layers is shallower than in the lowermost lava flow. This indicates **rotation** of the **principal stresses** at the contacts between these layers so that σ_1 is more gently dipping in the pyroclastic layers, as is the dike. The dike is well exposed and its vertical tip is located in a scoria layer at the contact with a stiff lava flow above. Thus, the dike tip is arrested exactly at the contact where the stiff lava flow takes over from a comparatively compliant scoria layer.

Another example is from the active Holocene rift zone in Southwest Iceland (Fig. 7.18). This basaltic dike is arrested only **5 m below the surface** at the contact between a compliant pyroclastic layer (a tuff) and a stiff basaltic lava flow. The dike thickness gradually decreases from 34 cm at the foot of the exposure, a sea cliff, to 10 cm at the tip. This dike tip appears very similar to the tip of the dike in Fig. 7.17. While this dike is clearly

Fig. 7.18 An arrested basaltic dike on the Reykjanes Peninsula in Southwest Iceland. View northeast, the dike is arrested at 5 m depth below the surface of the active volcanic zone, at the contact between a stiff basaltic lava flow and a soft pyroclastic (tuff) layer. At the bottom of the 8-m-tall exposure (point A) the dike is 0.34 m thick, but gradually thins to 0.1 m at the arrested tip (point B).

arrested in the section and thus is a non-feeder, a feeder-dike of similar dimensions is a short distance to the west of the arrested dike (cf. Gudmundsson, 2017).

Both these dikes (Figs. 7.17 and 7.18) are arrested at contacts between a stiff lava flow above and a compliant pyroclastic layer below the contact. This is a common type of arrest, and all the three mechanisms discussed above could theoretically contribute to dike arrest at such a contact. There are many similar arrests observed in other areas (Gudmundsson, 2002, 2003). But let us take these two dikes (Figs. 7.17 and 7.18) as examples and try to assess the importance of the three main mechanisms of arrest and deflection discussed above.

The Cook–Gordon **delamination mechanism** could possibly operate at both localities. The dike in Santorini is arrested at a depth of less than 200 m below the surface, and the dike in Iceland at just 5 m below the surface. Thus, the dike-induced tensile stress (Fig. 7.13) could theoretically have opened the contacts. The field evidence, however, does not indicate much effect of this mechanism as regards these two dikes. Both tips are somewhat **rounded** and located within the scoria/tuff layers. No sill emplacement is apparent at the contact, whereas some sill-like injection might be expected if the contact became open over a significant area. The contact in Santorini also does not appear to have been particularly weak; the scoria is **welded** to the lower margin of the lava flow above. The contact in Iceland was presumably weaker. However, the overpressure in both dikes may have been too low to generate a sufficiently high dike-parallel tension to open the contacts. Both dikes are thin, tens of centimetres at maximum, and become thinner towards their tips, suggesting that the overpressure in the top parts of the dike just below the contact was perhaps too low to induce tensile stresses for contact opening.

The mechanisms related to a **stress barrier** and an **elastic mismatch** are likely to have contributed to both dike arrests. The abrupt change in the stiffness of the rocks across the contact, from a soft scoria/tuff layer to a stiff lava flow, would encourage abrupt change in the orientation of σ_1, particularly its rotation by as much as, in places, 90° (cf. Figs. 7.6 and 7.9). Thus, in both cases, for this reason alone, the lava flows might have acted as stress barriers. And that conclusion is further supported, for the arrested dike in Santorini (Fig. 7.17), by several subvertical dikes that dissect the lava flow, tens of metres from the present dike (Fig. 2.34). These dikes, if emplaced earlier than the present arrested dike, may have generated temporarily high subhorizontal compressive stresses in the lava flow and thus stress barriers (as is also the case for the dike in Fig. 7.18; Gudmundsson, 2017).

The ideal elastic-mismatch mechanism operates when the layer above the contact is much **stiffer** than the layer below the contact and hosting the dike. That is exactly the situation for both these dikes, with stiff lava flows being in contact with soft scoria/tuff layers below the contact (Figs. 7.17 and 7.18). Thus, the field observations, together with the theoretical discussion and numerical models above, suggest that the mechanisms of the generation of stress barriers and elastic mismatch very likely both operated during the arrests of these two dikes.

7.5 Surface Effects of Dikes

On approaching the Earth's surface, sheets in general, and dikes in particular, generate stresses and associated deformation at the surface. When hosted by a homogeneous, isotropic crustal segment, a vertical dike generates stresses and displacement at the surface that are symmetric around the projection of the dike (Figs. 3.2, 3.37, 3.39, and 3.40). By contrast, the **stresses and displacement** above an inclined dike are asymmetric (Fig. 3.2). Such models as these are normally referred to as elastic half-space (or semi-infinite elastic plate) models because it is imagined that an infinite elastic body is cut in half and the deformation and stresses are analysed at the surface thus generated by the section cut through the centre.

In the solid-mechanics and fracture-mechanics literature results on the surface effects of vertical extension fractures have been well known for many years (Isida, 1955). The results of analytical or quasi-analytical studies are generally similar whether the loading is external tension or internal fluid overpressure. The tensile stress (and the associated deformation/displacement) **peaks symmetrically** around the imaginary projection of the dike plane to the surface, but the stress and displacement peaks are at different locations, as discussed below. There is normally no tensile-stress concentration at the surface directly above the dike itself, as is also shown by numerical model results (Figs. 3.37 and 3.40).

Early numerical models indicated large surface **subsidence** above arrested dikes, suggesting dike-induced graben formation in rift zones (Pollard and Holzhausen, 1979). Subsequent studies, however, did not yield any large subsidence directly above the arrested tips of vertical dikes (Pollard et al., 1983), in agreement with the analytical solutions of Isida (1955) and others. There is, of course, tensile stress at the dike tip (e.g. Fig. 3.40), namely the stress that generates the extension fracture at the tip to allow the dike to propagate (e.g. Eq. (5.6)). And when the dike is close to the surface, the extension fracture runs ahead of the magma front and reaches the surface before the magma starts to pour out (Fig. 7.19). Many photographs (Figs. 7.19 and 7.20) and detailed descriptions (Macdonald, 1972; Lockwood and Hazlett, 2010) confirm that the **dike-fracture** reaches the surface well **before** the **magma**. On rupturing the surface, the tension fracture can be metres to, perhaps, a few tens of metres **ahead** of the **magma front**. This is in accordance with theory and experiments which show that the magma front normally lags behind the fracture front during hydrofracture (including dike) propagation (Fig. 4.10).

Analogue models have also been used to model the surface deformation associated with propagating dikes (Mastin and Pollard, 1988; Trippanera et al., 2015; Acocella and Trippanera, 2016; Xu et al., 2016). The results again suggest that close to the surface a feeder-dike commonly generates a fracture at the surface, namely the fracture that the magma eventually uses if and when it reaches the surface. There is thus no question that in most cases, especially at divergent plate boundaries and in rift zones in general, an extension fracture propagates **faster** than the magma and may reach the surface as much as tens of metres ahead of the fluid/magma front. Indeed, experiments of hydraulic fracturing at shallow depth show exactly the same. At any particular moment the fracture tip may be metres or tens of metres **ahead** of the fluid front (Chapter 4; Warpinski, 1985; Bunger, 2009).

For many decades, however, large normal faults and **grabens** have commonly been attributed to **dikes** that failed to reach the surface, that is, to arrested dikes (Figs. 2.11, 2.29, and 2.30). Originally, this was suggested by Walker (1965) for the volcanic zones of Iceland. The idea sounded very plausible since the volcanic zones contain numerous dikes that do not reach the surface. For exploring these ideas, detailed studies of dikes and faults were made in Iceland (Forslund and Gudmundsson, 1991, 1992), but **no dike tips** were found connected with **normal faults** or tension fractures in that study. Studies of arrested dikes in other volcanic areas have also generally failed to find significant grabens connected with the tips (Figs. 2.11, 2.29, 2.30, and 7.17), even when the dike tip is as shallow as 5 m below the surface of the rift zone (Fig. 7.18). Nevertheless, there is no doubt that some dikes very close to the surface generate **tension fractures** (Fig. 7.19) and some

Fig. 7.19 Beginning phase of the March 2011 eruption in Kilauea, Hawaii. Part of the feeder-dike has reached the surface, issuing pahoehoe lava, but part is still seen as a magma front in the tension fracture closest to the viewer. The photograph demonstrates well that (1) during dike propagation the magma front normally lags behind the dike-fracture tip at any particular moment (Fig. 4.10), and (2) that very close to the surface the magmatic overpressure induces tensile fractures at the surface – fractures that, for a feeder-dike, are subsequently used as a channel for the magma to the surface. Photo: Tim Orr/USGS, 5 March 2011. A black and white version of this figure will appear in some formats. For the colour version, please refer to the plate section.

also **small grabens** (cf. Chapter 3, Figs. 3.41–3.44; Chapter 10, Figs. 10.16–10.20; Mastin and Pollard, 1988; Gudmundsson, 2003; Trippanera et al., 2015; Acocella and Trippanera, 2016; Xu et al., 2016).

As regards **large grabens** in rift zones, their mode of formation is not entirely clear. No doubt grabens in rift zones form like grabens anywhere, namely through two normal faults dipping towards each other. **Conjugate faults** are the normal results of rock-physics experiments generating shear failure. Shear failure under considerable normal stress, that is, significant crustal depth, when the maximum compressive principal stress σ_1 is vertical and the minimum one σ_3 is horizontal, results in normal faulting and, commonly, in graben formation. Large grabens in rift zones are most likely formed in the same way as grabens everywhere; σ_3 **is reduced** through extension until normal-fault failure occurs. No normal dike tips in rift zones could **accommodate** normal-fault displacements of the order of many tens or hundreds of metres. But small-scale normal faulting is common in rift zones, and dike-induced stresses are likely to play a role, either **directly** (through the magmatic pressure) or **indirectly**, through reducing the effective thickness of the crustal segment that is subject to horizontal (mostly relative) tension as the dike tip approaches the surface.

To throw further light on the **surface stresses** of arrested dikes, and the potential relation between arrested dikes and grabens, we turn now to numerical models of dikes in a layered crustal segment. In Chapter 3 we analysed the effects of layering on the variation in the

Fig. 7.20 Aerial view of the initiation, the first 'finger', of the volcanic fissure of the July 1980 eruption in the Krafla Volcanic System in North Iceland (located in Fig. 6.15). Apart from the feeder-dike-fracture itself, no surface fractures were generated by the subsurface part of the feeder-dike (the nearby fractures are old). At this stage the fissure is only tens of metres long, but subsequently other 'fingers' hit the surface, the initial fissures propagated laterally at the surface (at the rate of 0.1–0.2 m s^{-1}), and eventually the volcanic (offset and disconnected) fissure reached the length of many kilometres. Photo: Aevar Johannsson.

tensile stresses ahead of the dike tip. In particular, a comparison of the non-layered model in Fig. 3.39 and the layered model in Fig. 3.40 shows the effects of layering in modifying the stress field and encouraging dike arrest. In both models the loading is the same, an internal magmatic overpressure of 10 MPa. The maximum theoretical tensile stress at the dike tip in the non-layered model (Fig. 3.39), however, is about 150 MPa but only about 30 MPa at the tip of the dike in the layered model (Fig. 3.40). Both models show that the tensile stress decreases from its maximum at the dike tip to zero at the point on the surface right above the dike tip, in agreement with earlier analytical (Isida, 1955) and numerical (Pollard et al., 1983) results (Fig. 3.37). For the dike-induced tensile stress to have a chance of reaching the surface and generate fractures, the dike tip has to be very close to the surface and with an overpressure capable of propagating the tip to the surface, as is observed (Fig. 7.19).

The direct **effects of layering** on the surface stresses induced by the dike are seen on comparing Figs. 3.39 and 7.21. Both models have dikes of similar dimensions, magmatic overpressure (10 MPa) and depths below the surface. In Fig. 7.21, however, the crust is composed of alternating layers of stiff (100 GPa) thin layers and soft (1 GPa) thick layers. The result is that the maximum dike-induced tensile stress at the surface is here about 0.5 MPa in contrast to 4 MPa for dike-induced tensile stress in Fig. 3.39. Also, the shapes of the tensile-stress curves at the surface differ between these models.

When the Cook–Gordon **delamination mechanism** operates, the surface deformation induced by a dike can be widely different from that induced by the same dike in

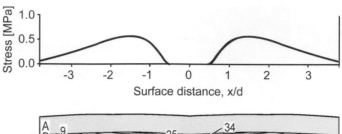

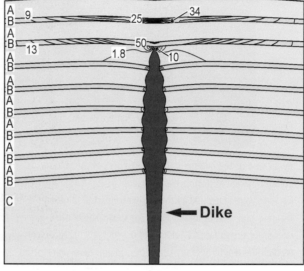

Fig. 7.21 Surface tensile stress σ_3, reaching a maximum of 0.5 MPa, induced by an arrested dike in a layered crustal segment. Here x is the horizontal distance along the surface whereas d is the depth to the tip (top) of the dike. In this numerical model, the thick layers have a stiffness of 1 GPa whereas the thin layers have a stiffness of 100 GPa. This is reflected in the shape of the dike; its aperture or opening is larger in the compliant thick layers and smaller in the stiff thin layers and also in the low stresses in the soft surface layer. The only loading is an internal magmatic overpressure of 10 MPa.

a homogeneous isotropic crustal segment. This is seen on comparing Fig. 3.39 with the model in Fig. 7.22. In the latter, the discontinuity or contact close to the surface is opened by the dike-parallel tensile stress. No tensile stresses can be transmitted through the open contact, but must go around the lateral ends of the open part of the contact. Similar results are obtained in other delamination models for dikes (Fig. 3.38). As a consequence, there, tensile stress peaks at the surface are **shifted** away from the surface point above the dike, and any attempt to infer the depth to the dike tip (or dike centre) from the stress/displacement peaks could be highly misleading. Thus, for a layered crustal segment, a standard inversion of the surface-deformation date during dike propagation to infer the depth to the dike tip is not a straightforward process. Using crack or dislocation models for elastic half-spaces may yield unreliable results, a conclusion that is of significant importance for hazard and risk assessments during unrest periods with dike injections. Further analysis of surface stresses and displacements, as well as internal crustal deformation, induced by arrested dikes is given in Chapter 10.

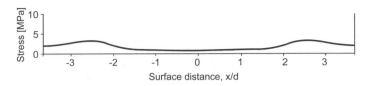

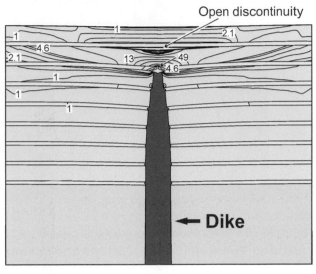

Fig. 7.22 Surface tensile stress σ_3, reaching a maximum of 3 MPa, induced by an arrested dike in a layered crustal segment. Here x is horizontal distance along the surface whereas d is the depth to the tip (top) of the dike. In this numerical model, the thick layers have a stiffness of 100 GPa whereas the thin layers have a stiffness of 1 GPa (opposite to that in Fig. 7.21). This is reflected in the shape of the dike; its aperture or opening is larger in the thin compliant layers and smaller in the stiff thick layers and also in the comparatively high stresses in the stiff surface layer. The only loading is an internal magmatic overpressure of 10 MPa. Because of a weak contact, with zero stiffness (a discontinuity), the peaks of the tensile stress are at a very different location in relation to the depth d to the top of the dike compared with the peaks in Fig. 7.21. This shows that when there are weak contacts that can open up (through the Cook–Gordon delamination mechanism) surface stresses and displacements cannot be used to infer the depth to the tip (cf. Fig. 3.38).

Field observations of arrested dike tips, as well as observations of beginning eruptions, generally support the analytical and numerical results discussed here. The arrested dike tips in Figs. 2.11, 2.29, 2.30, 7.17, and 7.18 all end **without** any significant tectonic fractures extending from the dike tip into the host rock. The one dike observed with faults in the host rock at the tip seems to have been captured by an existing graben (Fig. 2.31), although this and similar cases need to be explored further. Similarly, photographs from the beginning of eruptions commonly show the magma within an extension fracture that has already reached the surface **ahead** of the magma front (Figs. 7.19 and 7.20).

In the case of feeder-dikes that are comparatively thick, or that have a large overpressure when they reach the surface, the extension fracture above the tip of the dike at the surface may split into **two or more** parallel segments. Then, the strip of land, usually from several

Narrow, small grabens are common in rift zones. View south, a small graben in the northernmost part of the Krafla Volcanic System in North Iceland (cf. Fig. 6.15). The left (east) fault is vertical because it dissects a Holocene basaltic lava flow with a tensile strength of a few mega-pascals; the fault was thus initially a tension fracture at the surface. The right (west) fault dissects soil and sand with zero tensile strength. The fault thus does not change into a vertical tension fracture at the surface but rather maintains its shear-fracture dip. Small grabens of this type, however, are normally not related to dike emplacement but rather to fault movements (cf. Figs. 7.24 and 7.25).

metres to a maximum of a few tens of metres or so, may subside between the parallel fractures (cf. Al Shehri and Gudmundsson, 2018). These **graben-like structures** are further encouraged to form if there is already an absolute tension, associated with divergent plate movements or volcano spreading or other similar processes, in the uppermost part of the rift zone. Such structures, however, are common in many rift zones and need not be related to dike emplacement (Fig. 7.23). In fact, narrow 'grabens' of this type characterise most large normal faults in the rift zones of Iceland (Fig. 7.24), and are common in the rift zones of mid-ocean ridges. They can often be explained in simple terms as the result of the normal faults evolving from tension fractures that, once they reach a critical depth, must change into normal faults (Chapter 3). Also, in many such narrow grabens the hanging wall is **tilted**, a feature that can partly be related to **friction** along the fault (Fig. 7.25).

The models and observations presented here indicate that it is difficult for dikes in layered crustal segments and arrested at significant depths – hundreds of metres or more – to generate **large grabens** at the surface. Also, the **graben rule** holds; the width of the a dike-induced graben at the surface should be similar to the depth to its top/tip (Chapter 10; Al Shehri and Gudmundsson, 2018; Bazargan and Gudmundsson, 2019). During many rifting events, however, normal faulting and graben formation are likely to occur essentially **simultaneously**. Both are favoured during rifting events and, in addition, as the dike propagates towards the surface, the tensile stress related to the rifting episode becomes concentrated in a thinner and thinner crustal layer (cf. Fig. 7.26). This means that the tensile stress in that layer increases and may result in the formation of tension fractures,

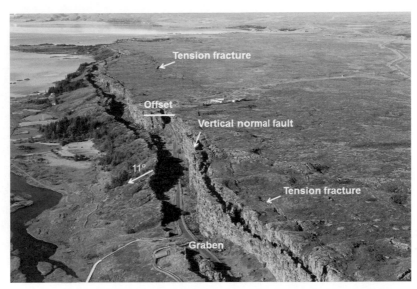

Fig. 7.24 Aerial view of part of the boundary fault of the active rift zone in Southwest Iceland. View southwest, the main boundary fault, Almannagja, is actually a narrow graben, mostly 20–60 m wide, whereas the total Holocene vertical displacement reaches a maximum of about 40 m. The east fault wall is tilted about 11° to the east, partly because of friction along the fault plane (Fig. 7.25; cf. Gudmundsson, 2017). A black and white version of this figure will appear in some formats. For the colour version, please refer to the plate section.

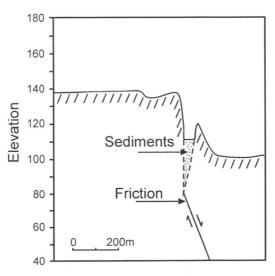

Fig. 7.25 East–west cross-section across the Almannagja boundary fault, roughly where seen in Fig. 7.25. The tilted east fault wall is partly so because of friction along the fault plane where the open (graben-like) fracture becomes closed. Normal faults with this geometry are very common in rift zones in general, but the geometry, including the narrow graben, is usually not directly related to arrested dikes. Modified from Gudmundsson (2017).

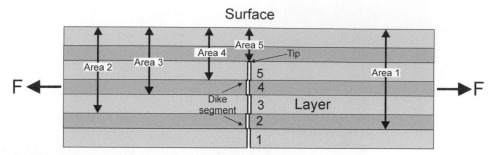

Fig. 7.26   The dike functions as an open fracture that concentrates stresses, particularly above its tip/top at any particular instance during its propagation. Here, the uppermost part of the dike is shown by five segments, 1–5, all of which are in physical contact. The dike propagates in a rift zone subject to extension, presented by the force F. The dike generates an open fracture, the tip of which advances closer and close to the surface, thereby reducing the area on which the force F acts and, for a constant force, increasing the (tensile and shear) stress concentration in the layers above the dike tip. The force F acts on area 1 (the area is regarded as having a unit distance or dimension into the illustration) when the dike tip is at the top of segment 1, on area 2 when the tip is at the top of segment 2, and so forth. Independently of the overpressure in the dike, the dike function as an open fracture while propagating gives rise to tensile- and shear-stress concentration in the layers above the dike tip at any instance, which can lead to fracture formation. Depending on the tensile/shear strength of the layers, the fracture formation does not have to be close to the dike but can be some distance away from the dike (or rather the dike projection to the surface).

normal faults, and grabens. If, however, there is no regional and absolute tension in the uppermost layer during the dike emplacement, no grabens will form or reactivate during the dike emplacement. In fact, rather the opposite; the dike may induce reverse faulting on an existing, nearby normal fault (Fig. 3.3).

By contrast, if there is already **regional tension** during the dike emplacement, tensile stresses concentrate in the uppermost layer. The resulting grabens, however, do not necessarily relate in any way to the depth to the dike tip; they can form anywhere in the layer subject to tensile-stress concentration, particularly in layered rocks where the layer subject to tension is stiff (say a thick lava flow) and is resting on soft sediments or pyroclastic rocks – as is common. In this mechanism, also, the graben faults (or other normal faults) would normally not meet with the tip of the dike – in fact, they would commonly be well outside the dike (Fig. 7.27). This is also in agreement with studies in palaeo-rift zones in Iceland and elsewhere which show that where normal faults and grabens are common, dikes are comparatively infrequent, and vice versa (Forslund and Gudmundsson, 1991, 1992; Gudmundsson, 1995). In the event of the dike reaching the surface, the resulting eruption is normally a fissure eruption, to which I turn now.

7.6 Fissure Eruption

When a **feeder-dike** hits the surface, an eruption starts. The details of the dynamics of an eruption, including effusion rates, are discussed in Chapter 8. Here, the focus is on the brief

Fig. 7.27 A graben and a dike dissecting 12-Ma basaltic lava pile in North Iceland. View southwest, the graben and the dike are side by side, as is common in rift zones. In fact, studies show that where normal faults and grabens are common, dikes are comparatively rare, and vice versa (Gudmundsson, 1995; Acocella and Trippanera, 2016). The dike is 6 m thick (the same as in Fig. 7.16) and the sea cliff is about 120 m high. The right boundary fault of the graben goes straight through the lava pile at a constant dip, whereas the left boundary fault changes dip when passing through layers of scoria and lava with different mechanical properties. No dike tip is visible at the bottom of the graben, which ends about 1500 m below the original surface of the volcanic zone. If the graben had reached the surface of the volcanic zone within which it formed, the graben width at the surface would have been 800–1000 m.

description of the feeder-dike and its evolution at the surface. Presumably, all dikes, including feeders, propagate as fingers (Fig. 7.28). The first finger to reach the surface (Fig. 7.29) then generates the first segment of the volcanic fissure. Initially, the volcanic fissures are thus normally discontinuous and sometimes offset and separated by distances that are large in comparison with the lengths of the initial finger-like fissure segments. Most or all of the segments, however, extend laterally until they reach their equilibrium aperture–length ratios. This does not mean that most segments become interconnected – many remain disconnected through the eruption – but that their final lateral length or strike dimension follows well-established fracture-mechanics principles.

For constant feeder-dike magma overpressure p_o, half the aperture or **opening u** of the volcanic fissure (or fissure segment, as discussed below) is related to half the surface length (half the strike-dimension) a of the fissure through the formula (Gudmundsson, 2011a):

$$u = \frac{2p_o(1 - v^2)}{E}(a^2 - x^2)^{1/2} \tag{7.12}$$

where v is the Poisson's ratio and E is the Young's modulus of the host rock, and x is the coordinate axis along the strike-dimension. This is a plane-strain formula for a two-dimensional elliptical through crack, a so-called tunnel crack (Fig. 5.16), subject to constant

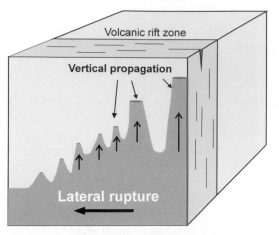

Schematic illustration showing how dikes propagate through discontinuous 'fingers'. Subsequently, the fingers may or may not become connected. Here, the overall propagation or rupture direction of the fracture is lateral, while the dike segments themselves, and the 'fingers', propagate primarily vertical. Generally, fractures, including dikes, propagate in all directions. For example, the surface fractures hosting the very short dike 'fingers' in Fig. 7.20 subsequently propagated or ruptured laterally into kilometre-long fissures as deeper (more slowly vertically propagating) dike segments reached the surface. For example, dike propagation in the Bardarbunga 2014–2015 volcanotectonic episode was very likely similar to that shown here (Gudmundsson et al., 2014).

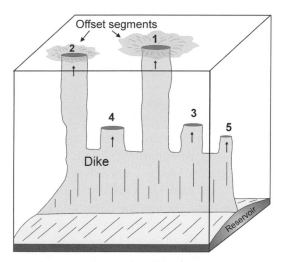

The first dike 'finger' to reach the surface initiates the resulting fissure eruption (Figs. 7.19 and 7.20). The first segments at the surface are normally short, tens of metres, but they commonly propagate laterally at the surface and may eventually link up into larger fissure segments (cf. Figs. 7.20 and 7.28; Gudmundsson, 2017).

fluid (here magma) overpressure p_o. The **strike-dimension** of the fracture is here the length of the volcanic fissure at the surface, or the length of the feeder-dike below the surface. The **dip-dimension** is the distance along the dike from the source magma chamber to the surface. The plane-strain model in Eq. (7.12) assumes that the dip- and strike-dimensions of the feeder-dike are similar, or that the dip-dimension is larger than the strike-dimension. Consequently, the opening displacement or aperture (here shown as half the aperture u) depends only on the smaller of the dip- and strike-dimensions, namely on half the strike-dimension a. If, however, the strike-dimension of the feeder-dike (or the dike segment being modelled) is much larger than the dip-dimension of the feeder-dike, then a plane-stress model is used in which case the factor $(1 - v^2) = 1$. Generally, plane-strain formulas are used when the crustal segment under consideration has lateral dimensions that are similar to, or less than that of, the vertical dimension. Similarly, plane-stress formulas are used when the lateral dimensions of the crustal segment under consideration are much larger than the vertical dimension. Thus, for equal-dimension rock bodies, plane strain is used; and for comparatively thin rock bodies, plane stress.

When modelling the opening of a volcanic fissure or a feeder-dike through Eqs. (7.12) and (7.13), we can either use the total length or strike-dimension of the fissure/dike or the strike-dimensions of individual segments. For a **segmented fissure/dike**, the total strike-dimension is normally used when (i) the segments are physically connected or (ii) when they are not connected but the lateral distances between the nearby segment tips is less than about 10% of strike-dimensions of the segments (Gudmundsson, 2011a). This follows because when the distance between the nearby segments of extension (mode I) fractures becomes small, the segmented fracture functions mechanically as a single fracture.

Most extension fractures, including dikes, open up into **flat ellipses** when the loading is constant – overpressure p_o for dikes and other hydrofractures and tensile stress $-\sigma$ for tension fractures. This means that the opening (or thickness of a solidified dike) is similar along the greater part of the strike-dimension. It is only close to the lateral ends that the opening/thickness decreases rapidly to zero. These types of thickness/opening variations are commonly observed in hydrofractures, both mineral veins (Kusumoto et al., 2013a; Kusumoto and Gudmundsson, 2014) and dikes (Gudmundsson et al., 2012; Kusumoto et al., 2013b). In vertical sections of layered rocks with widely different mechanical properties, however, the variation in dike thickness may deviate significantly from that of a flat ellipse. This applies particularly to feeder-dikes, whereas some non-feeders approximate flat-ellipse thickness variations in vertical sections (Fig. 2.33; Geshi et al., 2010, 2012). It follows that so long as the measurements made are not of the opening/thickness close to the lateral ends, they will yield a value that is generally approximately constant and representative of the opening/thickness for almost the entire strike- (or dip-) dimension of the extension fracture. Using this information, it is often convenient to use the **ratio** between the maximum opening/thickness (assumed to be the value measured in the field) and the total strike-dimension of the fracture to infer the overpressure/tensile stress associated with fracture formation. Alternatively, if the strike-dimension of, say, a volcanic fissure is measured and the overpressure can be estimated, the likely (strictly maximum) opening/aperture can be calculated.

Let us denote the total strike-dimension by $L = 2a$ and assume it to be smaller than (or roughly equal to) the dip-dimension of the dike. Then the maximum (central) **opening**

displacement of the volcanic fissure/feeder-dike Δu_I, that is, the maximum aperture of the **dike-fracture**, is obtained by using $x = 0$ in Eq. (7.12), thus:

$$\Delta u_I = \frac{2p_o(1 - v^2)L}{E} \qquad (7.13)$$

If the tensile stress $-\sigma$ (or $-\sigma_3$) is substituted for the overpressure p_o, then the corresponding equation for a **tension fracture** is obtained, namely:

$$\Delta u_I = \frac{-2\sigma(1 - v^2)L}{E} \qquad (7.14)$$

For tension fractures in rift zones, particularly those hosted by smooth pahoehoe lava flows (Fig. 1.6), both the strike-dimension and the opening can normally be measured (Figs. 1.6 and 1.9), in which case Eq. (7.14) can be used to calculate the tensile stress $-\sigma$ responsible for the fracture formation. For volcanic fissures supplying magma during eruptions (Figs. 7.19 and 7.20), the opening or aperture of the fissure can rarely be measured directly in the field, but sometimes may be **estimated from GPS and/or InSAR** and other geodetic studies. Also, for volcanic fissures that formed before modern geodetic instruments were available, that is, volcanic fissures older than a few tens of years, the opening can hardly ever be measured in the field, except in the rare cases where the feeder-dike is exposed in vertical sections. By contrast, the strike-dimensions of young volcanic fissures (those not buried by later eruptive materials) can almost always be measured. Equation (7.13) can then be used to calculated the likely (maximum) opening or aperture of the fissure (the feeder-dike close to the surface) while it was active.

The **aperture** of a volcanic fissure is of fundamental importance for understanding volumetric flow (effusion) rates. In particular, the volumetric flow rate along the vertical z-coordinate Q_z through a volcanic fissure supplied with magma by a vertical feeder-dike (Figs. 3.3 and 7.19) is given by (Fig. 5.25; Gudmundsson, 2011a):

$$Q_z = \frac{\Delta u^3 W}{12\mu_m}\left[(\rho_r - \rho_m)g - \frac{\partial p_e}{\partial z}\right] \qquad (7.15)$$

where Δu is the opening or aperture (the subscript for mode I is omitted here but implied since feeders are mostly mode I cracks), and W is its width in a direction that is perpendicular to the magma-flow direction and thus corresponds to the length or strike-dimension of the fissure/feeder at the surface. It follows therefore that the cross-sectional area of the volcanic fissure perpendicular to the magma flow is $A = \Delta uW$. Other symbols are the dynamic (absolute) viscosity of the magma μ_m, the average host-rock density ρ_r, the average magma density ρ_m, the acceleration due to gravity g, and the excess-pressure gradient along the feeder-dike from the chamber to the surface $\partial p_e/\partial z$, which is measured in the direction of the flow, that is, along the vertical z-coordinate. Notice that the z-coordinate is regarded as positive downwards, that is, in the direction of increasing depth in the crust, whereas the magma flow in the feeder is vertically upwards, towards the surface. It therefore follows that when using Eq. (7.15) for eruptions, there are two minus signs (–) in front of the term $\partial p_e/\partial z$, which become a plus sign.

A similar formula can be derived for a **dike of any dip**. For an inclined sheet with a dip α the dip-dimension L becomes the linear distance to the Earth's surface from the point of rupture and dike initiation at the boundary (the roof) of the shallow source chamber (Fig. 5.25). Then, Eq. (7.15) becomes modified to:

$$Q_L = \frac{\Delta u^3 W}{12\mu_m}\left[(\rho_r - \rho_f)g\sin\alpha - \frac{\partial p_e}{\partial L}\right] \tag{7.16}$$

where the subscript L on the volumetric flow rate Q_L indicates that the flow is up along the dip-dimension L which differs from the dip-dimension z of a vertical dike (Fig. 5.25). All the other symbols are the same as in Eq. (7. 15).

If the injected sheet is horizontal, and thus strictly **a sill**, the sheet dip is zero, or $\alpha = 0°$, so that $\sin\alpha = 0$ and the first term in the brackets in Eqs. (7.15) and (7.16) drops out. For a sill in the horizontal xy-plane, with the width W measured along the y-axis and the length L measured along the x-axis, we may substitute x for L in Eq. (7.16), in which case the volumetric flow rate Q_x becomes:

$$Q_x = -\frac{\Delta u^3 W}{12\mu_m}\frac{\partial p_e}{\partial x} \tag{7.17}$$

where all the symbols are as defined above. In Eq. (7.17), the only pressure driving the flow along the sill is the excess pressure p_e. Notice, however, that for a sill supplied with fluid from a dike or an inclined sheet, as is common (Figs. 5.2, 5.7, and 5.15), the excess pressure in the sill at its initiation is equal to the overpressure that develops in the dike or inclined sheet on its path towards the contact with the sill.

From Eqs. (7.15)–(7.17) we can calculate the **volumetric flow rate** through a volcanic fissure of any dimensions (Figs. 7.17, 7.18, 7.27). The dependence of the volumetric flow rate on the cube of the aperture of the volcanic fissure applies to the flow of any fluid through a rock fracture and is known as the **cubic law** (cf. Chapter 5). Commonly, the volumetric flow rate and the length of the fissure can be measured, whereas the unknown parameter is the opening or aperture of the volcanic fissure. These equations allow us to estimate the apertures of volcanic fissures, which can then be compared with actual measurements of the thicknesses of exposed feeder-dikes (Fig. 3.3; Galindo and Gudmundsson, 2012; Geshi et al., 2010, 2012; Kusumoto et al., 2013b).

7.7 Heat Transfer and Magma Solidification

During magma transport through a dike to the surface, heat is continuously transferred from the magma to the host rock and, at the surface, to the atmosphere (on dry land) or water (at the bottom of a lake or the sea). Similarly, arrested dikes and other intrusions that do not reach the surface, such as most sills and inclined sheets (and plutons in general), gradually transfer heat to the surrounding rock, solidify, and then eventually reach the same temperature as the host rock.

In thermal physics, **heat** is defined as disorganised thermal energy in transit. The **thermal energy** of a system such as an injected dike is its kinetic energy plus the latent or hidden energy, also referred to as latent heat. More specifically, **heat** is disorderly transfer of energy – through uncoordinated motion and collision of the particles that constitute the matter under consideration (solid, liquid, or gas). By contrast, **work** in physics is an orderly transfer of energy through coordinated motion of the particles that constitute the matter of interest. For instance, when magmatic overpressure ruptures the rock to form a vertical fracture – a dike – all the particles in the dike-fracture walls move as part of a coordinated opening displacement. Thus, the particles that constitute the left fracture wall move to the left, and those that constitute the right fracture wall move to the right (e.g. Figs. 3.3, 7.2, 7.16, and 7.18). Similarly, for a fault, all the particles in the walls on the same side of the fault plane move in the same direction (Figs. 7.24, 7.25, and 7.27). During displacement on a normal fault, for instance, all the particles in the hanging wall move down whereas all the particles in the footwall move up (Fig. 1.5). This movement or displacement of the walls of rock fractures constitutes work, defined as force times displacement in the direction of the force.

Work comes in many forms, such as mechanical, magnetic, electrical, and chemical, whereas heat comes in only one form, namely as heat. Thermal energy is transferred through **heat flow**. In turn, heat flows by virtue of temperature difference. More specifically, heat flows from one body, say a cooling dike, to another body, say the host rock of the dike, until the **temperature gradient** – the change in temperature with distance – disappears. **Temperature** is a measure of the average kinetic energy of the particles that constitute the matter; for example, the particles that constitute the magma of a chamber. We say that the heat 'flows' from one body to another down the temperature gradient even if, strictly speaking, nothing is flowing. What happens instead is a **transfer of thermal energy** (on an atomic scale) from the hot body to the cold one. The process of cooling and solidification of a dike involves a decrease in the average kinetic energy of the particles (molecules, atoms) that constitute the magma of the dike. When the average kinetic energy of the particles decreases, the temperature falls.

Heat flux, a common measure of heat transport in the crust, indicates the rate of transport of heat energy through a surface per unit time, the unit being $J\ m^{-2}\ s^{-1}$, that is, watt per square metre, $W\ m^{-2}$ (Blundell and Blundell, 2006). There are, however, other definitions of heat flux, some of which give the unit as watt (W) and refer to the above definition as **heat-flux density.** The concept is not used further in this book but it is commonly used in connection with mantle plumes, hot spots, and related large-scale features.

Heat can be transferred between bodies through conduction, convection, and radiation. For cooling of intrusions, **radiation** – transmission of energy through waves (such as electromagnetic waves) or particles – is not of importance. **Convection,** however, certainly plays a role in the cooling of intrusions, as is exemplified by natural geothermal or hydrothermal systems, in which convection of water plays a major role. However, dikes and other intrusions commonly develop a chilled selvage, a glassy margin (Chapter 2), through which heat transfer is primarily via conduction. **Conduction** is thus commonly the limiting and controlling factor in cooling and solidification of intrusions and magma chambers. Here, the focus is conduction. In what follows, a simple version of the cooling and solidification of a sheet-like intrusion (dike, inclined sheet, or a sill) is presented.

Solidification is here the phase transition whereby magma turns into solid (igneous) rock; solidification is also referred to as freezing. The theory of solidification of intrusions presented here is the simplest possible and is based on the general theory of heat conduction in solids (Carslaw and Jaeger, 1959).

Consider a sheet (e.g. a dike or a sill) located in a crustal segment. **Half the thickness** of the sheet is denoted by u, so that the **total thickness** is $2u = \Delta u$. In the analysis it is assumed that the sheet is located at a depth in the crust that is large in comparison with the sheet thickness. Thus, for a 4-m-thick vertical dike, for example, the analysis would be accurate if made, say, some tens of metres below the surface, but not a few metres below the surface. The magma in the sheet is initially at the temperature T_i, which for basaltic magma would commonly be between 1100 °C and 1400 °C. The initial temperature of the host rock is assumed 0 °C, which is not far from common host-rock temperatures at shallow depths. We can then calculate the **temperature T** within the sheet at any given time t since the sheet emplacement took place and at any **distance x** from the mid-plane or centre of the sheet using the formula (Carslaw and Jaeger, 1959):

$$T = \tfrac{1}{2}T_i \left(erf\, \frac{u-x}{2(\kappa t)^{1/2}} + erf\, \frac{u+x}{2(\kappa t)^{1/2}} \right) \tag{7.18}$$

Here $erf\, w$ is the **error function**, which is defined as:

$$erf\, w = \frac{2}{\sqrt{\pi}} \int_0^w e^{-z^2}\, dz \tag{7.19}$$

and κ is the **thermal diffusivity**, given by the formula:

$$\kappa = \frac{K_t}{\rho_r c} \tag{7.20}$$

Here K_t is the thermal conductivity of the host rock and ρ_r its density, and c is the specific heat capacity. The new concepts introduced in Eqs. (7.18) – (7.20) will now be explained further.

The **error function ($erf\, w$)** given by Eq. (7.19) yields the probability that a given value or quantity will be within given limits, provided the measurements of that particular value follow a normal (Gaussian) curve or distribution. The error function occurs in statistics, as well as in physics and engineering problems. For arbitrary values of w, the error function (Eq. (7.19)) is the area under the curve defined by the equation, from the origin 0 to the value w. The function $erf\, w$ can only be evaluated numerically, and is usually obtained, for various values of w, from tables.

The **thermal diffusivity κ** is a measure of the capacity or ability of a material to conduct, in proportion to its ability to store, thermal energy. More specifically, thermal diffusivity has the unit of $m^2\, s^{-1}$ and is the quantity of heat passing through a unit area (m^2) per unit time (s^{-1}) divided by the product of material density and specific heat capacity. Diffusion of any kind is driven by a gradient; thermal diffusion is, for example, driven by a temperature gradient. Thermal diffusivity is thus also a measure of how rapidly the temperature difference in a given body approaches zero, that is, how quickly the body reaches thermal equilibrium.

The **thermal conductivity K_t** or heat conductivity is a measure of the ability of the material to conduct heat. More specifically, thermal conductivity measures the rate of thermal energy transfer through a unit area of a body and per unit temperature gradient. Its unit is $J\,s^{-1}\,m^{-1}\,K^{-1}$ or $W\,m^{-1}\,K^{-1}$, where J is the energy unit joule, s is second, m is metre, K is kelvin, and W is the power unit watt ($W = J\,s^{-1}$). Thus, if p is the rate of thermal energy transfer per unit area A per unit temperature difference dT/dx, where x is a coordinate, then the thermal conductivity K_t may be given by the formula:

$$K_t = \frac{p}{A\frac{dT}{dx}} = \frac{p}{A}\frac{dx}{dT} \tag{7.21}$$

The **specific heat capacity c** or specific thermal capacity is a measure of the quantity of heat that must be supplied to a unit mass of substance to raise its temperature by one degree. The unit of specific heat capacity is joule per kilogram per kelvin, or $J\,kg^{-1}\,K^{-1}$.

Coming back to the sheet emplacement, it follows from Eq. (7.18) that the temperature T_c at the contact between the sheet and the host rock is given by:

$$T_c = \tfrac{1}{2}T erf \frac{u}{(\kappa t)^{1/2}} \tag{7.22}$$

At the exact time of sheet emplacement, $t = 0$, the denominator in Eq. (7.22) becomes zero and the value of the fraction is thus infinite (∞). Since $erf(\infty) = 1$ it follows from Eq. (7.22) that immediately at emplacement the contact temperature $T_c = \tfrac{1}{2}T_i$ and subsequently falls slowly with time.

The conclusion that the contact temperature is half the initial magma temperature is based on several assumptions, not all of which are necessarily valid under the various conditions in the crust. The main assumptions are:

- When the sheet is emplaced the **host-rock temperature is 0 °C**. This may be a reasonable assumption close to the surface, and to certain depths in active rift zones, particularly in highly porous and fractured rocks where water circulation keeps the rock temperature low (similar to the groundwater temperature, commonly 5–20 °C). However, earlier intrusions and shallow magma chambers may bring the host-rock temperature to much higher values, as high as a few hundred degrees at the depths of several kilometres. A high host-rock temperature increases the sheet host-rock contact temperature and slows the rate of cooling and, therefore, increases the time for sheet solidification.
- The **size of the sheet** does not affect the contact temperature. However, in large sheets such as thick sills – some of which reach thicknesses of hundreds of metres, even kilometres, and act as magma chambers (Chapter 5) – large-scale convection is thought to be common. Such convection may contribute significantly to raising the contact temperature, perhaps to melting temperatures of the host rock, as many have suggested. Anatexis or partial melting of the host rock at the contacts with

large sills are likely to be common, and is one mechanism by which shallow magma chambers change their shape to become more smooth or close to ideal ellipsoids (Fig. 5.11).

- The **contacts are smooth**. While this is commonly the case, many contacts, particularly those of large sheets, are initially irregular. Where the contact projects into the sheet, that is into the magma, partial or total melting is most likely to occur. By contrast, where the contact projects into the host rock, as a sort of dike-let or an apophyse, the magma is likely to solidify quickly and little or no melting takes place. As a consequence, the contact tends to become smoother with time because of the joint effect of melting host-rock parts that project into the magma and rapid solidification of magma parts that project into the host rock (Fig. 5.11).

This last point, that of smoothing of originally irregular sheet (or magma-chamber) contacts, has been analysed by Jaeger (1961). The results suggest that at a **corner** where the magma in the dike is bounded by two plane surfaces of the host rock at an angle α (Fig. 7.30), the contact temperature T_c becomes:

$$T_c = \frac{\alpha T_i}{2\pi} \tag{7.23}$$

where, as before, T_i is the initial temperature of the magma in the dike. For a flat contact between the dike and the host rock, $\alpha = \pi$, and Eq. (7.23) yields $T_c = \frac{1}{2}T_i$, as we got before. However, if at the corners of a rectangular sill (Fig. 7.30), $\alpha = 90^\circ = \pi/2$,

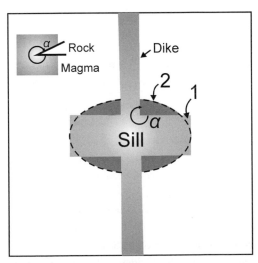

Fig. 7.30 At the contact between a dike and a sill, the latter being a potential magma chamber, the magma is bounded by two plane surfaces at an angle a. Where the host rock projects into the magma, as illustrated on the inset figure, partial or complete melting is likely to occur. By contrast, magma solidifies most rapidly where it projects into the host rock. The initial angular contact between the dike and the sill (stage 1) gradually becomes more smooth or, here, elliptical (stage 2). For the same reason, irregular contacts between magma chambers and their host rocks gradually become smooth (Fig. 5.11). Modified from Gudmundsson (1990).

in which case Eq. (7.23) gives $T_c = \frac{1}{4}T_i$, a much lower contact temperature is obtained, and thus there is no host-rock melting. By contrast, at the upper contact between the sill and dike injected from it (Fig. 7.30), $\alpha = 270° = 3\pi/2$, in which case Eq. (7.23) gives $T_c = \frac{3}{4}T_i$, or a much higher contact temperature and most certainly partial melting of the host rock. Thus, for static conditions of this type where the dike has irregular contacts, partial melting of the host rock is likely to occur in jags that project into the dike and solidification of the magma is likely to be most rapid in apophyses that project into the host rock. The general effect is to make originally **irregular contacts** gradually more **smooth**.

The time that it takes the dike to solidify, that is, to change from a fluid state or phase to a solid state, can also be calculated. To do so, we must consider the (specific) **latent heat of solidification**. Latent heat is the quantity of heat absorbed or released (or transformed) when, at a constant temperature, a substance changes its state. The name derives from the fact that the energy involved in the change of state of matter is referred to as latent or 'hidden' energy, that is, as latent heat. The change of state is also named **phase transition**. Latent heat or energy is, for example, the heat absorbed or released during a change from solid to liquid, or from liquid to gas. The **specific latent heat** L measures the quantity of energy as heat needed for a complete phase transition or change of state of a unit of mass of the substance and has the unit of J kg^{-1}.

Proceeding as above, and taking the specific latent heat into consideration, **the time t_s for the solidification** of a sheet is given by (Carslaw and Jaeger, 1959; Jaeger, 1964):

$$t_s = \frac{u^2}{4\kappa\lambda^2} \tag{7.24}$$

Here λ is the root or solution of the equation (Carslaw and Jaeger, 1959):

$$\lambda(1 + erf\lambda)e^{\lambda^2} = \frac{cT_m}{L\pi^{1/2}} \tag{7.25}$$

where again c is the specific heat capacity of the host rock, T_m is the melting-point temperature of the magma in the dike, and L is the specific latent heat of solidification. Tables of λ for various ranges of solidification of magma (felsic, intermediate, mafic) and different specific latent heat values are provided by Jaeger (1957) and specific heat capacities and other thermal properties of rocks are given by Robertson (1988) and Eppelbaum et al. (2014).

Using Eqs. (7.22) and (7.23), the contact temperature at the time of sheet emplacement can be calculated and, from Eq. (7.24), the time it takes for the sheet to solidify. As indicated above, these equations make various assumptions which may be well or less-well justified depending on thermal and other conditions that prevail in the host rock at the time and location of the sheet emplacement. Generally, however, these equations can be used for rough **estimates** of the contact temperatures and the **time of solidification** of sills, inclined sheets, and dikes, including feeders to volcanic eruptions.

7.8 Summary

- Dikes and other sheet intrusions are formed when the excess pressure in a chamber or a reservoir becomes so high that the magma-chamber walls or, normally, the roof ruptures, and magma flows into the resulting fracture. Dikes, inclined sheets, and sills are extension fractures driven open (formed) by the magma overpressure. The general conditions for chamber rupture and dike initiation is that the excess pressure in the chamber or reservoir reaches the tensile strength of the host rock. The *in situ* tensile strength of rocks is low: the measured range is 0.5–9 MPa, and the most common values are 2–4 MPa. The *in situ* tensile strength of solid rocks is thus close to being a constant.
- Rupturing of the host rock and the propagation of a dike generates earthquakes. Most of the resulting earthquakes are the result of slip on existing fractures in the host rock as the magma pressure in the dike induces shear stresses that cause slip on favourably oriented fractures. The dike propagation thus induces an earthquake swarm, the migration of which through the host rock can be used to monitor the propagation of the dike.
- The selected path of a dike depends on the local mechanical properties and stress fields of the layers that constitute the host volcano. Dike paths in basaltic edifices have generally simpler geometries than those in stratovolcanoes. In the latter, dike deflection into sills or inclined sheets are common, resulting in an irregular path. Some dikes also use steeply dipping faults as channels for parts of their paths. The overall path of all dikes, however, is perpendicular to the general direction of the minimum compressive principal stress σ_3.
- Many dikes and sheets become arrested on their paths and never reach the surface to supply magma to eruptions. They are thus non-feeders. In every long-lived polygenetic volcano (stratovolcano or basaltic edifice) the majority of dikes and sheets become arrested, most commonly at contacts between mechanically dissimilar layers in the volcano. General considerations suggest that arrested dikes – non-feeders – constitute a higher proportion of injected dikes in stratovolcanoes than in basaltic edifices. Few systematic studies, however, have been made as to the exact proportion of feeders versus non-feeders in different tectonic regimes and volcano settings.
- There are three basic mechanisms by which dikes and sheets become arrested (halted or stopped) on their paths. The first is the Cook–Gordon delamination mechanism, whereby a weak contact opens ahead of the dike that arrests the dike or deflects it into a sill. The second is the stress-barrier mechanism, whereby the dike meets a layer with an unfavourable local stress field. For instance, if a vertically propagating dike meets a layer that favours sill emplacement, the dike either changes into a sill or stops altogether. The third is the elastic-mismatch mechanism, where the layer ahead of (above) a dike tip is much stiffer than the layer below the contact, hosting the dike tip. The result is either a deflection of the dike to form a sill along the contact between the mismatching layers or dike-tip arrest. All the three mechanisms can operate together, but the Cook–Gordon mechanism is presumably primarily common at shallow crustal depths whereas the other two can operate at any depth.

- Ideally, the surface stresses and displacements induced by a dike close to the surface peak (form 'ridges') to the sides of the projection of the dike tip onto the surface, and reach a minimum (form a 'valley' or 'depression') right above the dike tip. These ideal stress and displacement fields apply to vertical dikes in a homogeneous isotropic crustal segment (an elastic half-space). In a real layered crustal segment, the stress and displacement peaks become modified in various ways. In particular, the location of the stress peaks may be shifted laterally, usually away from the dike, and the peak magnitudes (sizes) may increase or decrease depending on the layering. It is thus not a simple matter to relate the surface stresses and displacements either to the depth or the magma overpressure of the associated dike.

- When the dike is close to the surface, it may induce surface fractures. Some of these may be small normal faults, whereas the main fractures ahead of the dike tip are extension fractures. These fractures subsequently become part of the dike-fracture in the case of a feeder-dike.

- The effusion or volumetric flow rate up through a feeder-dike (a volcanic fissure at the surface) depends on the aperture or opening or thickness of the dike-fracture to the third power. This relationship, which applies to flow of any fluid through any type of fracture, is referred to as the cubic law. Since the effusion rates of fissure eruptions, as well as the fissure length or strike-dimension, can usually be measured, the unknown opening or aperture of the fissure (the feeder-dike thickness) can be estimated from the cubic law.

- The time it takes sheet intrusions to solidify depends on their thickness to the second power. In commonly used equations for the solidification of sheets it is assumed that the temperature of the host rock at the time of sheet emplacement is 0 °C and that the heat transfer is primarily through conduction. These are generally reasonable assumptions. Rapid solidification of irregularities (magma fingers, apophyses) that project into the host rock, and melting of irregularities (host-rock fingers) that project into the magma are among the main reasons why magma-chamber margins or contacts with the host rock are not highly irregular but rather smooth.

7.9 Main Symbols Used

A	area
a	half length (strike-dimension) of a crack
c	specific heat capacity
E	Young's modulus (modulus of elasticity)
E_1	plane-strain Young's modulus of layer 1
E_2	plane-strain Young's modulus of layer 2
e	2.7183
$erf\,w$	the error function
G_{Ic}	critical strain energy release rate for mode I crack, material toughness

G_I	strain energy release rate during the extension of a mode I crack
G_{II}	strain energy release rate during the extension of a mode II crack
G_{III}	strain energy release rate during the extension of a mode III crack
g	acceleration due to gravity
h	height or dip-dimension of a dike (measured up from its magma source)
K_t	thermal conductivity
K_I	stress-intensity factor for a mode I crack
K_{II}	stress-intensity factor for a mode II crack
K_{III}	stress-intensity factor for a mode III crack
K_{Ic}	plane-strain fracture toughness of a mode I crack
L	strike-dimension of a dike or sheet, equal to $2a$
p_e	fluid excess pressure
p_l	lithostatic pressure (or stress)
p_o	fluid overpressure
Q	volumetric (effusion) flow rate of magma
T	temperature
T_c	contact temperature
T_i	initial temperature in a sheet
t	time since sheet emplacement
T_0	tensile strength
t_s	time of solidification
u	normal opening displacement of a sheet, half the thickness of a sheet or dike
Δu	maximum aperture (total opening) or thickness of a sheet
Δu_I	maximum aperture (total opening) or thickness of a sheet modelled as a mode I crack
W	dimension perpendicular to fluid (magma) flow; length of a volcanic fissure
x	coordinate axis, horizontal coordinate
z	vertical coordinate, positive down from the Earth's surface
α	dip of an inclined sheet (Eq. (7.16))
α	angle between host-rock surfaces or walls in contact with magma (Eq. (7.23))
α_D	Dundurs uniaxial (extensional) elastic-mismatch parameter
κ	thermal diffusivity
λ	root (solution) of Eq. (7.25)
μ_m	dynamic or absolute viscosity of magma
v	Poisson's ratio
π	$3.1417\ldots.$
ρ_m	magma density
ρ_r	rock or crustal density
σ_d	differential stress $(\sigma_1 - \sigma_3)$
σ_3	minimum compressive (maximum tensile) principal stress

7.10 Worked Examples

Example 7.1

Problem

Dike propagation is driven by magmatic overpressure. Use the information and equations in this chapter to answer the following questions:

(a) If the excess magma pressure at the time of chamber/reservoir rupture and dike injection into the roof is equal to the typical *in situ* tensile strength, that is, 2–4 MPa, how can the overpressure in a dike, as calculated from its aspect (length–thickness) ratio, reach 10–20 MPa closer to or at the surface?

(b) At what depth in a typical volcano or volcanic zone would the magmatic overpressure in a dike reach its maximum value? Discuss your results in view of the idea of dikes being arrested or deflected into sills at 'levels of neutral buoyancy'.

Solution

(a) From Eq. (7.5) the overpressure in a dike in relation to its height h above the source magma chamber/reservoir is given by:

$$p_o = p_e + (\rho_r - \rho_m)gh + \sigma_d$$

This equation shows that the overpressure p_o depends partly on the excess pressure in the chamber p_e, and also on the buoyancy term $(\rho_r - \rho_m)gh$, in addition to the stress difference σ_d. Thus, even if the initial driving pressure of the dike is only the excess pressure, as the dike propagates vertically, its dip-dimension or height h increases and so does the buoyancy contribution to the overpressure so long as the average density of the rock layers through which the dike propagates is greater than the density of the magma. Consider, for example, a basaltic feeder-dike propagating vertically through 10 km of host rock with an average density of 2800 kg m^{-3}. If the average magma density is 2700 kg m^{-3}, the buoyancy term in Eq. (7.5) would yield:

$$(\rho_r - \rho_m)gh = (2800 \text{ kg m}^{-3} - 2700 \text{ kg m}^{-3}) \times 9.81 \text{ m s}^{-2} \times 10\,000 \text{ m}$$
$$= 9.8 \times 10^6 \text{ Pa} = 9.8 \text{ MPa}$$

Thus, if we add to this $p_e = 3$ MPa we have close to 13 MPa overpressure in the dike at this crustal depth. Furthermore, during extension, as is common during volcanotectonic episodes, σ_d may be of the order of several mega-pascals, in which case the overpressure in the dike at a certain crustal level could reach about 15 MPa.

(b) Generally, the density of crustal layers increases with depth. The density of individual rock layers or units varies much, so that at any crustal depth we may find a low-density layer, that is, a layer with much lower density than many of the layers above it in the pile. Nevertheless, statistically the density of layers increases with depth, partly because of compaction (particularly of pyroclastic and sedimentary layers) and filling of their pore spaces (e.g. vesicles and joints in the lava flows) with secondary minerals. It follows that, generally, the buoyancy term in Eq. (7.5) may be positive below a certain crustal depth, namely the depth where the magma density equals the density of the crustal layers, and negative above that depth.

For felsic magma, say with a density of 2200 kg m^{-3}, the buoyancy term in Eq. (7.5) may be positive right up to the Earth's surface. For mafic magmas, say with densities of 2700–2800 kg m^{-3}, the buoyancy term becomes zero (neutral) and then negative at certain crustal depths. Typical rock densities in the uppermost few kilometres of the crust are 2300–2500 kg m^{-3}, and in many sedimentary basins less than this. Thus, no typical basaltic eruptions would occur anywhere in the world if dikes were always arrested or deflected into sills at a 'level of neutral buoyancy'. Since dike-fed basaltic eruptions are the most common on the planet, it is clear that this mechanism generally does not operate. In fact, there is no reason why it should. From Eq. (7.5) the highest overpressure in a basaltic dike is reached at the level of neutral buoyancy, and this overpressure is, theoretically, normally high enough to propagate the dike to the surface. If the dike does not make it to the surface, it is thus not because of neutral buoyancy, but because of the operation of one or all of the three mechanisms of dike arrest discussed in this chapter and amply supported by direct field observations.

Example 7.2

Problem

The surface expression of a feeder-dike injected from a magma chamber at 2 km depth below the surface of the associated volcano is a 3-km-long volcanic fissure. For the uppermost 2 km of the crust, the Young's modulus is 10 GPa and the Poisson's ratio is 0.25. The estimated magmatic overpressure in the dike at the time of emplacement is 3 MPa. Calculate the opening or aperture of the volcanic fissure (roughly the thickness of the dike). If the dike is basaltic, what would be the main source of the overpressure?

Solution

Since the dike is a feeder, we can use the through-crack model (Eq. (7.13)), which yields:

$$\Delta u_I = \frac{2p_o(1-v^2)L}{E} = \frac{2 \times 3 \times 10^6 \text{ Pa} \times (1-0.25^2) \times 3000 \text{ m}}{1 \times 10^{10} \text{ Pa}} = 1.7 \text{ m}$$

This is a very common thickness of dikes in local sheet swarms associated with shallow magma chambers, such as in Iceland and elsewhere (Fig. 2.18). For a typical basaltic magma, the buoyancy term in Eq. (7.5) would be zero or, normally, somewhat negative. Thus, the main source of the magmatic overpressure in this case would be the excess pressure p_e in the magma chamber.

Example 7.3

Problem

A basaltic feeder-dike injected from a sill-like magma chamber at the depth of 2 km forms a 1.5-km-long volcanic fissure at the surface. The maximum aperture or opening of the fissure, as determined by measurements, is 1 m. The magma viscosity is 100 Pa s, its density is 2650 kg m^{-3} whereas the estimated average density of the uppermost 2 km of the host rock, between the surface and the top of the shallow magma chamber, is 2550 kg m^{-3}. Assume the typical *in situ* tensile strength of 3 MPa for the host rock of the magma chamber and the dike to be vertical. Calculate the volumetric flow or effusion rate Q_z at the beginning of the eruption. Comment on the implication of the densities of the magma and the host rock for the effusion rate.

Solution

Since the dike is assumed to be vertical we can use Eq. (7.15). When using that equation we assume that the excess pressure in the chamber at the time of rupture and dike injection is equal to the in situ tensile strength of the host rock in the roof of the chamber, that is, 3 MPa. From Eq. (7.15):

$$Q_z = \frac{\Delta u^3 W}{12\mu_m}\left[(\rho_r - \rho_m)\, g - \frac{\partial p_e}{\partial z}\right] = \frac{(1\ \text{m})^3 \times 1500\ \text{m}}{12 \times 100\ \text{Pa s}}$$

$$\left[(2550\ \text{kg m}^{-3} - 2650\ \text{kg m}^{-3}) \times 9.81\ \text{m s}^{-2} + \frac{3 \times 10^6\ \text{Pa}}{2000\ \text{m}}\right] = 650\ \text{m}^3\text{s}^{-1}$$

The calculation method and the results may be clarified further as follows:

- The sign before the last term, the pressure gradient, is plus because the z-coordinate is positive downwards, from the surface of the Earth, whereas the magma flow in the dike is positive upwards, towards the surface. The gradient is thus negative, here -1500 Pa m^{-1}, so we have $-- = +$, that is, the two minuses become plus.
- Because the density of the host rock between the upper part of the chamber and the surface is less than that of the basaltic magma, the buoyancy term in Eq. (7.15) is negative. As discussed above, it is common that the buoyancy term for very shallow magma chambers is, as regards typical basaltic magmas, either neutral (equal density of rock and magma) or negative (magma denser than the rock). This means that the buoyancy term decreases the overpressure available to drive the magma to the surface. More specifically, the only positive contribution to the overpressure or driving pressure of the magma on its path to the surface is then the excess pressure in the chamber. This is one reason why basaltic eruptions from shallow magma chambers are normally not very powerful.
- The calculated volumetric flow rate applies only to the earliest stages of the eruption. As soon as the chamber ruptures and dike propagation towards the surface begins, the volume of magma in the chamber starts to fall, and so does the excess pressure. While an inflow of magma from the deeper source reservoir may occur at the same time, that inflow is normally at

a much lower rate than the outflow from the shallow chamber during dike injection and eruption, and thus does not significantly affect the most common inflation–deflation–eruption scenarios. The calculated maximum effusion rate, 650 m^3 s^{-1}, is similar to that estimated in common small basaltic eruptions in Icelandic central volcanoes (e.g. Galindo and Gudmundsson, 2012).

Example 7.4

Problem

A basaltic dike is emplaced in a rift zone. The initial temperature of the magma in the dike is 1200 °C and the range of solidification is taken as 1200–1000 °C (Williams and McBirney, 1979). The latent heat of solidification L for basaltic magma is about 0.42 MJ kg^{-1} (Jaeger, 1968; Fukuyama, 1985). For this value of L and the above range of solidification, the value of the root λ in Eqs. (7.24) and (7.25) is 0.349 (Jaeger, 1957). For basalts, the thermal conductivity K_t is in the range of 1.3–2.9 W m^{-1} K^{-1}, with an average of 2.1 W m^{-1} K^{-1} (Oxburgh, 1980), the specific heat capacity is around 850 J kg^{-1} K^{-1} (Robertson, 1988), and the rock density for the uppermost several kilometres of the crust may be taken as 2700 kg m^{-3} (Gudmundsson, 2011a). Using these data, calculate the time of solidification of a dike as a function of its thickness in metres.

Solution

In order to use Eq. (7.24) we must first estimate the thermal diffusivity κ. From Eq. (7.20):

$$\kappa = \frac{K_t}{\rho_r c} = \frac{2.1 \text{ Js}^{-1}\text{m}^{-1}\text{K}^{-1}}{2700 \text{ kgm}^{-3} \times 850 \text{ Jkg}^{-1}\text{K}^{-1}} = 9.1 \times 10^{-7} \text{ m}^2\text{s}^{-1}$$

where we have used the unit J s^{-1} for W (watt) in connection with thermal conductivity. Using this value in addition to those above, Eq. (7.24) gives the time for solidification in seconds t_s as:

$$t_s = \frac{u^2}{4\kappa\lambda^2} = \frac{u^2}{4 \times 9.1 \times 10^{-7} \text{ m}^2\text{s}^{-1} \times 0.349^2} = 2.3 \times 10^6 u^2 \tag{7.26}$$

Since this is a useful approximate relation for the time of solidification in seconds of a basaltic sheet of half thickness u in metres, we number the equation. Perhaps a more appropriate time unit here would be days, in which case we obtain the approximate relation:

$$t_d = 26.6 \, u^2 \tag{7.27}$$

where t_d is the approximate time in days for solidification of a sheet of half thickness u in metres. For example, if an emplaced basaltic dike is 1 m thick, it takes approximately 6.6 days for it to solidify, that is, for the magma to cool down from an initial temperature of 1200 °C to 1000 °C. Similarly, for a 4-m-thick dike, the time for solidification would be about 106 days.

Example 7.5

Problem

The range of solidification of basaltic magma used in Example 7.4 is only one of several possibilities. The temperatures of basaltic magmas may be as high as 1300 °C (and of ultramafic magmas as high as 1500 °C). Also, some basaltic magmas may still be somewhat fluid below 1000 °C. In both cases, the temperature range for solidification would be different from that in Example 7.4, and the calculated time would also be different. Here we consider the case where the magma has definitely become brittle and therefore solid rock, namely the temperature at which cooling or columnar joints begin to develop (Fig. 2.8). This temperature is generally regarded as 60% of the initial magma temperature, that is, $0.6T_i$, which for 1200–1300 °C magma would be 720–780 °C. Here we take the temperature for columnar joint initiation as 800 °C and the range of solidification as being from 1200 °C to 800 °C. For this range, the root $\lambda = 0.108$ (Jaeger, 1957). If the latent heat of solidification, the specific heat capacity, and the rock density are the same as in Example, 7.4, calculate the time from dike or sill emplacement to columnar joint initiation as a function of the dike/sill thickness.

Solution

From Eq. (7.20), the thermal diffusivity 9.1×10^{-7} m^2 s^{-1}, as in Example 7.4. Using this result and Eq. (7.24) we obtain the time for the initiation of columnar joints in a dike or sills as:

$$t_s = \frac{u^2}{4\kappa\lambda^2} = \frac{u^2}{4 \times 9.1 \times 10^{-7} \text{ m}^2\text{s}^{-1} \times 0.108^2} = 2.4 \times 10^7 u^2 \qquad (7.28)$$

This result is for the time in seconds. If we recalculate this as time in days, then we get:

$$t_d = 278\, u^2 \qquad (7.29)$$

For example, if the dike is 2 m thick, so that $u = 1$ m, then it takes about 278 days for the first columnar joints to begin to form, and possibly more because we rounded up the temperature at which such joints start to form (to 800 °C from 720 °C for the initial magma temperature of 1200 °C). Similarly, for a 5-m-thick dike (Fig. 2.8), columnar joints would only start to form some 1737 days or about 5 years after the dike emplacement.

7.11 Exercises

7.1 What mechanical type of fracture is a dike?

7.2 How is the dike-fracture formed; what kind of loading generates a dike-fracture?

7.3 State (using an equation) the mechanical condition for magma-chamber rupture and dike or inclined-sheet or sill injection. Explain all the symbols in the equation.

7.4 What is material toughness? Define and provide the unit.

7.5 What is fracture toughness? Define and provide the unit.

7.6 What pressure drives dike propagation while the dike is very close to its magma chamber/reservoir? What is the typical magnitude of this pressure and how do we know?

7.7 Explain, using an equation (explaining all the symbols), how the overpressure or driving pressure in a dike can increase as the dike propagates towards the surface. For a basaltic dike injected from a deep-seated reservoir, at which depth in the crust would the overpressure normally reach its maximum value?

7.8 What main factors control dike (or inclined-sheet) propagation paths?

7.9 What are the three main mechanisms or factors that contribute to dike arrest?

7.10 Illustrate the geometries of typical arrested dike tips. What principal factors determine if the tip is narrow and crack-like or rounded?

7.11 Illustrate and explain the typical surface stresses and displacement induced by a dike approaching the surface.

7.12 What are the main methods for monitoring active dike propagation?

7.13 What is the main geometric parameter that controls the volumetric or effusion rate of a volcanic fissure or feeder-dike?

7.14 Define and provide the unit of thermal conductivity.

7.15 Define and provide the unit of specific heat capacity.

7.16 Why do large magma bodies, such as chambers, tend to have comparatively smooth (in contrast to angular or irregular) boundaries or contacts with the host rock?

7.17 A magma chamber beneath a caldera injects an acid dike that reaches the surface to supply magma to a 1.5-km-long volcanic fissure. The depth of the chamber roof below the surface at the location of the dike injection is at 2 km. The average Young's modulus of the host rock that the dike dissects is 7 GPa and its Poisson's ratio is 0.27. If the maximum opening or aperture of the fissure is 2 m, what was the magmatic overpressure in the feeder-dike when the fissure formed?

7.18 The velocity v of magma flow in a dike or, in general, the velocity of any fluid flow in a fracture, can be obtained from the formula:

$$v = \frac{Q}{A} \tag{7.30}$$

where v is the velocity of the magma (or other fluid) in m s^{-1}, Q is the volumetric flow rate (denoted by Q_z if along the vertical coordinate axis) in m^3 s^{-1}, and A is the cross-sectional area in map view (plan view) of the volcanic fissure (or feeder-dike, or, in general, the fluid-transporting fracture) in m^2. In Example 7.3 the initial volumetric flow or effusion rate is 650 m^3 s^{-1}. Using this and other information in Example 7.3, calculate the magma-flow velocity during the initial stages of the eruption.

7.19 Many dikes and sills are multiple intrusions, that is, formed in many injections of essentially the same magma (Figs. 2.4 and 5.5). Some of those composed of basaltic magma have individual injections (columnar rows) that are, on average, about 4 m thick. Clearly, the columnar joints mostly formed before the next injection took place, but we do not know if all earlier injections had cooled down to host-rock temperatures before the next magma injection (and columnar-row formation) took place. Use the equations developed in this chapter to estimate, crudely, the minimum time between injections of 4-m-thick columnar rocks (forming parts of multiple dikes or sills), if they had developed at least some columnar joints before the next injection took place.

7.20　Among the thickest sills in the world are mafic sills in Antartica, some of which reach thicknesses of about 400 m. Take the range of solidification as being from 1200 °C to 1000 °C and calculate the time it would take for such a sill to solidify – assuming that it does not receive new magma during its solidification.

References and Suggested Reading

Acocella, V., Trippanera, D., 2016. How diking affects the tectonomagmatic evolution of slow spreading plate boundaries: overview and models. *Geosphere*, **12**, 867–883.

Al Shehri, A., Gudmundsson, A., 2018. Modelling of surface stresses and fracturing during dyke emplacement: application to the 2009 episode at Harrat Lunayyir, Saudi Arabia. *Journal of Volcanology and Geothermal Research*, **356**, 278–303.

Bazargan, M., Gudmundsson, A., 2019. Dike-induced stresses and displacements in layered volcanic zones. *Journal of Volcanology and Geothermal Research*, **384**, 189–205.

Becerril, L., Galindo, I., Gudmundsson, A., Morales, J. M., 2013. Depth of origin of magma in eruptions. *Scientific Reports*, **3**, 2762, doi:10.1038/srep02762.

Blundell, S. J., Blundell, K. M., 2006. *Concepts in Thermal Physics*. Oxford: Oxford University Press.

Bunger, A., 2009. *Near-Surface Hydraulic Fracture: Laboratory Experimentation and Modeling of Shallow Hydraulic Fracture Growth*. Saarbrucken: Lambert Academic Publishing.

Carslaw, H., Jaeger, J.C., 1959. *Conduction of Heat in Solids*. Oxford: Oxford University Press.

Eppelbaum, L., Kutasov, I., Pilchin, A., 2014. *Applied Geothermics*. Berlin: Springer Verlag.

Forslund, T., Gudmundsson, A. 1991. Crustal spreading due to dikes and faults in Southwest Iceland. *Journal of Structural Geology*, **13**, 443–457.

Forslund, T., Gudmundsson, A. 1992. Structure of Tertiary and Pleistocene normal faults in Iceland. *Tectonics*, **11**, 57–68.

Freund, L. B., Suresh, S., 2004. *Thin Film Materials: Stress, Defect Formation and Surface Evolution*. Cambridge: Cambridge University Press.

Fukuyama, H., 1985. Heat of fusion of basaltic magma. *Earth and Planetary Science Letters*, **73**, 407–414.

Galindo, I., Gudmundsson, A., 2012. Basaltic feeder dykes in rift zones: geometry, emplacement, and effusion rates. *Natural Hazards and Earth System Sciences*, **12**, 3683–3700.

Geshi, N., Kusumoto, S., Gudmundsson, A., 2010. The geometric difference between non-feeders and feeder dikes. *Geology*, **38**, 195–198.

Geshi, N., Kusumoto, S., Gudmundsson, A., 2012. Effects of mechanical layering of host rocks on dike growth and arrest. *Journal of Volcanology and Geothermal Research*, **223–224**, 74–82.

Gudmundsson, A., 1990. Emplacement of dikes, sills and crustal magma chambers at divergent plate boundaries. *Tectonophysics*, **176**, 257–275.

Gudmundsson, A., 1995. Infrastructure and mechanics of volcanic systems in Iceland. *Journal of Volcanology and Geothermal Research*, **64**, 1–22.

Gudmundsson, A., 2002. Emplacement and arrest of sheets and dykes in central volcanoes. *Journal of Volcanology and Geothermal Research*, **116**, 279–298.

Gudmundsson, A., 2003. Surface stresses associated with arrested dykes in rift zones. *Bulletin of Volcanology*, **65**, 606–619.

Gudmundsson, A., 2006. How local stresses control magma-chamber ruptures, dyke injections, and eruptions in composite volcanoes. *Earth-Science Reviews*, **79**, 1–31.

Gudmundsson, A., 2011a. *Rock Fractures in Geological Processes*. Cambridge: Cambridge University Press.

Gudmundsson, A., 2011b. Deflection of dykes into sills at discontinuities and magma-chamber formation. *Tectonophysics*, **500**, 50–64.

Gudmundsson, A., 2017. *The Glorious Geology of Iceland's Golden Circle*. Berlin: Springer Verlag.

Gudmundsson, A., Philipp, S. L., 2006. How local stress fields prevent volcanic eruptions. *Journal of Volcanology and Geothermal Research*, **158**, 257–268.

Gudmundsson, A., Kusumoto, S., Simmenes, T. H., et al., 2012. Effects of overpressure variations on fracture apertures and fluid transport. *Tectonophysics*, **581**, 220–230.

Gudmundsson, A., Lecoeur, N., Mohajeri, N., Thordarson, T., 2014. Dike emplacement at Bardarbunga, Iceland, induces unusual stress changes, caldera deformation, and earthquakes. *Bulletin of Volcanology*, **76**, 869, doi:10.1007/s00445-014-0869-8.

He, M. Y., Hutchison, J. W., 1989. Crack deflection at an interface between dissimilar elastic materials. *International Journal of Solids and Structures*, **25**, 1053–1067.

He, M. Y., Evans, A. G., Hutchinson, J. W., 1994. Crack deflection at an interface between dissimilar elastic materials: role of residual stresses. *International Journal of Solids and Structures*, **31**, 3443–3455.

Hutchinson, J. W., 1996. Stresses and failure modes in thin films and multilayers. Notes for a Dcamm Course. Technical University of Denmark, Lyngby, pp. 1–45.

Isida, M., 1955. On the tension of a semi-infinite plate with an elliptic hole. *Scientific Papers of the Faculty of Engineering, Tokushima University*, **5**, 75–95.

Jaeger, J. C., 1957. Temperature in the neighbourhood of a cooling intrusive sheet. *American Journal of Science*, **255**, 306–318.

Jaeger, J. C., 1961. The cooling of irregularly shaped igneous bodies. *American Journal of Science*, **259**, 721–734.

Jaeger, J. C., 1964. Thermal effects of intrusions. *Reviews of Geophysics*, **2**, 443–466.

Jaeger, J. C., 1968. Cooling and solidification of igneous rocks. In Hess, H. H. and Poldervaart, A. (eds.), *Basalts, Volume 2*. New York, NY: Interscience, pp. 503–536.

Jaupart, C., Mareschal, J. C., 2011. *Heat Generation and Transport in the Earth*. Cambridge: Cambridge University Press.

Kavanagh, J., Menand, T., Sparks, R. S. J., 2006. An experimental investigation of sill formation and propagation in layered elastic media. *Earth and Planetary Science Letters*, **245**, 799–813.

Kim, J. W., Bhowmick, S., Hermann, I., Lawn, B. R., 2006. Transverse fracture of brittle bilayers: relevance to failure of all-ceramic dental crowns. *Journal of Biomedical Materials Research*, **79B**, 58–65.

Kusumoto, S., Gudmundsson, A., 2014. Displacement and stress fields around rock fractures opened by irregular overpressure variations. *Frontiers in Earth Science*, **2**, doi:10.3389/feart.2014.00007.

Kusumoto, S., Geshi, N., Gudmundsson, A., 2013a. Inverse modeling for estimating fluid-overpressure distributions and stress intensity factors from arbitrary open-fracture geometry. *Journal of Structural Geology*, **46**, 92–98.

Kusumoto, S., Geshi, N., Gudmundsson, A., 2013b. Aspect ratios and magma overpressure of non-feeder dikes observed in the Miyakejima volcano (Japan), and fracture toughness of its upper part. *Geophysical Research Letters*, **40**, doi.org/10.1002/grl.50284.

Lockwood, J. P., Hazlett, R. W., 2010. *Volcanoes: Global Perspectives*. London: Wiley-Blackwell.

Macdonald, G. A., 1972. *Volcanoes*. Upper Saddle River, NJ: Prentice Hall.

Mastin, L. G., Pollard, D. D., 1988. Surface deformation and shallow dike intrusion processes at Inyo Craters, Long Valley, California. *Journal of Geophysical Research*, **93**, 13 221–13 235.

Menand, T., 2008. The mechanics and dynamics of sills in layered elastic rocks and their implications for the growth of laccoliths and other igneous complexes. *Earth and Planetary Science Letters*, **267**, 93–99.

Oxburgh, E. R., 1980. Heat flow and magma genesis. In Hargraves, R.B. (ed.), *Physics of Magmatic Processes*. Princeton, NJ: Princeton University Press, pp. 161–199.

Philipp, S. L., 2008. Geometry and growth of gypsum veins in mudstones at Watchet, Somerset, SW England. *Geological Magazine*, **145**, 831–844.

Philipp, S. L., 2012. Fluid overpressure estimates from the aspect ratios of mineral veins. *Tectonophysics*, **581**, 35–47.

Pollard, D. D., Holzhausen, G., 1979. On the mechanical interaction between a fluid-filled fracture and the earth's surface. *Tectonophysics*, **53**, 27–57.

Pollard, D. D., Delaney, P. T., Duffield, W. A., Endo, E. T., Okamura, A. T., 1983. Surface deformation in volcanic rift zones. *Tectonophysics*, **94**, 541–584.

Robertson, E. C., 1988. Thermal properties of rocks. *US Geological Survey, Open-File Report* **88–441**, 1–106.

Sun, C. T., Jin, Z. H., 2011. *Fracture Mechanics*. New York, NY: Academic Press.

Tibaldi, A., 2015. Structure of volcano plumbing systems: A review of multi-parametric effects. *Journal of Volcanology and Geothermal Research*, **298**, 85–135.

Trippanera, D., Ruch, J., Acocella, V., Rivalta, E., 2015. Experiments of dike-induced deformation: insights on the long-term evolution of divergent plate boundaries. *Journal of Geophysical Research*, **120**, 6913–6942.

Valko, P., Economides, M. J., 1995. *Hydraulic Fracture Mechanics*. New York, NY: Wiley.

Walker, G. P. L., 1965. Some aspects of Quaternary volcanism in Iceland. *Quaternary Journal of the Geological Society*, **49**, 25–40.

Warpinski, H. R. 1985. Measurement of width and pressure in a propagating hydraulic fracture. *Journal of the Society of Petroleum Engineers*, February, 46–54.

Williams, H., McBirney, A. R., 1979. *Volcanology*. San Francisco, CA: Freeman.

Xu, W., Jonsson, S., Corbi, F., Rivalta, E., 2016. Graben formation and dike arrest during the 2009 Harrat Lunayyir dike intrusion in Saudi Arabia: insights from InSAR, stress calculations and analog experiments. *Journal of Geophysical Research*, **121**, doi:10.1002/2015JB012505.

Yew, C. H., Weng, X., 2014. *Mechanics of Hydraulic Fracturing*, 2nd edn. Houston, TX: Gulf Publishing.

Dynamics of Volcanic Eruptions

8.1 Aims

When a magma-filled fracture reaches the surface, a volcanic eruption occurs. In Chapters 5 and 7 we have discussed the conditions under which this may happen. Here, the focus is on the likely course of events once an eruption has started. Among the main questions facing scientists and civil authorities during a beginning eruption are: (1) What is the likely size or magnitude of the eruption? (2) What is its likely duration? (3) Is it going to be primarily effusive or explosive or both? All these questions ultimately relate to the hazards and associated risks posed by the particular volcano. For a more reliable assessment of the hazards associated with volcanoes, the frequencies and sizes of their eruptions need to be known and related to a general understanding of the dynamics of eruptions. The main aims of this chapter are to:

- Provide data on and discuss the size distribution of eruptions.
- Discuss the frequencies of eruptions in different types of volcanoes.
- Present a general framework for ordinary and large effusive eruptions.
- Present a general framework for ordinary and large explosive eruptions.
- Provide a framework for understanding magma-transport variation during an eruption.
- Discuss and explain the duration of different types of eruptions.

8.2 Eruption Sizes and Frequencies

How we classify the sizes of eruptions is to some extent arbitrary. As earlier in the book, here eruptions with volumes of 0.1 km^3 or less are regarded as small. Eruptions with volumes in the range 0.1–10 km^3 are regarded as moderate, and eruptions larger than 10 km^3 as large. It could certainly be argued that any eruption exceeding 1 km^3 should be called large. However, given that some eruptions reach volumes of the order of 1000 km^3, I think it is better to classify eruptions of the order of 1 km^3 as moderate.

So how does this grouping into small, moderate, and large eruptions compare with the recorded history of eruption sizes? If we first look at small eruptions, then their number must be very large. There are no complete records available of the number of small eruptions on dry land (and hardly any records at all for eruptions on the sea bottom). Nevertheless, the general relation between the number or frequency of eruptions and the volumes erupted suggests a power-law size or magnitude distribution (Fig. 8.1). This implies that, for any given (not-too-short) period of time, say a decade or century, the number of small eruptions is very large whereas the number of large eruptions is very small.

This general observation as to the relative frequency of small and large eruptions is in agreement with observations. There are perhaps about **60 volcanoes** erupting on average each year, of a total of about **1500 active volcanoes.** (All figures refer to subaerial eruptions – many more occur on the sea bottom, particularly at mid-ocean ridges, but there is little information available on submarine eruptions.) Some 150–160 volcanoes erupt once or more every decade, and some **550 volcanoes** are known to have erupted in historical times, that is, have documented historical eruptions. It is estimated that around 1500 volcanoes have erupted in the Holocene and are then regarded as **active**. Some of these volcanoes have erupted almost **continuously** for hundreds of years – in the case of Stromboli in Italy probably for some 2000 years. At any one time, there are some 20 volcanoes erupting on land.

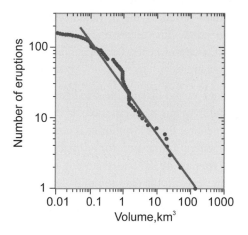

Fig. 8.1 Size distribution of volcanic eruptions in terms of eruptive volumes in cubic kilometres (cf. Table 8.1); modified from Pisarenko and Rodkin (2010). The bi-logarithmic (log–log) plot means that both the cumulative number on the vertical axis (y-axis) and the eruptive volumes on the horizontal axis (x-axis) are given on logarithmic scales (cf. Chapters 2 and 4). The straight line fits much of the distribution reasonably well. Since the only frequency or probability distribution that generates a straight line on a log–log plot is the power law, the fit suggests that the volume distribution of eruptive materials follows approximately this law. There are many methods for testing how well a power law (which forms a part of the larger set of heavy-tailed distributions) fits a distribution such as this one in comparison with other functions, and also to check if more than one straight line (power law) fit the distribution better than a single line (cf. Clauset et al., 2009; Mohajeri and Gudmundsson, 2012; Gudmundsson and Mohajeri, 2013).

The largest instrumentally documented subaerial eruption so far is the 1912 **Novarupta eruption** in Alaska, which produced about 30 km^3 of acid to intermediate tephra (Hildreth and Fierstein, 2012). Calculated as dense magma, this eruptive volume corresponds to some 13–14 km^3 of magma being transported out of the source chamber during the eruption (including the volume of the feeder-dike). The second-largest and much better documented subaerial eruption, also in the twentieth century, was the 1991 **Pinatubo eruption** in the Philippines (Newhall et al., 1997), which produced some 10 km^3 of eruptive materials. This eruptive volume corresponds to about 4 km^3 of dense magma being transported out of the source chamber (the feeder-dike volume is assumed to be included in all these numbers; cf. Chapter 6). Many earthquakes, including large ones of M6–7.8, were associated with these eruptions, both of which were associated with caldera collapses. The caldera collapse associated with the Novarupta eruption occurred in the nearby Katmai Volcano, 10 km from the eruption site. The caldera has a diameter of 3–4 km and a volume of 5–6 km^3 (much smaller than the eruptive volume). The caldera associated with the Pinatubo eruption has a diameter of 2.5 km. Both these eruptions are large based on the above classification and both are explosive. In terms of magma volume leaving the chamber, however, the 1991 Pinatubo eruption would be classified as moderate. No large subaerial effusive eruptions occurred in the twentieth century.

Going further back than the twentieth century, there have been many large subaerial eruptions in historical time. The largest one is presumably the 1815 **Tambora eruption** in Indonesia (Self et al., 1984; Rosi et al., 2003). The eruption produced mainly tephra, the total estimated volume of the eruptive materials being 160–170 km^3. This is regarded as equivalent to about 50 km^3 of dense magma being transported out of the shallow chamber during the eruption. The caldera associated with the eruption is 6–7 km in diameter and has a volume of about 36 km^3 (Fig. 9.11), thus considerably less than the magma volume leaving the chamber in the eruption. Other large historical explosive eruptions include the 1883 **Krakatoa eruption** in Indonesia (about 20 km^3 of tephra), the 934–1085 **Tianchi** (or Baitoushan) **eruption,** the volcano being at the boundary between China and North Korea (some 96–172 km^3 of tephra, or around 30 km^3 of dense magma leaving the chamber), and the 186 (or 230 – the eruption year is unclear) **Taupo** (Hatepe) **eruption** in New Zealand (some 120 km^3 of tephra). The last eruption occurred before humans settled in New Zealand and is thus hardly historical in an ordinary sense, but it had effects that were noticed in other areas such as in China and in the Roman Empire. The largest historical effusive eruptions are presumably the 934 **Eldgja eruption** (lava volume about 20 km^3) and the 1783 **Laki eruption** (lava volume about 15 km^3), both of which occurred in South Iceland. Given that both these lava flows were fed by feeder-dikes, tens of kilometres long, it is likely that the dense magma being transported out of the source reservoir during the Eldgja eruption was 25–35 km^3, and that during the Laki eruption, 18–20 km^3.

The large explosive and effusive eruptions above are much smaller than the largest known eruptions on Earth. The largest of these exceed 1000 km^3, some reaching about **5000 km^3** of eruptive materials, and thus one to two orders of magnitude larger than the largest historical eruptions. The most recent of these very large eruptions took place in Sumatra (Indonesia) some 75 000 years ago at the site of the present-day Lake Toba. The **Toba eruption** produced around 2800 km^3 of eruptive materials and a collapse caldera with major and minor axes of 100 km and 30 km, respectively. The present lake occupies only a part of the original caldera since a resurgent dome forms a large island in the central part of the caldera (Fig. 9.13b). The Toba eruption is the largest known subaerial eruption in the past 25 million years, but 27–28 million years ago an even larger eruption occurred in Colorado (the United States). This is the **La Garita caldera eruption**, which produced eruptive volumes of 4000–5000 km^3, that is, the Fish Canyon Tuff, during which an elongated caldera formed. The caldera, with major and minor axes of 75 km and 35 km, respectively, has dimensions similar to those of Lake Toba. Some effusive eruptions also reach volumes of 4000–5000 km^3, such as several of the Columbia River basalt lava flows, which were erupted 15–16 million years ago.

The bi-logarithmic plot in Fig. 8.1 shows that the great majority of the listed eruptions recorded are less than 1 km^3. In fact, the plot suggests that most of the eruptions are small, that is, between 0.01 km^3 and 0.1 km^3. None of the very large or great eruptions, with eruptive volumes of the order 1000 km^3, are listed here, because none of them is 'historical' in the proper sense. These great eruptions – sometimes referred to as super-eruptions – are very rare; perhaps one eruption every 50 000 years on average (Table 8.1). Even if rare, the very large eruptions are extremely important since they are among the **greatest threats** to the survival of our civilisation and, possibly, mankind. Understanding the conditions for their formation, hence how to forecast them, is one of the main themes of this chapter.

8.3 Magma Transport to the Surface

8.3.1 Cylindrical Conduits

Magma transport to the surface, through a dike or a cylindrical or an elliptical conduit, can be modelled in various ways. Here, the basic models are reviewed and those recommended for use are indicated. The simplest model is a cylindrical or pipe-like vertical conduit where the magma is driven by a fluid-pressure gradient to the surface (Figs. 8.2 and 8.3). The general formula for fluid flow through a circular pipe, in relation to pressure drop or pressure gradient, is referred to as the Hagen–Poiseuille equation, which in its simplest form may be given by:

$$Q = -\frac{\pi R^4}{8\mu_m}\frac{\Delta p}{\Delta z} = -\frac{\pi R^4}{8\mu_m}\frac{\partial p}{\partial z} \tag{8.1}$$

Here, Q is the volumetric flow rate of magma (and gas), R is the radius of the cylindrical conduit, μ_m is the dynamic (absolute) viscosity of the magma, $\Delta p = p_b - p_t$ is the fluid pressure difference between the bottom and top of the conduit, Δz is the height (vertical dimension or dip-dimension) of the conduit, and $\partial p/\partial z$ is the fluid-pressure gradient. In all the calculations here, as earlier (Section 7.6) when dealing with fluid or magma flow through conduits or dikes and inclined sheets, the coordinate z is regarded as positive measured downwards from the Earth's surface (the surface of the volcano). As discussed further below, the pressure gradient $\partial p/\partial z$ is thus negative – the pressure decreasing up

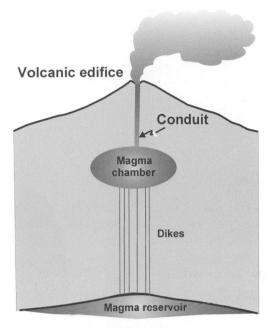

Volcanic edifice

Conduit

Magma chamber

Dikes

Magma reservoir

Fig. 8.2 Some polygenetic (central) volcanoes are supplied with magma during most eruptions through cylindrical (or somewhat elliptical) conduits. These conduits normally connect with a shallow magma chamber, which, in turn, is supplied with magma from a deeper, and normally much larger, reservoir. Drilling into some active elliptical conduits suggests, however, that they are largely composed of dense swarms of dikes (Nakada et al., 2005). Some eroded conduits, exposed as volcanic necks or plugs, are also composed of dense swarms of dikes and sheets (Fig. 8.3).

along the z-axis. Since the terms on the right-hand side of the equations have a minus sign and the pressure gradient itself is negative, it follows that the volumetric flow rate Q becomes a positive quantity.

Equation (8.1) assumes that the behaviour of the crustal segment hosting the conduit, including that of the associated volcano and magma chamber, is rigid. Recall that rigid rock behaviour means that however much fluid pressure or stress is applied to the rock body, there is no change in its internal configuration, that is, there is no strain. Measurements indicate, however, that the crustal segments hosting magma chambers behave as elastic, to a first approximation, so that when the magma pressure changes, in the chamber or in a conduit, the host rock is strained. When the host rock behaves as elastic, then the buoyancy term must be added to the pressure term in Eq. (8.1). The general equation for magma overpressure in a conduit or feeder-dike where the buoyancy term is included is (Chapter 5):

$$p_o = p_e + (\rho_r - \rho_m)gh + \sigma_d \tag{8.2}$$

Here, p_e is the excess pressure in the chamber (or reservoir) at the time of rupture and the start of magma flow up to the surface. The other symbols are ρ_r, which is the average host-rock density, ρ_m, which is the average magma density, g, which is the acceleration due to gravity, and h, which is the dip-dimension (height) of the conduit as measured from its point

Fig. 8.3 Volcanic necks/plugs in Tenerife, Canary Islands. (a) A volcanic plug, some 50 m in diameter, in Anaga, Tenerife. The plug is largely made of breccia which is dissected by many dikes. (b) The 'Cathedral', a plug in the Las Canadas caldera in Tenerife. The plug is about 100 m in diameter and composed primarily of intrusions, dikes, and sheets. Both plugs are somewhat elliptic in map view.

of contact with the magma chamber. The symbol σ_d denotes the differential stress ($\sigma_d = \sigma_1 - \sigma_3$) in the crustal layer (at the crustal level) which the magma in the conduit has reached when examined during its flow towards the surface.

In some conduits there is **continuous magma flow** to the surface from the chamber over long periods of time. Well-known volcanoes where this appears to be the case are Stromboli in Italy and, in particular, volcanoes with continuously active lava lakes such as Erta Ale in Ethiopia (Fig. 3.35), Mount Nyiragongo in the Democratic Republic of the Congo, and Puu Oo (until 2018) in Kilauea, Hawaii. For these volcanoes, so long as they are continuously erupting, no significant excess pressure can build up in the source magma chamber so that p_e is close to zero and does not contribute to the overpressure in Eq. (8.2). But volcanoes with continuously open conduits, and thus continuously erupting, are rare and should be regarded as special cases. Thus, normally, before an eruption occurs excess pressure builds up in the chamber, so that at the time of chamber rupture $p_e > 0$.

For circular (or somewhat elliptical) conduits, the effects of the variation in the differential stress σ_d are less important than is commonly the case for dikes. Also, when considering the volumetric flow rate (the effusion rate) at the surface the potential variation in σ_d is limited. This follows because for an eruption on dry land, particularly in an area undergoing local extension (as applies to many, perhaps most, volcanoes), σ_1 is normally vertical and equal to the atmospheric pressure (about 0.1 MPa or 1 bar), whereas σ_3 is either the same as the vertical stress (isotropic state of stress in the atmosphere where the magma reaches the surface) or zero – and possibly somewhat negative just below the surface. In our calculations for the effusion rates at the surface, we thus normally ignore σ_d, and do so in this chapter. Note, however, that when overpressure is calculated for a dike exposed at significant depth below the original surface of the volcano (and thus in an eroded crustal segment), σ_d may have to be considered for accurate results (Fig. 5.14).

In Eq. (8.1), as normally in this book (and in solid-earth geoscience in general), the vertical coordinate z is regarded as positive downwards, that is, in the direction of increasing crustal depth from the Earth's surface. Since the height or dip-dimension of the conduit is regarded as positive upwards, with its origin at the contact with the source magma chamber, it follows that h in Eq. (8.2) is positive along negative z (cf. Section 7.6, Eq. (7.15)). The magma flow up the conduit (or up a feeder-dike) is driven from higher pressure (in the chamber) to lower pressure (atmospheric pressure at the Earth's surface, for example) so that, as indicated above, the magma-pressure gradient must also be negative. Based on these considerations, and using Eq. (8.2), for a crustal segment that behaves as elastic, Eq. (8.1) for vertical magma transport up through a cylindrical conduit (Fig. 8.3) becomes:

$$Q = -\frac{\pi R^4}{8\mu_m} \frac{\partial}{\partial z} [p_e - (\rho_r - \rho_m)gz] = \frac{\pi R^4}{8\mu_m} \left[(\rho_r - \rho_m)g - \frac{\partial p_e}{\partial z} \right] \qquad (8.3)$$

(Notice the rearrangement of the terms and signs in the final version of Eq. (8.3) and recall from Eqs. (8.1) and (7.15) that for magma flow vertically upwards there will be two minus signs ($- -$) in front of the term $\partial p_e/\partial z$, which thereby becomes a plus sign.) Equation (8.3), sometimes written in somewhat different form, has commonly been used to model magma transport to the surface in cylindrical conduits. Some authors drop the last term, $\partial p_e/\partial z$, in which case the entire driving pressure or overpressure is attributable to the buoyancy term $(\rho_r - \rho_m)g$. But then it is implicitly assumed that the conduit is continuously open to the surface; that no magma excess pressure can build up in the chamber before eruption. As indicated above, an open conduit of this kind is a very special situation and primarily found in lava lakes, which are very few. Also, for basaltic magmas injected from shallow chambers, the buoyancy term is commonly zero or negative, in which case omitting the term $\partial p_e/\partial z$ would make it impossible to explain how the basaltic magma is able to reach the surface (Chapter 7; cf. Example 7.3). Normally, therefore, the equation should include the term $\partial p_e/\partial z$. Approximately cylindrical conduits exist, as exemplified by many necks and plugs (Fig. 8.3). More commonly, however, the conduit is elliptical in shape.

8.3.2 Elliptical Conduits and Dikes

Many conduits are elliptical in plan view, that is, have elliptical horizontal (map-view) cross-sections (Fig. 8.4). The elliptical model can also be used for magma flow in many necks and plugs, but (using elongated ellipses) also for dikes, inclined sheets, and sills – although, as we shall see, the parallel-plate model is more common for dikes, sills, and sheets. An elliptical plan-view cross-section of a conduit is given by:

$$\frac{x^2}{a^2} + \frac{y^2}{b^2} = 1 \tag{8.4}$$

where a and b are the semi-major and semi-minor axes of the ellipse, respectively. By analogy with Eq. (8.1), the volumetric flow rate Q up through a vertical **elliptical conduit** (Fig. 8.4) is given by (White, 2005):

$$Q = -\frac{\pi}{4\mu_m} \frac{(ab)^3}{(a^2 + b^2)} \frac{\partial p_e}{\partial z} \tag{8.5}$$

where $\partial p_e / \partial z$ is the pressure gradient driving the flow. The mean velocity $\bar{q}$ of the flow up through the conduit is then:

$$\bar{q} = \frac{Q}{A} = \frac{Q}{\pi ab} \tag{8.6}$$

where $A = \pi ab$ is the horizontal cross-sectional area of the elliptical conduit. Combining Eqs. (8.5) and (8.6), the mean magma-flow velocity is obtained:

$$\bar{q} = -\frac{a^2 b^2}{4\mu_m(a^2 + b^2)} \frac{\partial p_e}{\partial z} \tag{8.7}$$

where, again, the minus sign is because the flow is along the negative direction of the z-coordinate axis. For the circular conduit discussed above (Eq. (8.1)), $a = b = R$ and the cross-sectional area $A = R^2\pi$, in which case Eq. (8.7) reduces to:

$$\bar{q} - \frac{Q}{R^2\pi} = -\frac{R^2}{8\mu_m} \frac{\partial p_e}{\partial z} \tag{8.8}$$

Equations (8.5) to (8.8), however, apply to a rigid crustal segment, the rock behaviour discussed in connection with Eq. (8.1) above. The crust hosting shallow chambers normally behaves as elastic to a first approximation. For an elastic crust, by analogy with Eq. (8.3), Eq. (8.5) may be rewritten in the form:

$$Q = -\frac{\pi}{4\mu_m} \frac{(ab)^3}{(a^2 + b^2)} \frac{\partial}{\partial z} [p_e - (\rho_r - \rho_m)gz] = \frac{\pi}{4\mu_m} \frac{(ab)^3}{(a^2 + b^2)} \left[(\rho_r - \rho_m)g - \frac{\partial p_e}{\partial z} \right] \tag{8.9}$$

where all the symbols are as defined above. Equation (8.9) reduces to Eq. (8.3) for a circular conduit, where $a = b = R$. Equation (8.9) can be used also for dikes, particularly for feeder-dikes and the associated volcanic fissures. Because most dikes are much longer than they are thick, it implies that the strike-dimension of a feeder-dike is normally many times the size of its aperture at the time of eruption. Also, theoretical considerations and

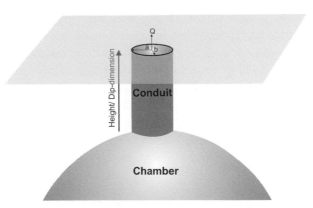

Fig. 8.4   Vertical conduit supplied with magma from a shallow chamber, only the uppermost part of which is seen here. Here a and b denote the semi-axes of the cross-section of the conduit and Q denotes the volumetric flow rate through the cross-section. When $a = b$, the cross-section is circular and the conduit is a vertical cylinder. When $a \neq b$ the cross-section is elliptical and as a becomes many times larger than b, the conduit becomes a volcanic fissure (fed by a vertical feeder-dike). The velocity of flow, $\overline{q}$, is obtained by dividing the volumetric flow Q rate by the cross-sectional area A of the conduit.

measurements of volcanic fissures and dikes indicate that they are commonly approximately flat ellipses where the aperture/thickness varies little along the greater part of the strike-dimension or surface length (Chapters 5 and 7). It follows that a fluid transport up through a dike can be modelled using the parallel-plate model, to which I turn now.

8.3.3 Parallel-Plate Models of Dikes, Inclined Sheets, and Sills

The parallel-plate model derives, like the equations above, from the Navier–Stokes equations and is the basis of the so-called **cubic law** in hydrogeology, volcanology, and related fields. The name 'cubic law' stems from the volumetric flow rate or **effusion rate** being related to the aperture or opening of the fluid-conducting fracture to the third power, that is, to the cube of the aperture. This is the standard model in hydrogeology, and is also the most common model for dikes, inclined sheets, and sills in volcanotectonics.

The cubic law is also the basis for **flow channelling.** More specifically, because the cubic law implies that the volumetric or effusion flow rate depends on the aperture to the third power any slight variations in the aperture or opening of a fluid-conducting fracture can give rise to great variations in the effusion rate. It follows that there is a **channelling** (focusing) of the flow on the parts of the fracture – such as a volcanic fissure – where the aperture is larger than in the neighbouring parts. Where, because of aperture variation, the effusion rate is high in comparison with that of the adjacent parts of the dike/fissure, a crater cone forms (cf. Chapter 9).

As for circular and **elliptical conduits**, the host-rock behaviour is normally modelled in one of two basic ways: either as being rigid or elastic. Both assumptions have been used in volcanotectonics, and the assumption of rigid host rock is common in hydrogeology. However, measurements show that volcanoes inflate and deflate during unrest periods and eruptions, and that this deformation can, to a first approximation, be modelled as (linear) elastic.

Thus, models assuming elastic behaviour are much more realistic and are recommended here. However, since rigid and elastic models are both used in volcanotectonics, here the results of both modelling approaches will be discussed and compared. Of primary consideration will be feeders, either vertical dikes or inclined sheets (Fig. 8.5). But at the end of this section, sills will briefly be considered.

Let us consider first a vertical feeder-dike in a rigid host rock (Fig. 8.6). From the Navier–Stokes equations (Lamb, 1932; Milne-Thompson, 1996) it follows that the volumetric flow or effusion rate Q up through the dike is given by:

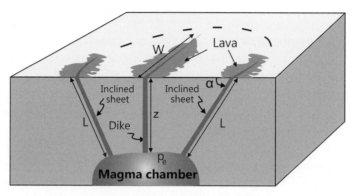

Fig. 8.5   Magma transport from a shallow magma chamber to the surface is primarily through vertical (radial) dikes and inclined sheets. Here a dike and two inclined sheets are shown schematically (cf. Fig. 5.25). The height or dip-dimension of the dike is denoted by z, that of the inclined sheets (dipping at (angle) a) is denoted by L. The length or strike-dimension of the dike is shown here and denoted by W. The excess pressure in the chamber at the time of rupture is denoted by p_e.

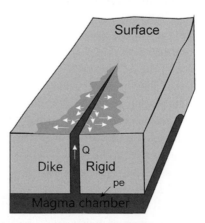

Fig. 8.6   A vertical feeder-dike hosted by rigid rock. Rocks or other solids modelled as rigid do not deform, however much stress or pressure is applied to them. It follows that in this model the rock does not respond to any pressure changes in the magma chamber and there is no buoyancy term contributing to the magmatic overpressure or driving pressure. This model is represented by Eq. (8.10).

$$Q = \frac{\Delta u^3 W}{12\mu_m}\left[\rho_m g - \frac{\partial p_e}{\partial z}\right] \tag{8.10}$$

Here Δu is the opening or aperture of the dike (roughly equal to the thickness of a solidified dike as measured in the field, Chapters 5 and 7) and W is its width perpendicular to the flow direction, that is, the strike-dimension of the dike or the length of the volcanic fissure at the surface. We know that the dip-dimension and the aperture or opening of a dike change with depth in the crust (Fig. 2.22, Chapter 2), but at the surface, the length or strike-dimension of the volcanic fissure that the dike supplies magma to may be used to estimate W. It follows that the dike cross-sectional area perpendicular to the magma flow is $A = \Delta u W$. It is always assumed that the strike-dimension is many times (commonly about 1000 times, Chapter 2) the opening, so that $W \gg \Delta u$. The other symbols in Eq. (8.10) are μ_m and ρ_m, which are the dynamic viscosity and the density of the magma, respectively; g, which is the acceleration due to gravity, and $\partial p_e/\partial z$, which is the vertical pressure gradient in the direction of the magma flow. Appendix F gives data on the densities and viscosities of some common magmas and other crustal fluids.

When the walls of the dike and the source magma chamber are free to deform elastically as fluid is transported towards the surface (Fig. 8.7), the weight of the rock above the chamber must be supported by its internal magma (and gas) pressure in the chamber. Since the host-rock density is normally different from the magma density ρ_m, a buoyancy term becomes added to the excess-pressure gradient. Thus, for a vertical dike in a crustal segment that behaves as elastic, and by analogy with Eq. (8.9), then Eq. (8.10) may be rewritten in the form:

$$Q = \frac{\Delta u^3 W}{12\mu_m}\left[(\rho_r - \rho_m)g - \frac{\partial p_e}{\partial z}\right] \tag{8.11}$$

where all the symbols are as defined above.

Equations (8.10) and (8.11) apply for magma transport up through a vertical feeder-dike. Vertical dikes are most common in regions such as rift zones that are dominated by external

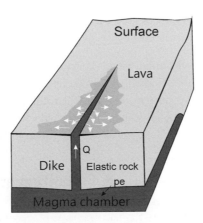

Fig. 8.7   Vertical feeder-dike hosted by elastic rock. Elastic rocks and other solids deform when stress or pressure is applied to them. In this model the rock deforms as the pressure in the chamber and dike changes. It follows that the buoyancy contributes to the magmatic overpressure or driving pressure during dike propagation and eruption. This model is represented by Eqs. (8.11) to (8.13).

tensile forces. For divergent plate boundaries, for example, the external tensile forces derive from the plate pull associated with the spreading vector. Regional dikes, that is, dikes formed in rift zones away from polygenetic (or central) volcanoes (Chapter 2), are controlled by the regional stress field and are most commonly close to vertical (Chapters 3 and 6). By contrast, dikes with an attitude controlled by the local stress fields associated with shallow magma chambers are commonly inclined and accordingly referred to as inclined sheets.

Many shallow magma chambers inject inclined sheets (Fig. 7.5, Chapter 7). For parallel-plate models of inclined sheets and other non-vertical hydrofractures, Eqs. (8.10) and (8.11) must be modified so as to take the dip of the sheet into account. Let us denote the dip-dimension (the path of the magma from its rupture site in the roof of the chamber and, for a feeder-sheet, to the surface) by L and its dip by α (Fig. 8.5). For a crustal segment that behaves as rigid, the volumetric flow rate Q up the inclined sheet is then given by:

$$Q = \frac{\Delta u^3 W}{12\mu_m}\left[\rho_m g \sin\alpha - \frac{\partial p_e}{\partial L}\right] \qquad (8.12)$$

Consider next an inclined sheet located in a crustal segment that behaves as elastic. By analogy with Eq. (8.12), the volumetric flow or effusion rate is then given by:

$$Q = \frac{\Delta u^3 W}{12\mu_m}\left[(\rho_r - \rho_m)g \sin\alpha - \frac{\partial p_e}{\partial L}\right] \qquad (8.13)$$

Inspection of Eqs. (8.10) and (8.11) shows that they are special cases of the more general Eqs. (8.12) and (8.13), namely the cases when the sheet (the dike) is vertical, so that $\alpha = 90°$, in which case $\sin\alpha = 1$.

For fluid transport along a horizontal **sill,** Eqs. (8.12) and (8.13) become modified for the special case where the dip is zero, so that $\alpha = 0°$ and $\sin\alpha = 0$. It follows that the first term in the brackets of Eq. (8.13) drops out. For a sill in the horizontal xy-plane with width W measured along the y-axis and the length L measured along the x-axis, x may be substituted for L to obtain, from Eqs (8.12) and (8.13), the volumetric flow rate through the sill as:

$$Q = -\frac{\Delta u^3 W}{12\mu_m}\frac{\partial p_e}{\partial x} \qquad (8.14)$$

In Eq. (8.14), the only 'force' driving the magma flow along the sill is the excess pressure gradient $\partial p_e/\partial x$. Normally, a sill is supplied with magma through a dike (or, sometimes, an inclined sheet connecting a lower sill with, and supplying magma to, an upper sill). When a vertical dike supplies magma to a horizontal sill (Chapter 5), then the excess pressure in the sill is equal to the overpressure that develops in the dike on its path towards the contact with the sill.

8.4 Duration of Eruptions

Once an eruption has started, supplied with magma through a dike, an inclined sheet, or a circular-elliptical conduit, the volumetric flow rate or effusion rate can be estimated. This estimate can be obtained from direct measurements of the advancing lava flow field and/or of the eruption column (volcanic plume). The measurements may then be compared with the theoretical models and the associated equations provided in Section 8.3. The flow properties of lavas are now reasonably well understood and can be modelled, so as to assess the likely flow direction and velocities (e.g. Kilburn, 2000; Harris, 2013). Similarly, the formation, maintenance, and eventual collapse of eruption columns (volcanic plumes) have been studied extensively in the past decades and are reasonably well understood (Sparks et al., 1997; Clarke et al., 2002; Parfitt and Wilson, 2008; Bonadonna and Costa, 2013; Gonnermann and Manga, 2013; James et al., 2013; Roche et al., 2013).

The effects of an eruption on human society and climate, among other things, depend not only on the volume of eruptive materials and the height of the eruption column, but also on the duration of the eruption. Reliable forecasting of the duration of an eruption, once it has started, is of fundamental importance in hazard studies and risk assessments (Chapter 10). Furthermore, together with the volumetric flow rate, the duration is a measure of the volume of magma available to erupt in the source magma chamber or reservoir (Chapters 6 and 10).

Here, we focus on the duration of an eruption from a shallow magma chamber. The analysis can easily be generalised to include deep-seated magma reservoirs. However, one major point in the analysis is that the volumetric rate of flow of magma from the source reservoir to the shallow magma chamber is assumed to be **much lower** than that of flow out of the chamber during an eruption (Chapter 6). This assumption is supported by observations which show that there is normally deflation and thus **shrinkage** of the shallow magma chamber during an eruption. This implies that magma is leaving the chamber at a higher rate during the eruption than it receives new magma from its source reservoir. In particular, geodetic data indicate that during basaltic eruptions the volumetric flow rate into the source chamber is much lower than the rate of outflow through the feeder-dike (e.g. Stasiuk et al., 1993; Woods and Huppert, 2003). The main exception to this scenario is continuously erupting volcanoes, in particular lava lakes. As discussed above, continuously erupting volcanoes are few and do not alter the general picture of what happens during an ordinary eruption.

A second major point in the analysis is that the magma-chamber behaviour is assumed to be **elastic.** This is the general assumption for exploring and explaining the dynamics of ordinary or common eruptions. The analysis thus does not include the effects of major non-elastic events that may coincide with the eruption. Such events include lateral and vertical collapses and graben subsidences. The effects of these on eruption volumes and durations are here regarded as different mechanisms from those that control ordinary eruptions. In particular, caldera and graben collapses are here proposed as a mechanism for generating large eruptions and are treated accordingly later in this chapter (Sections 8.5 and 8.6).

In an ordinary eruption, a **tiny fraction** of the magma in the chamber flows out before the eruption comes to an end (Chapter 6). The crude sizes of many shallow chambers are known from geophysical, mainly seismic, studies. And even if they are partially molten, it is clear that small eruptions of the order of 0.01–0.1 km^3 as well as moderate eruptions of the order of 1 km^3 are normally very small fractions of the total volume of magma in the chambers. Here, the process that decreases the excess pressure in the chamber during an eruption and, eventually, results in closing of the feeder-dike or feeder-sheet at its contact with the chamber is analysed.

8.4.1 Theory

Consider the **transport of basaltic magma** through a feeder-dike (Fig. 8.8). It is assumed for convenience that the chamber is totally molten and that the average density of the rock hosting the dike (the rock between the surface and the magma chamber) is equal to that of the magma. Also, the analysis assumes that the flow of magma out of the chamber relates to its elastic shrinkage or reduction in volume, rather than to a caldera/graben collapse and piston-like subsidence of the roof. These assumptions can easily be relaxed in a more detailed analysis, whereby the average density of the host rock could be either higher or lower than that of the magma and the reduction in chamber volume may be related to caldera/graben subsidence. Also, the chamber need not be totally molten; the analysis is easily extended to a partially molten chamber or reservoir (Chapter 6). Here, we limit the analysis thus because the main points are changes in the effusion rate and the excess pressure in the chamber itself during the eruption. The **excess-pressure decline**, in turn, results in the gradual **closing of the feeder-dike** at its contact with the chamber, thereby bringing the eruption to an end. The assumption of equal density of magma and host rock implies that the buoyancy term in Eq. (8.11) is zero, that is, $(\rho_r - \rho_m)g = 0$. How quickly the excess pressure in the chamber approaches zero determines the duration of the eruption and, partly, its volume.

During magma transport to the surface along a feeder-dike (or other types of conduits) the excess pressure p_e in the source chamber decreases if one or more of the following is satisfied:

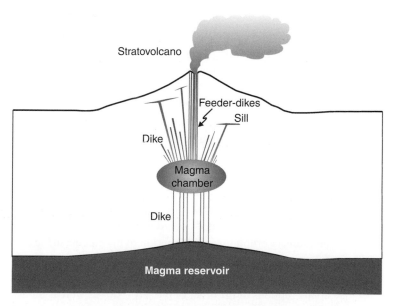

Fig. 8.8   A totally molten magma chamber supplies magma through a feeder-dike during an eruption in a volcano. It is assumed that during the eruption the volumetric flow rate of magma to the surface through the feeder is many times greater than the flow rate of magma from the source reservoir to the chamber, so that the latter may be ignored.

- There is not enough **gas exsolution** to maintain the excess pressure.
- The volumetric flow rate of new magma into the chamber from a source reservoir is **much lower** than flow rate out of the chamber and to the surface.
- The volume reduction of the chamber during the eruption is within roughly the **elastic limits**, that is, elastic shrinkage; thus no large volume reduction, related to caldera or graben collapse, occurs.

As for the first point, gas exsolution is unlikely to contribute significantly to excess pressure in shallow basaltic chambers – and even more so in deep-seated basaltic reservoirs. The second point is already assumed to hold, based on inflation/deflation observations of volcanoes worldwide. The third point is considered in Sections 8.5 and 8.6 on large eruptions.

Following these considerations, let us first recall Eq. (6.7) whereby the **total volume** V_{er} 'erupted' by, or rather transported out of, a totally fluid magma chamber through a feeder-dike (including the volume of the feeder itself) during an eruption is given by:

$$V_{er} = p_e(\beta_r + \beta_m)V_c \tag{8.15}$$

where p_e is the excess pressure in the chamber at rupture and dike injection, β_r is the host-rock compressibility, β_m is the magma compressibility, and V_c is the total volume of the totally fluid chamber. Using Eq. (8.15) we can define the ratio of excess magmatic pressure over the eruptive volume ψ (with the unit of Pa m^{-3}) as:

$$\psi = \frac{1}{(\beta_m + \beta_r)V_c} = \frac{p_e}{V_{er}} \tag{8.16}$$

From the considerations above, and using Eq. (8.16), it follows that the **excess pressure** in the source chamber of the feeder-dike at any instant $p(t)$ may be given by:

$$p(t) = p_e - \psi \int_0^t Q dt \tag{8.17}$$

where p_e is the excess pressure at the time of chamber rupture (as used in the equations above) and feeder-dike initiation, that is, at $t = 0$, ψ is defined in Eq. (8.16), and Q is, as before, the volumetric flow rate up through the feeder-dike (or, alternatively for measurements, the effusion rate of the associated volcanic fissure).

The **volumetric flow rate** $Q(t)$ through the volcanic fissure (fed by a dike), that is, the effusion rate, changes with time according to the equation (Machado, 1974):

$$Q(t) = Q_i - C \int_0^t Q dt \tag{8.18}$$

where Q_i is the initial volumetric flow rate and C is a constant that depends on the excess pressure, compressibility, and volume of the magma chamber, as well as the dimensions of the feeder-dike. Equation (8.18) has the solution:

$$Q(t) = Q_i e^{-Ct} \tag{8.19}$$

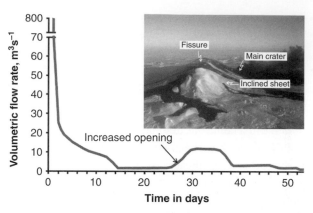

Fig. 8.9 Variation in the volumetric flow rate (the effusion rate) during the 1991 Hekla eruption in Iceland. The abrupt increase in the volumetric flow rate in the period from 26 to 38 days after the beginning of the eruption may be related to a small increase in the opening (aperture) of the feeder-dike. The inset is a photograph of the Hekla Volcano early in the eruption (data from Gudmundsson et al., 1992).

Equation (8.19) shows that the volumetric flow rate out of the chamber through the feeder-dike declines exponentially with time. A roughly **exponential** decrease in the volumetric flow or effusion rates is common during eruptions (Machado, 1974; Wadge, 1981; Stasiuk et al., 1993; Thordarson and Self, 1993; Thordarson and Larsen, 2007; Neri et al., 2011). One such example (with irregularities, however) is given in Fig. 8.9.

Using similar arguments, it can be shown, based on the above assumptions and by analogy with Eq. (8.19), that the excess pressure $p(t)$ in the magma chamber during the eruption is also a **negative exponential function** of time and given by (Gudmundsson, 2016):

$$p(t) = p_e \left(1 - e^{-\left(\frac{V_{er}}{Qt} - 1 \right)} \right) \tag{8.20}$$

where Qt is the magma volume (in m^3) flowing out of the chamber during the time interval t. The exponent has the units of m^3/m^3 and is thus dimensionless. Equation (8.20) is an approximation, but indicates an **exponential decrease** in excess pressure in the shallow source chamber during the eruption. One assumption, mentioned above, is that there is no significant inflow of magma into the chamber from a deeper source reservoir during the eruption. That assumption is commonly valid for eruptions of short duration, such as days or weeks, but less so for eruptions that last for months or years. An analysis of the flow of magma to the chamber during the eruption is presented by Woods and Huppert (2003). When the excess pressure eventually approaches zero, so that $p_e \to 0$, the bottom of the feeder-dike closes at its contact with the chamber and the eruption comes to an end. For volcanic hazard and risk assessments, one of the main points is therefore how long it will take for $p_e \to 0$, that is, how long the eruption will be.

8.4.2 Observations

Comprehensive reporting of volcanic eruptions on a global scale began only in the 1960s. The records before that time are incomplete and become gradually more so the further back in time we go. This is demonstrated in Fig. 8.10. There is a clear **increase** in the number of reported active volcanoes from the earliest records, from the end of the eighteenth century, and until about 1970. Since that time, when global recording of eruptions became reasonably well established, the reported number of volcanoes active each year has been more or less **steady**. This diagram, however, refers to all eruptions, including small ones, from volcanoes all over the world. Many active volcanoes are located in remote areas so that many of their small eruptions would not have been noticed before the advent of satellite monitoring.

Data obtained in the past decades for small eruptions indicate that they are very frequent. Table 8.1 presents a crude estimate of the **frequencies of eruptions** of different volumes (cf. Fig. 8.1). As the results suggest, small eruptions with eruptive volumes of $0.001–0.1$ km^3 are by far the most common; many such eruptions occur every year. By contrast, moderately large eruptions with eruptive volumes of $0.1–10$ km^3 occur, on average, once every $5–100$ years. Large eruptions, with eruptive volumes of $10–1000$ km^3 occur, on average, once every $1000–10\ 000$ years. Finally, the largest eruptions, of the order of several thousand cubic kilometres ($1–5 \times 10^3$ km^3) occur, on average, once every 50 000 years.

As regards moderate to large eruptions the record appears fairly complete. The number of reported active volcanoes issuing eruptive volumes in **excess of 0.1 km^3** has been

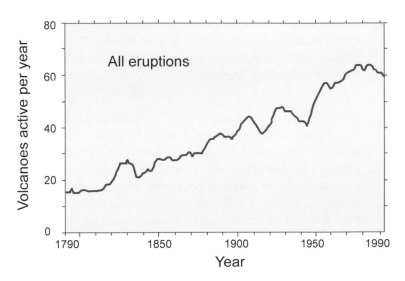

Fig. 8.10 Number of volcanoes recorded as erupting each year, presented as a 10-year running mean number, as a function of time since 1790. The data suggest that the number of volcanoes erupting each year has gradually increased over this period, but presumably this reflects mainly the improved recording of eruptions during the same period (cf. Fig. 8.11). Modified from Simkin and Siebert (2000).

Table 8.1 Frequency of eruptions as a function of eruptive volume. Modified from sciences.sdsu.edu (accessed September 2019).

Eruptive volume (km^3)	Frequency (every)
0.001–0.01	Several months
0.01–0.1	5 years
0.1–1.0	10 years
1–10	100 years
10–100	1000 years
100–1000	10 000 years
>1000	50 000 years

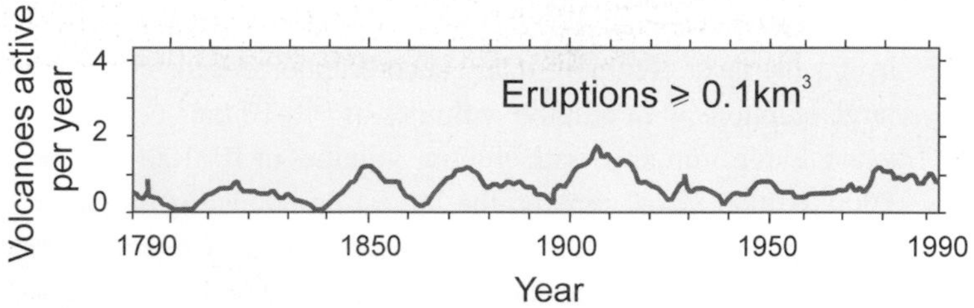

Fig. 8.11 Number of volcanoes recorded as producing eruptive materials of volume 0.1 km^3 or more each year, presented as a 10-year running mean number, as a function of time since 1790. In contrast to the graph in Fig. 8.10, where all recorded eruptions (irrespective of volume produced) are listed, this graph of the number of volcanoes producing moderate to larger eruptions each year does not show any increase with time. The difference between Figs. 8.10 and 8.11 is presumably primarily due to better recording of moderate to large eruptions (Fig. 8.11) than of small eruptions – the improved recording of which is the main reason for the gradual rise of the graph in Fig. 8.10. Modified from Simkin and Siebert (2000).

comparatively steady since the end of the eighteenth century (Fig. 8.11). This suggests that the eruption activity in terms of eruption frequency has been close to **constant** in the past centuries, even though the number of small eruptions seems to have increased – this being primarily because of improved recording with time. Again, it should be kept in mind that all these data refer to **subaerial eruptions**; our data on submarine eruptions, particularly at mid-ocean ridges, is still very limited and no statistical data of the types presented in Figs. 8.10 and 8.11 are available. These results thus suggest that, over tens or hundreds of years, the subaerial global volcanism in terms of frequency of eruptions has stayed roughly the same. The next task then is to obtain data on how long most eruptions last.

The durations of many eruptions have been measured (Simkin and Siebert, 2000). In the latest count, some **15 volcanoes** had been erupting more or less continuously for several decades. These include well-known volcanoes such as Sakurajima (Japan, Fig. 8.12a), and Stromboli (Italy, Fig. 8.12b), Erta Ale (Ethiopia, Fig. 3.35), and Erebus (Antarctica). In addition, there are several tens of eruptions in other active volcanoes every year. Over the

Fig. 8.12 Two volcanoes that erupt almost continuously. (a) Sakurajima in Japan has been almost continuously erupting since 1955, with as many as hundreds of small explosive eruptions each month. One such eruption is seen here (19 July 2013). View east, the stratovolcano reaches 1117 m above sea level and is located within the large Aria caldera (formed about 22 000 years ago) and close to the city of Kagoshima. (b) Stromboli, Italy, has been more or less continuously active for hundreds of years – possibly for 2500 years. Aerial view east, the stratovolcano reaches 924 m above sea level and, because of its continuous (strombolian) activity, is known as the 'Lighthouse of the Mediterranean'. Photo: Valerio Acocella.

past five decades there have normally been between 50 and 70 volcanoes active each year (Fig. 8.13). This value is essentially constant with an average of about **56 active volcanoes** every year.

The **duration** of 3301 eruptions is shown in Fig. 8.14. These data indicate that:

- The duration of close to 10% of the eruptions is less than 1 day.
- The duration of some 45% of the eruptions is less than 1 month.
- The median duration of an eruption is about 7 weeks.
- The duration of most eruptions is less than 3 months.
- The duration of about 85% of the eruptions is less than 1 year.

The results show that **most eruptions are very short**, several weeks, in comparison with the longest eruptions, the duration of which is several tens of years or more. A remarkable feature of the histogram in Fig. 8.14 is its shape, which is somewhere between **normal** and log-normal. By contrast, the lengths of volcanic fissures (Fig. 2.19) and eruption volumes

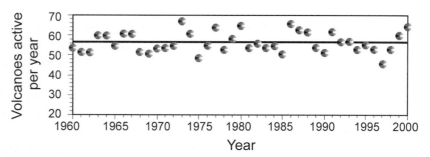

Fig. 8.13　Number of volcanoes erupting each year since accurate recording began in the 1960s. As is seen, the number of volcanoes erupting per year has been constant at an average of 56 during this time (cf. Figs. 8.10 and 8.11). Data from the Smithsonian Institution (www.volcano.si.edu; accessed September 2019).

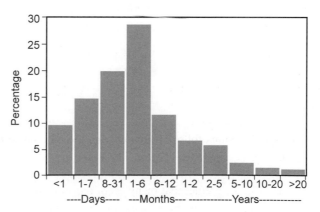

Fig. 8.14　Estimated durations of 3301 eruptions. The median duration is 7 weeks. About 85% of the eruptions lasted less than 1 year, and about 10%, less than 1 day. Modified from Simkin and Siebert (2000).

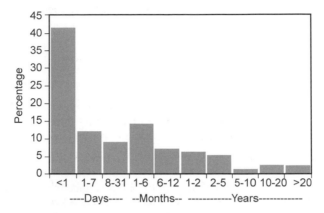

Fig. 8.15   Length of time from the beginning of an eruption to its peak. The data are for 252 explosive eruptions with a VEI ≥ 3. About 42% peak within 1 day and 52% within 1 week. However, 16% of the eruptions reach their peaks after 1 year or more. Modified from Simkin and Siebert (2000).

(Fig. 8.1) follow power-law size distributions. We might have thought that the duration, as the eruptive volumes, would also follow a power-law size distribution, but based on these data that is apparently not the case. Why the length distribution of eruptions is so different from the volume distribution is not known.

The time from the beginning of an eruption to its **peak** in terms of volumetric flow rate is an important measure of the eruption dynamics (Fig. 8.15). Here the data are limited to 252 explosive eruptions with a Volcanic Explosivity Index (VEI) larger than or equal to 3. Although comparable data on effusive eruptions are missing, the results are nevertheless interesting and indicate that:

- Some 42% of the eruptions peak within the first day.
- Some 52% of the eruptions peak within the first week.
- Some 16% of the eruptions peak after one year or more.
- Some 4% of the eruptions peak after more than 10 years.
- Some 2% of the eruptions peak after more than 20 years.

The first thing to mention about this histogram is, again, its shape. It is clearly different from the shape of that in Fig. 8.14. In fact, the histogram in Fig. 8.15 is much more similar to that obtained for the size distributions of the lengths of volcanic fissures and volumes of eruptive materials, namely a crude **power law**.

The second thing to be said about the **time-to-peak** results is that the statements in the first two bullet points follow logically from the fact that some 45% of eruptions are very short, or shorter than one month. The shortness of many eruptions, together with the exponential decline in volumetric flow rate and excess pressure in the chamber (Eqs. (8.19) and (8.20)) predict that many eruptions must peak within the first week. However, a small but significant number of explosive eruptions have times to peak of the order of years or tens of years. Some of these may be the most important explosive eruptions, namely the largest ones.

A case in point is the 1815 **Tambora eruption** (Indonesia). This is generally considered to be the **largest eruption** on Earth during historical time. As indicated in Section 8.2, the eruption produced 160–170 km^3 of eruptive materials (about 50 km^3 of magma) and a caldera 6–7 km in diameter, with a maximum depth of 1200 m and a volume of about 36 km^3 (Figs. 9.11 and 10.21). Following several centuries of apparently no volcanic activity, small eruptions began in the volcano in 1812, that is, about 3 years before the main eruption. The main eruptive phase started on 5 April 1815 and peaked 5 days later, namely on 10 April, when the main eruption occurred. On a much smaller scale, but instrumentally documented, the 1991 eruption of **Pinatubo** in the Philippines was also associated with a caldera collapse. The first precursor to that eruption may have been an M7.7 earthquake on 16 July 1990, but its connection with the eruption is uncertain since the earthquake occurred on a fault about 100 km from Pinatubo. On 15 March 1991, smaller earthquakes began, and the first of the small eruptions occurred on 2 April 1991. These continued until the main eruption on 16 July 1991, that is, about 3.5 months after the first eruptions. These and related observations suggest that precursors to moderate-to-large caldera eruptions may be months or years, a suggestion that has implications for the dynamics of large explosive eruptions in general.

8.5 Large Explosive Eruptions

8.5.1 Background

For understanding large eruptions in general, let us focus on a mechanism for squeezing out the magma from the chamber or reservoir during the eruption. The squeezing is achieved through **reducing** the chamber or reservoir **volume** as a result of caldera or graben collapse during the eruption so as to maintain the excess pressure. This implies that the exponential decrease in excess pressure indicated by Eq. (8.20) is delayed until very late in the eruption. In the model for large explosive eruptions, a piston-like caldera collapse is assumed to be the common driving force for maintaining the excess pressure in the chamber, hence for generating the large eruption. In Section 8.6, I turn to large effusive eruptions where, I suggest, graben subsidence commonly plays a role analogous to that of a caldera collapse.

As indicated in Section 8.2 no really large eruptions have been recorded instrumentally. The largest eruption in the twentieth century was the 1912 **Novarupta eruption** in Alaska, which produced some 30 km^3 of acid-to-intermediate tephra, calculated as equivalent to about 13–14 km^3 of magma leaving the magma chamber. The part of the chamber that generated the main caldera collapse, however, is beneath the nearby Katmai Volcano. There, a collapse caldera of 3–4 km in diameter, 0.6 km deep, and a volume of 5–6 km^3, developed during the eruption. The magma came partly from the chamber, or the part of the chamber that is beneath Katmai Volcano (possibly the same chamber is beneath Katmai and Novarupta), which is at a distance of about 10 km from Novarupta. The chamber was clearly layered with acid (rhyolite) magma on the top and less evolved andesite magma at the bottom. The second, and much better documented, reasonably large (but classified as moderate) explosive eruption in the twentieth century was the 1991 **Pinatubo** eruption in the Philippines, discussed above (Newhall et al., 1997). This eruption produced 10 km^3 of erupted materials, equivalent to about 4 km^3 of magma leaving the chamber. The eruption was associated with a caldera collapse, the caldera being 2.5 km in diameter.

Other moderate to large explosive eruptions in the past centuries include the 1815 **Tambora eruption** in Indonesia, which produced 160–170 km^3 of eruptive materials (mainly tephra), as mentioned above. Another similar, but smaller, explosive caldera eruption occurred in the **Krakatau** (or Krakatoa) Volcano in Indonesia, in 1883. Small eruptions began on 20 May that year and continued intermittently until the eruption peak occurred on 26–27 August. The eruption was associated with the collapse of a roughly circular caldera, about 6 km in diameter and with a maximum depth of about 1000 m. The estimated volume of eruptive materials is 18–21 km^3, corresponding to roughly 6–7 km^3 of dacite magma flowing out of the chamber (Simkin and Fiske, 1980; Rosi et al., 2003).

The largest explosive eruptions, however, produce several thousand cubic kilometres of eruptive materials. The most recent such eruption is the **Toba eruption** in Sumatra (Fig. 9.13b), some 75 000 years ago, which produced some 2800 km^3 of eruptive materials. Other eruptions on that scale include the **La Garita caldera** (in Colorado, United States) eruption, which generated the Fish Canyon Tuff some 27–28 million years ago, with an estimated volume of 4000–5000 km^3 (Lipman et al., 1997; Mason et al., 2004). The associated caldera itself is elongated with a minor axis of 35 km and a major axis of 75 km and is thus among the largest calderas on Earth.

For explosive eruptions that happened long before historical time, perhaps tens of millions of years ago or more, the volume estimates of the eruptive materials will always be inaccurate. Part of the material may have fallen on the sea or on lakes, been carried away by erosion, buried by materials from later eruptions, or affected in other ways that make accurate area (and therefore volume) estimates impossible. Yet the estimates for these old pyroclastic rocks, such as those associated with the Toba eruption and the La Garita eruption (the Fish Canyon Tuff) refer to the 'dry-rock-equivalent' volumes (Mason et al., 2004), which are similar in density to their source magmas. Thus, despite the somewhat limited accuracy, these volumes may be regarded as **crude measures** of the actual magma volumes that left the source chambers during the eruptions.

Explosive eruptions that start as 'ordinary' ones may develop into large eruptions in connection with a caldera collapse. In most reasonably well-documented caldera-collapse eruptions – such as the 1815 Tambora eruption, the 1883 Krakatau eruption, and the 1991 Pinatubo eruption – the **eruption intensity,** that is, the volumetric flow rate, reaches its peak at a later stage than in most 'ordinary' eruptions. Thus, in the Tambora eruption, the time from the eruption initiation to the peak may have been about 3 years, and in the Krakatau and Pinatubo eruptions just over 3 months. By contrast, over 50% of ordinary explosive eruptions reach their peak within a week (Fig. 8.15). It is likely that for many large explosive eruptions the late peak or plateau in the eruption is partly attributable to the maintenance of the excess pressure as a result of the development or reactivation of

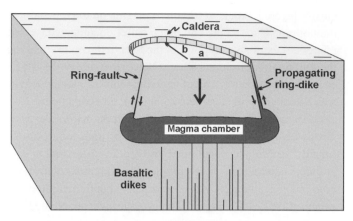

Fig. 8.16   As the caldera block subsides (vertical arrow) on an outward-dipping reverse ring-fault, the fault must open up and gradually increases its aperture (cf. Fig. 5.24). The ring-fault would thus normally be injected by magma to form a ring-dike, which commonly reaches the surface to erupt. There is thus essentially no friction along the outward-dipping ring-fault so that the caldera block may subside to the bottom of the chamber. This type of ring-fault displacement is referred to as unstable. It follows that (i) the volumetric flow or effusion rate should increase greatly with increasing subsidence of the caldera block, and (ii) most or all the magma in the chamber may be squeezed out during the eruption. The eruption is commonly large and the collapse may all but destroy the chamber (cf. Gudmundsson, 2015).

a caldera ring-fault, as will be explained in detail below. First, however, we need to consider the geometry and development of the ring-fault itself (cf. Chapter 5).

There are two primary ideas or models as to the ring-fault strike and dip, or attitude. One model assumes that the caldera faults are outward-dipping (Fig. 8.16), the other that the faults are inward-dipping (Fig. 8.17). The outward-dipping model is well established in the caldera literature, and is the original idea behind ring-fault formation – dating back to Anderson's (1936) model. This ring-fault attitude is also suggested by many analogue experiments. The field and analogue experimental evidence for outward-dipping ring-faults is discussed in Chapters 5 and 9.

The **outward-dipping model** has several important mechanical implications for the course of the caldera-forming eruption, many of which relate to the gradual increase in opening or aperture of the ring-fault (that is, the 'gap' between the piston-like caldera block and the walls of the host rock) as the caldera subsidence increases (Figs. 5.24 and 8.16). The implications are as follows:

1. A ring-dike would be expected to form, and much of the eruption should normally occur with the ring-dike as the main feeder. This follows because such an opening, which extends from the chamber to the surface, would normally be injected by magma – a ring-dike – that reaches the surface.

2. The volumetric flow rate through the ring-dike should normally increase with increasing subsidence of the caldera floor. As the piston-like caldera block subsides into the magma chamber, the opening or aperture of the fracture between the host rock and the subsiding

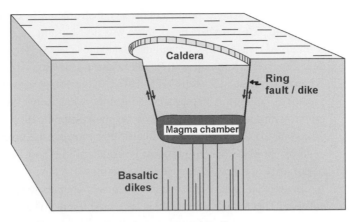

Fig. 8.17   When caldera subsidence occurs on an inward-dipping normal ring-fault, the caldera walls gradually become subject to higher horizontal compressive stresses (cf. Fig. 5.26). If no ring-dike is formed the friction, together with the normal stress on the ring-fault, ensures that the fault remains closed and the displacement stable. If a ring-dike feeder forms, collapse on an inward-dipping normal ring-fault is thus less likely than that on an outward-dipping ring-fault (Fig. 8.16) to result in increasing volumetric flow rate as the subsidence increases. The magma chamber is also less likely to be destroyed during collapse on an inward-dipping ring-fault than on an outward dipping one (cf. Gudmundsson, 2015).

block gradually increases so that the aperture Δu in Eqs. (8.11) and (8.13) increases. The volumetric flow rate through the feeder ring-dike depends on Δu^3, that is, on the cube of the aperture. Thus, the aperture increase with caldera subsidence should result in a very great increase in the volumetric flow rate Q.

3. There is essentially no friction along the ring-fault. Since the fault is occupied by the ring-dike, the only resistance to slip is the magma viscosity of the ring-dike. Also, the fracture or ring-dike opening increases as the caldera subsidence progresses. There is thus hardly anything to stop the piston-like block from sinking essentially to the bottom of the chamber, unless the piston rock is lighter than the magma in the lower part of the chamber. This means that for an outward-dipping ring-fault, a large fraction of the magma chamber should erupt – very little if any magma should be left in the chamber following the collapse. This means that outward-dipping ring-faults encourage large eruptions, the actual size of which depends on the size of the chamber.

The data available today allow only for partial tests of these three implications. Briefly, these data indicate the following:

- As to the first point, many calderas have very thin or non-existing ring-dikes (Chapters 3 and 9).
- As for the second point, it is not clear if the volumetric flow rate increases as the ring-fault subsidence continues. Nor is it known if many caldera eruptions are primarily or entirely fed by ring-dikes. In some calderas the eruption is partly along a ring-dike, in others the eruption appears to be primarily in the central part – from the main conduit of the stratovolcano that subsequently collapses or forms a central part of an already existing caldera.

- Further to the second point. Of the six instrumentally documented caldera collapses during the past 100 years or so, there was no ring-dike eruption in at least five, namely: Katmai, Alaska (1912), Fernandina, Galapagos (1968), Miyakejima, Japan (2000), Piton de la Fournaise, Reunion (2007), and Kilauea, Hawaii (2018). The collapse of the Plosky Tolbachik pit crater, Russia (1975–1976), may perhaps also be included as an example of a caldera collapse (Chapter 5); it did not include any ring-dike eruption. Part of the eruption during the collapse in Pinatubo, the Philippines (1991), may have been on a ring-dike, but the documentation indicates that most of the eruption was along the central conduit. Similarly, the eruptions of Krakatau (1883) and Tambora (1815) appear to have been partly or mainly along central conduits.
- As for the third point, some magma chambers are clearly largely destroyed by the caldera collapse. This follows since there is very little if any volcanic activity at the location of the caldera subsequent to its formation or collapse. Most calderas, however, are highly active after collapse (Newhall and Dzurisin, 1988; Gottsmann et al., 2008), suggesting that the magma chamber is not destroyed during the collapse. As for the recent examples discussed above, there have been eruptions in Fernandina, Pinatubo, Miyakejima, and Piton de la Fournaise since their recent collapses (13 to 52 years ago), showing that the associated magma chambers are still active. Going further back in time, Krakatau has erupted around 40 times since the 1883 eruption, the first one after the 1883 eruption being in 1927, that is, some 44 years after the eruption. Tambora did not erupt again for some 60 years following the 1812–1819 eruption. This assumes that the 1819 eruption was a continuation of the eruption that started in 1812 and reached its peak in 1815. Then, again, there was an eruption in Tambora in 1967, although the exact year is poorly constrained. It thus appears that there are three possibilities: (1) Magma chambers are not completely destroyed during caldera collapses, even in large eruptions such as the 1815 Tambora eruption. (2) New magma chambers form quickly (over tens of years) at roughly the same location as the destroyed ones. (3) The new eruptions in the collapse calderas are supplied with magma from deeper reservoirs that acted as sources for the (presumably destroyed and caldera-related) shallow chambers.

As for the **inward-dipping model** of collapse calderas (Fig. 8.17), the requirement for slip is an extension across the ring-fault. This extension can be generated either through as little as centimetre-scale doming of the crustal segment hosting the shallow chamber, through pressure increase in the deeper source reservoir (Chapter 3), rift-related extension (Chapter 6), gravity sliding resulting in volcano spreading, or any other similar mechanism that can produce extension. Because of the existing shallow magma chamber as a cavity in the crustal segment, and particularly when the ring-fault starts to develop (or an existing one slips), there will be a tensile-stress concentration in and around the piston-like block that will reduce the friction (Fig. 8.17) and promote caldera subsidence. When the ring-fault and/or ring-dike are close to vertical, there can be a total collapse so that the magma chamber is to a large degree emptied. For other conditions, the friction limits the subsidence to a maximum of, in most cases, 1–2 km. Also, where the friction is highest, for example where there is no ring-dike injected during the eruption, or where the dip of the ring-fault is unusually gentle (say 40-60°), there is little or no slip during the collapse, thereby generating trapdoor calderas (Amelung et al., 2000).

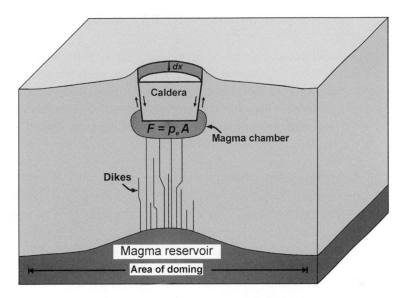

Fig. 8.18   As the caldera block subsides by the differential distance dx into the associated magma chamber, the chamber volume is reduced (the chamber shrinks). The excess pressure p_e in the chamber may thus be maintained, thereby allowing a much larger fraction of the magma to be squeezed out of the chamber than during an ordinary poroelastic eruption. The excess pressure p_e can be related to the average force F on the cross-sectional area of the subsiding caldera block A through Eqs. (8.21) and (8.22). The formation or reactivation of a collapse caldera can occur through an extension or a small doming (uplift) of the crustal segment hosting the caldera, as indicated (Chapter 5; cf. Gudmundsson, 2015).

Thus, for **inward-dipping** ring-faults, the slips are more controlled and **restricted** than those of **outward-dipping** faults. All piston-like caldera collapses, however, reduce the volume of the magma chamber, thereby offering the possibility of maintaining the excess pressure for a longer time and squeezing out an unusually large fraction of the magma in the chamber. Observational support for this mechanism was provided by the 2018 Kilauea collapse. Following many of the inward-dipping ring-fault slips and a reduction in the associated magma-chamber volume (Fig. 8.18), magma-pressure pulses propagate from the chamber and along the 40-km-long feeder-dike, resulting in a temporary increase in the volumetric flow (effusion) rate at the eruption site (Neal et al., 2019).

The squeezing of magma out of the chamber during ring-fault slips offers a possible mechanism of generating large eruptions – eruptions of many tens or hundreds of cubic kilometres from ordinary-sized chambers, the latter being normally in the range of 20–500 km^3. And, for exceptionally large magma chambers, such as may reach 5000–10 000 km^3, based on the areas of the associated calderas and reasonable thicknesses of the magma bodies, the eruptive volumes may reach several thousand cubic kilometres. Let us now analyse a simple model to explain large explosive eruptions generated by caldera collapses.

8.5.2 Model

Consider the magma chamber in Fig. 8.18. Recall first that pressure has the unit of N m^{-2}, that is, force per unit area. On multiplying the **pressure by area** (m^2), the **force** F (unit newton, N) is obtained. The excess pressure in the chamber p_e times the cross-sectional area of the caldera A (assumed to be roughly the same as the horizontal or map-view cross-sectional area of the associated chamber) then gives the force F, namely:

$$F = p_e A \tag{8.21}$$

As the **piston-like caldera roof** moves the differential distance dx (Fig. 8.18), the corresponding very small (infinitesimal) work dW done by the surroundings on the magma chamber system is:

$$dW = F dx = p_e A dx \tag{8.22}$$

and the **change in the volume** of the chamber dV_c is given by:

$$-dV_c = A dx \tag{8.23}$$

The minus sign in Eq. (8.23) indicates that chamber **volume decreases** during the piston-like caldera subsidence, as seen in Fig. 8.18. Here, I follow the thermodynamic convention that work done on the system is regarded as positive. Next Eqs. (8.22) and (8.23) are combined using the relation $dW = dU$, which represents the first law of thermodynamics when no heat is added to the system. Here dU denotes the change in internal energy of the magma chamber during its shrinkage or contraction. From this procedure the **change in internal energy** is obtained:

$$dU = -p_e dV_c \tag{8.24}$$

Equation (8.24) is then a measure of the **elastic potential energy** dU transformed during a caldera eruption as a function of the excess pressure p_e and the **contraction** or shrinkage of the chamber $-dV_c$.

The excess pressure p_e at the time of magma-chamber rupture and feeder-dike injection is roughly equal to the *in situ* tensile strength, T_0 (Chapter 7), which, in the crust, is close to being a constant, mostly 1–6 MPa and typically around 4 MPa. It follows that the excess pressure p_e at the time of chamber rupture is essentially constant, here taken as 4 MPa.

In the present model of a **large explosive eruption**, this excess pressure is largely maintained until close to the end of the caldera-forming eruption. It then follows from Eq. (8.24) that the elastic potential energy of the eruption is directly related to the volume change dV_c, here the contraction or shrinkage of the chamber during the eruption. This volume change dV_c corresponds roughly to the volume of material, that is, magma, which **leaves the chamber** during the eruption. Feeder-dikes associated with shallow magma chambers have volumes that are normally a fraction of a cubic kilometre and, at most, several cubic kilometres (Chapter 6). For large eruptions of the order of tens or hundreds of cubic kilometres or more, the feeder-dike volume may thus be regarded as being included in the estimated eruptive volume V_{er}.

From these considerations it follows that V_{er} can be equated with $-dV_c$ to obtain the **elastic potential energy** U_{er} driving the caldera-forming eruption, from the following equation (Gudmundsson, 2014):

$$U_{er} = p_e V_{er} \tag{8.25}$$

From Eq. (8.25) the elastic energy driving an explosive caldera-forming eruption of a given volume can be estimated, which is very useful for hazard assessments. Let us now extend the analysis to large effusive eruptions.

8.6 Large Effusive Eruptions

Numerous basaltic lava flows reach tens or hundreds of cubic kilometres, and some reach thousands of cubic kilometres. It is likely that many, perhaps most, large basaltic fissure eruptions – as indicated by their volumes and primitive composition – are supplied with magma from large deep-seated reservoirs (Fig. 8.2; cf. Figs. 6.1, 6.7, and 6.8). Here, let us recall the main equations for calculating the eruptive volumes during 'ordinary' eruptions from such reservoirs. Thus, the **eruptive volume** V_{er} is, from Eq. (6.24), given by:

$$V_{er} = \eta p_e (\beta_p + \beta_m) V_b \tag{8.26}$$

where η is the porosity or melt fraction of the reservoir, p_e is the magma excess pressure, β_p is the bulk compressibility of the reservoir, β_m is the magma compressibility, and V_b is the bulk or total volume of the reservoir. Inserting the appropriate values obtained in Chapter 6 into Eq. (8.26) yields the following ratio between the bulk volume of the reservoir V_b and the eruptive volume in an 'ordinary' eruption V_{er}:

$$V_{er} \approx 2 \times 10^{-4} V_b \tag{8.27}$$

Equation (8.27) indicates that there are limits to how large a lava flow can be generated in a single eruption by an ordinary mechanism. For example, a lava flow of 100 km^3 would, according to Eq. (8.27), require a reservoir volume of 5×10^5 km^3. If we consider a reservoir beneath a typical divergent plate boundary, with a width (lateral dimension measured perpendicular to the rift-zone axis) of 20 km, the along-axis dimension (the length) of the reservoir would have to be 1000 km and its thickness 25 km to supply magma to such an eruption by the ordinary mechanism. A magma body of such dimensions is **unlikely to exist** at any particular time and location so as to be able to supply magma to a single eruption. In the largest effusive eruptions, such as those of the comparatively recent (16–17 Ma-year-old) Columbia River Basalts, the individual lava flows reach estimated volumes from several hundred cubic kilometres to as much as several thousand cubic kilometres. It is very unlikely that such volumes could be generated by the ordinary mechanism that results in the ratio given in Eq. (8.27).

Grabens are in many ways analogous to collapse calderas (Fig. 8.19). Both occur on faults and involve subsidence of the crustal segment between the faults. Large-scale graben subsidence in a volcanic field or zone or system normally results in temporary **reduction** in

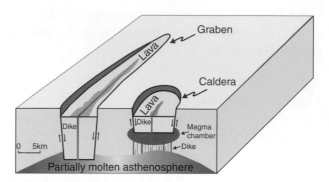

Fig. 8.19 Grabens and collapse calderas are mechanically analogous structures. Both relate to faults along which subsidence of the crustal block in-between the faults occurs. As for caldera subsidence, major graben subsidence results in a temporary reduction in the volume of the reservoir below the graben and may thus contribute to squeezing out a comparatively high proportion of its magma during an eruption (cf. Gudmundsson, 2016).

the volume of the magma reservoir below the graben. Some grabens also dissect central volcanoes with shallow magma chambers, in which case graben subsidence may temporarily reduce the chamber volume in a similar way as a caldera collapse does, but usually the graben subsidence in a central volcano is much smaller than a caldera subsidence. Here, however, we focus on large-scale graben subsidences above deep-seated reservoirs.

The mechanism proposed here for generating a large effusive eruption, in particular basaltic lava flows with volumes of tens to hundreds or even thousands of cubic kilometres, is essentially that of large-scale **graben subsidence** or collapse in the roof of the associated reservoir during the eruption. This mechanism is similar to that for generating large explosive eruptions, namely the subsidence of a **crustal block** into the reservoir during the eruption (Fig. 8.19). The subsidence of the graben block into the top part of the reservoir largely maintains its excess pressure during the eruption and dramatically increases the V_{er} /V_b ratio (Eq. (8.27)). More specifically, as the crustal segment within the graben – the strip of land between the boundary faults of the graben – subsides, the reservoir volume is reduced and the excess pressure in the reservoir is maintained in a manner analogous to that during caldera subsidence into shallow chambers, and expressed in Eqs. (8.22) to (8.24).

While the mechanism is quite general, some specific examples may be proposed. Consider first the large effusive eruptions in Iceland. The postglacial Skjaldbreidur lava shield has an estimated volume of about 15 km^3 (Fig. 8.20; Rossi, 1996). The age of Skjaldbreidur is not known exactly, but based on mapping it is in the range of 6000–9000 B.P., and probably closer to 9000 B.P. (Sinton et al., 2005). The northern part of the Thingvellir Graben is mostly located in an older lava flow, a pahoehoe flow referred to as the Thingvallahraun (Thingvellir Lava). It had previously been estimated to have formed in about 10 200 B.P. (Sinton et al., 2005), and thus perhaps is as much as 1000 years older than the Skjaldbreidur shield. No large normal faults and tension fractures dissect the main shield of Skjaldbreidur itself. In the **Thingvellir Graben**, southwest of the shield (Fig. 8.20), the early Holocene subsidence reaches at least 40 m and possibly 70 m. The graben extends into the lake and south to the Hengill Volcanic System (Figs. 4.8 and 8.20) where the displacement in Pleistocene rocks

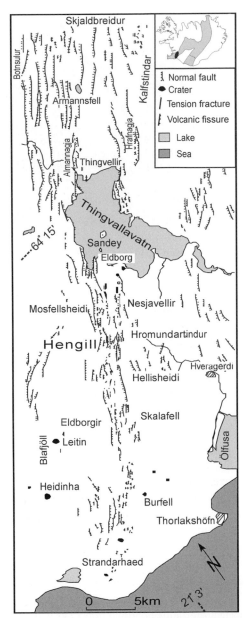

Fig. 8.20 Part of the West Volcanic Zone in Iceland, including the Hengill Volcanic System. Also shown are the Holocene Thingvellir Graben, the Hengill central volcano (Hengill), Lake Thingvallavatn (occupying the main graben), the hyaloclastite mountains Armannsfell and Botnsulur, and the Holocene lava shield Skjaldbreidur. The inset shows the location of the area.

reaches at least 200 m (Fig. 1.5). To the west, the graben structure extends to normal faults in the hyaloclastite mountain Botnsulur, where the displacements reach about 400 m (Fig. 8.20). It is not known if these faults participated in the postglacial subsidence, but those in Hengill

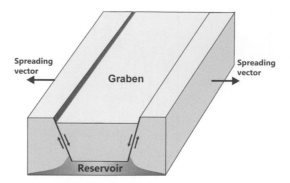

Fig. 8.21   Comparatively rapid subsidence of a graben within an active volcanic zone or system during an eruption reduces the volume of the associated reservoir. The temporary reduction in volume, if rapid enough, may maintain the excess pressure in the reservoir and thus allow a much higher proportion of the magma to be squeezed out of the reservoir during the eruption than is possible in ordinary eruptions (cf. Gudmundsson, 2016).

almost certainly did. It is well known that part of the displacement on a normal fault is due to the rise of its flanks, but the greater part is normally related to absolute subsidence.

If the extended Thingvellir Graben is taken to be 50 km long and 7 km wide (Fig. 8.20; Gudmundsson, 1987), then its area is about 350 km^2. An overall **subsidence of 30–40 m** for the extended graben as a whole (Fig. 8.21) corresponds to a volume of $10–14 \text{ km}^3$, which is roughly the volume of the Skjaldbreidur lava shield. If the entire graben system of the West Volcanic Zone was involved, then Skjaldbreidur is roughly in its centre (Fig. 8.22). The total length of the subsided strip of land is then about 85 km, the total width is about 18 km, and the area is 1500 km^2. For such a graben system, an average subsidence of a mere 10 m would be sufficient to correspond to the entire volume of Skjaldbreidur, 15 km^3. For such a simple analysis, however, several factors must be considered, including the following:

- Part of the subsidence of the graben must be **younger** than the formation of the Skjaldbreidur shield and thus cannot have contributed to the maintenance of the excess pressure in the reservoir. This follows because fault-displacement occurred in AD 1789. However, it is widely accepted that a large fraction of the subsidence along the faults occurred in the early Holocene (Sonnette et al., 2010), as indicated above.

- The maximum displacement on the faults where they dissect Holocene lava flows is about 40 m. The average displacement of 30–40 m may then look like an overestimate. However, as indicated, the faults continue into Lake Thingvallavatn (Bull et al., 2003) and south to Hengill Volcano where they reach displacements of at least 200 m. It is not known how large a fraction of the 200 m displacement was generated during the formation of the Skjaldbreidur lava shield, but it may be in excess of 40 m. Furthermore, the **total subsidence** of the Thingvellir Graben may be much larger; Tryggvason (1982) estimates the maximum subsidence in the graben within the Holocene lava flows to be 70 m.

- The eruption possibly took **many decades**. The volumetric flow rates are of course not known, but, by analogy with other eruptions, Sinton et al. (2005) suggest that the eruption may have taken 40–80 years. If true, then inflow into the reservoir could also contribute to maintain its excess pressure during the eruption.

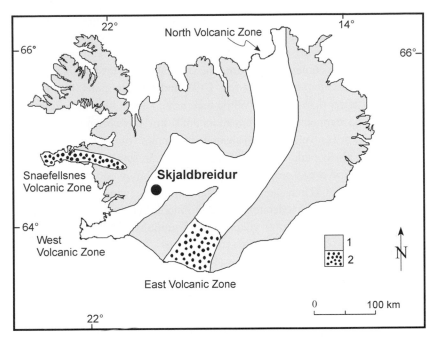

Fig. 8.22   The active volcanic zones of Iceland. The West Volcanic Zone is indicated. The location of the Skjaldbreidur lava shield is also indicated.

Large grabens are normally associated with flood basalt provinces or traps, such as **Columbia River** Basalts and the Deccan Traps. There are, for example, very **large grabens** in the Columbia River Plateau, including the Oregon–Idaho Graben. This particular graben is 50–60 km wide and 100 km long (Cummings et al., 2000), and thus has an area of at least 5000–6000 km^2. With a maximum graben subsidence of up to 800 m, there is a clear potential for squeezing out large volumes of magma during the graben development. Indeed, the authors of the study of the Oregon–Idaho Graben conclude that its formation coincided with large eruptions, but consider that the flows were primarily of rhyolite and that the graben itself (of age 15.3–15.5 million years B.P.) developed after the main basaltic lava flows of the Columbia River formed. Be that as it may, this and other grabens of similar size would certainly be candidates for squeezing out magmas from deep-seated reservoirs, in accordance with the model presented here.

8.7 Summary

- Eruptive volumes follow roughly power-law size distributions. Such distributions form part of heavy-tailed distributions and, for eruption volumes, mean that there are very many small-volume eruptions and very few large-volume eruptions. The size distribution applies to volcanoes worldwide and is likely to apply to many individual polygenetic (central volcanoes, stratovolcanoes, basaltic edifices) volcanoes as well.

• Eruptions are divided into three classes based on volumes of eruptive materials given in cubic kilometres (including the volumes of the feeder-dikes and feeder-sheets): (1) small eruptions (less than 0.1 km^3), (2) moderate eruptions (0.1–10 km^3), and large eruptions (larger than 10 km^3). While there is no known limit as to how small an eruption can be, those much smaller than about 0.001 km^3 have often not been reported. Similarly, while there is no clear known absolute maximum volume that can be erupted, the largest known explosive and effusive volumes formed in single eruptions are about 5000 km^3.

• The average eruption frequencies in relation to volumes can be further specified as follows. There are many small eruptions, with eruptive volumes of 0.001–0.1 km^3, each year. Moderately large eruptions, with eruptive volumes of 0.1–10 km^3 occur once every 5–100 years. Large eruptions, with eruptive volumes of 10–1000 km^3, occur once every 1000–10 000 years. The largest eruptions, of the order of several thousand cubic kilometres (1–5 × 10^3 km^3) occur, on average, once every 50 000 years.

• Magma transport to the surface from a chamber or reservoir is modelled primarily using three geometric types of conduits. The first model is a circular conduit, a vertical cylinder, which is most suitable for some stratovolcanoes with conduits that later solidify as cylindrical necks or plugs. The second model is that of an elliptical conduit. This is a flexible model, which can be applied to somewhat elliptical (in horizontal or map-view section) conduits that later solidify as elliptical necks or plugs, but also to dikes and inclined sheets, particularly when the length/aperture (or thickness) ratio is comparatively small. The third and most commonly used model is that of a flow between parallel plates. This is the standard model for magma transport through dikes, inclined sheets, and sills.

• The time from the beginning of an eruption to its end is known as the eruption duration. The duration depends on several factors, including the size of the source magma chamber or reservoir, the dimensions of the conduit/feeder-dike, and the excess-pressure variation in the chamber/reservoir. Simple theoretical considerations suggest that, for ordinary eruptions, that is, eruptions that are not associated with a caldera or graben collapse, the excess pressure in the chamber/reservoir decreases or falls crudely as an exponential function of time. Similarly, the volumetric or effusion flow rate decreases roughly exponentially with time until the pressure becomes too low in the source chamber/reservoir to keep the feeder-dike open, whereby the eruption comes to an end.

• The duration of 55% of eruptions is less than four weeks, the median (most common) duration being 7 weeks, and the duration of 85% of eruptions is less than 1 year. A few volcanoes, however, erupt continuously for decades, particularly those associated with lava lakes, and one or two (such as Stromboli) for hundreds of years. Some 52% of explosive eruptions (data are lacking for effusive eruptions) reach their maximum volumetric eruption rates, that is, the eruption peak, within the first week and 42% within the first day. However, 16% of explosive eruptions do not peak until more than 1 year after the beginning of the eruption, and 4% peak 10 years or more after the beginning of the eruption.

• Large explosive eruptions, erupting more than 10 km^3 and some more than 1000 km^3, are explained in terms of a piston-like subsidence of a collapsing caldera reducing the chamber volume and thereby maintaining close-to-constant excess pressure during most of the eruption. In this mechanism it is the caldera subsidence and associated chamber volume reduction that squeezes the magma out of the chamber. Thus, the caldera collapse is the cause of the large eruption, in agreement with all explosive eruptions exceeding about 25 km^3 in eruptive volumes being associated with collapse calderas. This mechanism explains how moderate-sized magma chambers can, during caldera subsidence, give rise to large-volume eruptions while the ordinary (non-caldera collapse) eruptions from the chamber are small to moderate in volume.

- Large effusive eruptions, again erupting more than 10 km^3 and some more than 1000 km^3, are explained in terms of associated graben subsidence or collapse. The graben subsidence reduces the reservoir volume and largely maintains the excess pressure for most of the eruption. This mechanism also applies to grabens dissecting central volcanoes and calderas, and may thus contribute to squeezing out magma from their shallow chambers. Here, however, the focus is on large effusive eruptions, most of which are fed by very large, deep-seated reservoirs. In this mechanism, the graben subsidence or collapse is the primary cause of the large eruption, in agreement with many large effusive eruptions being associated with large grabens. Thus, many graben subsidences or collapses generate large effusive eruptions, whereas many caldera collapses generate large explosive eruptions.

8.8 Main Symbols Used

A	area
a	semi-major axis of an ellipse
b	semi-minor axis of an ellipse
C	constant
e	2.71828 .... (Euler's number, a constant)
g	acceleration due to gravity
h	height or dip-dimension of a conduit measured from its contact with its chamber
p_b	fluid (magma) pressure at the bottom of a conduit
p_e	excess magma pressure
$p_e(t)$	excess pressure as a function of time
p_o	magmatic overpressure
p_t	fluid (magma) pressure at the top of a conduit
Δp	fluid (magma) pressure difference between top and bottom of vertical conduit
Q	volumetric flow rate (of a fluid, here magma)
$Q(t)$	volumetric magma flow rate as a function of time
$\bar{q}$	mean velocity of magma flow through a conduit
R	radius of cylindrical conduit
t	time
Δu	opening or aperture of a dike (or an inclined sheet or a sill)

V_c	total volume of a shallow chamber
V_{er}	'eruptive' volume of magma, including both intrusive and extrusive volumes
W	length of dike (strike-dimension) perpendicular to the magma-flow direction
x	horizontal coordinate, horizontal distance
y	horizontal coordinate, horizontal distance
z	vertical coordinate, vertical distance
β_m	magma compressibility
β_r	host-rock compressibility
μ_m	dynamic viscosity of magma
π	3.1416
ρ_m	magma density
ρ_r	rock or crustal density
σ_d	differential stress or stress difference in a particular crustal layer
ψ	ratio of excess magmatic pressure over eruptive volume
∂	partial derivative

8.9 Worked Examples

Example 8.1

Problem

During eruption through a cylindrical conduit of a stratovolcano, the peak volumetric flow rate is measured at 1000 m^3 s^{-1}. The conduit extends from a magma chamber at an estimated depth of 4 km; the chamber excess pressure is assumed to be 4 MPa when the eruption started. Assume that the viscosity of the magma is of the order of 10^5 Pa s (intermediate magma) and calculate roughly the diameter of the conduit.

Solution

Since no information is given about the elastic properties of the host rock, we use the equation for rigid rock, namely Eq. (8.1), which is:

$$Q = -\frac{\pi R^4}{8\mu_m}\frac{\partial p}{\partial z}$$

The aim is to find first the radius R of the conduit and then its diameter $2R$. To do so we rewrite Eq. (8.1) and solve for R as follows:

$$R = \left(\frac{8Q\mu_m}{-\frac{\partial p_e}{\partial z}\pi}\right)^{1/4} \tag{8.28}$$

Since the magma flow is in the negative direction of the vertical coordinate z, it follows that the gradient $\partial p_e/\partial z$ is itself negative, and that the minus sign in front of it thus changes its value to a positive one. The pressure change from the magma chamber to the surface is then 4 MPa, since the pressure at the top of the conduit is atmospheric (0.1 MPa) and assumed, in this context, to be 0 MPa. Using the values given above, we have from Eq. (8.28):

$$R = \left(\frac{8Q\mu_m}{-\frac{\partial p_e}{\partial z}\pi}\right)^{1/4} = \left(\frac{8 \times 1000 \text{ m}^3\text{s}^{-1} \times 1 \times 10^5 \text{Pa s}}{1000 \text{Pa m}^{-1} \times 3.1416}\right)^{1/4} = 22.5 \text{ m}$$

It follows that the diameter of the conduit is about 45 m. Thus, a reasonably large conduit would be needed. On solidification, the conduit would become a volcanic plug. Plugs of this diameter are common in many areas (Fig. 8.3).

Example 8.2

Problem

Use Example 8.1, but now with the additional knowledge that the rock is elastic. The average magma density is 2550 kg m^{-3} and that of the rock hosting the conduit (the uppermost 4 km of the crustal segment, including the volcano) is 2650 kg m^{-3}. Calculate the diameter of the conduit and explain the difference between the results obtained here and in Example 8.1. Discuss the different approaches and, in particular, which one is more appropriate.

Solution

Since we now have information on the density of the magma and the host rock, we do not use Eq. (8.1) but rather Eq. (8.3). Rewriting and solving for the radius of the conduit, R, Eq. (8.3) becomes:

$$R = \left(\frac{8Q\mu_m}{\pi\left[(\rho_r - \rho_m)g - \frac{\partial p_e}{\partial z}\right]}\right)^{1/4} \tag{8.29}$$

It is important to remember, as in Example 8.1, that the magma flow is in the negative direction of the vertical coordinate z, so that the minus sign in front of the gradient $\partial p_e/\partial z$ makes its value positive. Using the given values, we have from Eq. (8.29):

$$R = \left(\frac{8 \times 1000 \text{ m}^3\text{s}^{-1} \times 1 \times 10^5 \text{ Pa s}}{3.1415 \times \left[(2650 \text{ kg m} - 2550 \text{ kg m}^{-3}) \times 9.81 \text{ m s}^{-2} + \frac{4\times10^6 \text{ Pa}}{4000 \text{ m}}\right]}\right)^{1/4} = 18.9 \text{ m}$$

It follows that the diameter of the conduit is close to 38 m, and thus is considerably smaller than that obtained in Example 8.1. The reason is that, since the density difference between the host rock and the magma is positive (100 kg m^{-3}), the buoyancy adds to the pressure gradient driving the magma up through the conduit. Thus, the magma flow velocity is higher here than in Example 8.1, and a conduit with a smaller radius can thus produce the same volumetric flow rate (Fig. 8.4), 1000 m^3 s^{-1}, as the larger conduit in Example 8.1.

Which is more appropriate to use, Eq. (8.28) or (8.29)? Solid rocks at comparatively shallow depths (the uppermost kilometres of the crust) behave to a first approximation as elastic (Chapter 3). The same applies to volcanoes. No rocks behave as rigid. Thus, Eq. (8.29) is the most appropriate one to use. In particular, when the density of the rock hosting the conduit is the same as, or less than, that of the magma – as is common for basaltic magma – then the buoyancy would be zero or negative and reduce the volumetric flow rate. That possible effect on the volumetric flow rate through a cylindrical conduit is totally ignored when using Eqs. (8.28) or (8.1) because buoyancy does not enter these equations. Thus, when modelling magma transport in conduits of any type, including dikes and inclined sills, the elastic behaviour of the host rock should be taken into account whenever possible.

Example 8.3

Problem

A volcanic fissure opens at the surface of a volcano. The fissure is initially 600 m long and has an estimated (from GPS measurements) maximum opening of 0.8 m. The peak effusion rate is 300 m^3 s^{-1}. The depth to the magma chamber is unknown. Assume that the volcanic fissure can be modelled as an elliptical opening. Estimate the velocity of the magma up through the volcanic fissure.

Solution

Since the depth to the magma source is unknown, then so is the pressure gradient that drives the magma up through the feeder-dike and the volcanic fissure. It follows that the velocity should be determined from Eq. (8.6), with the semi-major axis of 300 m and semi-minor axis of 0.4 m. Using this information we get:

$$\bar{q} = \frac{Q}{A} = \frac{Q}{\pi a b} = \frac{300 \text{ m}^3 \text{ s}^{-1}}{3.1416 \times 300 \text{ m} \times 0.4 \text{ m}} = 0.8 \text{ m s}^{-1}$$

This is a very reasonable velocity for magma migration as estimated from the migration of earthquakes associated with dike propagation (Chapter 4).

Example 8.4

Problem

A vertical dike reaches the surface to form a volcanic fissure that is 4 km long. The maximum opening or aperture of the feeder-dike is estimated at 3 m. The source magma chamber is at 5 km depth. The basaltic magma viscosity is 1000 Pa s and the magma density is 2650 kg m^{-3}, whereas the average crustal density of the host rock (the uppermost 5 km of the crustal segment) is 2600 kg m^{-3}. The excess magmatic pressure in the chamber at the time of rupture and dike injection is estimated at 4 MPa. Use the elliptical conduit model of a dike (Eq. (8.9)) to calculate the volumetric flow rate.

Solution

For convenience, we shall here omit the units of the parameters in the calculations, but show them for the volumetric flow rate at the end. From Eq. (8.9) we have:

$$Q = \frac{\pi}{4\mu_m} \frac{(ab)^3}{(a^2 + b^2)} \left[(\rho_r - \rho_m)g - \frac{\partial p_e}{\partial z} \right] = \frac{3.1416}{4 \times 10^3} \frac{(2000 \times 1.5)^3}{(2000^2 + 1.5^2)}$$
$$\left[(2600 - 2650) \times 9.81 + \frac{4 \times 10^6}{5 \times 10^3} \right] = 1640.8 \text{ m}^3 \text{ s}^{-1}$$

This effusion rate is similar to that during the beginning stages of many moderate to large basaltic eruptions, inside and outside central (polygenetic) volcanoes. Please notice the following:

(a) As before the excess pressure gradient itself is negative so that the negative sign in front of it in Eq. (8.9) makes the gradient positive.
(b) The buoyancy effect is negative so that it decreases the effusion rate. This is, as said earlier, very common for basaltic eruptions from shallow chambers. In fact, the crustal density used here is from the upper 5 km of the crust in Iceland and the magma density is typical for tholeiitic magma.
(c) It follows from point b that the excess pressure in the chamber is the only driving pressure of the eruption. In this case the excess pressure, and thus roughly the *in situ* tensile strength in the chamber roof, is assumed to be 4 MPa. But we know that the tensile strength can be lower, say 1–2 MPa. And if that were the tensile strength here, the dike would never be able to reach the surface. An excess pressure of 2 MPa is too low to allow the dike to reach the surface because the driving pressure inside the bracket above would then be: $-490.5 + 400 = -90.5$ Pa m^{-1}. A negative gradient cannot drive the magma to the surface (or anywhere else), so that the dike would become arrested and no eruption would take place.

Example 8.5

Problem

Repeat Example 8.4 using the parallel-plate model (Eq. (8.11)). Comment on the difference between the results obtained from that model and the elliptical model (Eq. (8.9)).

Solution

Here, again, we omit the units during the calculations, but show them at the end. From Eq. (8.11) we have:

$$Q = \frac{\Delta u^3 W}{12\mu_m} \left[(\rho_r - \rho_m)g - \frac{\partial p_e}{\partial z} \right] = \frac{3^3 \times 4000}{12 \times 1000} [(2600 - 2650) \times 9.81 + 800]$$
$$= 2785 \text{ m}^3 \text{ s}^{-1}$$

Clearly, the parallel-plate model gives a much higher effusion rate than the elliptical model – to be exact, close to a 70% higher effusion rate. The reason is the difference in the assumed cross-sectional area of the opening or aperture of the feeder-dike. The elliptical shape means that the opening is everywhere less than the maximum, which occurs in the centre. By contrast, the parallel-plate model assumes that opening to be constant throughout the entire length or strike-dimension of the dike, hence a much larger horizontal cross-sectional area. Both models can be used, but normally the parallel-plate model is closer to the actual thickness (aperture) variation in the feeder-dikes.

Example 8.6

Problem

The total volume of intrusive and extrusive material produced during an eruption is estimated at 2 km^3. The average volumetric flow rate during the eruption was 100 m^3 s^{-1}. Assume the excess pressure in the source chamber at the time of its rupture and dike injection to be 4 MPa. If no significant inflow of magma from a deeper reservoir occurred during the eruption, estimate (a) the excess pressure in the chamber after 5 months or 20 weeks and (b) crudely, the duration of the eruption. (c) Comment on the accuracy of the methods used to solve the problems.

Solution

(a) Since no inflow takes place, Eq. (8.20) can be used to obtain a crude estimate of the excess pressure in the chamber after 20 weeks. First we recast Eq. (8.20) in a more explicit form, namely:

$$p(t) = p_e - p_e e^{-\left(\frac{V_{er} - Qt}{Qt}\right)} \qquad (8.30)$$

where p_e is the excess pressure in the chamber at the time of its rupture, here 4 MPa, Q is the average volumetric flow rate, here 100 m^3 s^{-1}, and V_{er} is the total magma flow out of the chamber during the eruption, here 2 km^3 or 2×10^9 m^3. Since we use SI units, time must be given in seconds: 20 weeks are $20 \times 7 \times 24 \times 60 \times 60 = 1.2 \times 10^7$ s. Inserting these values in Eq. (8.30) we get:

$$p(1.2 \times 10^7 \text{s}) = 4 \times 10^6 \text{ Pa} - 4 \times 10^6 \text{ Pa} \times e^{-\left(\frac{2\times10^9 \text{m}^3 - 100 \text{ m}^3\text{s}^{-1}\times 1.2\times10^7\text{s}}{100 \text{ m}^3 \text{ s}^{-1}\times 1.2\times10^7\text{s}}\right)} = 1.3 \times 10^6 \text{ Pa}$$
$$= 1.3 \text{ MPa}$$

Thus, after 20 weeks, the excess pressure in the magma chamber has decreased from 4 MPa to about 1.3 MPa, or about 32% of its original value.

(b) Here we assume that the eruption comes to an end when the flow of magma out of the chamber and into the feeder-dike stops. We also assume that this happens when the excess pressure in the chamber approaches zero, so that $p_e \rightarrow 0$.

To simplify the problem, we set $p_e = 0$,

which means that the two terms containing p_e on the right side of Eq. (8.30) must be equal. For this to happen:

$$e^{-y} = 1 \tag{8.31}$$

It then follows, because $e^0 = 1$, that:

$$y = \left(\frac{V_{er} - Qt}{Qt}\right) = 0 \tag{8.32}$$

from which it also follows that, since then we must have $V_{er} - Qt = 0$, the duration of the eruption t becomes:

$$t = \frac{V_{er}}{Q} \tag{8.33}$$

From Eq. (8.33) and the volume V_{er} we then have $Qt = 2 \times 10^9$ m^3. Given that $Q = 100$ m^3 s^{-1} it follows that:

$$t = \frac{2 \times 10^9 \text{m}^3}{100 \text{ m}^3\text{s}^{-1}} = 2 \times 10^7\text{s} = 232 \text{ days or } 33 \text{ weeks}$$

So, based on the information given and the above assumptions, the eruption should have lasted 33 weeks or close to 8 months. This is therefore a rather long eruption (Fig. 8.14).

(c) There are many assumptions in this analysis and the associated equations, some of which are less justified than others. First, in the present analysis, and in Eq. (8.20), an average, constant volumetric flow rate Q is used, whereas during an eruption Q varies and commonly declines as a negative exponential function (Eq. (8.19)). Such a decline would be expected to be in harmony with the decline in the excess pressure in the chamber (Eq. (8.20)). However, volumetric flow rates depend strongly on the geometry of the volcanic fissure/conduit, and this may change during the eruption. For example, the volumetric flow rate may temporarily increase during an eruption, even if the excess pressure in the source chamber is decreasing. This is, for example, seen in Fig. 8.9. Such an increase in volumetric flow rate may relate to sudden changes in the dimensions of the volcanic fissure, particularly to a sudden increase in its aperture. This follows because of the cubic law (Eq. (8.11)), which makes the volumetric flow rate through a volcanic fissure/feeder-dike very sensitive to aperture changes.

Second, flow of magma from a deeper source reservoir to the chamber during the eruption is assumed to be insignificant. This is presumably a valid assumption for most eruptions, since the majority are less than 3 months long and the median length is only 7 weeks. For eruptions of longer duration, lasting many months or years, this assumption becomes gradually less justified since significant volumes of magma may flow into the chamber from the source reservoir. Even for some shorter eruptions, lasting a few weeks or less, the chamber may receive magma from its source. Much depends on the type of connection between the chamber and the source reservoir (Fig. 8.2) – whether the connection is continuously open (generally unlikely) or a new one is very easily formed – the distance between the chamber and the source reservoir, the state of stress in the crust (Chapter 3), and other factors. Woods and Huppert (2003) consider the case where the shallow chamber receives magma during chamber eruptions.

8.10 Exercises

8.1 What are the eruptive volume ranges for small, moderate, and large eruptions?

8.2 How many subaerial polygenetic volcanoes are thought to be active? How many volcanoes erupt, on average, each decade or each year, and how many are erupting at any moment (for example, each day)?

8.3 Which was the largest (in terms of eruptive volumes) eruption in the twentieth century? What was the eruptive volume?

8.4 Name and indicate the volumes of the two largest eruptions in the nineteenth century.

8.5 What are the estimated volumes of the largest effusive and explosive eruptions and how frequently do they occur?

8.6 What sort of probability size distributions do eruptive volumes follow? Name several other natural processes that give rise to similar size distributions.

8.7 What are the main factors or parameters that determine the volumetric flow rate of magma up through a cylindrical conduit? Which of these parameters has the greatest effect on the flow rate?

8.8 What volcanotectonic structures correspond to cylindrical conduits? What are the common dimensions of these structures?

8.9 Why is the magma-pressure gradient for flow to the surface in a volcano from a chamber/reservoir normally shown with a negative sign?

8.10 Some volcanoes apparently have continuously open conduits (over years or decades). Which are the normal surface expressions of continuously open conduits?

8.11 What is the name of the general equation used to describe magma flow in fractures such as in dikes, inclined sheets, and sills, and which factor in that equation has the greatest effect on the volumetric flow rate?

8.12 How does the dip of a feeder-dike or feeder-sheet affect the volumetric flow rate?

8.13 Explain under what conditions a fracture (a dike) containing high-density basaltic magma could propagate to the surface through rocks of an average density less than that of the magma. What does the result imply for the concept of 'neutral buoyancy'?

8.14 What is the median duration of a volcanic eruption in weeks?

8.15 What is the overall percentage of eruptions that peak within (a) the first day and (b) the first week, and (c) after 10 years?

8.16 In which way do outward- and inward-dipping ring-faults differ as regards their eruptive potential?

8.17 Provide an explanation for large explosive eruptions from moderately sized shallow magma chambers.

8.18 Provide an explanation for large effusive eruptions from moderately sized deep-seated reservoirs.

8.19 A volcanic fissure has a peak effusion rate of about 600 m^3 s^{-1}. The initial fissure is 1800 m long and has a maximum aperture of 2.2 m. Model the fissure as an elliptical opening and calculate the velocity of the magma flow up through the fissure during the eruption peak.

8.20 Further information about the volcanic fissure in Exercise 8.19 yields that the source magma chamber is at the depth of 1.9 km and the magma viscosity is 1000 Pa s. Also, the magma density is 2650 kg m^{-3}, whereas the average crustal density of the host rock (the uppermost 3 km of the crustal segment) is 2500 kg m^{-3}. The excess magmatic pressure in the chamber at the time of rupture and dike injection is estimated to be 4 MPa. Use the elliptical conduit model of a dike (Eq. (8.9)) to calculate the volumetric flow rate.

8.21 Use the parallel-plate model to calculate the maximum opening of a volcanic fissure with a length of 1500 m at the peak effusion rate of 800 m^3 s^{-1}. The source magma chamber of the vertical feeder-dike is at a depth of 3 km, the average host-rock density is 2600 kg m^{-3}, the host-rock tensile strength is 3 MPa, the magma density is 2500 kg m^{-3}, and the magma viscosity 10 000 Pa s.

8.22 During an eruption in a stratovolcano, the peak volumetric flow rate is measured at 700 m^3 s^{-1}. The source chamber is at 3 km depth and the excess pressure in the chamber before eruption is assumed to be 3 MPa. If the magma viscosity is 10 000 Pa s, estimate the diameter of the cylindrical conduit, assuming the host rock to behave as rigid.

References and Suggested Reading

Amelung, F., Jonsson, S., Zebker, H., Segall, P., 2000. Widespread uplift and 'trapdoor' faulting on Galapagos volcanoes observed with radar interferometry. *Nature*, **407**, 993–996.

Anderson, E.M., 1936. The dynamics of formation of cone sheets, ring dykes and cauldron subsidences. *Proceedings of the Royal Society of Edinburgh*, **56**, 128–163.

Bonadonna, C., Costa, A., 2013. Plume height, volume, and classification of explosive volcanic eruptions based on the Weibull function. *Bulletin of Volcanology*, **75**, doi:10.1007/s00445-013-0742-1.

Bull, J. M., Minshull, T. A., Mitchell, N. C., et al., 2003. Fault and magmatic interaction within Iceland's western rift over the last 9 kyr. *Geophysical Journal International*, **154**, F1–F8.

Cashman, K. V., Sparks, R. S. J., 2013. How volcanoes work: a 25 year perspective. *Geological Society of America Bulletin*, **125**, 664–690.

Clarke, A. B., Voight, B., Neri, A., Macedonio, G., 2002. Transient dynamics of vulcanian explosions and column collapse. *Nature*, **415**, 897–901.

Clauset, A., Chalizi, R. C., Newman, M. E. J., 2009. Power-law distributions in empirical data. *Society for Industrial and Applied Mathematics*, **51**, 661–703.

Cummings, M. L., Evans, J. G., Ferns, M. L., Lees, K. R., 2000. Stratigraphic and structural evolution of the middle Miocene synvolcanic Oregon-Idaho graben. *Bulletin of the Geological Society of America*, **112**, 668–682.

Fagents, S. A., Gregg, T. K. P., Lopes, R. M. C. (eds.), 2013. *Modeling Volcanic Processes: The Physics and Mathematics of Volcanism*. Cambridge: Cambridge University Press.

Fedotov, S. A., Chirkov, A. M., Gusev, N. A., Kovalev, G. N., Slezin, Yu. B., 1980. The large fissure eruption in the region of Plosky Tolbachik Volcano in Kamchatka, 1975–1976. *Bulletin of Volcanology*, **43**, 47–60.

Gonnermann, H. M., Manga, M., 2013. Dynamics of magma ascent in the volcanic conduit. In Fagents, S. A., Gregg, T. K. P., Lopes, R. M. C. (eds.), *Modeling Volcanic Processes*. Cambridge: Cambridge University Press, pp. 55–84.

Gottsmann, J., Camacho, A. G., Marti, J., et al., 2008. Shallow structure beneath the Central Volcanic Complex of Tenerife from new gravity data: implications for its evolution and recent reactivation. *Physics of the Earth and Planetary Interiors*, **168**, 212–230.

Gudmundsson, A. 1987. Tectonics of the Thingvellir fissure swarm, SW Iceland. *Journal of Structural Geology*, **9**, 61–69.

Gudmundsson, A., 2014. Energy release in great earthquakes and eruptions. *Frontiers in Earth Science*, **2**, doi:10.3389/feart.2014.00010.

Gudmundsson, A., 2015. Collapse-driven large eruptions. *Journal of Volcanology and Geothermal Research*, **304**, 1–10.

Gudmundsson, A., 2016. The mechanics of large volcanic eruptions. *Earth-Science Reviews*, **163**, 72–93.

Gudmundsson, A., Mohajeri, N., 2013. Relations between the scaling exponents, entropies, and energies of fracture networks. *Geological Society of France Bulletin*, **184**, 377–387.

Gudmundsson, A., Oskarsson, N., Gronvold, K., et al., 1992. The 1991 eruption of Hekla, Iceland. *Bulletin of Volcanology*, **54**, 238–246.

Harris, A. J. L., 2013. Lava flows. In Fagents, S. A., Gregg, T. K. P., Lopes, R. M. C. (eds.), *Modeling Volcanic Processes*. Cambridge: Cambridge University Press, pp. 85–106.

Hildreth, W., Fierstein, J., 2012. The Novarupta–Katmai eruption of 1912: largest eruption of the twentieth century – centennial perspectives. *US Geological Survey Professional Paper*, 1791. Denver, CO: US Geological Survey, pp. 1–278.

James, M. R., Lane, S. J., Houghton, B. F., 2013. Unsteady explosive activity: strombolian eruptions. In Fagents, S. A., Gregg, T. K. P., Lopes, R. M. C. (eds.), *Modeling Volcanic Processes*. Cambridge: Cambridge University Press, pp. 107–128.

Kilburn, C. J., 2000. Lava flows and flow fields. In Sigurdsson, H. (ed.), *Encyclopedia of Volcanoes*. New York, NY: Academic Press, pp. 291–305.

Lamb, H., 1932. *Hydrodynamics*, 6th edn. Cambridge: Cambridge University Press.

Lipman, P. W. 1997. Subsidence of ash-flow calderas: relation to caldera size and magma chamber geometry. *Bulletin of Volcanology*, **59**, 198–218.

Lipman, P. W., Dungan, M. A., Bachmann, O., 1997. Comagmatic granophyric granite in the Fish Canyon Tuff, Colorado: implications for magma-chamber processes during a large ash-flow eruption. *Geology*, **25**, 915–918.

Machado, F., 1974. The search for magmatic reservoirs. In Civetta, L., Gasparini, P., Luongo, G., Rapolla, A. (eds.), *Physical Volcanology*. Amsterdam: Elsevier, pp. 255–273.

Mason, B. G., Pyle, D. M., Oppenheimer, C., 2004. The size and frequency of the largest explosive eruptions on Earth. *Bulletin of Volcanology*, **66**, 735–748, doi:10.1007/s00445-004-0355-9.

Milne-Thompson, L. M., 1996. *Theoretical Hydrodynamics*, 5th edn. New York, NY: Dover.

Mohajeri, N., Gudmundsson, A., 2012. Entropies and scaling exponents of street and fracture networks. *Entropy*, **14**, 800–833.

Murase, T., McBirney, A. R., 1973. Properties of some common igneous rocks and their melts at high temperatures. *Geological Society of America Bulletin*, **84**, 3563–3592.

Nakada, S., Uto, K., Sakuma, S., Eichelberger, J. C., Shimizu, H., 2005. Scientific results of conduit drilling in the Unzen Scientific Drilling Project (USDP). *Scientific Drilling*, **1**, 18–22, doi:10.2204/iodp.sd.1.03.2005.

Neal, C. A., Brantley, S. R., Antolik, J. L., et al., 2019. The 2018 rift eruption and summit collapse of Kilauea Volcano. *Science*, **363**, 367–374.

Neri, M., Acocella, V., Behncke, B., et al., 2011. Structural analysis of the eruptive fissures at Mount Etna (Italy). *Annals of Geophysics*, **54**, 464–479.

Newhall, C. G., Dzurisin, D., 1988. Historical unrest of large calderas of the world. *US Geological Survey Bulletin*, **1855**, Reston, VA.

Newhall, C., Hendley, J. W., Stauffer, P. H., 1997. *The Cataclysmic 1991 Eruption of Mount Pinatubo, Philippines*. US Geological Survey Fact Sheet-113–97.

Parfitt, L., Wilson, L., 2008. *Fundamentals of Physical Volcanology*. New York, NY: Wiley.

Pisarenko, V., Rodkin, M., 2010. *Heavy-Tailed Distributions in Disaster Analysis*. Berlin: Springer Verlag.

Poland, M. P., Takahashi, T. J., Landowski, C. M. (eds.), 2014. *Characteristics of Hawaiian Volcanoes. US Geological Survey Professional Paper, 1801*. Denver, CO: US Geological Survey.

Roche, O., Phillips, J. C., Kelfoun, K., 2013. Pyroclastic density currents. In Fagents, S. A., Gregg, T. K. P., Lopes, R. M. C. (eds.), *Modeling Volcanic Processes*. Cambridge: Cambridge University Press, pp. 203–229.

Rosi, M., Papale, P., Lupi, L., Stoppato, M., 2003. *Volcanoes*. Buffalo (USA): Firefly Books.

Rossi, M. J., 1996. Morphology and mechanism of eruption of postglacial lava shields in Iceland. *Bulletin of Volcanology*, **57**, 530–540.

Self, S., Rampino, M. R., Newton, M. S., Wolff, J. A., 1984. Volcanological study of the great Tambora eruption of 1815. *Geology*, **12**, 659–663.

Sigurdsson, H., Houghton, B. F., McNutt, S. R., Rymer, H., Stix, J. (eds.), 2000. *Encylopedia of Volcanoes*. New York, NY: Academic Press.

Simkin, T., Fiske, R. S. (eds.), 1980. *Krakatau, 1983: The Volcanic Eruption and Its Effects*. Washington DC.: Smithsonian Books.

Simkin, T., Siebert, L., 2000. Earth's volcanoes and eruptions: an overview. In Sigurdsson, H. (ed.), *Encyclopedia of Volcanoes*. New York, NY: Academic Press, pp. 249–261.

Sinton, J., Gronvold, K., Saemundsson, K., 2005. Postglacial eruptive history of the western Volcanic Zone, Iceland. *Geochemistry, Geophysics, Geosystems*, **6**, doi:10.1029/2005GC001021.

Sonnette, L., Angelier, J., Villemin, T., Bergerat, F., 2010. Faulting and fissuring in active oceanic rift: Surface expression, distribution and tectonic-volcanic interaction in the Thingvellir Fissure Swarm, Iceland. *Journal of Structural Geology*, **32**, 407–422.

Sparks, R. S. J., Bursik, M. I., Carey, S. N., et al., 1997. *Volcanic Plumes*. New York, NY: Wiley.

Stasiuk, M. V., Jaupart, C., Sparks, R. S. J., 1993. On the variation of flow rate in non-explosive lava eruptions. *Earth and Planetary Science Letters*, **114**, 505–516.

Thordarson, T., Larsen, G., 2007. Volcanism in Iceland in historical time: Volcano types, eruption styles and eruptive history. *Journal of Geodynamics*, **43**, 118–152.

Thordarson, T., Self, S., 1993. The Laki (Skaftar Fires) and Grimsvotn eruptions in 1783–1785. *Bulletin of Volcanology*, **55**, 233–263.

Tryggvason, E., 1982. Recent ground deformation in continental and oceanic rift zones. *American Geophysical Union Geodynamic Series*, **8**, 17–29.

Wadge, G., 1981. The variation of magma discharge during basaltic eruptions. *Journal of Volcanology and Geothermal Research*, **11**, 139–168.

White, F. M., 2005. *Viscous Fluid Flow*, 3rd edn. New York, NY: McGraw-Hill.

Woods, A. W., Huppert, H. E., 2003. On magma chamber evolution during slow effusive eruptions. *Journal of Geophysical Research*, **108**, 2403, doi:10.1029/2002JB002019.

9 Formation and Evolution of Volcanoes

9.1 Aims

How are volcanoes born, in which way do they live, and why and when, eventually, do they become extinct? More specifically, why are there any polygenetic central volcanoes at all? Why is the volcanism not simply evenly distributed through the volcanic zone or field? There are certainly volcanic areas where the volcanism is more or less evenly distributed – such as at fast-spreading ridges. But for most other volcanic areas one or more locations within the area dominates the volcanism, resulting in the formation of a specific central volcano. The central volcano erupts much more frequently than its surroundings and, therefore, normally forms an edifice. The edifice is a structure with a certain mechanical strength, most of which derives from the internal structure of the volcano. Later in the evolution of the volcano, it may form a collapse caldera, as well as being subject to lateral collapses, all of which affect its shape. When, eventually, the supply of magma to the volcano is cut off, it becomes extinct, that is, dies. The aim of this chapter is to use the results of the previous chapters to explain:

- Why and how central volcanoes, basaltic edifices, and stratovolcanoes, are born and form edifices.
- The eruption and intrusion frequencies of typical central volcanoes and how these depend on the mechanical interaction between shallow chambers and deeper reservoirs.
- The overall shapes of central volcanoes and how they depend on vertical and lateral collapses.
- The internal structure and mechanical strength of central volcanoes, including the main structural elements that constitute the volcanoes.
- How volcanoes become extinct, that is, cease to be active.

9.2 Why Are There Any Volcanic Edifices?

By a volcano, here and elsewhere in the book (unless specified otherwise) we mean a polygenetic volcano, that is, a **central volcano**. Such volcanoes have various names, depending on their geometry and composition. Here, as elsewhere in the book, mainly the

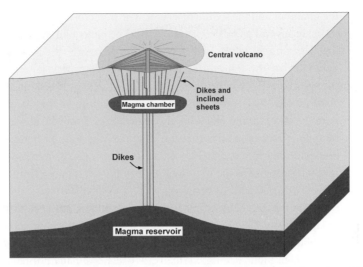

Fig. 9.1   A shallow magma chamber is a necessary and, normally, a sufficient condition for the formation of the edifice of a central/polygenetic volcano. The shallow magma chamber acts as a sink for magma from the associated deep-seated reservoir and as a source for the intrusions and extrusions of the central volcano. The reservoir is always much larger than the shallow chamber to which the reservoir supplies magma (cf. Figs. 1.12, 6.1, 6.7, 6.8).

following names are used: stratovolcanoes (composite volcanoes), basaltic edifices, and collapse calderas. While some calderas do not belong to collapsed edifices and may be graben-like (Aguirre-Diaz, 2008), the discussion here focuses on calderas as a part of the evolution of volcanic edifices.

So why are there any **volcanic edifices**? Why is the eruption frequency much **higher** in one part of a volcanic field or zone than in the adjacent parts? The answer to both questions is: a shallow magma chamber. The formation of a shallow magma chamber is a **necessary** and, normally, a **sufficient** condition for the formation of a volcanic edifice (Fig. 9.1). The chamber acts as a **sink**, that is, collects magma from a deeper reservoir, and as a **source** for eruptions at the surface above the chamber. Thus, a shallow magma chamber is both a sink and a source. If there were no shallow magma chambers, there would be hardly any volcanic edifices.

The formation of a shallow magma chamber has already been discussed (Chapter 5), so let us start from that point in the history of the volcano. A magma chamber gradually generates a volcanic edifice, as illustrated schematically in Fig. 5.2. The source reservoir of the shallow chamber is always **much larger** than the chamber. This follows from general considerations of the flow of magma or melt and information about the geometries of the volcanic fields/zones and systems. Once the shallow chamber has formed, it modifies the local stress field, depending on the chamber shape, size, and depth.

Let us consider how magma chambers function as sinks – collect magma, from the source reservoirs – in relation to the chamber shape. Here, only two common chamber **shapes** are considered, and the mechanical layering of the crust is ignored while the main principles are discussed. For a sill-like chamber – an oblate ellipsoid – the trajectories of σ_1 indicate the most likely direction of the propagation of magma paths, dikes, or inclined sheets, between the chamber and the source reservoir. The trajectories, calculated using numerical software, indicate several possible magma paths. The paths will of course be different when mechanical layering is taken into account. Furthermore, the loading has great effects on the geometry of the trajectories and therefore from how large a fraction of the top part of the reservoir the chamber draws or collects magma. When internal magmatic pressure in the shallow chamber is the only or primary loading, as may be the case during some unrest periods, the fanning shape of the trajectories from the lower margin or the floor of the magma chamber implies that the chamber can draw in magma from an area much wider than the chamber itself (Fig. 9.2a). In other words, a chamber of this shape subject to internal pressure as the only loading can tap off magma from a very **large fraction** of the upper part or roof of its reservoir.

While the geometry of most magma chambers is likely to be approximately that of an oblate ellipsoid, some chambers may have a circular vertical cross-section, and they may even be of the shape of a prolate ellipsoid with a vertical long axis (Gudmundsson, 2006, 2012). Those with circular vertical cross-sections may be either cylindrical, with a horizontal long axis, or spherical. Magma chambers are most likely to reach the approximate shape of a sphere at the end stages of their lifetimes. This follows from thermal and mechanical considerations (Gudmundsson, 2012). The stress models presented below are two-dimensional, so that they can apply either to a cylindrical magma chamber or a spherical one.

The stress trajectories around a chamber with a **circular vertical cross-section** subject to an internal excess pressure as the only loading are more widely spread than those around an oblate ellipsoid (Fig. 9.3a). Many of the potential dike paths are gently inclined, so that magma can be collected from a large fraction of the top part or the roof of the source reservoir. However, the dips of the stress trajectories, and thus the **magma paths**, depend on the loading. Thus, if the main or only loading is an external tension, such as related to rifting, rather than an internal excess pressure, then the stress trajectories and the related magma paths become very different (Fig. 9.3b). In particular, the paths become much more clustered – similar to those indicated schematically in Fig. 9.1 – in which case the chamber is only able to draw magma from a small fraction of the upper part of the reservoir. Depending on the loading, a chamber of a given shape may thus draw magma from different parts of the source reservoir, that is, from different compartments, some of which may contain magmas of widely **different composition** (Gudmundsson, 2012). If the reservoir has a completely flat upper margin or roof, the trajectories tend to be vertical on approaching the margin of the reservoir. This means that the trajectories change their dips between the chamber and the reservoir – forming curves (Figs. 9.2, 9.3).

Some reservoirs may have essentially flat upper margins, but many are likely to be somewhat **dome-shaped** (Fig. 9.1), particularly those in areas of extension, such as rift zones and divergent plate boundaries. This follows partly because of thinning of the crust in those areas, resulting in a sort of neck or notch. For a dome-shaped reservoir, the stress trajectories are also curved. But because they must meet the upper dome-shaped margin of the reservoir at right angles, the direction of the dip must change between the reservoir and the chamber.

The above considerations indicate how the chamber acts as a sink, collecting magma paths from a large part of the underlying reservoir. To see how the chamber acts as a source, let us consider again the σ_1 stress trajectories, in this case those between the chamber and the surface. It

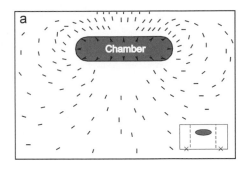

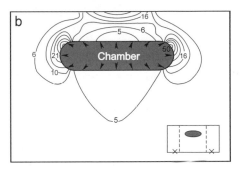

Fig. 9.2   Numerical models show (a) the trajectories of the maximum compressive principal stress σ_1 and (b) the magnitude (shown as contours) of the maximum tensile principal (minimum compressive principal) stress σ_3 around a shallow, sill-like magma chamber. The dikes/inclined sheets follow the σ_1 trajectories (Chapter 10), which become vertical on meeting free surfaces, namely the surface of the rift zone and the upper margin (roof/ceiling) of the magma chamber. Depending on the loading and the chamber and reservoir shape, and other factors, the chamber may draw in magma from a large part of the reservoir. Here the only loading is a magmatic excess pressure of 5 MPa in the chamber and the fanning shape of the stress trajectories indicate that the chamber could get magma through paths through a large part of the reservoir roof (the lower margin of the model). The crustal segment is homogeneous and isotropic with a Young's modulus of 40 GPa and a Poisson's ratio of 0.25. The horizontal width (lateral dimension in this vertical section) of the chamber is about four times the depth of its roof below the surface. Thus, the roof of a chamber, the width of which in this cross-section is 8 km, would be at about 2 km depth below the surface – as the roofs of many shallow chambers are (Figs. 1.17, 6.17, 6.18, 6.19, 7.5).

is clear from the distribution of the trajectories (Figs. 9.2, 9.3) that the magma paths from the chamber are commonly channelled to a limited area at the surface (Fig. 9.1). Again, for a chamber with a circular cross-section, the surface area to which the magma paths are channelled is larger, for a given lateral diameter of a magma-chamber subject to an internal excess pressure as the only loading, than for a sill-like or oblate-ellipsoidal magma chamber. The net result, however, is similar: magma is channelled to a comparatively small area at the surface.

Because of the stress concentration around the chamber (Fig. 9.2; Chapter 3) and the collection of magma from the source reservoir, **magma-path formation** (dike/sheet injection and propagation) is much more frequent above the chamber than elsewhere in the

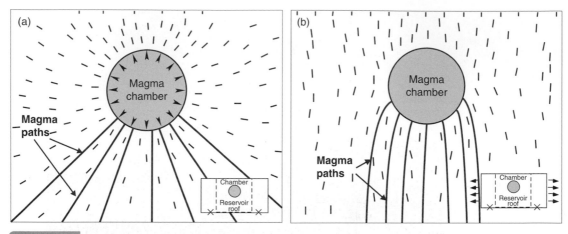

Fig. 9.3 Stress trajectories of σ_1 and likely magma paths for a shallow chamber with a circular vertical cross-section. (a) When the only loading is an internal magmatic excess pressure in the chamber (see the inset), the trajectories are fan-shaped and would allow the chamber to draw magma from a large fraction of the upper part of the source reservoir (the roof of which is the lower edge of the model). (b) When the only loading is external tension (see the inset), the trajectories between the floor of the chamber and the roof of the reservoir (the lower edge of the model) are close to vertical so that the chamber is able to collect magma from a much smaller fraction of the upper part of the reservoir than in (a).

associated volcanic system or volcanic field. It follows that, even if most of the injected paths/dikes become arrested, **dike-fed eruptions** are much more frequent at the surface above the magma chamber than in the volcanic field/system outside the chamber. Furthermore, because of chemical changes of the magma in the chamber, primarily through anatexis and crystal fractionation, intermediate and, in particular, acid magma develops more easily and in greater volumes in the chamber than in the deep-seated reservoir.

The main result of **frequent eruptions** and comparatively large volumes of **acid and intermediate** magmas in addition to the basaltic magmas at the surface above the magma chamber is that a polygenetic volcanic edifice, a stratovolcano or, if little or no intermediate and acid magma is generated (when the basaltic magma dominates due to a short time lag in the shallow chamber), a basaltic edifice builds up. Thus, the formation of a volcanic edifice (Fig. 9.4) is a **direct consequence** of the shallow magma chamber acting as a sink and a source. The reason for the more frequent eruptions in the polygenetic edifice than in its surrounding is also related to the interaction between the magma chamber and the magma reservoir, primarily to their size or volume differences, as will now be discussed.

9.3 Intrusion and Eruption Frequencies of Central Volcanoes

The high eruption frequency of central volcanoes, in comparison with the eruption frequency of the adjacent volcanic system/field, is due to two main factors. One factor is the size or **volume ratio** between the chamber and its source reservoir. The other factor is

Fig. 9.4 Mayon Volcano, a central/polygenetic volcano in the Philippines. Mayon is a highly active stratovolcano, with many tens of eruptions in the past 500 years, producing intermediate and basaltic lava flows and pyroclastics. The Mayon's peak reaches an elevation of 2463 m, or about 2447 m above its surroundings (the volcano's prominence), with slopes as steep as 40°. The volcano has an ideal cone shape with a base diameter of about 20 km. The location of the source magma chamber is not well constrained but its roof may be as shallow as 2–3 km beneath the summit of the volcano with a source reservoir perhaps at 10 km depth (cf. Rosi et al., 2003). Photo: Tomas Tam/Wikipedia Commons. A black and white version of this figure will appear in some formats. For the colour version, please refer to the plate section.

the **stress concentration** around the shallow magma chamber due to its being a cavity in an elastic crustal segment, subject to external and internal loading. Both factors have been discussed in earlier chapters (e.g. Chapter 6), but in a different context. Here, we explore both factors further, starting with the volume ratio of the chamber and its source.

9.3.1 Effects of Chamber–Reservoir Volume Ratio

As indicated above and in Section 6.7 the source magma reservoir is normally much larger than the shallow magma chamber (Fig. 9.1). A dike from the reservoir, meeting with the shallow chamber, may thus trigger many dike/sheet injections from the shallow chamber, and hence, potentially, **many eruptions** from the volcano to which the chamber supplies magma. The interaction between the shallow chamber and its source reservoir is a classic example of a **double magma chamber**. In volcanic areas with a thick crust, such as at convergent plate boundaries and above mantle plumes – for example, the central parts of Iceland – the plumbing system may be composed of at least three chambers/reservoirs, giving rise to the triple magma chamber. Here, however, the focus is on the double magma chamber.

From Eq. (6.27) the **number of dikes** (or inclined sheets or both) N_s from the shallow chamber that can be triggered by a single magma flow along a dike path from the source reservoir is:

$$N_s = \frac{\eta p_e q (\beta_p + \beta_m)}{f p_e \lambda (\beta_r + \beta_m)} = \frac{\eta q (\beta_p + \beta_m)}{f \lambda (\beta_r + \beta_m)} \qquad (9.1)$$

Here η is the porosity of the source reservoir, p_e is the excess magmatic pressure in the shallow chamber at the time of rupture and dike injection, β_m is the magma compressibility, β_p is the pore compressibility of the reservoir, and β_r is the compressibility of the rock hosting the chamber. The magma compressibility β_m and the excess magmatic pressure at rupture and dike/sheet injection p_e are assumed the same for the chamber and the reservoir. By contrast, the host-rock compressibility of the chamber β_r is likely to be different from the pore compressibility of the reservoir β_p, one reason being that the temperature of porous matrix of the reservoir is higher than that of the host rock of the chamber. In addition to these elastic constants and factors, there are three other factors in Eq. (9.1), namely q, λ, and f. They all range from zero to one. The factor q denotes the fraction of the total flow of magma from the reservoir during its rupture and formation of a magma path (a dike or inclined sheet) received by the shallow chamber. The factor λ denotes the volume ratio of the magma chamber V_c and the reservoir V_r, that is, $V_c = \lambda V_r$ (Eq. (6.26)). The factor f denotes the magma fraction or porosity of the shallow chamber, and is equal to one when the chamber is totally molten.

Some chambers are totally molten, for a while at least, so to explore the double-chamber effect on the intrusion frequency, let us begin with a totally molten chamber (Fig. 9.1). Common **shallow-chamber volumes** are estimated to be in the range of 20–500 km^3 while some – namely those issuing the largest caldera-driven explosive eruptions – reach at least 5000 km^3 (Gudmundsson, 2016). Based on the above estimates and the common areas of calderas (Chapter 5), a volume of 50 km^3 may be regarded as typical for a shallow chamber. For example, a totally molten sill-like chamber, circular in plan view and 8 km in diameter and 1 km thick would be of this volume.

It is more difficult to determine typical **volumes of source reservoirs**. Perhaps the best information is on reservoirs that supply magma to shallow chambers in Iceland. The reason is partly that the horizontal cross-sectional area of the reservoirs beneath the rift zones can be roughly determined from the areas of the associated volcanic systems/fissure swarms. The largest volcanic systems in Iceland reach areas of 2300–2500 km^2 (Thordarson and Höskuldsson, 2008), which could thus be similar to, but is normally somewhat smaller than, the lateral cross-sectional areas of the associated reservoirs. The thicknesses of the reservoirs are not known with any certainty, but are perhaps between 5 km and 15 km. If the **average porosity** or melt fraction of this part is 0.25 (25%) – perhaps with a totally fluid top and then gradually decreasing fluid fraction with depth –the volume of the **largest reservoirs** would be from about 3000 km^3 to over 9000 km^3. Since the common cross-sectional area of a volcanic system in Iceland is much smaller than this, or about 1000 km^2, it follows that a **typical reservoir** might have a volume from just over 1000 km^3 to about 3700 km^3. If a reservoir with a volume between these extremes, say 2000 km^3, supplies magma to a shallow chamber of 50 km^3, clearly the volume ratio $\lambda = 0.025$.

The **factor q** may be about 0.5, based on some crude estimates for double magma chambers in Iceland (Chapter 6). However, q may also be zero, because many regional dikes and some inclined sheets from the reservoir may not reach the shallow chamber at all.

By contrast, q may possibly be as great as 0.9. Given that the vertical distance between the source reservoir and the shallow chamber is commonly 10 km or more, and that the dike/sheet may be many kilometres in strike-dimension and have a thickness of several metres, it is clear, however, that much of the magma volume would normally go into making the dike/sheet connection. It is thus assumed here, for the present purpose, that q is 0.3 (30%), even if, as indicated, it may be much higher and also much lower.

The same values are used for the elastic properties as in Chapter 6, that is, $\beta_m = 1.3 \times 10^{-10}$ Pa^{-1}, $\beta_p = 9 \times 10^{-11}$ Pa^{-1}, and $\beta_r = 3 \times 10^{-11}$ Pa^{-1} (Gudmundsson, 2006), $\lambda = 0.025$, $\eta = 0.25$, and $q = 0.3$; then, if the chamber is totally molten ($f = 1$), Eq. (9.1) yields N_s of about 4. That is, for the given conditions, a magma flow from the reservoir could trigger **four dike/sheet injections** from the associated chamber. Most or all of these injections would become arrested, so that triggering dike/sheet injections is not the same as triggering eruptions.

If the proportion of magma received by the chamber was less than 30% (0.3), then the magma flow from the reservoir would trigger fewer dike/sheet injections. For example, if $q = 0.1$, then only one dike/sheet injection could be triggered. By contrast, if the proportion of magma received by the chamber was as great as 90%, so that $q = 0.9$, then twelve dike/sheet injections could be triggered. Other parameters can also vary. For example, the ratio λ between the volume of the chamber and that of the reservoir may be from close to zero to close to one (but it is unlikely to be greater than one). Similarly, the porosity or melt fraction in the reservoir may differ from 0.25, either being less or greater, and the shallow magma chamber need not be totally molten, so that the parameter f can be less than one. Thus, while the elastic constants are reasonably well constrained, the parameters η, q, f, and λ can all vary from close to zero to close to one. It follows that the number of dike/sheet injections triggered by a single magma flow from a reservoir to an associated shallow chamber of a double-chamber system can **vary widely**.

The main point here, however, is that a source reservoir normally has magma/melt volumes many times greater than that of the shallow magma chamber to which the reservoir supplies magma (Fig. 9.1). It follows that a single magma flow from the reservoir, provided the magma path meets the shallow chamber, can commonly trigger a few to many dike/sheet injections from the chamber. Some of the dikes/sheets become arrested at various crustal depths, whereas others reach the surface to supply magma to eruptions. Thus, one primary reason for the high eruption frequency of a central volcano is that its source magma chamber ruptures much more frequently than the source reservoir, which also supplies magma to eruptions in the volcanic system/field outside the central volcano. And the **high rupture frequency** of the chamber is partly because it is normally much smaller than the source reservoir. Additionally, this high rupture frequency of the shallow chamber is because it concentrates stresses, a topic to which I turn now.

9.3.2 Effects of Stress Concentrations

When subject to loading, all cavities in solid bodies concentrate stress. It follows that magma chambers in a solid crust subject to internal and/or external loading concentrate stresses. A **stress concentration** around the shallow source **chamber** of a central volcano is

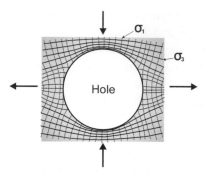

Fig. 9.5 Stress concentration around a hole located in a biaxial stress field. The trajectories are the orientations of the maximum and the minimum principal stresses in the plane of the illustration. Since stress is force per unit area, and the trajectories may also be thought of as 'lines of force', it follows that the greater the number of lines of force per unit area, the higher is the stress. More specifically, the lines may be regarded as representing the 'flow of force' around the hole, comparable to that of horizontal flow of liquid around a solid with a circular horizontal cross-section and represented by streamlines (Gudmundsson, 2011a). It follows that the stresses are high (are raised or concentrated) where there are many lines of force per unit area; here at the top and bottom of the image. For the application to magma chambers with a circular lateral cross-section or a vertical circular cross-section, the chamber must be assumed to be in lithostatic equilibrium (same stress/pressure on the walls from the inside as from the outside – Chapter 3) before the stress difference between σ_1 and σ_3 develops. For the application to the stress field around a circular collapse caldera, the hole is assumed to contain no liquid, that is, to be subject only to atmospheric pressure (equal to 0.1 MPa and ignored). The orientation of the trajectories of σ_1 and σ_3 are shown. Further details on stress fields around holes of various geometries are provided by Savin (1961).

one of the reasons for a high dike/sheet injection frequency and the associated eruption frequency – and, thereby, for the formation of volcanic edifices (Gudmundsson, 1988, 2006).

The stress concentration around magma chambers have already been discussed in Chapter 3, so only a brief summary is needed here. Stresses cannot be transmitted through holes or cavities, empty or filled with fluids. Instead, the principal stresses, represented by the **stress trajectories**, become clustered around the hole/cavity (Fig. 9.5). The trajectories are analogous to lines of force – and, indeed, to **streamlines** in a flowing liquid (Gudmundsson, 2011a). Because stress is force per unit area (Eq. (3.1)), it follows that when the number of lines of force passing through a unit area increases close to the hole/cavity, then the **stress increases**, that is, is raised or concentrated. And that is, exactly, what happens at the boundary of, and close to, any hole/cavity in a solid, such as a magma chamber in a crustal segment, subject to external or internal loading.

For a shallow magma chamber, the loading can be internal, external, or both. Recall that here the term loading includes applied **force, stress, pressure, and displacement**. For a shallow magma chamber, the common loading is the internal fluid pressure; more specifically, the magmatic excess pressure. For chambers located at divergent plate boundaries or, in general, in rift zones or volcanic fields subject to crustal doming, the main loading may be an external tension. To illustrate the effects of stress concentration, several numerical models are shown here in addition to those in Chapter 3. All the models are two-dimensional unless stated otherwise.

Consider first a **sill-like magma chamber**, which are common (Fig. 9.2). Here, the only loading is an internal excess magmatic pressure of 5 MPa. The chamber is located in a volcanic system/zone that is 15 km wide, the chamber itself being 8 km wide in a direction perpendicular to the strike of the volcanic system, which is then the horizontal dimension of the chamber in Fig. 9.2. The crustal segment hosting the chamber is 12 km thick, the chamber itself being 2 km thick and with a roof located at 2 km below the free surface of the volcanic system to which it supplies magma. All these numbers are common for shallow magma chambers worldwide and, in particular, for active and fossil chambers in Iceland (Gudmundsson, 2012; cf. Chaussard and Amelung, 2014). In this model, the crustal segment is homogeneous and isotropic (non-layered), and should be regarded as illustrative rather than representing actual volcanic systems. Also, the surface of the volcanic system above the chamber is flat, so that a volcanic edifice has not formed as yet. The model may thus be regarded as representing a young magma chamber, during the beginning stages of volcanic-edifice formation.

In this model, the highest tensile stresses occur close to and at the **upper corners** of the chamber (Fig. 9.2b). The model corners are rounded, because angular corners (and other irregularities) are mechanically and thermally unstable, that is, **short-lived**, in comparison with the lifetime of the chamber itself – here of the order of hundred thousand years (Fig. 5.11; Gudmundsson, 2012). The theoretical tensile stress, however, exceeds 5 MPa along the greater part of the upper margin of the chamber so that dike/sheet injection is possible (Fig. 9.2b). For this magma-chamber geometry and loading, most of the ruptures would occur close to and at the **upper edges** or corners of the chamber, because that is where the theoretical tensile stresses are highest. Also, the trajectories of the maximum compressive principal stress σ_1 are mostly inclined, suggesting that many of the injected dike paths would actually be **inclined sheets**. In a realistic, mechanically layered model, most of the sheet paths would become arrested at contacts between dissimilar layers. Yet, many sheets would reach the surface (Chapter 7), thereby supplying magma to the formation of a volcanic edifice.

For a magma chamber with a circular vertical cross-sectional area, there are basically two models to be considered. One is the spherical chamber, the other is the cylindrical chamber with a long axis parallel with the axis of the associated volcanic system/volcanic zone. For a **spherical chamber** with a radius many times smaller than the depth to its roof – and thus a spherical cavity in an infinite elastic body (Fig. 3.8) – the maximum theoretical tensile stress at the margin of the chamber is only half the (absolute value of) the magmatic excess pressure (Fig. 3.9). However, due to **geometric irregularities** as well as the effects of layering and near-surface effects on the stress concentration around a real shallow magma chamber, the theoretical tensile stresses are normally considerably higher than this. For a **cylindrical chamber** with a long axis parallel with the volcanic system, the model is essentially a two-dimensional circular hole. For such a chamber, far away (in relation to the chamber size) from the surface of the associated volcanic system/zone, the maximum tensile stress at the boundary of the chamber is equal in magnitude to the magmatic excess pressure (Fig. 3.16). Again, however, for real shallow magma chambers in layered crustal

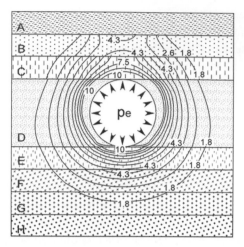

Numerical model showing the magnitude, in mega-pascals, of the maximum principal tensile stress σ_3 around a shallow magma chamber of circular vertical cross-section. The only loading is a magmatic excess pressure p_e of 10 MPa, similar to the maximum *in situ* tensile strength of crustal rocks (Gudmundsson, 2011a). The chamber is located in a layered crustal segment. All the layers/units have a Poisson's ratio of 0.25, whereas they have the following Young's moduli: A = 5 GPa, B = 20 GPa, C = 30 GPa, D = 50 GPa, E = 70 GPa, F = 80 GPa, G = 90 GPa, and H = 100 GPa. Since no tensile stress is transferred to layer A, the surface layer, it would be expected to act as a stress barrier to dike propagation (cf. Gudmundsson and Brenner, 2004).

segments and comparatively close to the surface, the theoretical tensile stresses may largely exceed the magnitude of the excess pressure.

An example of a magma chamber of a circular vertical cross-sectional area in a layered crust is shown in Fig. 9.6. Here, the magma chamber is 6 km in vertical diameter, with a top at 6 km below the free surface of a volcanic zone with a crustal thickness of 20 km. For a spherical chamber, the volume would be about 113 km^3, so that the chamber is reasonably large. The mechanical layers or units increase in stiffness with depth, and the loading is a magmatic excess pressure of 10 MPa.

The results show that the **tensile stress σ_3** exceeds 10 MPa close to the magma chamber, and is highest around the upper margin of the chamber. Thus, dike or sheet injection would be most likely from the upper part of the chamber. However, while dike/sheet intrusions are thus encouraged by this local stress field, there is no stress concentration in the uppermost unit/ layer, so that most of the dikes/sheets become **arrested**. Thus, there may be numerous unrest periods with dikes/sheets injected into the central volcano, but few will reach the surface, so long as the surface unit/layer is comparatively compliant.

Similar results are obtained even if there is, in addition to the internal excess magmatic pressure of 10 MPa, an **external tension** of 3 MPa. This is seen in the model in Fig. 9.7, where the mechanical layering is exactly the same as in the model in Fig. 9.6, but a horizontal tension of 3 MPa has been added. The tension could correspond to that generated by the spreading vector at divergent plate boundaries, or to the effect of volcano spreading, for example. The concentration of the tensile stress σ_3 is here, understandably,

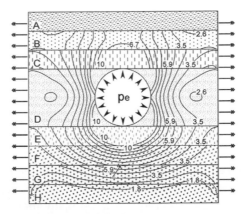

Fig. 9.7 Numerical model showing the magnitude, in mega-pascals, of the maximum principal tensile stress σ_3 around a shallow magma chamber of circular vertical cross-section. The layering is the same as in the model in Fig. 9.6. The difference in loading between the models is that in addition to the magmatic excess pressure p_e of 10 MPa in the shallow chamber, there is here an additional loading through horizontal tensile stress of 3 MPa, as indicated by the arrows. So this model is more appropriate than the one in Fig. 9.6 when the shallow chamber is subject to a rifting episode while also receiving new magma that generates an excess pressure of 10 MPa. Despite the horizontal tensile stress, layer A, the surface layer, is still free of tensile stress and might be expected to act as a barrier to the propagation of dikes to the surface (cf. Gudmundsson and Brenner, 2004).

higher than in the model in Fig. 9.6. This follows from the much larger areas around the magma chamber in Fig. 9.7 than in Fig. 9.6 where σ_3 exceeds 10 MPa. Thus, magma-chamber ruptures with dikes or inclined sheet injections are likely to be common. However, again, there is no tensile-stress concentration in the surface layer, so that dikes/sheets will have difficulty in reaching the surface. These models thus demonstrate the fact, which is also discussed in Chapter 7, that many dikes/sheets become **arrested** at contacts between mechanically dissimilar layers, so that most of them, even when supplied with magma from very shallow magma chambers, do not reach the surface to erupt.

The same result is seen in the oceanic floor, in particular in **ophiolites**. The sheeted dike complexes – which correspond to the swarms of inclined sheets and radial dikes in Iceland, Scotland, Tenerife, and other similar areas – reach close to **100% dikes** (Coleman, 1977; Lippard et al., 1986; Nicolas, 2013). That is, nearly all the original host rock has disappeared and the entire crustal layer is composed of intrusions, namely dikes. In the Semail Ophiolite in Oman, the thickness of the sheeted dike complex, the layer composed of 100% dikes, varies from 600 m to 2000 m, with an average of about 1500 m (Lippard et al., 1986). The dikes are mostly subvertical and 0.1–4.4 m thick, but most are between 0.5 m and 1 m. Thus, the thicknesses are very similar to those observed in the **sheet swarms** in Iceland and elsewhere.

The most remarkable thing about the dike complexes in the ophiolites, however, is the comparatively abrupt change in **dike intensity** or frequency at the upper margin of the sheeted dike complex. There is a transition zone from the dike complex into the dike swarm in the zone of extrusive rocks (mainly lava flows) above. The thickness of the transition zone depends on the type of lava flows that constitute the bulk of the extrusive zone. Where the lava flows are mostly sheet flows, the zone is less than 50 m thick in Oman and also in the Troodos Ophiolite (Baragar et al., 1987). However, where the lava flows are mostly pillow lavas in the Semail Ophiolite, the zone is about 100 m thick. Above the transition zones, the dikes constitute only a few per cent of the rock. Thus, across the transition zone, the percentage of dikes decreases from close to 100%, below the zone, to a few percent, above the zone. Clearly, therefore, the zone acts as a **dike arrester**, in a similar way as many comparatively compliant layers do in central volcanoes and volcanic zones worldwide.

There is no doubt that magma chambers act as cavities/elastic inclusions that concentrate or raise stresses (Gudmundsson, 2006; Andrew and Gudmundsson, 2008). The stress concentration results in the magma chamber **rupturing more frequently** and injecting dikes/sheets than the surrounding parts of the volcanic zone/field in which they are located. This is the main reason why there are **many more** inclined sheets and radial dikes in the local swarms of central volcanoes than there are regional dikes in the swarms outside the central volcanoes. While most of the injected dikes/sheets become arrested, many do reach the surface. The effect of stress concentration and associated frequent ruptures of the shallow magma chamber is thus, together with its acting as a sink for magma from the deeper reservoir, the primary reason for the formation of a **volcanic edifice** – a central volcano.

9.4 Shape (Geometry) of a Volcano

9.4.1 Differences between Stratovolcanoes and Basaltic Edifices

Polygenetic/central volcanoes generally form **edifices**, that is, mountains that rise high above their surroundings. The overall vertical cross-sectional shapes or profiles of many **stratovolcanoes** are rather similar (Fig. 9.8). They gradually become steeper towards the top part, where some of the slopes are as steep as 35–42° (Francis, 1993; Kilburn and McGuire, 2001; Gudmundsson, 2009). The profiles or shapes of **basaltic edifices** tend to be widely different from those of stratovolcanoes. For comparison, consider the profiles of Mauna Loa and Mauna Kea in Hawaii (Fig. 9.9). More specifically, the stratovolcanoes have, on average, much steeper slopes than the basaltic edifices, particularly in the upper-most parts of the volcanoes. Thus, while the top parts of some stratovolcanoes are, as indicated, so steep as 42° basaltic edifices are rarely steeper than 12°. In fact, the slopes of the Hawaiian shield volcanoes are mostly between 2° and 12° (Macdonald, 1972).

Why do the stratovolcanoes and basaltic edifices differ so markedly in shape? More specifically, why are basaltic edifices much more gently sloping than stratovolcanoes? Several explanations have been offered, the most common one being the difference in **viscosity** and **yield strength**. The basaltic lava flows that constitute the bulk of basaltic edifices have, while flowing, a low viscosity – commonly 10–100 Pa s (Appendix F; Kilburn, 2000; Spera, 2000). By contrast, intermediate and acid lava flows have viscosities of 10^4–10^7 Pa s (andesite) and 10^9–10^{13} cf. Pa s (rhyolite), that is, many orders of magnitude larger than typical viscosities of basaltic lava. The viscosities depend much on temperature.

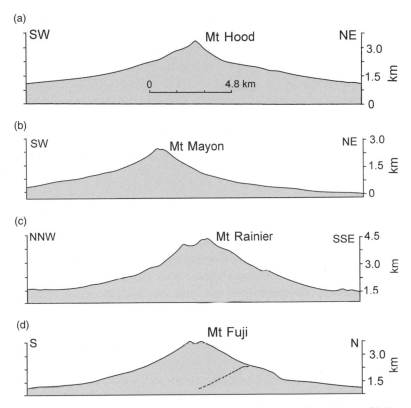

Fig. 9.8 Schematic profiles of four stratovolcanoes (composite volcanoes). (a) Hood, the United States. (b) Mayon, the Philippines, (c) Rainier, the United States, and (d) Fuji, in Japan. Hood has no major top crater, whereas the other three have. For Mayon see also Fig. 9.4 and for Fuji see also Fig. 9.10 (cf. Rosi et al., 2003). Modified from Macdonald (1972).

Here the eruption temperatures used are 1050–1200 °C for basaltic lava, 950–1170 °C for andesitic lava, and 700–900 °C for rhyolitic lava (Appendix F; Kilburn, 2000).

Most lava flows behave as **Bingham plastics**. They therefore do not flow until the yield strength is reached. The yield strengths of active (erupting) hot lava flows are considerably less than their *in situ* tensile strengths of 0.5–9 MPa. Humle (1974) estimated the yield strength of active basaltic lava flows to be in the range $0.1–8 \times 10^3$ Pa s, that of intermediate lava flows to be $0.1–1 \times 10^5$ Pa s, and that of acid lava flows to be $0.5–5 \times 10^6$ Pa s. Thus, the yield strength of acid lava flows is similar to the common *in situ* tensile strength of rocks in general.

While viscosity and yield strength may restrict the **flow length** of many lava flows, particularly intermediate and acid flows, neither can explain why stratovolcanoes are much steeper than basaltic edifices. This follows because the steepness of stratovolcanoes is a long-term feature, maintained through most of their lifetimes. This means that they must be strong enough to maintain the steep slopes over long periods of time. More specifically, the stratovolcanoes must, as structures, be much stronger than basaltic edifices.

The reason why stratovolcanoes are generally **stronger structures** than basaltic edifices is because **fractures** are more easily **arrested** in stratovolcanoes (Chapters 5 and 7). This follows because the variation in mechanical properties is much greater in stratovolcanoes than in basaltic edifices, which means that the local stresses within the stratovolcanoes are also very variable. In turn, this implies that stress-field homogenisation along the potential path of a fracture is reached only rarely. Such a homogenisation is a necessary condition for a fracture – a dike, a fault, a landslide fracture – to propagate through the edifice. In particular, for large landslides (lateral collapses) to occur, there must be stress homogenisation, that is, a favourable stress field for a landslide fracture (a fault) along the path of the fracture.

9.4.2 Top Crater

Many volcanoes have a noticeable crater at their tops (Figs. 9.8, 9.10), although some do not (Figs. 9.8, 9.9). Generally, there is no particular reason why the top of a stratovolcano should be occupied by a **crater** (Fig. 9.10). This follows because stratovolcanoes are primarily constructive or positive landscape features – edifices – and most are fed by dikes. Dikes may generate lava flows with little or no specific crater formation. However, many dikes feed crater rows – volcanic fissures – with several or many crater cones. But the **crater cones** are mostly positive landforms; there is a depression in the top, but the cone as a whole stands above its surroundings (Fig. 1.1b).

The formation of crater cones is not difficult to understand: they are a direct consequence of the **cubic law** or, more specifically, of **flow channelling** whereby much of the magma flow is through the large-aperture parts of the feeder-dike/volcanic fissure (Chapter 8). Since the volumetric or effusion flow rate depends on the aperture to the third power (the cubic law), it follows that slight variations in the aperture can result in great variations in the effusion rate and thereby in the channelling (focusing) of the flow on the parts of the fissure

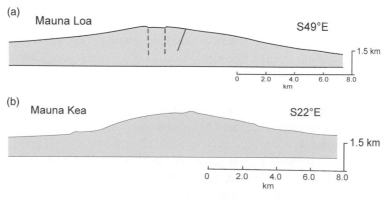

Fig. 9.9   Schematic profiles of two basaltic edifices (shield volcanoes). (a) Mauna Loa, in Hawaii, the United States, and (b) Mauna Kea, in Hawaii, the United States. Mauna Loa has a collapse caldera, as indicated, but Mauna Kea has no caldera (cf. Rosi et al., 2003). Modified from Macdonald (1972).

Fig. 9.10 Aerial image of the stratovolcano Fuji in Japan (cf. Fig. 9.8d). The clear depression in the top of the volcano is the top crater, about 700 m in diameter and about 250 m deep. Fuji is an active stratovolcano, composed of lava flows and pyroclastic layers ranging in composition from basalt to dacite, with a base diameter of some 50 km. The most recent eruption was in 1707–1708. The peak of Fuji reaches an elevation of 3776 m, and its prominence is the same. The location of the source magma chamber is not well constrained but its roof may be as deep as 8 km, with a deeper source reservoir at perhaps 20 km depth (Kaneko et al., 2010; cf. Rosi et al., 2003). Photo: NASA.

where the aperture is, by chance, larger than in the neighbouring parts. Where, because of aperture variation, the effusion rate is high in comparison with that of the adjacent parts of the dike/fissure, a **crater cone forms**. The depression in the centre of the cone is primarily the result of the particles – the scoria/spatter – being thrown out along trajectories that encourage their falling onto the surface at a certain distance from the dike/fissure opening. When the fissure segment is short, as is common during the later stages of an eruption, a crater cone forms, the plan view of which is circular or somewhat elliptical (Fig. 1.1b).

The formation of craters that are **negative landscape features**, however, is more difficult to explain. This is because the craters, being negative landscape features, form depressions with floors that are normally much lower than those of the adjacent summit area of the volcano. The craters are not collapse calderas – they are normally much less than a kilometre in diameter, and are not directly related to a shallow magma chamber (Chapter 5). Here, I propose that the formation of large top craters in many volcanic edifices is primarily attributable to the shape of the conduit. More specifically, I suggest that many stratovolcanoes containing such a crater have **cylindrical conduits** below the crater, modelled as prolate ellipsoids – as is, indeed, supported by the common observations of volcanic plugs/necks in many volcanoes (Section 9.5; cf. Chapter 8) and the formation of radial dikes.

So how does the cylindrical conduit contribute to the formation of a large crater? The answer is: through loading changes, primarily changes in magmatic pressure. The crater formation may be the result of either magmatic-pressure decreases or increases. The

pressure decrease normally occurs at the end of an eruption, or because the magma finds another path, out of the conduit, usually at a lower altitude. For example, a radial dike from the conduit normally lowers the pressure in the magma column above and can, thereby, result in a subsidence in the top part, forming a depression, namely a crater. In this way, the top crater will be somewhat analogous to a pit crater.

The **pressure-increase** mechanism is perhaps less intuitive and therefore more interesting. The main point is that for certain aspect ratios of a conduit, magmatic overpressure can generate subsidence at the surface above the conduit. When modelling the conduit as a prolate ellipsoid, with a vertical long axis, for certain length/width ratios there is subsidence at the surface above the top of the conduit. Thus, for a prolate ellipsoidal conduit, an increase in magmatic pressure can, theoretically, generate subsidence at the surface above the conduit (Bonaccorso and Davis, 1999; Dzurisin, 2006), thereby forming a top crater with a negative landscape form.

9.4.3 How Lateral and Vertical Collapses Shape Volcanoes

The mechanics of vertical and lateral collapses have already been discussed in Chapters 3 and 5, and the effects of vertical collapses, namely caldera collapses, on eruption sizes have been detailed in Chapter 8. Here, the focus is on how vertical and lateral collapses affect the shapes of the volcanoes within which they occur. Let us begin with the vertical collapses, namely collapse calderas, and then move on to lateral collapses, that is, landslides.

Vertical Collapses. Caldera collapses are mainly along close-to-vertical ring-faults (Figs. 3.33, 3.34, 5.21, 5.22, 5.23, 5.24, 5.26, 8.16, 8.17, 8.19). Clearly, when a collapse occurs and part of a volcano subsides into an associated shallow magma chamber, the overall shape of the volcano changes. More specifically, the upper part of the volcano that previously may have been a sharp cone, with or without a crater (Figs. 9.4, 9.8, 9.10), becomes a large depression (Fig. 9.11), so that, from a distance, the volcano looks **flat-topped**. In the case of a single caldera, much smaller than the volcano, the flat top is clear, and it looks like the top has been cut off – which, in a way it has, during the collapse (Fig. 9.9a). For a multiple collapse or, in general, when the collapse includes a large part of the volcano profile, the effect becomes even greater (Fig. 9.12). In an extreme case, the entire volcano collapses so that where there once was a topographic high – the volcanic edifice – there is now just a depression, normally filled with water to form a lake or part of the sea (Fig. 9.13).

Clearly, therefore, vertical collapses have great effects on the shape of the associated volcanoes. These effects range from being essentially a large depression in the top of the volcano, the structure and geometry of which is otherwise intact (Fig. 9.11), to major changes in the shape of the volcano (Fig. 9.12), where a significant part of the volcano has been destroyed, to the end member where the entire volcano basically disappears during the collapse (Fig. 9.13). The latter type of calderas is common in the case of very large explosive eruptions, such as those associated with the formation of the calderas of Santorini (Fig. 9.13a) in Greece and of Toba (Fig. 9.13b) in Indonesia (on Sumatra), as well as some of the very large calderas in North America, such as Yellowstone and Long

Fig. 9.11 Aerial image of the stratovolcano Tambora, Sumbawa Island, Indonesia. (a) Image of the entire volcano. The diameter at sea level is 60 km and before the 1815 eruption its elevation was about 4300 m but is now 2850 m (Gertisser et al., 2012). (b) Close-up of the collapse caldera formed in the 1815 eruption. The caldera has a diameter of 6–7 km and a maximum depth of about 1200 m (Self et al., 1984; cf. Rosi et al., 2003). The estimated volume of the eruptive materials during the collapse is between 95 km^3 and 175 km^3 (Gertisser et al., 2012), which corresponds to between 28 km^3 and 50 km^3 of magma flowing out of the shallow chamber as the caldera subsided, making it probably the largest eruption in historical time. The estimated volume of the caldera itself is 36 km^3 (Self et al., 1984). The depth of the shallow chamber is poorly constrained, but it may be between about 1.5 km and 7.5 km (Foden, 1986; Gertisser et al., 2012). There was an unrest period for at least 3 years prior to the caldera collapse and associated eruption. Photos: NASA.

Fig. 9.12 Las Canadas caldera and Teide in Tenerife, the Canary Islands (Spain). (a) View southwest, the caldera is elliptical in shape with a major axis of about 17 km and a minor axis of about 9 km. The caldera perimeter is some 80 km, and the maximum height of the caldera wall about 500 m. The caldera is multiple/overlapping and was formed in three major collapses (Marti and Gudmundsson, 2000), the oldest caldera (Ucanca, 1.02 Ma) being the southwestern part of the Las Canadas caldera. The second oldest caldera (Guajara, 0.57 Ma) forms the middle part, and the youngest caldera (Diego Hernandez, 0.17 Ma) is the northeastern part of the Las Canadas caldera. This age difference is clearly reflected in the weathering and erosion of the walls, those of the Ucanca caldera being heavily eroded whereas those of the Diego Hernandez caldera have hardly been eroded at all. Also seen is the stratovolcano Teide, which has formed during the past 0.16 million years, after the formation of the Diego Hernandez caldera. Photo: NASA. (b) View southwest, close-up of the upper part of Teide, which has an elevation of 3718 m; its most recent eruption was in 1909. The magma chamber of Teide is poorly constrained. One chamber may be at about 8 km below sea level, and then primarily basaltic, but a shallower acid (phonolitic) chamber is likely to be located directly beneath Teide itself, perhaps at around sea-level depth (Pina-Varas et al., 2018). A black and white version of this figure will appear in some formats. For the colour version, please refer to the plate section.

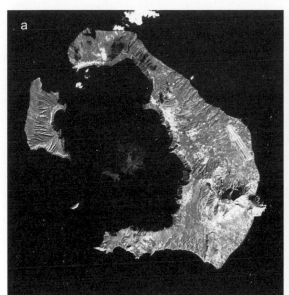

Fig. 9.13   Caldera collapses that have largely destroyed their volcanoes. (a) The Santorini caldera is elliptical, with a major axis of 12 km and a minor axis of 7 km and formed in the Minoan eruption which occurred some 3600 years ago (producing about 60 km^3 of eruptive materials). The most recent eruptions in Santorini occurred in 1950 (cf. Browning et al., 2015). (b) The Toba caldera in Sumatra, Indonesia, formed during the Toba eruption, some 75 000 years ago. This eruption produced about 2800 km^3 of eruptive materials and is the largest subaerial eruption in Earth in the past 25 million years (Chesner, 2012). The caldera is elliptical, with a major axis of about 100 km and a minor axis of 30 km. The present lake occupies only a part of the original caldera since a resurgent dome forms a large island in the central part of the caldera (cf. Chapter 8). Photos: NASA.

Valley (Lipman, 1997). While the exact relationship between a large caldera and the associated large eruption is not always clear-cut, it seems likely that, in many such collapses, it is the piston-like collapse itself which drives out the magma and is therefore the main reason for the large eruption (Chapter 8; Gudmundsson, 2016).

Lateral Collapse. Lateral collapses – landslides – have great effects on the geometries of volcanoes. Many, and probably most, volcanic islands are subject to frequent landslides, some of which are so large as to make significant changes to the geometries of the islands. Thus, volcanic islands (and their associated volcanoes) such as Hawaii and the Canary Islands have been subject to many large landslides that have had significant effects of their overall shapes.

While lateral collapses are common in both stratovolcanoes and basaltic edifices, there is considerable evidence that the **large ones** are more common in basaltic edifices (Gudmundsson, 2009). Large landslides are common in basaltic edifices and, as indicated above, they are very common around the Big Island of Hawaii (Moore et al., 1989, 1994; Morgan et al., 2003; Garcia et al., 2006), the Canary Islands (Mitchell et al., 2002; Walter and Schmincke, 2002; Acosta et al., 2003; Hurlimann et al., 2004), and Reunion (Oehler et al., 2005, 2008). While most large landslides of basaltic edifices occur on the ocean floor, landslides are also very common in subaerial parts of these edifices (Acocella et al., 2003; Oehler et al., 2005; Rust et al., 2005; del Potro and Hurlimann, 2008).

Landslides are also common in stratovolcanoes (e.g. Siebert, 1984; Voight and Elsworth, 1997; Reid, 2004; Pinel and Jaupart, 2005), as follows from the steep slopes in the upper parts of the volcanoes. However, landslides in stratovolcanoes are mostly very small (Ponomreva et al., 2006; Boudon et al., 2007), large ones being less common than might be expected from the steep slopes (Boudon et al., 2007). Large landslides or lateral collapses in stratovolcanoes seem to require special mechanical conditions, such as hydrothermally altered, weak and outward-dipping layers, or external loading, such as shallow magmatic intrusions (Kerle and de Vries, 2001; Tibaldi, 2001; Wooller et al., 2004; Tibaldi et al., 2006) or earthquakes (Lipman and Mullineaux, 1981; Vinciguerra et al., 2005). For example, Apuani et al. (2005) made a limit-equilibrium study of Stromboli, a stratovolcano composed primarily of (andesitic and basaltic) lava flows, breccias, and pyroclastic flows, and they concluded that, in the absence of external loading (primarily overpressured dikes), the volcano is mechanically stable.

Overall Shape Effects

Lateral and vertical collapses have great effects on the shapes of volcanoes. These effects, however, are to a considerable extent different. Broadly speaking, the differences may be summarised as follows:

- Lateral collapses contribute to keeping the slopes of the volcanoes comparatively gentle. While slopes of over 40° do occur, for the most part the slopes are much gentler. Thus, while lateral collapses keep the slopes within reasonable limits – collapse occurs when the slope of large parts of the volcano become too steep and thus mechanically unstable – they do not change the overall cone shape of the volcano. While tectonic elements and weaknesses play a role in lateral collapses, and dike intrusion may occasionally contribute to the triggering of them, such collapses are primarily driven by loads due to **gravity**.
- Vertical collapses, by contrast, tend to change the overall shape of the volcano. The profile of the volcano becomes flat-topped, if the collapse is comparatively small (Figs. 9.9a, 9.11). By comparatively small I mean in relation to the overall dimension of the volcanic edifice. Alternatively, most of the volcanic edifice may disappear (Fig. 9.13a) and the profile becomes essentially flat, as in the case of some of the largest collapse calderas (Fig. 9.13b). In contrast to lateral collapses, caldera collapses are primarily driven by **volcanotectonic loading** – the effects of a tectonically driven stress concentration above the lateral margins of an associated shallow magma chamber.

9.5 Internal Structure of a Volcano

In the earlier chapters all the main elements that constitute a central volcano have been discussed. Briefly, these are mechanical/lithological layers and fractures of various types. More specifically, the layers include lava flows, pyroclastic and breccia layers, sedimentary and soil layers, sills, dikes, inclined sheets and, occasionally, ring-dikes, as well as normal faults, reverse faults, strike-slip faults, joints, and, occasionally, ring-faults. While all these structures have been described before, we shall here try to understand their implications for processes that take place inside volcanoes – although those processes are primarily modelled and discussed in Chapter 10. The focus in this chapter is on the mechanical properties and function of the structures rather than their composition and chemistry.

9.5.1 Layers and Contacts

The main layers that constitute volcanoes are **lava** flows and **pyroclastic** layers. In basaltic edifices, the lava flows dominate, whereas in some stratovolcanoes and volcanoes erupted partly under water, pyroclastic layers/basaltic breccias may be as abundant as lava flows. Mechanically, the layering and their contacts imply that a volcano behaves mechanically as a composite material.

The main mechanical characteristic of structures made of such composite materials is the great variation in their local internal properties and stresses. The variation in local properties is primarily due to changes in the **Young's modulus** between layers, as well as in changes in the contact properties themselves. The variation in the Young's modulus between layers in volcanoes is by many orders of magnitude. Stiff lava flows and intrusions with high Young's moduli may reach values of more than 100 GPa, whereas compliant or soft sedimentary, pyroclastic, and soil layers between lava flows and intrusions may be as soft as 0.1 GPa or softer (Appendix D), a variation by three orders of magnitude. For comparison, the Poisson's ratio of most rocks is between 0.1 and 0.3, a variation of a factor of 3 and thus much less than that of Young's modulus.

The commonly abrupt changes in Young's moduli across contacts results in a reorientation of the principal stresses and changes in their magnitudes (Fig. 9.14; Chapters 7 and 10). The magnitude changes are primarily because the stiffer layers take on much more of the loading than the compliant layers. This means that when a volcano or a crustal segment composed of layers of different stiffnesses is subject to loading (force, pressure, stress, displacement), the stiff layers concentrate **most of the stress** and the soft, much less, if any (Chapters 7 and 10). This applies for any kind of loading. Thus, when there is compressive loading, the stiff layers concentrate **compressive stress**; where there is tensile loading or extension, the stiff layers concentrate **tensile stress**. Similarly, there is commonly an abrupt change in the **orientation** of the principal stresses between adjacent layers of dissimilar stiffness. These changes determine the fracture-propagation path and, eventually, constitute one of the main conditions for fracture arrest at contacts between layers (Chapters 7 and 10).

The **contact properties** themselves also have important effects on the propagation paths of fractures inside volcanoes, particularly the paths of sheet intrusions such as dikes. The stiffness of the contact material – the interface – in relation to that of the layers above and below the contact is a main part of the **elastic-mismatch** mechanism of fracture deflection and arrest (Chapters 7 and 10). Many contacts are real discontinuities in the sense that the (tensile and shear) strength across and along the contact is less than that of contact-parallel zones in the adjacent layers. Other contacts, however, are essentially '**welded**', meaning that their strengths are very similar to that of contact-parallel zones in the nearby layers.

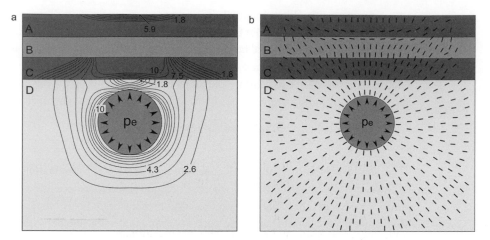

Fig. 9.14 Numerical model showing the effects of abrupt changes in the Young's modulus across contacts on the magnitudes and directions of the principal stresses around a shallow magma chamber of circular vertical cross-section. The only loading is an internal magmatic excess pressure p_e of 10 MPa. Each of the layers A, B, and C has a thickness of 1 km, whereas layer D has a thickness of 7 km. The magma chamber itself has a vertical diameter of 2.5 km and its top is at 3.75 km below the surface. Here, laboratory values are used for the Young's moduli of the rock layers, with a total range from 0.03 GPa to 150 GPa (Appendix D). The Young's moduli are as follows: for layer A, 100 GPa; layer B, 1 GPa; layer C, 100 GPa; and layer D, 10 GPa. Thus, layers A and C are very stiff but layer B is very compliant. (a) Maximum principal tensile (minimum compressive) stress σ_3 in mega-pascals. The compliant layer B suppresses the tensile stresses, whereas the very stiff layers A and C raise (concentrate) the stresses. (b) Ticks (short lines) showing the direction (trajectories) of the maximum compressive principal stress σ_1. The 90° flip to horizontal in the direction of σ_1 (which sheet intrusions follow) at the contact between layers A and B implies that this contact will act as a stress barrier to vertically propagating dikes, that is, it will tend to arrest the dikes or deflect them into sills (cf. Gudmundsson and Brenner, 2005).

Mechanical layering, including contact or interface properties, to a large degree determines the resistance to fracture propagation in volcanoes. Since the strength of a volcano depends on this resistance, it follows that mechanical layering largely controls the **strength of volcanoes**. The strength or **fracture resistance**, in turn, has a great effect on the evolution of a volcano through the following factors and processes:

- Strength, as reflected in mechanical layering and internal structure, determines the equilibrium slope of the volcanic edifice. More specifically, stratovolcanoes appear to be more fracture resistant, because of the variation in their mechanical properties, than basaltic edifices. It follows that stratovolcanoes are stronger structures than basaltic edifices.
- Since stratovolcanoes are stronger structures they can sustain steeper slopes than basaltic edifices. The difference in slopes is confirmed by direct measurements (Gudmundsson, 2009). Basaltic edifices have gentle slopes, mostly in the range of 2–12°. By contrast, stratovolcanoes have much steeper slopes, particularly in their uppermost parts where the slopes may reach 35–42°. Both types of volcanic edifices may have close-to-vertical sea cliffs, but the cliffs are mechanically unstable and they are subject to frequent rock falls and landslides.

- For an essentially large-scale uniform tensile and shear strength of a given volcanic edifice, the channelling effects of the shallow magma chamber results in most of the dikes/sheets reaching the surface in the upper part of the edifice. This is the main reason that the volcanoes maintain their cone shape (Figs. 9.4, 9.8, 9.10). The details of dike/sheet paths are discussed further in Chapter 10.

9.5.2 Sills, Dikes, and Inclined Sheets

Sheet intrusions are among the most common structures in volcanic edifices. They are mostly injected from the shallow magma chamber which constitutes the heart of the volcano (Fig. 9.1; cf. Figs. 6.8, 7.5). The intrusion forms, mechanisms, and mechanics are discussed elsewhere in the book (Chapters 2, 5, 7, and 10). Here, the focus is on their contributions as building blocks of the volcanic edifices.

Sills and gently dipping sheets can add much to the strength of a volcano (Fig. 9.15; cf. Figs. 1.16, 2.29). Because sills are normally subhorizontal and stiffer (with a higher Young's modulus) than the adjacent layers (Fig. 2.35), many propagating fractures are either deflected or vertically arrested on meeting a sill. Since fracture arrest is the main measure of fracture resistance and thus strength, it follows that sills contribute to making volcanic edifices stronger. Reasonably thick sills may also function as shallow magma chambers, and sill formation is the main mechanism of magma-chamber formation (Chapters 5 and 6).

Dikes and inclined sheets also add to the strength of volcanic edifices (Figs. 1.20, 2.1, 2.34, 6.13). Commonly, they form a **framework** that reinforces or **strengthens** the edifice as a whole. This is particularly the case when numerous inclined sheets, sills, and dikes form a clear and coherent framework (Figs. 9.16 and 9.17). While systematic quantitative studies are lacking as to the relative frequencies of sills in stratovolcanoes and basaltic edifices, it is likely that sills are much more common in stratovolcanoes. This follows because the conditions for dike and sheet deflection into sill are more widely satisfied, particularly through elastic mismatch or stress barriers (Section 7.4), in stratovolcanoes than in basaltic edifices. Since dikes and inclined sheets are also common in stratovolcanoes, it follows that the intrusive framework in stratovolcanoes has normally more **reinforcing effects** in stratovolcanoes than in basaltic edifices.

While sheet intrusions generally increase the strength of a volcanic edifice, they may occasionally contribute to **decreasing the strength** when there is extensive geothermal alteration (Fig. 9.17). Intense geothermal alteration generates clay, which can significantly reduce the shear strength of rock bodies in volcanoes. The clay is normally very local, however, and occurs primarily close to large magma bodies rather than comparatively thin sheet intrusions. Also, the reduction in strength through clay formation would be similar in basaltic edifices and stratovolcanoes, on the assumption that both are supplied with magma through shallow chambers. Also, clay formation does not necessarily reduce the overall strength of the rock body within which a clay layer or unit occurs. Due to its very low Young's modulus, which may be 0.05 GPa or lower (Appendix D), clay also suppresses stresses generated through loading, and it commonly contributes to fracture arrest, thereby occasionally increasing the **fracture resistance** and strength of the rock body.

Fig. 9.15 Sills and gently dipping inclined sheets generally make volcanic edifices stronger. This follows because many vertically propagating fractures such as dikes are either deflected or arrested on meeting the sills/sheets. Here, a vertically propagating basaltic dike (in Tenerife, Canary Islands) has become arrested on meeting with a gently dipping stiff sheet, marked as a stiff layer. View north–northeast, the maximum thickness of the dike is about 0.8 m. A black and white version of this figure will appear in some formats. For the colour version, please refer to the plate section.

Once it has formed, a framework of sheet intrusions increases the strength and stiffness of volcanic edifices in two basic ways. First, the framework acts as a **resistance** against any external and internal general loading. For example, during magma-chamber expansion, the resulting inflation or surface doming becomes less because of the intrusion framework than it would normally be without the framework. When the potential inflation is reduced, so are the chances of surface-fracture formation, and related failures. Second, the framework also acts as a resistance to local fracture propagation and associated failure. This is because the

Fig. 9.16 A network of dikes and sills/sheets can increase the mechanical strength of a volcanic edifice. The photo shows a network of dikes and sills/sheets in Southeast Iceland. View northeast, the thickness of most of the intrusions is about 1 m. The sheets/sills dissect the dikes and are therefore younger than the dikes.

Fig. 9.17 When sheet intrusions produce extensive geothermal alteration with the formation of clay, the shear strength of a volcanic edifice may be locally reduced. Here, geothermal alteration is shown in the mountain Esja in West Iceland, a fossil central volcano. View east, the geothermal alteration is associated with a swarm of inclined sheets (cf. Gudmundsson, 2017). A black and white version of this figure will appear in some formats. For the colour version, please refer to the plate section.

great variety in the attitude of the sheets means that fracture propagation in almost any direction is likely to meet one or more sheet intrusions. Because of the high stiffness of such intrusions, they are likely to arrest or deflect the propagating fracture (Fig. 9.15; Gudmundsson, 2009). This arrest mechanism applies particularly to stratovolcanoes because they have many soft sedimentary and pyroclastic layers to which the stiff sheets form a strong mechanical contrast, thereby encouraging fracture arrest or deflection according to the mechanisms of stress barriers or elastic mismatch (Chapter 7).

In addition to their effects on the strength and stability of volcanic edifices, sheet intrusions are the main carriers of magma to the surface in volcanic eruptions. The internal structure exposed in deeply eroded volcanoes shows the paths of sheet intrusions and, in particular, how and where they become arrested (Chapters 7 and 10). In detail, the intrusions paths depend on the availability of joints and, occasionally, faults in the host rock, as discussed in the next subsection. Studies of the **sheet-intrusion paths** and arrest conditions are of fundamental importance for understanding volcanic unrest periods, and as tests of models for forecasting dike/sheet-fed volcanic eruptions.

9.5.3 Joints and Faults

Joints are the most common outcrop-scale structures in volcanoes and associated crustal segments. In fact, in most outcrops, of any kind of rock, joints are the dominant structures. While **tectonic joints** are common in volcanic areas, as in other parts of the brittle crust, the characteristic joints of volcanic areas are columnar or **cooling joints**. Recall that these are normally very narrow fractures generated during the solidification and contraction (shrinkage) of an intrusion, lava flow, or pyroclastic layer. The joints occur in lava flows but are normally best developed in sheet intrusions, such as dikes and sills, where they commonly form dense networks, characterised by hexagonal joints (Figs. 9.18, 9.19; cf. Figs. 2.8, 2.35, 5.5, 5.17).

While mostly small, joints are very important structures in volcanoes for several reasons. First, joints affect the mechanical **properties** of the layers that host them. In particular, joints have strong effects on the effective Young's modulus (stiffness) of the rock hosting rock layers. More specifically, as the joint frequency increases in a rock body, its effective Young's modulus decreases. It follows that to get correct estimates of the *in situ* Young's modulus for analytical and numerical modelling of tectonic processes in a volcano, the joint frequency in the layers which constitute that volcano needs to be considered. There are analytical methods for calculating the effective Young's modulus based on the joint (or, generally, fracture) frequency (cf. Gudmundsson, 2011a).

Second, jointed rock bodies store and **transport fluids**. In volcanoes, the fluids are primarily groundwater and geothermal water. Intrusions, in particular, commonly have a very low matrix porosity. Yet, their fracture porosity may be considerable (Figs. 9.18, 9.19: cf. Figs. 2.8, 2.35, 5.5, 5.17) and they can thus act as reservoirs for geothermal fluids or, at shallower depths, as fractured aquifers. It is known that in some sills the chilled selvage at the lower boundary may be partly ruptured while the selvage at the upper boundary remains intact. Such sills can act as **reservoirs** and aquifers (Gudmundsson and Lotveit, 2012). The joints also conduct fluids. Much of the permeability in intrusions and lava flows is due to columnar joints.

Fig. 9.18 Columnar (cooling) joints are common in lava flows, pyroclastic layers, and, particularly, in sheet intrusions such as the sills, part of one which is seen here. Columnar joints are the main weaknesses in a lava pile used to form the paths of dikes (Chapter 10) and for the generation of faults (Gudmundsson, 2011a). View east, these columns form a part of a basaltic sill in Reynisfjall, South Iceland.

Third, joints are the main **weaknesses** or discontinuities used during the formation of larger fractures. The larger fractures include faults, inclined sheets, and dikes. It is well established how joints in lava flows and intrusions link up to form large-scale tension fractures and normal faults (Gudmundsson, 2011a). Also, field studies of dikes show clearly that they commonly use columnar joints as weaknesses to form parts of their propagation paths (Fig. 9.20: cf. Fig. 7.3). Since dikes normally propagate along a path that is parallel to the maximum compressive principal stress σ_1, it follows that they mostly use joints that are parallel to σ_1, that is, in a principal stress plane. Dikes can, however, also occasionally use joints that are oblique to σ_1 and thus potential or actual shear fractures, that is, faults. The energy conditions that encourage dikes to use oblique joints/faults as parts of their paths are, however, rarely met, as is discussed in detail in Chapter 10.

Fig. 9.19 Top part of columnar joints in a basaltic lava flow at Kirkjubaejarklaustur in South Iceland. The outcrop is known as the 'Church Floor', but has never been the site of a church. Most of the columns are either hexagonal or pentagons. (a) Overview of a large part of the outcrop, the total area of which is about 80 m². (b) Close-up of a part of the outcrop. Most of the columns are 20–30 cm in diameter. The darker rock in-between the columns is also basalt but it is more vesicular than that of the columns themselves (cf. A.T. Gudmundsson, 1990).

This brings us to the faults themselves. The most common **large faults** in active volcanoes/volcanic fields are **normal faults** (Fig. 9.21: cf. Fig. 7.24). This is as expected from the stress conditions in volcanic fields/volcanoes that favour dike or sheet emplacement. These stress conditions are mostly such that σ_1 is either vertical (for vertical dikes) or

Fig. 9.20 Dikes use joints as weaknesses when forming their paths. View northwest, a segmented basaltic dike partly follows cooling joints in a small basaltic intrusion (a laccolith) on the peninsula of Anaga in Tenerife (Canary Islands). The dike ends vertically – thins outs or tapers away to its tip. Cf. Fig. 7.3. Photo: Kevin D'Souza.

inclined at various angles (for inclined sheets). A vertical or steeply dipping σ_1 favours the formation of normal faults.

However, **strike-slip** and **reverse faults** do also occur in volcanoes, primarily close to intrusions. Thus, small strike-slip faults are common in volcanic regions of Iceland (Gudmundsson et al., 1992) and strike-slip faulting has been recorded during present-day dike propagation in volcanic rift zones in Iceland (e.g. Agustsdottir et al., 2016). Some reverse faulting is also associated with dike emplacement, as recorded during many current dike-emplacement events as well as by direct field observations (Gudmundsson et al., 2008). For example, dikes emplaced close to a normal fault may induce a reverse slip on that fault (Fig. 3.3).

Strike-slip faulting and reverse faulting in volcanoes is, as indicated, commonly associated with dike emplacement. Because of its overpressure or driving pressure, the dike temporarily

Fig. 9.21 Aerial view of a large normal fault in the rift zone of Southwest Iceland. View south, the fault (known as Jorukleif) has a total vertical displacement of about 200 m (Gudmundsson, 2017). The displacement is on two steps, the main upper one, seen here, has a vertical displacement of 130 m. The fault is a part of a segmented fault, tens of kilometres long, that constitutes the western boundary fault of the Thingvellir Graben (cf. Fig. 7.24). A black and white version of this figure will appear in some formats. For the colour version, please refer to the plate section.

increases the horizontal compressive stress so as to make it the largest principal stress σ_1. This is particularly likely to happen at shallow depths (Fig. 3.3). If σ_2 and σ_1 are originally similar in magnitude – where σ_1 is vertical for normal faulting – then with the new σ_1 being horizontal and perpendicular to the strike of the existing normal fault, a reverse displacement may be triggered on the normal fault. The reverse slip in Fig. 3.3 is at least 5 m, but millimetre- and centimetre-scale reverse slips are common during dike emplacement.

Theorctically, reverse faulting can also occur in volcanoes during subsidence of the surface, that is, deflation. Most deflation periods, however, do not induce high horizontal compressive stresses. This follows because the deflation normally stops when the excess pressure in the chamber p_e has reduced to zero (Chapter 5). When that happens, the magma chamber is again in a **lithostatic equilibrium** so that it is not generating any tectonic stresses (tensile or compressive) in the host rock.

9.5.4 Ring-Faults and Ring-Dikes

Ring-faults as seen at the surface of volcanoes and associated eruptions have been discussed above and earlier in the book (Chapters 5 and 8). When the eruptions occur on the ring-faults themselves, they are supplied with magma by ring-dikes. So far, however, not much information has been provided on what the ring-faults and ring-dikes look like at depth in volcanoes. More specifically, it has not been indicated how these structures can be recognised in the field in deeply eroded, extinct volcanoes.

As discussed in Chapter 5 (Section 5.7), ring-faults are generally **dip-slip faults**. Some ring-faults have a strike-slip component, while others may have segments that are primarily strike-slip faults. The main ring-fault displacement is vertical, however, so that the faults are mainly dip-slip. This means that they are either normal faults or reverse faults. The main displacement along a ring-fault is thus **vertical shear** displacement. When such a fault is occupied by a ring-dike, there is also a significant **opening** displacement (extension), as discussed below, but the main displacement is still that of a shear fracture – a fault – rather than an extension fracture – a dike.

Well-exposed ring-faults are rare in deeply eroded volcanoes. Where a ring-fault is seen in cross-sections such as steep walls and cliffs, normally only a small segment of the fault is exposed. One of the best-exposed ring-fault segments is the one in the Hafnarfjall Volcano in West Iceland (Fig. 5.22). The ring-fault in Hafnarfjall is a steeply dipping **normal fault**. This appears to be the most common ring-fault attitude seen in the field, namely steeply (inward-) dipping **normal** faults (Simkin and Howard, 1970; Macdonald, 1972; Filson et al., 1973; Fedotov et al., 1980; Aramaki, 1984; Lipman, 1984, 1997; Newhall and Dzurisin, 1988; Gudmundsson, 1998; Geshi et al., 2002; Browning and Gudmundsson, 2015a,b). There is also some field evidence for outward-dipping **reverse** ring-faults (e.g. Williams et al., 1970; Branney, 1995; Cole et al., 2005; cf. Geyer and Marti, 2008; Marti et al. 2009), but less so than for the inward-dipping normal ring-faults. The outward-dipping reverse fault was assumed in the original model on ring-fault formation (Anderson, 1936) and is commonly observed in analogue experiments (Marti et al., 1994, 2008; Acocella, 2007; Geyer and Marti, 2014). But, as said, field observations indicate that inward-dipping normal ring-faults are more common. The attitude of the ring-fault has important implications for the associated eruption mechanics, as is discussed in Chapter 8. In particular, the largest explosive eruptions are very likely associated with displacements on outward-dipping ring-faults (Gudmundsson, 2016).

Ring-faults are commonly occupied by **ring-dikes**, many of which are vertical or dip steeply inward (Oftedahl, 1953; Almond, 1977), while some may be outward-dipping (Anderson, 1936). Although a ring-dike is not a necessary condition for a ring-fault to form or slip, such a dike (while fluid) reduces the friction along the fault and therefore makes a ring-fault slip much easier. In fact, there is no evidence of a ring-dike formation in five of the six main caldera collapses instrumentally recorded during the past hundred years, that is, in Katmai (1912), in Fernandina (1968), in Miyakejima (2000), in Piton de la Fournaise (2007), and in Kilauea (2018). A ring-dyke may, however, have formed during the collapse in Pinatubo (1991). Nevertheless, if a ring-dike forms during the collapse, it is more likely that the collapse becomes comparatively large.

As structures, ring-dikes are arcuate or oval – **circular** or part-circular – in plan view with **steep** to vertical **contacts** with the host rock. Some ring-dikes can be traced almost to the entire circle, whereas others are exposed only for a small part of the circle (Fig. 9.22). Most ring dykes are felsic, particularly the thick ones, but some are mafic or intermediate.

The main mechanical difference between ring-dikes and other dikes – as well as inclined sheets and sills, all of which are primarily extension fractures – is that ring-dikes occupy **shear fractures**, namely, ring-faults. More specifically, most ring-dikes are thought to be generated during, or soon after, caldera collapses, and to be injected into the caldera fault – the ring-fault (Figs. 9.23 and 9.24). This is also what distinguishes ring-dikes from

Fig. 9.22   Part of the ring-dike in Ketillaugarfjall in Southeast Iceland. View northeast, the ring-dike is tens of metres thick here, but it can be traced only over a small arc.

otherwise geometrically similar concentric dikes (in plan view) such as inclined sheets; the former occupy shear fractures (faults), the latter occupy extension fractures.

Ring-dikes vary widely in **thickness**. Some of the best-known ring-dikes are felsic and may reach thicknesses of hundreds of metres (Billings, 1972; Johnson et al., 2002). For example, some of the ring-dikes in the Oslo area of Norway are as thick as 500 m (Segalstad, 1975). Many ring-dikes, however, are much thinner, commonly several metres or less (Segalstad, 1975; Browning and Gudmundsson, 2015b). While most ring-dikes, particularly thick ones, are generated through magma being injected into the ring-fault directly from the magma chamber – especially if the ring-fault is outward-dipping (Fig. 9.24) – some ring-dikes, particularly thin ones, may form by the ring-fault capturing inclined sheets and deflecting them upwards along the ring-fault (Browning and Gudmundsson, 2015b).

Most studied ring-dikes are poorly exposed in the vertical dimension (Fig. 9.22), so that deep vertical sections of tens or hundreds of metres through the dikes are rare. It follows that accurate estimates of the ring-dike dip is difficult. The **dip measurements**, so far as they go, indicate that most ring-dikes are close to vertical, as indicated above, but the debate is whether they dip steeply outward or inward. There is a body of evidence indicating that some dip steeply outward whereas others dip steeply inward. A recent study of some of the best-exposed ring-dikes in Ardnamurchan in Scotland, for example, indicates that the dike may be dipping inward at 50–70° (O'Driscoll et al., 2006). Also, studies of an unusually well-exposed, but thin, ring-dike in the fossil central volcano Hafnarfjall in West Iceland shows that the ring-dike (and the ring-fault) is overall very steeply inward dipping (Fig. 5.22). The importance of the dip of ring-dikes is because outward-dipping ring-dikes/ring-faults are much more likely to give rise to large eruptions than inward-dipping ones (Chapter 8; Gudmundsson, 2016).

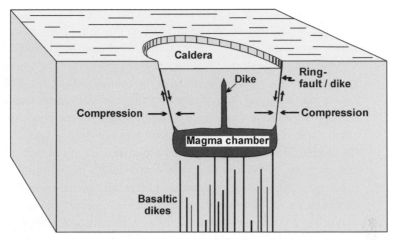

Fig. 9.23   An inward-dipping ring-fault and a central dike (as in Fig. 9.24). Collapse on an inward-dipping, normal ring-fault is less likely to generate a large explosive eruption than collapse on an outward-dipping ring-fault and ring-dike (Chapter 8; cf. Gudmundsson, 2015, 2016).

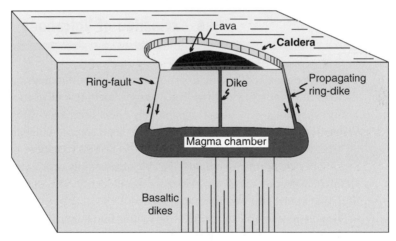

Fig. 9.24   An outward-dipping ring-fault – a ring-dike – following an initial central eruption on an ordinary dike (producing a lava flow). Collapse on an outward-dipping, reverse ring-fault may result in the piston-like rock body subsiding to the floor of the associated chamber and squeezing out much of the magma in the chamber. Outward-dipping ring-faults are thus more likely to generate large explosive eruptions than inward-dipping faults. Also, the ring-dikes are likely to become thicker, for a given magma composition, for outward- than for inward-dipping faults (Chapter 8; cf. Gudmundsson, 2015, 2016).

9.5.5 Plugs and Necks

Earlier (Section 8.3), I discussed volcanic eruptions supplied with magma through cylindrical conduits. Here, I briefly outline the resulting structures, as seen in inactive and somewhat eroded volcanoes. A solidified cylindrical or somewhat elliptical (in plan view) conduit is referred to as a **volcanic plug** or **neck** (Fig. 8.3). Plugs are common, but most are small, that is, less than 1 km in diameter. They may, however, stand tens or hundreds of metres above their surroundings (Fig. 8.3). They are discordant intrusions.

Some plugs are primarily composed of tephra, breccia, and other **pyroclastic rocks** formed through explosive activity in the conduit (Fig. 8.3a). These are later sealed and healed into hard rock. Other plugs are primarily, or almost exclusively, formed by fluid magma that solidifies to form the plug. Commonly, the plugs are then to a large degree composed of a **cluster of dikes** (Fig. 8.3b). In fact, drilling into active conduits, particularly the conduit of the active Unzen Volcano in Japan, showed that at 1.3 km depth below the summit crater the conduit is largely composed of a cluster of dikes (Nakada et al., 2005).

Plugs are normally more **resistant to erosion** than the surrounding parts of the volcano, so that they stand as columns – plugs or necks – above the surroundings when a significant part of the rest of an extinct volcano has been destroyed through erosion (Fig. 8.3). Well-known volcanic plugs include the Castle Rock on which Edinburgh Castle in Scotland stands, and the Shiprock in New Mexico (the United States). The Castle Rock rises some 80 m above its surroundings and is mostly of basalt (dolerite) whereas the Shiprock rises to 480 m above its surroundings and is mostly of breccia.

9.6 Extinction of Volcanoes

On Mars, polygenetic/central volcanoes are active for hundreds of millions of years, but on Earth at most for a few million years. Why is there this difference? Why do volcanoes on Earth become extinct after comparatively short lifespans? The answer is that on Earth volcanoes become disconnected from their main sources comparatively quickly, whereas on Mars that does not happen. The reason that central volcanoes become thus disconnected is primarily the **plate movements**: the volcanoes simply drift away from their main sources. Before that process is discussed in more detail, let us briefly summarise the active time, the life, of a typical polygenetic/central volcano.

All polygenetic volcanoes are born, evolve into maturity, and then become extinct, that is, they die. The volcano is **born** when a shallow magma chamber is formed and begins to channel magma to a limited area at the surface (Fig. 9.1). This focusing of eruptions and eruptive materials on a comparatively small part of the associated volcanic field/system/ zone results in the formation of a volcanic edifice, namely a polygenetic volcano (Section 9.2; Gudmundsson, 1998, 2000). While the shallow chamber is the necessary condition for the formation of a polygenetic volcano, the deep-seated reservoir (or reservoirs, in case there is more than one) is the one that primarily controls the longevity of the volcano and its eventual extinction (Fig. 9.1). More specifically, it is the interplay between the shallow chamber and the source reservoir(s) that largely determines how long a volcano remains active.

Following its initiation, a shallow **magma chamber** grows in size. The maximum eruptive volume from a chamber during a single eruption depends on the magma volume that the chamber holds – which normally varies positively with the overall size of the chamber. It follows that so long as the chamber volume is increasing with the growth of the chamber, the **maximum eruptive volume** from a chamber (the combined intrusive and extrusive volume in a single event) also increases. The chamber growth comes to an end, however, when the influx of magma from the source reservoir equals the volume of magma that solidifies within the chamber or flows out of the chamber during dike/sheet intrusion and/or eruption. Normally, the shallow chamber spends most of its lifetime at this equilibrium stage, which is thus also the mature and main stage of the associated volcano (Section 9.3).

The volcano enters the **extinction stage** when the shallow chamber stops receiving magma from its source reservoir. For its supply of magma, the chamber depends on the reservoir, so that when that supply is much diminished and eventually ceases altogether, the chamber and its volcano gradually become extinct. After the supply of magma from the reservoir is partly or entirely **cut off**, the chamber may continue some activity for perhaps a few tens of thousands of years, depending on its size, shape, and depth (Gudmundsson, 2000; cf. Spera, 1980). One consequence of the reduction in the rate of magma received and thus in the volume of the shallow chamber is that the chamber's shape changes, whatever its mature shape, to that of a **sphere** so as to prolong its life.

The main reason why the supply of magma from the deep-seated reservoir to the shallow magma chamber becomes cut off is the lateral displacement of the shallow chamber. While the deep-seated reservoir is commonly **stationary** below the central part of the plate boundary, the shallow chamber is displaced or **moves laterally** with the moving plate. This applies to many divergent plate boundaries and probably many convergent boundaries as well. At divergent plate boundaries, and in rift zones in general, the deep-seated reservoir is commonly located at the top of the upper mantle, at the axis of the plate boundary (Fig. 6.3). While some reservoirs may shift their location due to changes in the location of the regions of minimum potential energy (Chapter 6), many reservoirs are likely to be essentially stationary for much longer times than the lifetimes of shallow chambers and associated volcanoes.

By contrast, the shallow chamber normally becomes a **permanent part** of one or the other of the two plates that meet at the boundary. This happens when a new spreading centre is formed to either side of the old one; for example, when tilting of the lava pile results in shifting of the main paths of dikes (Figs. 6.9 and 6.10). Also, when the spreading rate is asymmetric so that one plate has a higher spreading rate than the other, the shallow chamber becomes a fixed part of one of the plates. The shallow chambers becoming attached to one plate, rather than both, is the reason that they are found in, say, East Iceland (a part of the Eurasian Plate) rather than West Iceland (a part of the North American Plate), and not split in half between these parts/plates. A shallow chamber thus eventually becomes part of one plate and moves laterally with it away from the deep-seated and more stationary source reservoir. For example, plate movements in Iceland move a shallow magma chamber laterally by as much as 20 km in a million years. If the source reservoir remains essentially stationary during that time, progressively **fewer dikes** will meet the chamber to supply magma to it, so that the chamber gradually becomes extinct.

Once the shallow chamber is largely or wholly cut off from its source reservoir, the chamber gradually solidifies and becomes extinct. It follows that the polygenetic volcano for which the chamber acts as a source also becomes extinct. Many polygenetic volcanoes are active for 0.5–1 million years, while some, particularly in the western United States, appear to have been active (sometimes intermittently) for several million years.

9.7 Summary

- All volcanoes (here: polygenetic/central volcanoes) are born, reach maturity, and then become extinct (die). A necessary condition for a volcano to be born is the formation of a shallow magma chamber. The chamber acts as a sink for magma (collects magma) from one or more deeper source reservoirs and as a source for dike/sheet injections and eruptions. That is, the shallow chamber channels magma to a comparatively small area at the surface, namely the area where the volcanic edifice builds up. If there were no local shallow chambers, then no volcanic edifices would form because the volcanism would be evenly distributed throughout the volcanic system/field/zone.
- Because the deep-seated reservoirs are generally much larger than the shallow chambers they supply magma to, a single magma flow from a reservoir, lasting perhaps many years, may trigger tens of magma flows from the chamber. Each magma flow from the chamber results in dike/sheet injection, some of which may reach the surface to supply magma to an eruption. Other dike/sheet injections become deflected into sills, mostly at contacts between mechanically dissimilar layers. Also, the stress concentration around the shallow chamber contributes to its rupturing more frequently than the deep-seated reservoir. Thus, the frequency of dike/sheet injections and eruptions is much higher in the volcano than in the surrounding volcanic system/zone/field, which is the primary reason that the volcanic edifice forms.
- Stratovolcanoes have generally much steeper slopes than basaltic edifices (shield volcanoes). The uppermost parts of some stratovolcanoes may be as steep as $42°$; basaltic edifices are rarely anywhere steeper than $12°$. To maintain their steepness throughout most of their lifetimes, stratovolcanoes must be mechanically stronger than basaltic edifices. The stratovolcanoes are mechanically stronger, and can maintain steeper slopes than basaltic edifices, primarily because the great contrast in the properties of the layers that constitute a stratovolcano encourages fracture arrest at layer contacts.
- The top crater cones on volcanoes, like crater cones in general, are a direct consequence of the cubic law and flow channelling, whereby much of the magma flow is through the large-aperture parts of the conduit/fissure. Top craters that are negative landscape features, that is, depressions, are normally related to cylindrical conduits where pressure changes, a decrease or an increase, can result in craters where the floor is much lower than that of the adjacent part of the volcano.
- Vertical (caldera) collapses and lateral (landslide) collapses both contribute much to the overall shapes of many volcanoes. Following caldera collapse, the top part of the volcano

that previously may have been a sharp cone becomes a depression so that, from a distance, the volcano looks flat-topped. Often, much more than just the top part of the volcano subsides during the collapse, but rarely less than that. In contrast to caldera collapses, which change the overall shape of the volcano, lateral collapses primarily contribute to keeping the slopes of the volcanoes comparatively gentle. Lateral collapses do not normally change the overall shape of the volcano.

- The main structural elements that constitute volcanoes, namely those that occur inside volcanoes, are the following: (1) mechanical layers and contacts; (2) sheet intrusions, that is, dikes, sills, and inclined sheets; (3) joints and faults; (4) ring-faults and ring-dikes; and (5) plugs and necks. The intrusions, such as dikes, ring-dikes, sills, inclined sheets, and plugs/necks, normally make the volcano stronger; they provide a strong framework inside the volcano. But the intrusions are also 'elastic inclusions' that concentrate stresses and may sometimes encourage failure at their contacts with the host rock. Joints and faults, including ring-faults, by contrast make the volcano normally mechanically weaker. Joints, in particular, provide the weaknesses along which many faults develop and dikes and sheet intrusions in general develop their paths.

- Volcanoes become extinct – dic – when the supply of magma is cut off. This normally happens gradually as the volcano and its shallow magma chamber are carried by plate movements away from the deep-seated reservoir which acts as the primary magma source for the shallow chamber and thus the volcano. The deep-seated reservoir may be essentially stationary for a long time, while the shallow chamber and the volcano become permanent features of one of the plates at a boundary and therefore move with that plate away from the boundary. When the shallow chamber becomes cut off from its source reservoir, the chamber gradually solidifies into a pluton, and the volcano to which the chamber supplied magma becomes extinct.

9.8 Main Symbols Used

f porosity (magma fraction) of a shallow chamber

N_s number of dikes/inclined sheets injected from a shallow chamber

p_e excess magmatic pressure in the chamber (equal to the host-rock tensile strength)

q fraction of magma volume 'erupted' by a reservoir that is received by its chamber

V_c volume of a shallow chamber

V_r volume of a deep-seated reservoir

β_m magma compressibility

β_p pore compressibility

β_r host-rock compressibility

η porosity (magma or melt fraction) of a reservoir

λ ratio of the volume of a shallow chamber and that of its source reservoir

9.9 Worked Examples

Example 9.1

Problem

A shallow magma chamber has an estimated total volume of 90 km^3 whereas the total volume of the source reservoir is 9000 km^3. As total volumes (in contrast to the volume of magma, which is normally much smaller), both values are within the typical range for chambers and reservoirs (Chapter 6; Gudmundsson, 2016). Assume that the fraction of the magma volume received by the shallow chamber during flow from the reservoir is 0.5. Assume also that the melt fraction of the reservoir is 0.25, that of the shallow chamber 0.40, the *in situ* tensile strength of the crust is 3 MPa, and that the compressibilities of the magma and the host rock are as follows: $\beta_m = 1.3 \times 10^{-10}$ Pa^{-1}, $\beta_p = 9 \times 10^{-11}$ Pa^{-1}, and $\beta_r = 3 \times 10^{-11}$ Pa^{-1}. Use this information to calculate the number of dikes and inclined sheets injected by the shallow chamber during a single magma flow from the source reservoir.

Solution

From Eq. (9.1) we have:

$$N_s = \frac{\eta p_e q (\beta_p + \beta_m)}{f p_e \lambda (\beta_r + \beta_m)} = \frac{\eta q (\beta_p + \beta_m)}{f \lambda (\beta_r + \beta_m)}$$

The values of the terms β_m, β_p, and β_r in the equation are provided directly. Let us now estimate the values of the other terms. The melt fraction (or porosity) of the magma chamber is given as 0.4 (40%) and that of the reservoir as 0.25 (25%), so that f is 0.40 and η is 0.25. The excess pressure p_e is equal to the *in situ* tensile strength, and is here 3×10^6 Pa. Notice, however, that since p_e occurs both in the nominator and the denominator in the first expression on the right-hand side of Eq. (9.1), the term cancels out and does not occur in the second expression on the right-hand side. Since the volume of the shallow chamber is 90 km^3 and that of the source reservoir 9000 km^3, it follows that their ratio λ is 0.01. Also, since the fraction of the magma volume received by the shallow chamber during flow from the reservoir is 0.5 it follows that q is 0.5.

Using these values, from the second term on the right-hand side of Eq. (9.1) we get:

$$N_s = \frac{0.25 \times 0.5 \times (9 \times 10^{-11} \text{ Pa}^{-1} + 1.3 \times 10^{-10} \text{ Pa}^{-1})}{0.40 \times 0.01 \times (3 \times 10^{-11} \text{ Pa}^{-1} + 1.3 \times 10^{-10} \text{ Pa}^{-1})} = \frac{2.75 \times 10^{-11} \text{ Pa}^{-1}}{6.4 \times 10^{-13} \text{ Pa}^{-1}} = 43$$

Thus, theoretically, a single magma flow from the reservoir could trigger 43 dike and inclined-sheet injections from the connected shallow magma chamber. Given the number of injections, it is likely that one or several would reach the surface to feed eruptions.

Example 9.2

Problem

Stress measurements show that the state of stress in a layer at 2 km depth in a rift zone at a divergent plate boundary is such that the maximum compressive principal stress σ_1 is vertical (its magnitude is obtained below), σ_2 is 47 MPa, and σ_3 is 45 MPa. The average density of the uppermost 2 km of the crust is 2500 kg m^{-3}.

(a) Calculate the vertical stress σ_1 in the layer at 2 km depth.
(b) How would the stress field change if a regional dike with a magmatic overpressure of 10 MPa penetrated the layer at 2 km depth?
(c) What are the main volcanotectonic implications of the dike-induced change in the stress field in (b)?

Solution

(a) The vertical stress is normally calculated either from Eq. (3.5) or Eq. (3.6), with Eq. (3.5) being used when only the average density of the crustal segment (rather than densities of the individual layers) is known. From Eq. (3.5) the vertical stress σ_v is given by:

$$\sigma_v = \rho_r g z$$

where ρ_r is the average density of the rock layers, g is the acceleration due to gravity (9.81 m s^{-2}) and z is the depth measured as positive down from the free surface of the Earth. Using the provided values, Eq. (3.5) gives the vertical stress as:

$$\sigma_v = 2500 \text{ kg m}^{-3} \times 9.81 \text{ m s}^{-2} \times 2000 \text{ m} = 4.9 \times 10^7 \text{Pa} = 49 \text{ MPa}$$

The calculations thus confirm that the vertical stress is the first principal stress, that is, $\sigma_v = \sigma_1$, as is normal in rift zones and at divergent plate boundaries in general.

(b) If a dike with a magmatic overpressure of 10 MPa propagated through the layer at 2 km depth, the overpressure should be added to the magnitude of σ_3 to find the new stress magnitude in a direction perpendicular to the strike of the dike. This new stress is 45 MPa + 10 MPa = 55 MPa and it is thus higher than the vertical stress, 49 MPa. It follows that this new stress of 55 MPa becomes the maximum compressive principal stress, so that, subsequently to the dike emplacement, σ_1, becomes, for a while, horizontal (it was vertical) and has a direction perpendicular to the strike of the injected dike.

(c) When the stress field in a volcanic rift zone is such that in one or more layers the maximum compressive principal stress σ_1 is, for a while, horizontal and perpendicular to the common strike of dikes, then there is a tendency for the subsequent dikes to either become arrested on meeting the contact with that layer, or to become deflected into a sill along the contact. Dike deflection into a sill due to rotation of the principal stresses (generating a stress barrier) is one of the common mechanisms for magma-chamber initiation (Chapters 5 and 7). A local stress of this type also favours reverse faulting, a not uncommon process during dike emplacement (Chapter 4).

Example 9.3

Problem

A basaltic dike is injected from a shallow chamber at 4 km depth. The chamber-roof (host-rock) tensile strength is 3 MPa and the host-rock density is 2600 kg m^{-3} (Appendix E) whereas the magma density is 2750 kg m^{-3} (Appendix F). Assume that the stress difference σ_d is everywhere 1 MPa in the uppermost 2 km of the crust. Calculate the theoretical dike overpressure at the depth of (a) 2 km and (b) 0.1 km below the surface. (c) Would this dike reach the surface to erupt? If not, at which depth would the dike propagation stop?

Solution

The overpressure or driving pressure p_o of a dike as a function of its height h above the source chamber (the dip-dimension) is, from Eq. (5.3), given by:

$$p_o = p_e + (\rho_r - \rho_m)gh + \sigma_d$$

(a) We recall that the excess pressure in the chamber p_e is equal to the tensile strength of the roof (3 MPa). From Eq. (5.3) above:

$$p_o = 3 \times 10^6 \text{Pa} + (2600 \text{ kg m}^{-3} - 2750 \text{ kg m}^{-3}) \times 9.81 \text{ m s}^{-2}$$
$$\times 2000 \text{ m} + 1 \times 10^6 \text{Pa} = 1.0 \text{ MPa}$$

(b) From Eq. (5.3) we now get:

$$p_o = 3 \times 10^6 \text{Pa} + (2600 \text{ kg m}^{-3} - 2750 \text{ kg m}^{-3}) \times 9.81 \text{ m s}^{-2}$$
$$\times 3900 \text{ m} + 1 \times 10^6 \text{Pa} = -1.7 \text{ MPa}$$

(c) At the depth of 2000 m, the theoretical overpressure (1 MPa) is still positive and high enough to propagate the dike, but at 100 m depth the theoretical over-pressure has become negative, that is, –1.7 MPa. A negative overpressure cannot exist. Thus, as soon as the overpressure becomes zero, the dike propagation would stop. We see from the results that the overpressure becomes zero when the absolute value of the second term (the buoyancy term), becomes equal to 4 MPa. Thus, the theoretical maximum propagation height or straight-line path length h_{max} that the dike can reach before its overpressure becomes zero and the propagation must stop is:

$$h_{max} = \frac{4 \times 10^6}{1471.5} = 2718 \text{ m}$$

This means that the dike propagation would stop and the dike become arrested at 4000–2718 = 1282 m below the surface. Here we are discussing arrest as the over-pressure becomes zero. Based on the discussion in Chapters 5 and 7, the dike could become arrested at greater depths than this if the dike meets with a stress barrier, that is, is affected by Cook–Gordon delamination, or through the process of elastic mismatch.

Example 9.4

Problem

Explain how a shallow magma chamber acts both as a sink and a source for magma.

Solution

The chamber acts as a sink, that is, collects or draws in magma, for magma from the deeper source reservoir. Dikes or other magma paths from the source reservoir are, largely through the effects of stress concentration and the associated local stress field, drawn into the shallow chamber since the paths follow the trajectories of σ_1. It follows that the chamber can draw magma from a wide area, the upper part of the source reservoir. The chamber also acts as a source of magma, namely for the volcanic edifice, the stratovolcano or basaltic edifice, that forms at the surface above the chamber. The chamber channels magma, through dikes and inclined sheets, to a limited area at the surface where the volcanic edifice builds up. Thus, the magma chamber acts both as a sink and a source, and is a necessary condition for the formation of a polygenetic/central volcano.

9.10 Exercises

9.1 What is the primary reason for the formation of polygenetic/central volcanic edifices: what is the necessary volcanotectonic condition for such structures to form?

9.2 Illustrate, using σ_1 trajectories or ticks for a chamber with a circular vertical cross-section in a homogeneous, isotropic (non-layered) crustal segment, how a shallow magma chamber can act as a sink and source for magma. Why are the σ_1 trajectories appropriate to illustrate this function of a chamber?

9.3 Why are eruptions much more frequent above a shallow magma chamber than elsewhere in the volcanic system/zone/field given that the majority of dikes/inclined sheets normally becomes arrested on their path to the surface?

9.4 In the book double magma chambers are commonly discussed and illustrated as being the sources for extrusive and intrusive activity, that is, a shallow magma chamber supplied with magma from a deeper reservoir. Under what conditions is the plumbing system likely to consist of three (or more) chambers/reservoirs, forming triple magma chambers?

9.5 What is thought to be the normal range for shallow magma-chamber volumes?

9.6 How can caldera sizes help us estimate shallow magma-chamber volumes? Based on common caldera sizes, what would be the common magma-chamber volumes?

9.7 What would be the likely maximum volumes of shallow magma chambers? How are these volumes inferred?

9.8 What are the likely maximum volumes of deep-seated reservoirs in Iceland? How are these volumes estimated?

9.9 Explain briefly why stress concentrations play an important role in the eruption frequency of polygenetic volcanoes.

9.10 Explain briefly, how mechanical layering affects eruption frequencies of polygenetic volcanoes. Why is the effect of layering normally stronger in stratovolcanoes than in basaltic edifices?

9.11 What is the general difference in maximum slope between basaltic edifices and stratovolcanoes? What is likely to be the primary reason for this difference?

9.12 What are the main physical reasons for the formation of crater cones, at the top of stratovolcanoes or along volcanic fissures?

9.13 Mention and explain briefly one or more possible mechanisms for the formation of a top crater, that is, circular or somewhat elliptical depressions (a negative landscape form) at the top of a stratovolcano.

9.14 How do vertical collapses primarily modify the shape or geometry of a volcanic edifice?

9.15 How do lateral collapses primarily modify the shape or general geometry of a volcanic edifice?

9.16 What are the main outcrop-scale structural elements that characterise the internal configuration or structure of a polygenetic volcano?

9.17 Briefly, what are the main effects of (a) layers and contacts, and (b) of joints and faults on the paths of dikes and inclined sheets?

9.18 What are the main mechanical and geometric differences between ordinary dikes and ring-dikes?

9.19 What is the typical lifespan of a polygenetic volcano (length of time that the volcano remains active)? How do such volcanoes normally become inactive, extinct?

9.20 What are the main mechanical reasons for a single magma flow from a deep-seated source reservoir being able to trigger many magma flows (dike/inclined sheet injections) from an associated shallow magma chamber?

References and Suggested Reading

Acocella, V., 2007. Understanding caldera structure and development: an overview of analogue models compared to natural calderas. *Earth-Science Reviews*, **85**, 125–160.

Acocella, V., Behncke, B., Neri, M., D'Amico, S., 2003. Link between major flank slip and eruptions at Mt. Etna (Italy). *Geophysical Research Letters*, **30**, 2286, doi:10.1029/2003GL018642.

Acosta, J., Uchupi, E., Munoz, A., et al., 2003. Geologic evolution of the Canarian Islands of Lanzarote, Fuerteventura, Gran Canaria and La Gomera and comparison of landslides at these islands and those at Tenerife, La Palma and El Hierro. *Marine Geophysical Research*, **24**, 1–40.

Aguirre-Diaz, G., 2008. Types of collapse calderas. *Institute of Physics Conference Series: Earth and Environmental Sciences*, **3**, 012021.

Agustsdottir, T., Woods, J., Greenfield, T., 2016. Strike-slip faulting during the 2014 Bardarbunga–Holuhraun dike intrusion, central Iceland. *Geophysical Research Letters*, **43**, 1495–1503.

Almond, D. C. 1977. The Sabaloka igneous complex, Sudan. *Philosophical Transactions of the Royal Society of London*, **A287**, 595–633.

Anderson, E. M., 1936. The dynamics of formation of cone sheets, ring dykes and cauldron subsidences. *Proceedings of the Royal Society of Edinburgh*, **56**, 128–163.

Andrew, R. E. B., Gudmundsson, A., 2008. Volcanoes as elastic inclusions: their effects on the propagation of dykes, volcanic fissures, and volcanic zones in Iceland. *Journal of Volcanology and Geothermal Research*, **177**, 1045–1054.

Apuani, T., Corazzato, C., Cancelli, C., Tibaldi, A., 2005. Stability of a collapsing volcano (Stromboli, Italy): limit equilibrium analysis and numerical modelling. *Journal of Volcanology and Geothermal Research*, **144**, 191–210.

Aramaki, S. 1984. Formation of the Aira caldera, southern Kyushu, ~ 22,000 years ago. *Journal of Geophysical Research*, **89**, 8484–8501.

Baragar, W. R. A., Lambert, M. B., Baglow, N., Gibson, I. L., 1987. Sheeted dykes of the Troodos Ophiolite, Cyprus. *Geological Association of Canada Special Paper*, **34**, 257–272.

Billings, M. P., 1972. *Structural Geology*, 3rd edn. Upper Saddle River, NJ: Prentice-Hall.

Bonaccorso, A., Davis, P., 1999. Models of ground deformation from vertical volcanic conduits with application to eruptions of Mount St. Helens and Mount Etna. *Journal of Geophysical Research*, **104**, 10 531–10 542.

Boudon, G., Le Friant, A., Komorowski, J. C., Deplus, C., Semet, M. P., 2007. Volcano flank instability in the lesser Antilles arc: diversity of scale, processes, and temporal recurrence. *Journal of Geophysical Research*, **112** (B8), Art. No. B08205.

Branney, M. J. 1995. Downsag and extension at calderas: new perspectives on collapse geometries from ice-melt, mining, and volcanic subsidence. *Bulletin of Volcanology*, **57**, 303–318.

Browning, J., Gudmundsson, A., 2015a. Surface displacements resulting from magma-chamber roof subsidence, with application to the 2014–2015 Bardarbunga–Holuhraun volcanotectonic episode in Iceland. *Journal of Volcanology and Geothermal Research*, **308**, 82–98.

Browning, J., Gudmundsson, A., 2015b. Caldera faults capture and deflect inclined sheets: an alternative mechanism of ring-dike formation. *Bulletin of Volcanology*, **77**, 889, doi:10.1007/s00445-014-0889-4.

Browning, J., Drymoni, K., Gudmundsson, A., 2015. Forecasting magma-chamber rupture at Santorini volcano, Greece. *Scientific Reports*, **5**, doi:10.1038/srep15785.

Chaussard, E., Amelung, F., 2014. Regional controls on magma ascent and storage in volcanic arcs. *Geochemistry, Geophysics, Geosystems*, **15**, doi:10.1002/2013GC005216.

Chesner, C. A., 2012. The Toba Caldera Complex. *Quaternary International*, **258**, 5–18.

Cole, J. W., Milner, D. M., Spinks, K. D. 2005. Calderas and caldera structures: a review. *Earth-Science Reviews*, **69**, 1–26.

Coleman, R. G., 1977. *Ophiolites. Ancient Oceanic Lithosphere?* Berlin: Springer Verlag.

Del Potro, R., Hurlimann, M., 2008. Geotechnical classification and characterisation of materials for stability analyses of large volcanic slopes. *Engineering Geology*, **98**, 1–17.

Dzurisin, D., 2006. *Volcano Deformation: New Geodetic Monitoring Techniques*. Berlin: Springer Verlag.

Fedotov, S. A., Chirkov, A. M., Gusev, N. A., Kovalev, G. N., Slezin, Yu. B., 1980. The large fissure eruption in the region of Plosky Tolbachik Volcano in Kamchatka, 1975–1976. *Bulletin of Volcanology*, **43**, 47–60.

Filson, J., Simkin, T., Leu, L. 1973. Seismicity of a caldera collapse: Galapagos Islands 1968. *Journal of Geophysical Research*, **78**, 8591–8622.

Foden, J., 1986. The petrology of the Tambora volcano, Indonesia: a model for the 1815 eruption. *Journal of Volcanology and Geothermal Research*, **27**, 1–41.

Francis, P., 1993. *Volcanoes: A Planetary Perspective*. Oxford: Oxford University Press.

Garcia, M. O., Sherman, S. B., Moore, G. F., et al., 2006. Frequent landslides from Koolau Volcano: results from ODP Hole 1223A. *Journal of Volcanology and Geothermal Research*, **151**, 251–268.

Gertisser, R., Self, S., Thomas, L. E., et al., 2012. Processes and timescales of magma genesis and differentiation leading to the great Tambora eruption in 1815. *Journal of Petrology*, **53**, 271–297.

Geshi, N., Shimano, T., Chiba, T., Nakada S., 2002. Caldera collapse during the 2000 eruption of Miyakejima volcano, Japan. *Bulletin of Volcanology*, **64**, 55–68.

Geyer, A., Marti, J., 2008. The new worldwide collapse caldera database (CCDB): a tool for studying and understanding caldera processes. *Journal of Volcanology and Geothermal Research*, **175**, 334–354.

Geyer, A., Marti, J., 2014. A short review of our current understanding of the development of ring faults during collapse caldera formation. *Frontiers in Earth Science*, **2**, doi:10.3389/feart.2014.00022.

Gudmundsson, A. 1988. Effect of tensile-stress concentration around magma chambers on intrusion and extrusion frequencies. *Journal of Volcanology and Geothermal Research*, **35**, 179–194.

Gudmundsson, A., 1998. Magma chambers modeled as cavities explain the formation of rift zone central volcanoes and their eruption and intrusion statistics. *Journal of Geophysical Research*, **103**, 7401–7412.

Gudmundsson, A. 2000. Dynamics of volcanic systems in Iceland: example of tectonism and volcanism at juxtaposed hot spot and mid-ocean ridge systems. *Annual Review of Earth and Planetary Sciences*, **28**, 107–140.

Gudmundsson, A., 2006. How local stresses control magma-chamber ruptures, dyke injections, and eruptions in composite volcanoes. *Earth-Science Reviews*, **79**, 1–31.

Gudmundsson, A., 2009. Toughness and failure of volcanic edifices. *Tectonophysics*, **471**, 27–35.

Gudmundsson, A., 2011a. *Rock Fractures in Geological Processes*. Cambridge: Cambridge University Press.

Gudmundsson, A., 2011b. Deflection of dykes into sills at discontinuities and magma-chamber formation. *Tectonophysics*, **500**, 50–64.

Gudmundsson, A., 2012. Magma chambers: formation, local stresses, excess pressures, and compartments. *Journal of Volcanology and Geothermal Research*, **237–238**, 19–41.

Gudmundsson, A., 2015. Collapse-driven large eruptions. *Journal of Volcanology and Geothermal Research*, **304**, 1–10

Gudmundsson, A., 2016. The mechanics of large volcanic eruptions. *Earth-Science Reviews*, **163**, 72–93.

Gudmundsson, A., 2017. *The Glorious Geology of Iceland's Golden Circle*. Berlin: Springer Verlag.

Gudmundsson, A., Brenner, S. L., 2004. Local stresses, dyke arrest, and surface deformation in volcanic edifices and rift zones. *Annals of Geophysics*, **47**, 1433–1454.

Gudmundsson, A., Brenner, S. L., 2005. On the conditions for sheet injections and eruptions in stratovolcanoes. *Bulletin of Volcanology*, **67**, 768–782.

Gudmundsson, A., Lotveit, I. F., 2012. Sills as fractured hydrocarbon reservoirs: examples and models. In Spence, G. H., Redfern, J., Aguilera, R., et al. (eds.), *Advances in the Study of Fractured Reservoirs. Geological Society of London Special Publications, 374*. London: Geological Society of London, pp. 251–271.

Gudmundsson, A., Bergerat, F., Angelier, J., Villemin, T. 1992. Extensional tectonics of Southwest Iceland. *Bulletin of the Geological Society of France*, **163**, 561–570.

Gudmundsson, A., Friese, N., Galindo, I., Philipp, S. L., 2008. Dike-induced reverse faulting in a graben. *Geology*, **36**, 123–126.

Gudmundsson, A. T., 1990. *On the Ringroad*. Reykjavik: Lif og Saga (in Icelandic).

Humle, G., 1974. The interpretation of lava flow morphology. *Geophysical Journal of the Royal Astronomical Society*, **39**, 361–383.

Hurlimann, A., Marti, J., Ledesma, A., 2004. Morphological and geological aspects related to large slope failures on oceanic islands – the huge La Orotava landslides on Tenerife, Canary Islands. *Geomorphology*, **62**, 143–158.

Johnson, S. E., Schmidth, K. L., Tate, M. C., 2002. Ring complexes in the Peninsular Ranges Batholith, Mexico and the USA: magma plumbing system in the middle and upper crust. *Lithos*, **61**, 187–208.

Kaneko, T., Yasuda, A., Fujii, T., Yoshimoto, M., 2010. Crypto-magma chambers beneath Mt. Fuji. *Journal of Volcanology and Geothermal Research*, **193**, 161–170.

Kerle, N., de Vries, B. V. W., 2001. The 1998 debris avalanche at Casita volcano, Nicaragua: investigation of structural deformation as the cause of slope instability using remote sensing. *Journal of Volcanology and Geothermal Research*, **105**, 49–63.

Kilburn, C. J., 2000. Lava flows and flow fields. In Sigurdsson, H. (ed.), *Encyclopedia of Volcanoes*. New York, NY: Academic Press, pp. 291–305.

Kilburn, C. J., McGuire, W.J., 2001. *Italian Volcanoes*. Harpenden: Terra Publishing.

Lipman, P. W. 1984. The roots of ash flow calderas in western North America: windows into the tops of granitic batholiths. *Journal of Geophysical Research*, **89**, 8801–8841.

Lipman, P. W. 1997. Subsidence of ash-flow calderas: relation to caldera size and magma chamber geometry. *Bulletin of Volcanology*, **59**, 198–218.

Lipman, P. W., Mullineaux, D. R. (eds.), 1981. *The 1980 Eruptions of Mount St. Helens, Washington. US Geological Survey Professional Paper, 1250*. Denver, CO: US Geological Survey.

Lipman, P. W., Dungan, M. A., Bachmann, O., 1997. Comagmatic granophyric granite in the Fish Canyon Tuff, Colorado: implications for magma-chamber processes during a large ash-flow eruption. *Geology*, **25**, 915–918.

Lippard, S. J., Shelton, A. W., Gass, I. (eds.), 1986. *The Ophiolite of Northern Oman*. Oxford: Blackwell.

Macdonald, G. A., 1972. *Volcanoes*. Upper Saddle River, NJ: Prentice Hall.

Marti, J., Gudmundsson, A., 2000. The Las Canadas caldera (Tenerife, Canary Islands): an overlapping collapse caldera generated by magma-chamber migration. *Journal of Volcanology and Geothermal Research*, **103**, 161–173.

Marti, J., Ablay, G. J., Redshaw, L. T., Sparks, R. S. J., 1994. Experimental studies of collapse calderas. *Journal Geological Society London*, **151**, 919–929.

Marti, J., Geyer, A., Folch, A., Gottsmann, J., 2008. A review on collapse caldera modelling. In Gottsmann, J. and Marti, J. (eds.), *Caldera Volcanism: Analysis, Modelling and Response*. Amsterdam: Elsevier, pp. 233–283.

Marti, J., Geyer, A., Folch, A., 2009. A genetic classification of collapse calderas based on field studies, and analogue and theoretical modelling. In Thordarson, T., Self, S. (eds.), *Volcanology: The Legacy of GPL Walker*. London: IAVCEI-Geological Society of London, pp. 249–266.

Mitchell, N. C., Masson, D. G., Watts, A. B., Gee, M. J. R., Urgeles, R., 2002. The morphology of submarine flanks of volcanic ocean islands – a comparative study of the Canary and Hawaiian hotspot islands. *Journal of Volcanology and Geothermal Research*, **115**, 83–107.

Moore, J. G., Clague, D. A., Holcomb, R. T., et al., 1989. Prodigious submarine landslides on the Hawaiian Ridge. *Journal of Geophysical Research*, **94**, 17465–17484.

Moore, J. G., Normark, W. R., Holcomb, R. T., 1994. Giant Hawaiian landslides. *Annual Review of Earth and Planetary Sciences*, **22**, 119–144.

Morgan, J. K., Moore, G. F., Clague, D. A., 2003. Slope failure and volcanic spreading along the submarine south flank of Kilauea volcano, Hawaii. *Journal of Geophysical Research*, **108**, 2415, doi.org/10.1029/2003JB002411.

Nakada, S., Uto, K., Sakuma, S., Eichelberger, J. C., Shimizu, H., 2005. Scientific results of conduit drilling in the Unzen Scientific Drilling Project (USDP). *Scientific Drilling*, **1**, 18–22, doi:10.2204/iodp.sd.1.03.2005.

Newhall, C. G., Dzurisin, D., 1988. *Historical Unrest of Large Calderas of the World*. Reston, VA: US Geological Survey.

Nicolas, A., 2013. *Structures of Ophiolites and Dynamics of Oceanic Lithosphere*. Berlin: Springer Verlag.

O'Driscoll, B., Troll, V.R., Reavy, R.J., Turner, P., 2006. The Great Eucrite intrusion of Ardnamurchan, Scotland: reevaluating the ring-dike concept. *Geology*, **34**, 189–192.

Oehler, J. F., de Vries, B. V. W., Labazuy, P., 2005. Landslides and spreading of oceanic hot-spot and arc basaltic edifices on low strength layers (LSLs): an analogue modelling approach. *Journal of Volcanology and Geothermal Research*, **144**, 169–189.

Oehler, J. F., Lenat, J. F., Labzuy, P., 2008. Growth and collapse of the Reunion Island volcanoes. *Bulletin of Volcanology*, **70**, 717–742.

Oftedahl, C. 1953. The igneous rock complex of the Oslo region: XIII. The cauldrons. *Skrifter Det Norske Videnskaps-Akademi i Oslo*, **3**, 1–108.

Pina-Varas, P., Ledo, J., Queralt, P., Marcuello, A., Perez, N., 2018. On the detectability of Teide volcano magma chambers (Tenerife, Canary Islands) with magnetotelluric data. *Earth, Planets and Space*, **70**, doi.org/10.1186/s40623-018-0783-y

Pinel, V., Jaupart, C., 2005. Some consequences of volcanic edifice destruction for eruption conditions. *Journal of Volcanology and Geothermal Research*, **145**, 68–80.

Ponomreva, V. V., Melekestev, I. V., Dirksen, O. V., 2006. Sector collapses and large landslides on late Pleistocene–Holocene volcanoes in Kamchatka, Russia. *Journal of Volcanology and Geothermal Research*, **158**, 117–138.

Reid, M. E., 2004. Massive collapse of volcano edifices triggered by hydrothermal pressurization. *Geology*, **32**, 373–376.

Rosi, M., Papale, P., Lupi, L., Stoppato, M., 2003. *Volcanoes*. Buffalo, NY: Firefly Books.

Rust, D., Behncke, B., Neri, M., Ciocanel, A., 2005. Nested zones of instability in the Mount Etna volcanic edifice, Italy. *Journal of Volcanology and Geothermal Research*, **144**, 137–153.

Savin, G. N., 1961. *Stress Concentration Around Holes*. New York, NY: Pergamon.

Segalstad, T. V., 1975. Cauldron subsidences, ring-structures, and major faults in the Skien district, Norway. *Norks Geologisk Tidsskrift*, **55**, 321–333.

Self, S., Rampino, M. R., Newton, M. S., Wolff, J. A., 1984. Volcanological study of the great Tambora eruption of 1815. *Geology*, **12**, 659–663.

Siebert, L., 1984. Large volcanic debris avalanches: characteristics of source areas, deposits and associated eruptions. *Journal of Volcanology and Geothermal Research*, **22**, 163–197.

Simkin, T., Howard, K.A. 1970. Caldera collapse in the Galapagos Islands, 1968. *Science*, **169**, 429–437.

Spera, F. J., 1980. Thermal evolution of plutons: a parameterized approach. *Science*, **207**, 299–301.

Spera, F. J., 2000. Physical properties of magmas. In Sigurdsson, H. (ed.), *Encyclopedia of Volcanoes*. New York, NY: Academic Press, pp. 171–190.

Thordarson, T., Hoskuldsson, A., 2008. Postglacial volcanism in Iceland. *Jokull*, **58**, 197–228.

Tibaldi, A., 2001. Multiple sector collapses at Stromboli volcano, Italy: how they work. *Bulletin of Volcanology*, **63**, 112–125.

Tibaldi, A., Bistacchi, A., Pasquare, F. A., Vezzoli, L., 2006. Extensional tectonics and volcano lateral collapses: insights from Ollague volcano (Chile–Bolivia) and analogue modelling. *Terra Nova*, **18**, 282–289.

Tsuchida, E., Yaegashi, A. 1981. Stresses in a thick elastic plate containing a prolate spheroidal cavity subjected to an axisymmetric pressure. *Theoretical and Applied Mechanics*, **31**, 85–96.

Tsuchida, E., Saito, Y., Nakahara, I., Kodama, M., 1982a. Stresses in a semi-infinite elastic body containing a prolate spheroidal cavity subjected to axisymmetric pressure. *Japan Society of Mechanical Engineers Bulletin*, **25**, 891–897.

Tsuchida, E., Saito, Y., Nakahara, I., Kodama, M., 1982b. Stress concentration around a prolate spheroidal cavity in a semi-infinite elastic body under all-around tension. *Japan Society of Mechanical Engineers Bulletin*, **25**, 493–500.

Vinciguerra, S., Elsworth, D., Malone, S., 2005. The 1980 pressure response and flank failure of Mount St. Helens (USA) inferred from seismic scaling exponents. *Journal of Volcanology and Geothermal Research*, **144**, 155–168.

Voight, B., Elsworth, D., 1997. Failure of volcano slopes. *Geotechnique*, **47**, 1–31.

Walter, T. R., Schmincke, H. U., 2002. Rifting, recurrent landsliding and Miocence structural reorganisation on NW-Tenerife (Canary Islands). *International Journal of Earth Sciences*, **91**, 615–628.

Williams, H., McBirney, A. R., Lorenz, V. 1970. *An Investigation of Volcanic Depressions. Part I. Calderas*. Houston, TX: Manned Spacecraft Center.

Wooller, L., de Vries, B. V. W, Murray, J. B., Rymer, H., Meyer, S., 2004. Volcano spreading controlled by dipping substrata. *Geology*, **32**, 573–576.

10 Understanding Unrest and Forecasting Eruptions

10.1 Aims

One of the main aims of the science of volcanology, and that of volcanotectonics in particular, is to understand volcanic unrest periods. By 'understanding' I mean that the signals coming from the volcano during the unrest can be interpreted in terms of plausible physical and chemical processes occurring inside the volcano. By volcanic 'unrest' I mean an increase in various physical and chemical signals, suggesting that associated processes within the volcano operate at different rates, intensities, or both. By interpreting the unrest period in terms of correct physical processes, there is a chance of assessing the volcanic hazard, namely the probability that the unrest period results in an eruption. Furthermore, when the understanding of the processes giving rise to the signals is accurate, not only the location of the eruption site but also the likely size (volume) of the eruption can be forecasted. The main aims of this chapter are to:

- Show how understanding of volcanotectonic processes can provide a correct interpretation of unrest periods.
- Show how the reliable interpretation of volcanotectonic processes provides a framework for accurate forecasting of eruptions.
- Demonstrate that the future of humankind depends partly on understanding the conditions for large eruptions.
- Discuss methods for preventing potentially devastating large eruptions.

10.2 Volcanic Unrest

Volcanic unrest is formally defined as a significant increase in geological, geophysical, and geochemical signals from a volcano, indicating changes in the modes, rates, or intensities of the associated physical and chemical processes within the volcano. In this definition, unrest means that the volcano shows a behaviour that deviates from its normal or background (baseline) behaviour. Thus, in order to decide if an unrest period has started it is necessary to have some minimum background information about the volcano, so as to be able to define its **baseline** behaviour. When the unrest is preceded by a long-term quiescence, perhaps lasting many decades or even centuries, it may be difficult to determine the baseline behaviour. This follows because the baseline behaviour is normally determined from instrumental data, or at least from professional volcanological observations, and these may not cover decades, not to speak of centuries.

10.2.1 The Main Signals

The signals that characterise volcanic unrest reflect many physical and chemical changes in the volcano. These include changes, and normally an increase, in the following (cf. Chapter 1):

- Frequency of earthquakes within the volcano.
- Ground elevation and tilting.
- Emission of volcanic gases (fumarole activity).
- Melting of snow and/or ice.
- Volumetric flow rate and chemistry of groundwater and geothermal water.

Perhaps the most noticeable and easily detected are the changes in frequency and location of **earthquakes**. When a volcano which for a long time has had little or no earthquake activity suddenly shows significant seismicity, it has entered an unrest period. The earthquake activity is initially normally widely distributed in the volcano, although most of the earthquakes occur close to the associated shallow magma chamber (Fig. 1.12). When the magma chamber ruptures and injects a dike or an inclined sheet, the earthquakes tend to concentrate around the dike/sheet, giving rise to an **earthquake swarm** (Figs. 1.13, 4.8, 4.9; Chapter 4).

The second most noticeable signal is normally changes in **ground elevation and tilting**. Ground-elevation changes are of two types: inflation and deflation. Both are also reflected in tilt changes. Inflation means uplift or doming of the surface of the volcano (Fig. 1.12). Since inflation implies an expansion of the associated magma chamber, it is more common during the early stages of an unrest period than its opposite process, namely deflation. As implied in the name, deflation means lowering of the surface of the volcano.

Inflation is primarily caused by an increased fluid pressure in the source magma chamber of the volcano (or the volcanic system or the volcanic field, as the case may be). The pressure increase may be due to gas expansion and accumulation, particularly in acid magmas, and occasionally to a pore-fluid-pressure increase in a geothermal reservoir above the magma chamber. Most commonly, however, the inflation is because of a flow of new magma (from a deeper reservoir, Chapters 8 and 9) into the shallow chamber. Deflation can be due to a horizontal extension in the crustal segment hosting the shallow chamber; and, over long periods of time, deflation may result partly from solidification of magma in a chamber, and its shrinkage. More commonly, however, deflation is attributable to the lowering of the magmatic pressure in the chamber as magma flows out of it along a newly formed dike/sheet.

Increased **fumarole activity** is often more difficult to interpret: it may be attributable to new magma coming into the shallow chamber, but it may also be due to a local permeability increase at the site of the fumaroles. The latter may be a result of an inflation and thus a magma-chamber expansion, but it does not have to be. Local mechanical/stress or chemical changes affect fracture apertures and thus the permeability and rate of degassing. How noticeable the fumarole activity is also depends on many external factors, including the weather. For example, fumaroles are generally more conspicuous during cold, clear, and still winter days than during warm and often cloudy summer days. Monitoring changes in the gas composition and volumetric flow rate, however, is important and complementary to that of monitoring volcanic earthquakes and deformation.

Changes in the **chemistry** of groundwater and/or geothermal water are also commonly monitored and detected during unrest periods. For example, many rivers that originate on the slopes of volcanoes, or from their ice caps, are monitored and show changes in chemistry during unrest periods. A related factor is the melting of ice/snow on volcanoes. Increased melting, normally reflecting increased geothermal activity, is common during unrest periods of ice-covered volcanoes.

10.2.2 Monitoring Techniques

Here follows a very brief overview of the main, modern monitoring techniques. These are discussed in more detail in Chapter 1. The focus, as elsewhere in the book, is on the geophysical and tectonic aspects of the methods and monitoring.

1. **Volcano seismicity** monitoring uses standard **seismometers** at specific seismic stations but also portable seismic networks. The latter make it possible to monitor in very great detail local earthquake swarms, such as those that occur during dike propagation (Agustsdottir et al., 2016). The technique is exactly the same as used for seismic monitoring of the propagation of fluid-driven fractures in general, such as hydraulic fractures used in the hydrocarbon industry (Fisher and Warpinski, 2011; Davis et al., 2012; Fisher, 2014; Shapiro, 2018). Some of the earthquakes are generated during rupture at the propagating tip of the dike (or inclined sheet), in the process zone around the fracture tip (Gudmundsson, 2011a); most, however, are triggered by the magmatic overpressure of the dike through slip on existing fractures in the walls of the dike, primarily joints such as columnar (cooling) joints. Using the seismic monitoring, many dikes have been inferred to deflect into sills (Sigmundsson et al., 2010). Similarly, many hydraulic fractures have been inferred to deflect into (water) sills (Fisher and Warpinski, 2011; Fisher, 2014).

2. **Ground elevation and tilting.** These monitoring techniques are partly old, and partly new. The old methods used **tiltmeters** (clinometers) to measure, in radians, very small changes in the slope of a part of the surface of a volcano. Standard **levelling** techniques serve a similar purpose. Some types of **strainmeters** are also old, such as those using metal rods. These are mostly extensometers, measuring the increase in the distance between two points and dividing the increase by the original distance. Fluid-filled **boreholes** can also function as strainmeters; the water level rises when the hole is subject to

compression, and falls when the hole is subject to extension. The most important elevation monitoring techniques today, however, use satellites. The use of the **GPS**, which began in the 1970s and was in complete use in the 1990s, is the older of the two main satellite methods. The satellites (32 in total) measure accurately the location of benchmarks or measurement points ('nails') at the Earth's surface. From changes in the configuration or location (due to horizontal and/or vertical displacements) of the points between successive measurements, it is possible to decide if the surface (of the volcano) is moving vertically and/or horizontally. The other main satellite method uses **InSAR** data, where the phase difference between the outgoing signal (sent by the satellite) and the returned signal (reflected from the Earth's surface) is a measure of the surface displacement.

3. **Chemical monitoring**. In addition to these geophysical/tectonic monitoring methods, there are various geochemical ones, as indicated above. The main methods are **gas monitoring**, which is a routine method used on many active volcanoes. Changes in the composition or the rate of flow of gas, primarily of CO_2 (carbon dioxide) and SO_2 (sulphur dioxide), both derived from magma, is an indication of processes taking place inside the volcano. Changes in the chemistry of the **springs and rivers** associated with active volcanoes are also commonly monitored.

10.2.3 Hazards

Understanding the physical processes associated with unrest periods is of fundamental importance for the reliable assessment of volcanic hazards in a given volcano or volcanic system/field. There are several definitions of a volcanic hazard, but the following is the one used here. A **volcanic hazard** is the probability that an eruption will occur in a given volcano or volcanic system/field within a specified time window.

There have been various attempts to assess the probability of an eruption in a given volcano over a given period of time based on purely statistical data (Sobradelo et al., 2013, 2015; Sparks et al., 2013). Such a purely probabilistic approach has been successful in some fields of science. For example, many aspects of statistical physics derive from purely probabilistic considerations of the items of interest falling into certain classes or bins in a given size distribution. Ultimately, the shape and range of the size distribution, however, can be related to the energy input into the system being analysed. For instance, the volumes of volcanic eruptions (Fig. 8.1), the thicknesses of dikes (Fig. 2.18), the lengths of volcanic fissures (Fig. 2.19), and the magnitudes of earthquakes (Fig. 4.7) are all crude power laws and therefore follow a specific size distribution. But these distributions are based on data obtained 'after the fact', that is, after the eruptions and the earthquakes occurred, or the fracture formed. The distributions do not tell us where and when the next earthquake in a given seismic zone will happen, or what magnitude it will have. Similarly, the distribution of eruptive volumes, either in a single volcano/volcanic field or for subaerial volcanoes in general, does not tell us when and where the next eruption will occur, or the volume/size it is likely to have.

It follows that in order to make realistic estimates of the hazards for individual volcanoes or volcanic fields during unrest periods, we must understand the physical and chemical processes responsible for the unrest signals. More specifically, there are **four principal questions** that we should, and in the future must, be able to answer when a volcano enters an unrest period, namely the following:

1. Is the magma chamber going to rupture and inject a dike (or a sheet)?
2. What is the likely propagation path of the dike/sheet?
3. Is the dike/sheet likely to reach the surface to erupt; if yes, then at which location?
4. If an eruption occurs, how large is it likely to be?

Until recently, we could not even provide the information needed to attempt an evidence-based answer to any of these questions. In recent years, there has been considerable progress in the direction of providing the data and models on which plausible answers can be given. Earlier chapters in the book discuss and present some of the relevant data and models. Yet, we still have a long way to go until we can provide reliable answers to these fundamental questions. In this last chapter I summarise some of the main findings relevant to the questions, and look ahead with a view to adding further details to the framework within which these questions must be answered.

10.3 Depth, Form, and Function of Magma Chambers

The magma chamber is the key concept in volcanotectonics. I have discussed how magma chambers form (Chapter 6) and their principal roles in the formation of volcanic edifices (Chapter 9). I have also discussed magma-chamber rupture and subsequent dike/sheet injection (Chapters 5 and 7) as well as their role in formation of caldera collapses (Chapters 5 and 9). Here, I discuss the conditions for magma-chamber rupture from a different perspective, with a focus on the unrest signals that can be taken as an indication that the rupture has occurred. But first I provide a short discussion about the chambers themselves.

10.3.1 Detecting Magma Chambers

Active shallow magma chambers have proved rather difficult to detect so as to leave little or no doubt about their existence. The main methods used for chamber detection are seismic and geodetic techniques. These are commonly complemented by petrological/geochemical methods.

The seismic methods rely on the so-called **S-wave shadow** effect, whereby the S-waves generated during earthquakes (natural or human-made) drop out at the location of the shallow chamber. This drop-out or disappearance of the S-waves follows from the fact that these waves cannot propagate through fluids. Thus, if part of or the entire shallow chamber is totally fluid, the S-waves will drop out, so that there will be an S-wave shadow at the location of the chamber. Similarly, even if the chamber is only partially molten, the S-wave and the P-wave velocities may decrease markedly (Fig. 4.1), so that the chamber can be crudely located (Chapter 4). Also, earthquakes tend to concentrate around the chamber during unrest periods (Fig. 6.16, Chapter 6). This does also provide an approximate indication of the depth and size of the chamber.

10.3.2 Depth

Geodetic methods, primarily using GPS and InSAR data, make it possible to locate the depth of the main source responsible for the inflation/deflation of the associated volcano. The method is crude, and is still primarily based on the nucleus-of-strain non-layered (elastic half-space) model referred to as the **Mogi model.** This model, discussed in detail in Chapter 3, can be used to obtain an estimate of the depth to the chamber during an unrest period. Combining geophysical (mainly seismological and geodetic) and petrological data, Chaussard and Amelung (2014) estimated the depths to the tops of magma chambers at subduction zones worldwide. Their results for 70 volcanoes in 15 volcanic regions (Fig. 6.17) show that the majority of the identified chambers/reservoirs are comparatively shallow, that is, with roofs at depths of less than about 5 km below the surface. Here, the surface is not the top of the associated volcano but rather that of the surrounding or adjacent area – or volcanic field/system. The definition of a **shallow chamber** is taken as one with a roof at **5 km or less** below the surface of the associated volcanic field/system.

These and other results strongly suggest that shallow magma chambers are common. The methods used are less applicable for detecting deep-seated reservoirs. This is partly because the crust between the deep-seated reservoir and the surface has many more layers and there is a greater chance of some of the layers being compliant and yielding unreliable information. And it is also partly because most deep-seated reservoirs are sources for shallow chambers (Chapters 6 and 9), and the local stress field of the shallow chamber may, to a degree, disturb or modify the stress and deformation fields of the deep-seated reservoir, making it more difficult to use surface deformation to infer the reservoir depth.

10.3.3 Form

The fact that shallow magma chambers are common is also strongly supported by direct **field observations** in deeply eroded volcanic systems and fields. All deeply eroded central/polygenetic volcanoes contain plutons. Depending on the depth of erosion, the plutons are either directly observed in the field, or inferred from geophysical (e.g. gravity and magnetic) measurements. For example, all the central volcanoes in Iceland that are eroded to depths of 1500 m or more contain microgabbro or gabbro bodies, many of which are in direct contact with inclined sheets and dikes (Fig. 10.1; cf. Figs. cf. 1.17, 7.3, 7.5; Chapter 7). Many plutons are found at much shallower crustal depths (Fig. 5.19; Gudmundsson, 2017). Similar plutons are seen in the deeply eroded Tertiary volcanoes of Scotland and, in fact, worldwide (Fig. 10.1; cf. Fig. 5.18). The roofs of these plutons are commonly ruptured at various locations to form dikes or inclined sheets (Figs. 10.1, 10.2; cf. Figs. 1.17, 6.19). Probability considerations suggest that some of the dikes very likely reached the surface to feed volcanic eruptions, which implies that the plutons acted as source chambers for volcanoes. The depths of the roofs of these plutons/chambers below the surface during their lifetimes were commonly from several hundred metres to several kilometres, that is,

Fig. 10.1 Part of the fossil magma chamber of Torres del Paine (cf. Fig. 5.18). The granitic peaks or horns themselves are named Cuernos del Paine. The 2-km-thick fossil shallow magma chamber, mostly of granite, was emplaced about 12 million years ago during many thick sill injections over a period of some 90 000 years (Michel et al., 2008; Lauthold et al., 2014) at a depth of 2–3 km below the top of the volcanic area at that time. View north, many granitic dikes and inclined sheets dissect the roof of the extinct chamber, some of which are seen here. Photo: Jim Ross/NASA.

similar to those of currently active shallow magma chambers (Fig. 6.17). For example, the roof of the chambers in Figs. 1.17, 6.18, 6.19, and 10.2 is at a depth of about 1.5 km (Gudmundsson, 2012), whereas that of the chamber in Fig. 10.1 is at a depth of about 2 km (Michel et al., 2008).

For many plutons, only the upper part is seen in the field. This applies, for example, to some of the exposed shallow magma chambers in Iceland (Figs. 5.6, 5.19, and 7.5). In some of these plutons/chambers, parts of the **roof and the walls** are well exposed (Figs. 1.17, 10.6, 10.7, 10.8), but not the floor. However, some plutons, particularly thick sills and laccoliths, have both part of the **roof and the floor** (lower contact) exposed. An excellent

Fig. 10.2 Part of the fossil magma chamber of Reydarartindur (height about 600 m) in Southeast Iceland. View north, the magma chamber is felsic (granophyre), and several felsic dikes are seen dissecting the roof, a basaltic lava pile. The top of the roof is at about 1300 m below the original top of the lava pile hosting the chamber.

example of a pluton/chamber with both the floor and the roof brilliantly exposed (Fig. 10.1; cf. Fig. 5.18) is the thick sill/laccolith Torres del Paine in Chile (Michel et al., 2008). That fossil chamber/pluton is primarily of granite, is formed in many magma injections (as most chambers appear to be, Chapter 5), and has a maximum thickness of about 2 km.

But there are many sills, some hundreds of metres thick, where parts of the **roof and the floor** are exposed (Fig. 10.3). Again, many of these have ruptured tops where dikes or inclined sheets propagate up from the sill, which thereby most likely acted as a magma chamber. Other sills have clearly exposed floors, even if the roofs are mostly eroded away (Fig. 2.35).

The field data presented here, as well as those discussed earlier (Chapters 1, 5, 7), indicate that shallow magma chambers are commonly of well-defined geometries. **Sill-like** chambers or oblate ellipsoids are presumably most common, while other geometries also exist. The function of a shallow magma chamber, however, depends on how much of it is really molten, a topic to which I turn now.

10.3.4 Function as Molten Bodies

What part of a magma chamber is really molten? There are widely different models and views on this aspect of magma chambers, the issue having been debated for a long time (Cashman et al., 2017). In the models discussed and elaborated in Chapters 5 and 6, the chambers are either partially molten or totally molten. It is plausible that many shallow chambers are partially molten during much of their lifetimes. The exact melt fraction is generally unknown, however, and it is likely to vary much during the lifetime of the chamber.

Fig. 10.3   Ferrar Dolerite sills in Antarctica are exceptionally well exposed, with both the roof and the floor visible. The sills, about 177 million years old, are commonly 100–500 m thick, and occasionally as thick as 1000 m (Fleming et al., 1997; Boudreau and Simon, 2007). Many dikes and, in particular, inclined sheets dissect the roof, suggesting that some of the sills acted as magma chambers (Arioldi et al., 2012). Photo: NASA.

Field observations provide some constraints on the melt content and general function of shallow chambers, as well as their potential to supply magma during eruptions and/or intrusions. If the **contacts** between individual layers or units that constitute the chamber/pluton are very clear (Fig. 5.5), the pluton was never totally molten and did not even function as a single partially molten body at any time. Such bodies are referred to as **multiple intrusions** and the only way that they can function as magma chambers is that the newest intrusion – a sill in this case – stays molten for a short while and is able to rupture the roof and inject a dike/sheet. Very thin sills, such as those in Fig. 5.5, are unlikely ever to have supplied magma to a dike/sheet. In fact, most such multiple sills, where each sill solidifies before the next one is injected, solidify as plutons that never function as magma chambers.

For other plutons, while there are some contacts seen between the individual intrusive bodies that constitute the pluton, it is clear that they acted as magma chambers. This follows because the roof is seen to have **ruptured** with dike/sheet injections (Figs. 10.1, 10.2; cf. Figs. 1.17, 6.19). These chambers were at least partially molten and may have been largely or totally molten during some stages. Examples of chambers that were totally molten include those where a single set of cooling or columnar joints extends through the entire exposed body. Such fossil chambers include those see in Figs. 10.1 and 10.3. When a single set of cooling joints extends through the entire pluton, it shows that no part had developed a joint set before the other parts. This implies that much of the chamber was at a similar temperature when the joints started to develop – which is at temperatures around 60% of the original temperature of the magma (Chapter 7; Jaeger, 1961). Thus, thermally, the chamber acted more or less as a single body, and therefore may have been totally molten at one stage.

The fact that large parts of shallow magma chambers are, from time to time, in a fluid state is supported by other observations. In particular, during eruptions associated with caldera collapses, the **subsidence** or vertical displacement of the piston-like caldera block may reach kilometres. For outward-dipping ring-faults (Chapter 8; Gudmundsson, 2016), the caldera block may subside to the floor of the associated chamber, implying that the entire chamber responded to the pressure changes due to the subsiding block as a fluid. It follows that during large caldera-forming eruptions much of and perhaps the **entire** magma chamber is in a **fluid state.**

10.4 Magma-Chamber Rupture

The cause of volcanic unrest is, most commonly, expansion or contraction of the associated magma chamber. The expansion results in an inflation of the volcano surface, whereas contraction results in a deflation of the surface. In many, and presumably most, unrest periods, the magma chamber does not rupture. When it does, however, the resulting dike or inclined sheet injection has the potential of reaching the surface to erupt. It is therefore very important to understand the conditions for magma-chamber rupture. These conditions have been discussed in Chapters 5 and 7, but here they are put into the context of understanding better the processes associated with volcanic unrest.

The **condition for rupture** and dike injection is given by (Eq. (5.1)):

$$p_l + p_e = \sigma_3 + T_0 \tag{10.1}$$

where p_l denotes the lithostatic stress at the rupture site at the boundary of the magma chamber, p_e is the magmatic pressure in the chamber in excess of σ_3, the minimum compressive (maximum tensile) principal stress, and T_0 is the local *in situ* tensile strength at the rupture site. Alternatively, Eq. (10.1) can be written in the form:

$$p_t = \sigma_3 + T_0 \tag{10.2}$$

where $p_t = p_l + p_e$ is the total fluid pressure, that is, the pressure of the magma and gas (if any) in the chamber at the time of rupture. Equations (10.1) and (10.2) state that when the total fluid pressure in the magma chamber becomes equal to the combined value of the minimum compressive (maximum tensile) principal stress and the *in situ* tensile strength, the chamber ruptures and a magma-filled fracture, an inclined sheet or a dike, becomes injected (Figs. 10.1, 10.2; cf. Fig. 1.17).

Field observations of thousands of cross-cutting relationships between dikes, between inclined sheets, between sheets and dikes, and between dikes/sheets and the layers that they dissect show that the great majority of dikes/sheets are **extension fractures.** More specifically, they are **hydrofractures,** that is, magma-driven fractures, where the magmatic pressure ruptures the rock. They can therefore be modelled as mode I cracks (Chapter 5). It also follows that the potential **paths** of the dikes/sheets are indicated crudely by the orientation of the stress trajectories of the maximum compressive principal stress σ_1 (Section 10.5; Chapter 7).

But how is the magma-chamber rupture noticed during an unrest period? How do we know that the chamber has ruptured? The answer is that there are **two** main geophysical signals that indicate magma-chamber rupture: surface **deflation** and an **earthquake swarm.** Both may take some time to become noticed. Let us first discuss the earthquakes.

As soon as the magma chamber ruptures, it forms a **magma-filled fracture**. Depending on the local stress field, the fracture is either a dike or an inclined sheet (Chapter 7). Here, the discussion focuses on the **dikes**, but the physical principles apply as well to inclined sheets (and sills). Not all earthquake swarms are generated by dikes (Chapter 4), but many are, and these latter are the centre of attention here.

When the dike begins to propagate up from the source chamber, earthquakes are induced by the dike primarily at two locations: at the dike tip and in the walls of the dike-fracture. Let us first consider the earthquakes induced or triggered at the **dike tip** (Figs. 1.11, 2.29–2.31, 5.9, 7.17, 7.18). Because a dike is an extension fracture, rather than a shear fracture, the earthquakes at the tip are partly related to extension fractures, that is, are non-double-couple earthquakes (Figs. 4.10, 4.11). This means that the earthquakes are generated during opening of fractures rather than slip (shear) along the fractures. Some fracturing in the **process zone** (Fig. 1.13) of the dike, however, is due to slip on shear fractures, that is, faults. Faults are commonly induced ahead of the dike tip, in the process zone, particularly at and close to the surface, once the dike tip reaches shallow levels in the crust (Section 10.6; Figs. 3.37–3.44, Chapter 3).

The dike also induces many earthquakes in the host rock, namely in the **walls** of the dike-fracture. Dike propagation is driven by magmatic driving pressure or **overpressure** p_o, given by (Eq. (5.3)):

$$p_o = p_e + (\rho_r - \rho_m)gh + \sigma_d \qquad (10.3)$$

where ρ_r is the average host-rock density, ρ_m is the average magma density, g is acceleration due to gravity, h is the height or the dip-dimension of the dike above its rupture site at a particular time during its propagation, and σ_d is the differential stress ($\sigma_d = \sigma_1 - \sigma_3$) at the crustal level (the layer/unit) which the dike has reached during its propagation. The overpressure is always with reference to a particular crustal level/layer, since it changes as the dip-dimension of the dike increases (the dike propagates towards the surface). In case of a feeder-dike, the dip-dimension h is the vertical distance from the dike initiation at the boundary of the chamber to the volcanic fissure at the Earth's surface. For a propagating dike, h is the vertical distance from the dike initiation at the boundary of the chamber to the tip of the propagating dike at a particular time. For an old dike exposed in an eroded volcano, h refers to the vertical distance from the dike initiation at the boundary of the fossil chamber to the layer/unit where the dike is measured.

Using appropriate values in Eq. (10.3), the overpressure in a propagating dike easily reaches 5–20 MPa (Section 10.5; Becerril et al., 2013). The earthquakes in the dike-fracture wall are due to shear stresses induced by this overpressure. Most of the earthquakes in the walls occur on **existing fractures**, primarily cooling (columnar) joints in the lava flows and intrusions (Figs. 4.11, 4.12), but also on joints in pyroclastic and sedimentary layers as well as on contacts between layers. Most of the joints are originally vertical and (to a lesser degree) horizontal whereas the contacts are mostly horizontal. When the layers/beds become buried, however, they gradually tilt, mostly towards the axis of the volcanic zone within which they form (Figs. 2.1 and 2.5; Gudmundsson, 2017). As a consequence, the joints and contacts also become tilted and thus **oblique** (as seen in a vertical section) to the horizontal compressive stress induced by a vertically propagating dike. This can thus trigger a slip on existing, oblique joints (acting as shear fractures), some of which may link up into larger faults (Fig. 4.12). The slip under these conditions would be primarily **reverse** in the walls (Fig. 3.3) but **normal** at and close to the upper tip of the dike (Fig. 1.13).

Some of the joints are also, from the beginning, oblique to the dike-induced horizontal compressive stress, and are thus subject to shear stresses. This follows because the strike of most of the joints, particularly the columnar joints, is oblique rather than parallel to the dike-induced direction of horizontal compressive stress (Fig. 10.4; cf. Figs. 9.18, 9.19). The dike-induced horizontal stress commonly becomes, for a while, the direction of the maximum compressive principal stress σ_1. Consequently, many of the joints slip in shear and generate earthquakes (Fig. 10.4; cf. Fig. 4.12). The movement is then **strike-slip**, so that the focal mechanisms of the resulting earthquakes are strike-slip. Studies of earthquake swarms associated with dike propagation in volcanic rift zones show that many, sometimes the majority, of the earthquakes occur on strike-slip faults (Agustsdottir et al., 2016). This is not surprising because strike-slip faulting is common inside and adjacent to rift zones, such as in Iceland, in many cases presumably because of dike-induced stresses and slip on existing joints (Gudmundsson et al., 1992; Villemin et al., 1994; Bergerat and Angelier, 1998; Karson, 2017).

A large fraction of the earthquakes induced by dikes are thus likely to be related to slips on oblique joints and contacts, some of which may eventually link up into larger faults (Fig. 4.12). Slips on favourably oriented existing major faults nearby the dike may also occur (Figs. 1.21 and 10.12). As indicated above, **all types of faults** may be induced by propagating dikes (Gudmundsson et al., 2008; Passarelli et al., 2015; Agustsdottir et al., 2016; Bonaccorso et al., 2017).

With seismic networks in place, dike-induced earthquake swarms are easily **monitored** (Figs. 4.9, 4.13, 6.16). From the migration of the swarm, the propagation of the associated dike can be inferred. As the swarm migrates closer to the surface, the chances of an **eruption** generally become greater. It should be remembered, however, that most dike-propagation paths become arrested (Chapters 5 and 7), in which case no eruption occurs. Thus, most earthquake swarms induced by dikes do not propagate to the surface. Nevertheless, earthquake swarms are the best empirical indications of **dike-propagation paths** during volcanic unrest periods that we have.

10.5 Dike Propagation

Dike-propagation paths have been discussed in general terms in Chapters 5 and 7. Here, the attention is on how to correlate dike propagation with other signals, primarily geophysical, received during the propagation; and, in particular, on the physical principles that control dike paths. Let us first look in more detail at Eq. (10.3), since that equation determines the

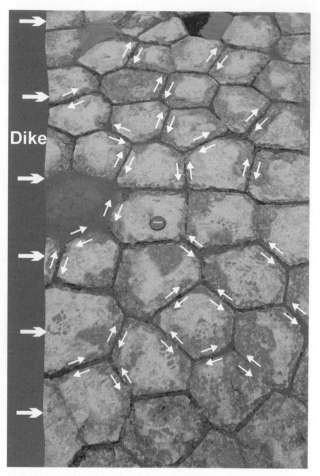

Fig. 10.4   When a dike propagates, its overpressure induces a fault slip along numerous joints in the walls of the dike-fracture. In lava flows and sills, most of the existing joints are columnar (cooling) joints of the type seen here in plan view (in a lava flow in South Iceland). The horizontal compressive stress generated by the overpressure of the dike (thick arrows) commonly induces strike-slip faulting on the existing joints (thin arrows).

overpressure, namely the fluid pressure available to drive the dike propagation. There are two sources of overpressure: the excess pressure and the buoyancy.

10.5.1 Excess Pressure

An **excess pressure** p_e is a necessary condition for magma-chamber rupture. If there is no excess pressure, there will be no magma-chamber rupture. It follows that for any magma-chamber rupture and dike injection to take place, there **must be** an excess pressure. As we see below (cf. Chapter 5), the value of the buoyancy term for dikes injected from shallow chambers can be zero or even negative, in which case the excess pressure is the only pressure available to drive the propagation of the dike.

The excess pressure at rupture is generally assumed to be equal to the *in situ* **tensile strength**. Measurements, primarily using hydraulic fractures injected from wells at various depths, indicate that the *in situ* tensile strength of rocks is generally from about 0.5 MPa to 9 MPa (Appendix E; Zang and Stephansson, 2010; Gudmundsson, 2011a). *In situ* hydraulic fracture measurement of the *in situ* tensile strength are the most accurate that exist and are particularly relevant to magma-chamber rupture and dike injection. This is because the hydraulic fractures used in the measurements are **hydrofractures**, so that the basic physics of their formation is the same as that of dike initiation.

But could the *in situ* tensile strength in the roof of a magma chamber not be significantly higher than this, perhaps because of the high temperature of the rock and **quasi-plastic behaviour**? High tensile strengths of rocks have, indeed, been proposed – some suggested *in situ* tensile strengths being as high as 100 MPa. This topic has been mentioned earlier in the book, but here I summarise the main reasons why tensile strengths are highly unlikely to be much higher in the roof or walls of magma chambers than elsewhere in the crust. The main points are as follows:

- The *in situ* tensile strength has been measured to a crustal depth of about 9 km. The values obtained at that depth are in the same range as above, that is, not greater than 9 MPa. The deepest measurements, at about 9 km, yielded tensile strengths of 8–9 MPa. These were made in the KTB drill hole in Germany. At this depth, the host-rock temperature is about 260 °C (Zoback and Harjes, 1997). For basalts and granites, plastic (ductile) behaviour is limited at temperatures as high as 600–900 °C and a confining pressure of 500 MPa (Paterson and Wong, 2005). For a confining pressure of 500 MPa, the corresponding crustal depth is about 16–18 km, **far deeper** than any shallow magma chambers.

- Direct observations of roofs of shallow magma chambers show that the fractures formed by the dikes are primarily extension fractures. If they were formed through plastic deformation, the fractures would normally follow slip lines, that is, they would be shear fractures. There is no evidence for dikes close to, or meeting, magma chambers generally occupying shear fractures. The **extension fractures** occupied by the dikes in the roofs of chambers are clearly formed by the magmatic overpressure, with no evidence that their formation was in any way different from that dictated by Eqs. (10.1) and (10.2).

- Focal mechanisms for earthquakes occurring in the roofs of magma chambers show the same range of **driving shear stresses** as are inferred from **stress drops**, mostly 1–10 MPa (Chapter 4), as anywhere else in the crust. The Modified Griffith criterion (Eq. (5.15)) indicates that the shear strength of the rock is equal to twice the tensile strength, that is, $\tau_0 = 2T_0$. If the tensile strength in the roof were tens of mega-pascals, or even 100 MPa, then, based on Eq. (5.15), the shear strength should be many tens or a couple of hundred mega-pascals. Such high driving shear stresses/stress drops are not observed.

- If the tensile strength of the rock hosting a magma chamber was as high as 100 MPa, then the excess pressure would be of the same order (Chapter 8). It follows, first, that the **effusion rate** (Eqs. 8.10–8.13) would be many times greater than is observed and, second, the fraction of magma leaving the chamber would, for basaltic chambers, be 10–20 times higher than is currently estimated in normal poroelastic-controlled eruptions (Chapter 8).

- The inferred **overpressure** of exposed dikes would also be greater by the same factor, that is, 10–20 times larger than is actually observed. The aspect ratios of many dikes have been measured (e.g. Gudmundsson, 1983; Geshi et al., 2010; Becerril et al., 2013; Kusumoto et al., 2013) and there is no evidence that the overpressure at the time of emplacement was of the order of 100 MPa.
- Similarly, if the tensile strength was as high as 100 MPa, the excess pressure before rupture would be of that order. The measured **inflation** (due to magma-chamber expansion as the excess pressure builds up) would also be 10–20 times greater than is normally observed (Chapter 3).

Other difficulties and mechanical inconsistencies that follow from the assumption of tensile strengths of the rocks hosting magma chambers being unusually high are discussed by Gudmundsson (2012). Thus, in the absence of any evidence for exceptionally high tensile strength in the host rocks of magma chambers, it can be concluded that the excess pressure at the time of the chamber rupture is most likely similar to that measured in hydraulic-fracture experiments, namely between 0.5 MPa and 9 MPa, and most commonly **2–4 MPa.**

10.5.2 Buoyancy

The second term on the right-hand side of Eq. (10.3) denotes the buoyancy, namely the term:

$$\text{Buoyancy} = (\rho_r - \rho_m)gh \tag{10.4}$$

Depending on the difference between the average rock density ρ_r and the magma density ρ_m, the buoyancy contribution to the overpressure can be (a) positive (rock denser than the magma), (b) zero or neutral (rock and magma density the same), and (c) negative (magma denser than the rock). Furthermore, the **buoyancy** contribution is always **close to zero** as long as h is small, that is, so long as the dike-propagation distance from the chamber is short. As an example, consider a dike with a magma density of 2500 kg m^{-3} (say, andesite) propagating into a crustal segment (chamber roof), the average density of which is 2600 kg m^{-3}, a typical value for the roofs of shallow chambers. For a vertical dike that has propagated 100 m, giving it a dip-dimension of 100 m, the buoyancy contribution to the overpressure is only 0.1 MPa. Earlier, when the dike was about 10 m tall (dip-dimension), the buoyancy contribution would be as small as 0.01 MPa, both values being totally negligible in comparison with the common excess pressure of 2–4 MPa. For such a dike, it is only when its dip-dimension has reached a kilometre or more that the contribution of the buoyancy to the overpressure becomes significant – here about 1 MPa.

More specifically, the density of most crustal rocks is generally between 2000 kg m^{-3} and 3000 kg m^{-3} whereas the density of the most common magmas is between 2250 kg m^{-3} and 2800 kg m^{-3}. The crustal rock density range given here applies to the entire crust. For the part of the crust that hosts shallow magma chambers – more specifically, forms the roofs of the chambers – and through which dikes propagate towards the surface, the range is much smaller. A typical range for a primarily basaltic (oceanic) crust with sedimentary rocks on the top would be $2000–2900 \text{ kg m}^{-3}$. The average density of the uppermost 5 km of the crust in Iceland – which may be regarded as largely oceanic – is 2670 kg m^{-3} (Gudmundsson,

1988). Iceland does not have thick sedimentary layers everywhere at the surface of the volcanic zones – although some thick glacial sedimentary layers do occur – but at the surface or at shallow depths (below the Holocene lava flows) there are thick low-density hyaloclastite (basaltic breccia) layers almost everywhere. For comparison, densities of basaltic magmas are between 2650 kg m^{-3} and 2800 kg m^{-3} (Appendix F).

From Eq. (10.4) and these density values it follows that, for common basaltic magma in a dike injected from a shallow magma chamber, the **buoyancy** makes **little or no** significant positive contribution to the overpressure. Even if we used the lowest magma density of basalt, 2650 kg m^{-3}, for an average crustal density of 2670 kg m^{3} for the crustal layers between the chamber and the surface (with the chamber roof assumed to be at the maximum depth of 5 km), the buoyancy contribution is only 1 MPa. For shallower chambers or higher magma densities, the buoyancy term is **zero or negative.** This is partly because average crustal densities everywhere are less at shallower depths. So if, for example, the chamber's upper margin or roof was at 3 km depth instead of 5 km depth, the average density of the crustal layers between the chamber and the surface would be 2575 kg m^{-3}, and thus lower than that of any basaltic magma. This would mean that the buoyancy effect would be **negative** for all basaltic dikes, that is, the most common dikes in most volcanoes.

The conclusion remains the same even when the effects of **gas expansion** on the magma density are taken into account. This is because a significant enough gas expansion in basaltic magma to greatly lower its density only occurs at very shallow depths, that is, at total pressures of about 25 MPa or less (Gonnermann and Manga, 2013), which corresponds to crustal depths of a kilometre or less. This is in agreement with results from Hawaii suggesting that most of the exsolution of gas in basaltic magmas occurs in the uppermost few hundred metres of the feeder/conduit (Greenland et al., 1985, 1988). These results are also consistent with direct field studies of basaltic dikes, which show that the vesicles (formed during gas expansion) are generally small and rare at crustal depths greater than several hundred metres below the original surface of the volcanic zone/central volcano. In the uppermost parts of many feeder-dikes, however, large vesicles are common and may form zones, with the largest vesicles in the dike centre (Galindo and Gudmundsson, 2012; Gudmundsson, 2017). Thus, the expected reduction in magma density due to gas expansion is unlikely to be of great significance except very **close to the surface**, particularly in feeder-dikes (cf. Chapters 7 and 8).

Many have suggested that dikes cannot penetrate layers with the same density as that of the magma, namely where the buoyancy contribution to the overpressure is zero. Such layers are said to constitute **levels of neutral buoyancy** – meaning levels of zero buoyancy – and they are supposed to act as traps for vertically propagating dikes, particularly basaltic dikes. The neutral-buoyancy level or layer is then assumed either to arrest the basaltic dike that meets it, or to deflect the dike into a sill (Bradley, 1965; Holmes, 1965; Gretener, 1969; Francis, 1982; Ryan, 1993; Chevallier and Woodford, 1999). This idea has also been used as an explanation for the formation of shallow magma chambers.

In earlier chapters, however, there have been many examples of basaltic dikes passing easily through layers of densities less than those of typical basaltic magmas (Figs. 1.11, 2.27, 2.34, 7.27). Layers of hyaloclastites, sediments, and acid or intermediate rocks cut by basaltic dikes are examples that come to mind. Similar examples are seen everywhere in the world where basaltic volcanism takes place (Figs. 10.5, 10.6). Most of this volcanism is

Fig. 10.5 A feeder-dike dissecting a pile of basaltic sills (thickness about 20 m), lava flows, and pyroclastic layers in the caldera wall of Las Canadas, Tenerife (Canary Islands; cf. Fig. 9.12a). View east, the basaltic dike passes easily through the thick pyroclastic layers, showing that local neutral buoyancy cannot stop propagating dikes (cf. Gudmundsson, 2012).

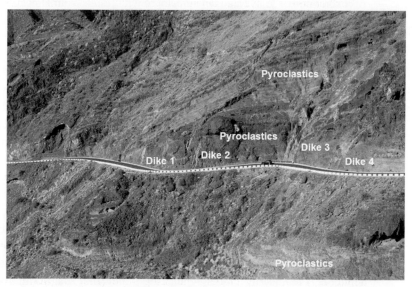

Fig. 10.6 High-density basaltic dikes normally propagate easily through low-density pyroclastic layers. Here, basaltic dikes are seen propagating through pyroclastic layers in the peninsula of Anaga, Tenerife (Canary Islands). Layers with densities similar to that of the magma ('neutral buoyancy') or less than that of the magma ('negative buoyancy') are normally not sufficient to arrest propagating basalt dikes. Dikes 1 and 2 are 1–2 m thick, and dikes 3 and 4 are both around 5 m thick. A black car close to dike 3 also provides a scale. A black and white version of this figure will appear in some formats. For the colour version, please refer to the plate section.

supplied with magma through dikes, all of which must **pass through** neutral (or negative) buoyancy layers on their paths to the surface. The basaltic feeder-dikes must normally propagate through **many layers** where the buoyancy is neutral or negative. This follows because typical volcanic zones are composed of numerous rock layers. And even a zone of predominantly basaltic rocks normally contains many layers of breccias, pyroclastics (hyaloclastites), sediments, intrusions, and other rocks with different densities (Fig. 10.5: cf. Figs. 1.11, 2.27, 2.34, 7.27). Thus, for a basaltic dike propagating up through any volcanic zone, there are normally many layers with a density equal to, or somewhat lower than, that of the magma at a particular crustal depth. Generally, therefore, for a typical basaltic dike-fed eruption to occur, the dike must propagate through many local layers or levels of neutral (or negative) buoyancy.

The average density of seismic crustal layers or units, however, **increases with depth**. As indicated above, the density of individual rock layers or units varies much, so that at any crustal depth we may find a layer with a much lower density than many of the layers above it in the pile, and also lower than that of typical basaltic magma. Nevertheless, statistically, the average density of the layers/units increases with depth. This increase is partly because of compaction (particularly of pyroclastic and sedimentary layers) and the filling of their pore spaces (e.g. vesicles and joints in the lava flows) with secondary minerals. Thus, the buoyancy term in Eq. (10.4) may, on average, be positive below a certain crustal depth, namely the depth where the magma density equals the density of the crustal layers, and negative above that depth.

Furthermore, the average density of the uppermost seismic crustal layers or units of an active volcanic zone is everywhere less than that of typical basaltic magma. Common basaltic magmas may have densities between 2650 kg m^{-3} and 2800 kg m^{-3} (Appendix F) whereas the uppermost crustal layers may have densities as low as 2500 kg cf. (and occasionally, 2300 kg m^{-3}) even in a predominately basaltic crust (e.g. Gudmundsson, 1988, 2011a). Thus, for **every single dike-fed basaltic eruption**, the dike must not only propagate through many local layers of neutral or negative buoyancy, but they must propagate through seismic layers or units (with thicknesses of hundreds of metres or more) of average densities, at shallow depths, of generally less than that of the magma. Neither the thick seismic units nor the local layers, as a rule, arrest the basaltic dikes or deflect them into sills based solely on their neutral or negative buoyancy. It can therefore be concluded that field observations show clearly that **neutral buoyancy** layers/units normally do **not stop** the vertical propagation of the dikes.

There are, in fact, **no mechanical reasons** why dikes should stop, change into sills, or propagate laterally at levels of neutral buoyancy. Equation (10.4) suggests that, for a gradually increasing average density of the host-rock units with increasing crustal depth (as is commonly, crudely, the case), the **highest magmatic overpressure** occurs at the 'regional' level of neutral buoyancy. It follows that, unless the local stress field, the tensile strength, or toughness of the rock change abruptly at the contact with the level of neutral buoyancy, there is every reason for the dike to continue its vertical propagation path and propagate though the level of neutral buoyancy. This is exactly what is observed in the field: as a rule, basaltic dikes propagate easily through layers of densities similar to, or lower than, that of the basaltic magma (Figs. 10.5, 10.6; cf. Figs. 1.11, 2.27, 2.34, 7.27).

So far, the focus has been on the buoyancy effect on basaltic magma. For a **felsic magma**, with a typical density of 2200 kg m^{-3} (Appendix F), the buoyancy term in Eq. (10.4) may be positive right up to the Earth's surface. By contrast, for an **intermediate magma** such as andesite, with a typical density of 2500 kg m^{-3}, the buoyancy term may be positive or neutral all the way to the surface or, alternatively, slightly negative in the uppermost several hundred metres of the crust. The gas expansion and density decrease is greater for intermediate and, particularly, felsic magma than for basaltic magma. So that even if the average density of the intermediate magma, at 2500 kg m^{-3}, may be somewhat greater than that of the uppermost part of the crust, the density of which may be 2300–2500 kg m^{-3}, and in many sedimentary basins it is less than this, the density decrease of the magma at shallow depths due to gas expansion may result in a positive or neutral buoyancy in this part of the crust, following the **positive buoyancy** at greater depths.

Dike-fed basaltic eruptions are by far the most common in the world. The considerations above indicate that **no** typical **basaltic eruptions** would occur **anywhere** in the world if dikes always become arrested or deflected into sills at a level of neutral buoyancy. Since dike-fed basaltic eruptions are the most common on the planet, it is clear that this **mechanism generally does not operate**. As indicated above there is, in fact, no reason why it should. From Eq. (10.4), the highest overpressure in a vertically propagating basaltic dike is reached at the regional level of neutral buoyancy. This overpressure (Eq. (10.3)) is, theoretically, normally high enough to propagate the dike to the surface. If the dike does not reach the surface to erupt, it is thus not because of neutral buoyancy, but because of the operation of one or all of the **three mechanisms of dike arrest** discussed earlier and amply supported by direct **field observations**.

10.5.3 Propagation Paths – Comparison with Other Hydrofractures

We have now considered the magmatic pressure, composed primarily of excess pressure in the chamber before rupture and buoyancy effects during dike propagation, which makes dike emplacement possible. Before discussing the principles that control dike-propagation paths in greater detail, it is worth summarising briefly the results on the propagation of human-made **hydraulic fractures.** These are used in the hydrocarbon industry to increase the permeability of reservoirs of oil and gas. Propagating hydraulic fractures (like propagating dikes) generate **earthquake swarms** (Shapiro, 2018), mostly of microseismicity, and their propagation paths can be monitored (Fisher and Warpinski, 2011; Davis et al., 2012; Flewelling et al., 2013; Fisher, 2014).

Hydraulic fractures are extension fractures (modelled as mode I cracks) where the rock is ruptured and the fracture is driven open by **water** under high pressure that is injected from a drill hole or well. **Conventional** hydraulic fracturing, used for the past 70 years in the hydrocarbon industry to increase the permeability of hydrocarbon reservoirs, generates a fracture that is **injected laterally** from a **vertical** well, arrested at the top and bottom. The aim is thus to confine the hydraulic fracture to the 'target layer', the layer or unit containing oil or gas. The fracture should thus not propagate much into the layers above and below the target layer (cf. Valko and Economides, 1995; Yew and Weng, 2014; Shapiro, 2018).

In the past decades, a new method has been developed whereby hydraulic fractures are **injected vertically** into the rock layers from a **horizontal** well (Wu, 2017; Shapiro, 2018). This technique has been extensively used for getting gas out of shales (gas shales) and is now well developed. The resulting (mostly) vertical hydraulic fractures are mechanically very similar to vertical dikes. Some of the hydraulic fractures are well over a kilometre tall, that is, with dip-dimensions similar to those of many radial dikes and inclined sheets injected from shallow magma chambers (Chapter 6; Gudmundsson, 2017).

Studies of the propagation paths of thousands of hydraulic fractures show that their paths are commonly complex. The microseismic studies indicate that vertical hydraulic fractures commonly deflect into contacts or other discontinuities, particularly at shallow depths (Fisher and Warpinski, 2011; Flewelling et al., 2013). Deflection of vertical hydraulic fractures into horizontal water-sills can occur at any depth in the sedimentary basins, but it is particularly common at depths of less than about 700–800 m (Fisher, 2014). Some hydraulic fractures also deflect into **existing faults**, and may propagate along them for a while (Davis et al., 2012, 2013; Lacazette and Geiser, 2013). This behaviour is entirely analogous to that of dikes and is discussed in greater detail in Section 10.5.4 below.

Hydraulic fractures have also been studied in **excavated sections** in the subsurface. Their geometric features are similar to those observed in dikes. Some fractures, or fracture segments, become arrested at contacts between mechanically dissimilar layers. Other segments become deflected into water-sills at such contacts or fractures (e.g. Fisher and Warpinski, 2011; Fisher, 2014). It is clear that the three principal mechanics of fracture arrest and deflection, namely Cook–Gordon delamination, stress barriers, and elastic mismatch, all operate on hydraulic fractures in the same way as they do on dikes.

In addition to human-made hydraulic fractures and dikes, hydrofractures also include many **joints** and **mineral veins**. These show very similar geometric structures to those of dikes and hydraulic fractures. The fractures are commonly arrested or deflected on meeting other fractures and, in particular, at contacts between mechanically dissimilar layers. It can be concluded that the same principles control the path formation, and eventual path arrest, of all hydrofractures. And now let us turn to the physical principles that control the actual paths selected by hydrofractures, focusing on the paths of dikes.

10.5.4 Propagation Paths – the Principles of Virtual Work and Least Action

One of the most important **unsolved problems** in earth sciences is to provide a theoretical framework that allows us to make reliable forecasts for rock-fracture propagation. The reason why the solution of this problem is so important is that brittle deformation, which dominates in the upper part of the crust, is through fracture initiation and propagation. Thus, fracture propagation controls earthquakes, landslides (lateral collapses), calderas (vertical collapses), the formation and development of all types of plate boundaries (divergent, convergent (including subduction-zone initiation), and transform boundaries), and, of course, dike propagation and eventual volcanic-fissure formation. In addition, fracture propagation has a large impact on various applied fields such as hydrogeology and petroleum geology.

Current theoretical understanding **does not make it possible** to forecast with any reliability the likely path of a rock fracture. This conclusion applies to all types of rock fractures, including faults, mineral veins, human-made hydraulic fractures, and dikes. Here, I focus on dikes (including inclined sheets and sills), and suggest a **new theoretical framework** for forecasting their paths. The focus is thus on fluid-driven **extension fractures** (hydrofractures). However, the theoretical framework is, with proper adjustment of the boundary conditions, suitable for all types of rock fractures.

From its point of initiation in the roof of a source magma chamber, there are theoretically **an infinite number of paths** that the dike may choose to follow to its point of arrest (Fig. 10.7). Here, the point of arrest is either within the crust – a true arrest – or at the surface, in the case of a feeder-dike. The reasons that the dike eventually selects a particular path among all those available are not known. The fundamental importance of being able to make reliable forecasts of a dike path is because the selected path determines **whether, where, and when** the dike reaches the surface to **erupt.**

Principle of Least Action for Discrete Systems

Here I propose that the eventual path selected by a dike (and any extension fracture) is the one of **least action** as determined by **Hamilton's principle.** In its simplest version

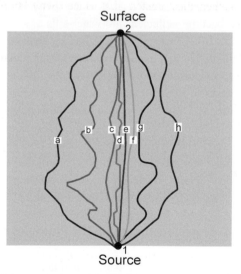

Fig. 10.7 A dike initiated at a source chamber can, theoretically, choose among an infinite number of paths to reach the surface (or its point of arrest within the crust). Here the point of initiation (at time t_1 in Eq. (10.5)) in the roof of the source chamber is denoted by 1 and the point of eruption at the surface, forming the volcanic fissure/crater cone (at time t_2 in Eq. (10.5)) is denoted by 2. Possible dike paths – only eight are shown here – are denoted by a–h. Hamilton's principle of least action implies that the dike selects the path along which the time integral of the difference between the kinetic and potential energies is stationary (is an extremum), and most commonly a minimum, relative to all other possible paths with the same points of initiation and arrest. In a homogeneous, isotropic crustal segment, the principle implies that the selected path would normally be close to path e. For a heterogeneous, anisotropic crustal segment, the path can be much more complex, depending on the mechanical layering and local stresses (Figs. 10.9, 10.10, and 10.12; cf. Figs. 6.10 and 7.10).

this principle states that the dike selects the path along which the time integral of the difference between the kinetic and potential energies is stationary (is an extremum) relative to all other possible paths with the same points of initiation and arrest. As you may recall from calculus, extremum or extreme values on a curve are its maximum, minimum, or points of inflection. For most processes to which the Hamilton's principle applies the extremum turns out to be a **minimum.** For dikes, which generally propagate slowly (commonly at 0.1–1 m s^{-1}) the kinetic energy is primarily associated with seismic waves of the induced earthquakes. By contrast, the potential energy is the strain energy stored in the volcano/volcanic zone during unrest plus the elastic energy supplied by the forces acting on the volcano/volcanic zone. When the kinetic energy is omitted, that is, when the forces associated with the dike propagation are conservative (explained below) and there are no constraints (also explained below), Hamilton's principle of least action reduces to the **principle of minimum potential energy.** This is a well-known principle in solid mechanics and was postulated as a basis for understanding dike propagation by Gudmundsson (1986). Here I provide a further development of these ideas with Hamilton's principle as the main guide.

Several concepts have now been introduced, which, for most readers, are likely to be unfamiliar. So let us go on to define and clarify them. To do so, the concepts must be presented in their mathematical context. Let us start with **Hamilton's principle**. Its simplest form is given by:

$$\delta S = \delta \int_{t_1}^{t_2} L \, dt = \delta \int_{t_1}^{t_2} (T - V) dt = 0 \qquad (10.5)$$

Here S is the action (explained below), L the Lagrangian, also called the Lagrangian function (also explained below), t_1 and t_2 are two specified and arbitrary chosen times in the evolution of the system, and δ is the **variational symbol**, which simply denotes a small change. In the second equation on the right hand side, T is the **kinetic energy** and V is the **potential energy**. It follows from Eq. (10.5) that the **Lagrangian** is equal to the difference between the kinetic energy and the potential energy, that is:

$$L = T - V \qquad (10.6)$$

Now that all the terms in Eq. (10.5) have been defined, it is time to discuss what Hamilton's principle really means. Briefly, it states that the **actual path** chosen by the system in moving from time t_1 to t_2 is such that the variation of the action, that is, δS is zero. This means that the actual path taken is the one for which the action integral (Eq. (10.5)) is an extremum. As indicated above, an **extremum** (or extreme value) means that the action integral (Eq. (10.5)) can be maximum or minimum or a point of inflection. Thus, Hamilton's principle of least action might more properly be referred to as the principle of stationary action. In practice, however, the action is normally **minimised**, made least, along the actual path taken. **Action** has the dimensions of **energy × time** (or linear momentum × distance) and the unit of J s (joule-second). Thus, the dike (fracture) path chosen (Fig. 10.7) is normally the one along which the **energy** transformed multiplied by the **time** taken for the propagation is the **least** (is a minimum).

Equation (10.5) applies to a **conservative system**, namely one in which the work done by a force is independent of path (reversible) and equal to the difference between the final and initial values of the potential energy (potential energy function). The associated force field can be expressed as a **gradient of the potential**. If all the given forces acting on a system are conservative, then the system is conservative. For many systems in solid mechanics (including rock mechanics), the external force system (the external loads) are such that the body force and the surface stresses are independent of the solid-body deformation (Fung and Tong, 2001). Such systems, thus, cannot involve **friction**. Later in the section, however, we consider friction. Before going further, several terms need to be explained that are referred to when using Hamilton's principle, particularly the principle of virtual work, which is also known as the principle of virtual displacements.

Principle of Virtual Work

Let us begin by defining the term **virtual displacements.** These are imagined (virtual) infinitesimal (tiny) changes in the coordinates that are consistent with the constraints of the system and assumed to take place instantaneously (Richards, 1977; Meirovitch, 2003; Goldstein et al., 2013). The displacements are named virtual because they are assumed or pretended to occur in **no time** at all (the system is 'frozen in time'; time is held constant), whereas all real displacements take time. In other words, a virtual displacement is the result of our imagining the system reaching a somewhat different position or, alternatively, a somewhat different configuration, while time is held constant.

Next is the term **constraints.** In the present context, constraints mean some restrictions to the freedom in the movements within or of the system (Richards, 1977; Gupta, 2015). More specifically, constraints imply that there are certain factors or parameters that partly control the movements of the system. Examples of constraints are the walls of a container of gas. Clearly, for a closed container the gas molecules can only move inside the container, the walls of which therefore constrain their movements. Another example is a body constrained to move along a smooth surface; for example, an inclined surface or a sphere. The movement of the body is then normally parallel with the surface, but the constraint forces act perpendicularly to the surface. It follows that the work done by the constraint forces (being at a right angle to the movement of the body) is zero. A well-known exception to this in geology is a **fault slip.** Then, the friction is a constraint force that acts parallel to the surface – the fault plane – in which case there is **work done** by the constraint force.

Now let us define **virtual work.** When a system is subject to virtual displacement the work done by any real force on the system or its particles is referred to as virtual work (Richards, 1977; Dym and Shames, 2013; Hamill, 2014; Gupta, 2015). More specifically, for a continuous system, such as a crustal segment or any elastic rock body, virtual work is the work done on the body by actual surface stresses and body forces during virtual displacements that are consistent with any constraints on the movement of the material points of the body (Reddy, 2002).

The **principle of virtual work** states that for a system in static equilibrium the virtual work done on the system due to reversible virtual displacements that are compatible with the constraints is zero (Richards, 1977; Meirovitch, 2003; Hamill, 2014; Gupta, 2015). An alternative name is the principle of virtual displacements (Washizu, 1975). In greater detail, the principle of virtual work may be stated as follows. For a system in equilibrium under given forces and constraints, the sum of the virtual work made by the internal and external forces associated with virtual displacements is zero.

The above statements are general and particularly appropriate for discrete systems. Let us therefore now state the principle of virtual work for a **continuous system** in a somewhat different manner, while the basic meaning is the same. If, for the given boundary conditions, the virtual work of all the external and internal forces associated with the virtual displacements in a continuous system, such as an elastic body, is zero, then the system is in equilibrium. This means that the following conditions are satisfied (Reddy, 2002):

$$\delta W = \delta W_I + \delta W_E = 0 \tag{10.7}$$

where δW is the total virtual work, δW_I is the virtual work attributable to the internal forces, and δW_E is the virtual work attributable to the external forces. But it has not yet been explained what is meant by internal and external forces, which we turn to now.

External and Internal Forces

Consider all the forces acting on a given volume of a solid body, for example a crustal segment. The solid body here corresponds to the system that was discussed earlier. **External forces** are outside the system, here the solid body, and so, generally, are their sources. By contrast, **internal forces** are inside the system, the solid body (Fig. 10.8). They are generally attributable to sources or processes inside the system (cf. Richards, 1977; Reddy, 2002; Meirovitch, 2003). More specifically, an internal force is a measure of the

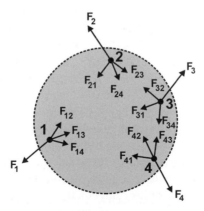

Fig. 10.8 External and internal forces. The shaded circular area represents the system (e.g. a crustal segment) under consideration and the points 1–4, particles (for a discrete system) or material points (for a continuous system, as here). External forces origin outside the system and are here given as F_1-F_4. These forces act across the boundary of the system (the edge of the circular shaded area) and they act on the system. Internal forces occur inside the system and are here presented by F_{12}, F_{23}, F_{32} and so forth. Internal forces act only inside the system, that is, they do not cross the boundary of the system. Modified from Richards (1977).

resistance of one part of the body to being separated from other parts. The internal force per unit area is referred to as the **stress**, which then causes strain. Internal forces cannot move the centre of mass of the body, they cannot move the body.

External forces, however, can move the centre of mass of the body and cause acceleration. They are divided into body (volume) forces and surface forces. **Body forces** act on all the elements or material points of a volume of a solid and are given as force per unit mass or, more commonly, force per unit volume. Body forces relate to the mass of the body, such as that of a crustal segment or a volcano. Examples include inertial, electromagnetic, and gravitational forces.

Surface forces act across surfaces of two types (Gudmundsson, 2011a). One is the physical surface, such as the walls of fractures, faults, and other discontinuities where the forces are due to the physical contact between the parts of the solid on either side of the discontinuity. The other surface is an imaginary one within the solid body (the system). The imaginary surface can have any attitude (orientation). The surface force acts on the imaginary surface as if it really existed. Because the surface can be of any orientation, and because stress is force per unit area, the imaginary surfaces allow the calculatation of the stress on a discontinuity (a fracture, a contact) of any orientation.

Virtual Work in Relation to Least Action and Minimum Potential Energy

Given the detailed discussion above about the principle of virtual work and related aspects of external and internal forces, you might wonder why all this detail is necessary. Why is this discussion needed when the focus is on the Hamilton's principle of least action and, in due course, the principle of minimum potential energy? The answer is that there are several reasons for the importance of the principle of virtual work in solid mechanics (and therefore for crustal segments), including those listed below (cf. Washizu, 1975; Richards, 1977; Tauchert, 1981; Wallerstein, 2002). The principle of virtual work can be summarised as follows:

- It applies to any material stress-strain relationships, that is, **any constitutive laws**. Thus, the principle is equally valid for materials that behave as elastic and inelastic (including plastic) materials, and to mechanical as well as thermal loading.
- It is valid both for **conservative** as well as **non-conservative** systems. This is particularly important in geology because systems with plastic (permanent) deformation and friction (such as faulting) are non-conservative.
- It can be used, with appropriate stress–strain (or stress–displacement) relationships, to set up mechanical **equilibrium conditions** for the system.
- It makes it possible to deal with **large actual displacements** and strains, even if the virtual displacements are supposed to be infinitesimal.
- It can be used to **derive** Hamilton's principle.
- It can be used to **derive** the principle of minimum potential energy.

It follows from these bullet points that the principle of virtual work is a fundamental principle in continuum mechanics (and rock mechanics) and is closely related to Hamilton's principle and the principle of minimum potential energy, both of which are used here to forecast dike-propagation paths. This brings us back to Hamilton's principle and, eventually, to the principle of minimum potential energy.

General Version of the Least-Action Principle

There are in fact several versions of Hamilton's principle. The version presented in Eq. (10.5) assumes that the system is conservative, that is, that the forces that operate on it are conservative. As said, that is commonly the situation in solid mechanics, but normally not during rock-fracture propagation; particularly not during shear-fraction propagation (an earthquake rupture), where friction plays an important role. The **most general** version of Hamilton's principle relaxes the assumption of the system being conservative and may be written as (Tauchert, 1981; Bedford, 1985):

$$\delta S = \int_{t_1}^{t_2} (\delta T - \delta V + \delta W + \delta C) = 0 \tag{10.8}$$

Here the system includes both conservative and non-conservative forces and is, in addition, subject to constraints. As before, T denotes the kinetic energy and V is the potential energy associated with the conservative forces. Additionally, δW is the virtual work of the generalised forces and δC is the constraint. Virtual work is not restricted to conservative forces, and thus includes non-conservative forces as well.

The **generalised forces Q_i**, when conservative and thus derivable from potential energy V, are defined by:

$$Q_i = -\frac{\partial V}{\partial q_i} \tag{10.9}$$

where q_i denotes generalised coordinates. The virtual work δW is related to generalised forces through the equation:

$$\delta W = Q_i \delta q_i \tag{10.10}$$

Equation (10.8) is the most general representation of Hamilton's principle, and applies in particular to **discrete systems.** Here, however, we are primarily interested in continuous systems and, in particular, elastic systems such as describe, to a first approximation, the behaviour of a crustal segment through which a dike propagates. So we now turn to continuous systems.

Least-Action Principle for Continuous (Elastic) Systems

As indicated, the Hamilton's principle of least action presented in Eqs. (10.5) and (10.8) applies to **discrete systems**, namely systems composed of (normally very many) particles. By contrast, for dike (or any rock-fracture) propagation, the system is normally not discrete (except in the unlikely case of dike propagation through unconsolidated pyroclastics or sediments) but rather is a solid rock and thus a **continuous system**. For a continuous **elastic**

system, as a crustal segment and associated volcano (through which the dikes propagate) are, to a first approximation, the following points need to be considered:

1. The degrees of freedom for a discrete system are finite, but for a continuous system they are **infinite**. By **degrees of freedom** I mean the minimum number of independent coordinates required to specify completely the position of each and every part of the system – the configuration of the system – which must also be compatible with any constraints on the system. **Configuration** here means the position, at a given moment, of all the particles (for a discrete system) or all the material points (for a continuous system; Reddy, 2002), whereas constraints are defined above. The degrees of freedom are thus the number of independent parameters needed to define the system configuration. For a discrete system of N particles without constraints there are $3N$ degrees of freedom. N may be very large, but it is always finite in a discrete system, whereas for a continuous system, N is infinite ($N \rightarrow \infty$).

2. **Hamilton's principle** for a continuous system takes on a form that is somewhat different from that of a discrete system. For a discrete mechanical system the potential energy V is only a function of the external forces (force field), such as gravity. For a continuous elastic system there is, in addition to the potential energy of the external forces (loading), an internal strain energy due to internal forces, which concentrates in the elastic rock before dike (fracture) propagation is initiated.

3. More specifically, the virtual work attributable to the **internal forces**, that is, the internal virtual work δW_I (Eq. (10.7)), may be considered as the variation of the **strain energy** in the deformed body or continuous system (Tauchert, 1981).

4. Similarly, the virtual work attributable to the **external forces**, that is, the external virtual work δW_E (Eq. (10.7)), may be considered as the work done by the body and surface forces during the variation in the displacement field of the deformed body or continuous system (Tauchert, 1981).

5. Most external forces may be regarded as being derived from a potential energy, as in Eq. (10.9). More specifically, the force components follow when the potential energy is differentiated with respect to the displacement components (Tauchert, 1981).

With these points clarified, a general presentation of the **Hamilton's principle** of least action for an **elastic solid** can now be provided as follows (Tauchert, 1981; Bedford, 1985; Reddy, 2002):

$$\delta S = \delta \int_{t_1}^{t_2} (T - V - U)dt = 0 \tag{10.11}$$

where, as before (Eq. (10.5)), S is the action and T is the kinetic energy. Here, however, V is the potential energy due to **generalised external forces** acting on the elastic rock body (Eqs. (10.9) and (10.10)), and U is the **strain energy** in the body. This formulation assumes that the external forces or loads that act on the rock body are independent of the elastic displacements that they generate, that is, are **conservative**, as is commonly the case in elastic deformation and discussed above (Tauchert, 1981; Fung and Tong, 2001). In this formulation, it is assumed that there are no constraints on the continuous elastic system, of the kind presented for the discrete system in Eq. (10.8).

Together, the strain energy stored in the body and the potential energy attributable to the external generalised forces acting on the body are referred to as the **total potential energy**, denoted by Π, where:

$$\Pi = V + U \tag{10.12}$$

in which case the **Lagrangian** (Eq. (10.6)) becomes:

$$L = T - \Pi \tag{10.13}$$

and **Hamilton's principle** (Eq. (10.11)) may be written as:

$$\delta S = \delta \int_{t_1}^{t_2} (T - \Pi) dt = 0 \tag{10.14}$$

where all the symbols have been defined above.

Principle of Minimum Potential Energy

This principle applies only to **elastic bodies**, both linear and non-linear elastic ones (Richards, 1977). As indicated, the brittle crust may be regarded as being elastic to a first approximation. From Eq. (10.7) we know that the sum of the internal and external forces is zero when the body is in equilibrium. It has also been mentioned that for many elastic systems, the forces may be regarded as being conservative. For a complete cycle of loading and unloading (removal of the load or forces), the original configuration is completely recovered and thus follows a closed path. The forces are thus conservative (Richards, 1977). More specifically, the internal forces, while doing no net work, provide a potential energy of strain, that is, **strain energy** (Eqs. (10.11) and (10.12)). The external forces (surface forces or stresses and body forces) are also assumed to be conservative, that is, they are only a function of position and do not depend on the deformation of the elastic body (Tauchert, 1981).

Assuming, therefore, that all the forces are conservative, we next need to consider the kinetic energy T in Eq. (10.14). Let us focus on dike propagation, in which we are mainly interested here. While there is normally earthquake activity associated with dike propagation (Chapter 4), the rate of dike-fracture propagation is slow in comparison with that of seismic ruptures. Thus, we use **static moduli** rather than dynamic ones when modelling dike emplacement (Gudmundsson, 2011a). The kinetic energy is therefore assumed to be zero, $T = 0$. From these premises, it can be shown (Richards, 1977; Tauchert, 1981; Reddy, 2002) that for an elastic body in equilibrium, the total potential energy is a minimum. Namely:

$$\delta(V + U) = \delta \Pi = 0 \tag{10.15}$$

where, as before, V is the potential energy due to the external forces, U is the strain energy due to the internal forces, and Π is the total potential energy.

In words, the **principle of minimum potential energy** means that of all the possible displacement fields or deformations (configurations) of an elastic body that satisfy the constraint conditions and the external and internal loads, the actual (true) displacements or deformations are those that make the total potential energy of the body a minimum. It is sometimes stated that the actual displacements or deformations make the potential energy stationary (an extremum). That is of course true. But it is also easily shown that the extremum value for a body in a stable equilibrium is actually a minimum.

Another way of expressing the principle of minimum total potential energy is as follows. For an elastic body to be in stable equilibrium, it is necessary and sufficient that the total potential energy of the body is a minimum. In contrast to the principle of virtual work, the principle of minimum total potential energy is valid only for **elastic bodies** or structures (linear or non-linear) that are subject to loads through conservative forces.

The principle of minimum potential energy has already been suggested as a basic mechanical framework for forecasting **dike paths** (Gudmundsson, 1986). The present analysis is thus a development of that general framework into a much more detailed one. Let us now use the present analysis to discuss possible dike paths, focusing first on general results and subsequently on the effects of discontinuities on the paths.

10.5.5 Propagation Paths in a Non-Fractured Elastic Crust

The classical Hamilton's principle implies that the 'path' along which a system 'moves' reflects changes in its configuration, rather than an actual movement of the system as a whole in three-dimensional space. Each point on the path or curve along which the system moves through time corresponds to one configuration of the system – one arrangement of the **particles** (for a concrete system) or the **material points** (for a continuous system). Here, however, the principle is extended so as to refer to actual paths in space and time, namely the propagation paths of rock fractures.

Quasi-Static Dike Propagation

This extension is particularly appropriate when dealing with dike (hydrofracture) propagation, because they advance in steps (Fig. 10.9). As discussed earlier (Chapter 4), the steps are partly a consequence of the **time lag** between the fracture front and the fluid (magma) front at any particular instant (Fig. 4.10). When the magmatic pressure in the dike-fracture reaches the conditions of Eqs. (10.1) or (10.2), the upper end (the tip) of the dike-fracture advances very quickly for a certain distance and then stops (becomes temporarily arrested). The rate of propagation of the fracture front (tip) during each advance (step) is similar to the velocities of S-waves, or of the order of kilometres per second. The high viscosity of the magma makes it impossible for it to flow as quickly as the dike-fracture front propagates. Consequently, following each fracture-front advance, the front will, for a while, be **empty** (Fig. 4.10); the magma front has to flow into the empty front, fill it, and build up a pressure so high that the dike-fracture tip can advance again. This process, filling the fracture front and building up the magma pressure for further rupture takes time, hence the time lag between the fracture front and the magma front.

It follows that dikes advance in **steps** through quasi-static fracture propagation, and each step may be regarded as following **Hamilton's principle** of least action (Eq. (10.14)). Alternatively, if the kinetic energy through earthquakes is omitted, then the principle of

10

9

8
7
6
5

4

3

2

1

Dike

Fig. 10.9 Dikes (and other rock fractures) propagate in steps. When a dike propagates through mechanical layers with different properties and sharp contacts (discontinuities), each step is likely to be similar in length (here height) as the thickness of the mechanical layer through which the dike is propagating. This is indicated schematically here, where the potential steps for the further vertical propagation of a dike through a lava pile (in Northwest Iceland) is indicated by the numbers 1–10. In the basaltic lava flows, the propagation steps would commonly coincide roughly with the existing columnar joints. While the steps are discrete, the resulting dike-fracture is normally physically continuous in that the segments/steps are in physical contact. If, with time and burial, the mechanical layers become 'welded together' so as to form thicker units, each composed of many lava flows, the steps/segments would become longer (higher), as seen in Fig. 10.10 (cf. Figs. 1.21, 2.34, and 7.16).

minimum potential energy (Eq. (10.15)) may be used. Here, we assume that the system (crustal segment) hosting the propagating dike-fracture reaches, temporarily, a mechanical **equilibrium** following each advance of the dike front. When the magma flows to the fracture front and the overpressure builds up again, the equilibrium becomes gradually unstable, until the overpressure becomes high enough for a new fracture-front advancement to occur.

The **size** of typical vertical fracture-front advancements during dike propagation depends much on the mechanical layering of the host rock. When the host rock is a pile of lava flows and pyroclastic or sedimentary/soil layers, the vertical advancements (the steps) may correspond to the thicknesses of the layers ahead of the fracture front (Fig. 10.9). This applies particularly to a young pile, the most common one in active volcanoes and volcanic systems, with sharp mechanical discontinuities at the contacts between layers. When the pile becomes older, there is commonly a gradual reduction in mechanical contrast between layers (partly due to secondary mineralisation and general compaction). Also, the contacts between layers may be irregular and partly welded together. In both these cases, several layers, sometimes many, may function as single mechanical layers during dike emplacement (Fig.10.10; cf. Figs. 4.11, 7.16). Also, the dike front

Fig. 10.10   For each dike-fracture step (Fig. 10.9) there is, theoretically, an infinite number of possible paths. Some of the potential paths that the propagating tip of the dike could have followed from segment A to segment C – so as to form segment B – are indicated by the numbers 1–7. The actual path taken to form segment B is path number 4 (the segment is about 1.5 m thick, cf. Fig. 1.7). Here, many lava flows and scoria layers are 'welded together' so as to function as single mechanical layers of thickness similar to the heights or dip-dimensions of the individual dike segments, such as segment B (cf. Figs. 1.21, 2.34, and 7.16).

cannot normally propagate through very thick layers, say many tens of metres or hundreds of metres, in a single step.

For each dike-fracture advance – each step generating a dike segment – there is, theoretically, an **infinite number of possible paths** (Figs. 10.10). And the same applies to the overall propagation path of the dike (Fig. 10.7). The **actual path** taken from the point of fracture initiation at time t_1 to the endpoint (the tip) of the fracture, reached at time t_2, is the one along which the **action S is minimised**. Since action is energy × time, it follows that its unit is J s. Let us now look at what energies are involved, to be minimised with time, during fluid-driven rock-fracture propagation.

Energy for Dike Propagation

All fracture formation needs an energy input for the hosting body. Fractures can be initiated and propagated only if energy is available to create the new fracture surfaces during the extension or growth of the fracture. The energy used to create new surfaces is referred to as **surface energy** (Anderson, 2005; Gudmundsson, 2011a). At an atomic level, surface energy is needed to rupture the solid and generate a fracture. More specifically, at this level, two atomic planes in the solid must be separated from each other to a distance where there are no longer any interacting forces between the planes. The separation requires work, namely surface energy. At the scale of rock fractures such as dikes, the rupture is rarely at an atomic level, but rather at the level of existing joints and other flaws in the host rock. The surface energy is denoted by W_s. For dike propagation, W_s represents energy that must be put into the hosting crustal segment or rock body (the continuous system) for the dike to be initiated and to propagate. Since the energy must be added to the continuous system, thermodynamically the surface energy W_s is regarded as positive.

For a dike to be initiated and to propagate, the **total energy** U_t of the hosting crustal segment must be large enough to overcome the surface energy W_s. The total energy may be regarded as being composed of two parts (Sanford, 2003; Anderson, 2005), namely:

$$U_t = \Pi + W_s \tag{10.16}$$

where Π is the **total potential energy** (Eq. (10.12)) of the crustal segment (the continuous system) hosting the propagating dike. From Eq. (10.12) we know that the total potential energy Π derives from two sources, namely the strain energy U and the potential energy V. The potential energy is due to the external applied load or generalised forces Q_i (Eq. (10.9)), which includes body forces such as gravity as well as surface forces/stresses. The external forces contribute to the **overpressure** of the magma that partly drives the dike-fracture propagation (Eq. (10.3)). The strain energy U, by contrast, is due to the internal forces between the material points in the deformed crustal segment. More specifically, strain energy is stored in the crustal segment, or any solid body, as a consequence of changes in the relative location of its material points, namely changes in the internal configuration of the body as it deforms and the material points become displaced. This happens before the dike-fracture is initiated, so that the strain energy is available to drive the fracture propagation, provided that certain conditions are satisfied.

Strain energy has been referred to several times, but so far it has only been defined qualitatively. A basic quantitative definition of **strain energy** U is stress × strain × volume and it has the unit of J (joule). **Strain energy per unit volume** U_0 may be defined as:

$$U_0 = \int_V \frac{\sigma_{ij}\varepsilon_{ij}}{2} dV_v \tag{10.17}$$

where σ is stress and ε is strain (the subscripts ij indicate the stress and strain components), and dV_v is the unit volume of the strained body or crustal segment. (Notice the subscript v for V_v for volume to distinguish it from the letter V which in this chapter denotes potential energy.) The factor ½ is because the force (or stress or pressure) that generates the strain energy, such as during the inflation (expansion) of a shallow magma chamber, varies linearly from zero (when the excess pressure begins to build up in the chamber) to its maximum value needed to reach the maximum displacement (expansion). It is thus the average value, namely half the value of the force that is used to calculate the strain energy. Dropping the subscripts ij to simplify the notation, and using Hooke's law, $\sigma = E\varepsilon$ (Chapter 3), where E is the Young's modulus, Eq. (10.17) can be rewritten in terms of **total strain energy** U and strains or stresses for the total volume V_v, thus:

$$U = \frac{\sigma \varepsilon V_v}{2} = \frac{E\varepsilon^2 V_v}{2} = \frac{\sigma^2}{2E} V_v \qquad (10.18)$$

Thus, any magma-chamber inflation or expansion prior to dike injection results in strain energy being stored in the crustal segment hosting the chamber, and this energy can be calculated either using strain or stress together with Young's moduli for the entire volume of the strained segment. This strain energy is partly used to form the two new fracture surfaces (used as surface energy), and partly for microcracking and plastic deformation in the process zone at the tip of the propagating dike (Gudmundsson, 2011a, 2012).

Once the magma chamber has ruptured, the dike-fracture propagates only if the total energy U_t in Eq. (10.16) either remains constant or decreases during each dike-front advancement; that is, for equilibrium conditions the dike-fracture propagates if $U_t = k$, where k is a constant. During dike-fracture propagation, a new surface area dA must be generated. From Eq. (10.16) and the condition $U_t = k$ we then have:

$$\frac{dU_t}{dA} = \frac{d\Pi}{dA} + \frac{dW_s}{dA} = 0 \qquad (10.19)$$

From Eq. (10.19) it follows that:

$$-\frac{d\Pi}{dA} = \frac{dW_s}{dA} \qquad (10.20)$$

which shows that the decrease in total potential energy Π during dike-fracture propagation equals the increase in surface energy W_s, namely that, when the dike grows, potential energy in the host rock is released and transformed into surface energy. The rate at which this release or transformation occurs is referred to as the **energy release rate**, denoted by G (Chapter 7), and, from Eq. (10.20), given by:

$$G = -\frac{d\Pi}{dA} \qquad (10.21)$$

Here, G is the energy available to drive the dike-fracture propagation (or any other extension, or mode I, fracture propagation). Thus, the dike-fracture will propagate if the energy release or transformation rate reaches the critical value on the right-hand side of Eq. (10.20), namely:

$$G_c = \frac{dW_s}{dA} \qquad (10.22)$$

G_c is referred to as the **material toughness** of the rock hosting the dike (cf. Chapter 7; Anderson, 2005; Gudmundsson, 2011a, 2012).

Common Dike Paths

Based on the considerations above, what can be said about likely dike-propagation paths? Using Hamilton's principle as a guide, the dikes will seek the path that minimises the action, that is, the used energy × time. Consider first the energy used during dike propagation. The crustal segment will rupture and the dike-fracture will propagate if the energy release rate reaches the critical value of the material toughness given by Eq. (10.22). But in addition to rupturing, a dike-fracture also **opens up** whereby the dike attains its thickness. While the final thickness of a dike is somewhat (about 10%) less than the opening of the dike-fracture, the thickness of old dikes may be taken as a good measure of their opening displacements and geodetic data give indications of the opening displacements of present-day dikes, particularly feeder-dikes.

When the dike-fracture opens up, **work is done** against a force, namely the normal force that acts on the dike-fracture walls. The greater the force, the larger the amount of work; that is, the greater is the energy needed or transformed for a given opening displacement. Since pressure (and stress) is force per unit area, it follows that when the magmatic overpressure first breaks the rock and then pushes the walls aside to open up the dike-fracture, more work is needed, for a given dike-fracture opening/dike thickness, when the push is against a high compressive stress/force per unit area than against a low one. Much more energy (work) is therefore required for the dike-fracture to open up (for the fracture walls to be displaced) against the maximum σ_1 or the intermediate σ_2 compressive principal stresses than against the minimum compressive (maximum tensile) principal stress σ_3. Dike propagation guided by Hamilton's principle of least action (Eq. (10.11)) or, if the kinetic energy is zero, the principle of minimum potential energy (Eq. (10.15)), should then be along a path that coincides with the trajectories (directions) of σ_1 and σ_2 and is thus perpendicular to the direction of σ_3. The **time constraints** in Hamilton's principle furthermore suggest that the dike will tend to follow the shortest path that is compatible with the other constraints.

Homogeneous, Isotropic Host Rock

So what dike paths can then be expected? Consider first the simplest case, namely dike propagation from a shallow chamber located in a **homogeneous and isotropic** crustal segment. Based on the least action/minimum principles discussed above, the dike path will follow the **trajectories** of σ_1. Furthermore, the path will be mostly **straight**, namely the shortest distance between t_1 and t_2 (Eq. (10.11)). That the path is straight follows from the arrangement of the σ_1 trajectories (Fig. 10.11; cf. Fig. 7.4) and is also a well-known result from the field of mathematical analysis known as **calculus of variations**, demonstrating the (intuitively well-known) fact that a straight line is the shortest distance between given points. Thus, provided the step-like average rate of dike propagation is roughly constant, the straight path also minimises the time needed for the propagation among a family of nearby curved or somewhat irregular paths with the same endpoints, t_1 and t_2 (marked as points 1 and 2 Fig. 10.7; cf. Fig. 10.10).

The question then arises: of the many straight potential paths from the chamber to the surface, which one does the dike choose? The answer is: **the path from point 1 (at time t_1),**

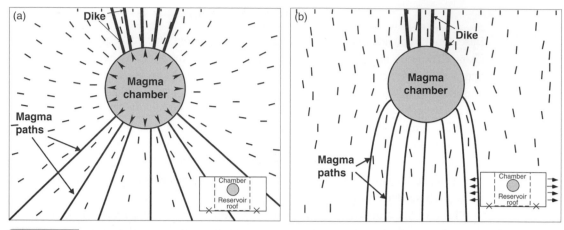

Fig. 10.11 Dike-propagation paths in a homogeneous, isotropic crustal segment. Several potential dike paths from the roof of the shallow chamber (of a circular, vertical cross-section) to the surface are indicated, as well as several potential magma paths from the source reservoir to the floor (lower margin) of the chamber (cf. Fig. 9.3 and Chapter 9). The ticks show the trajectories of σ_1, the likely dike paths (and magma paths) being parallel to these. When the loading of the chamber is (a) the internal magmatic excess pressure (see the inset), the potential dike paths are more spread (fan-shaped), than (b) when the only loading is external tension (see the inset).

parallel with σ_1, to point 2 (at time t_2) (for points 1 and 2 on Fig. 10.7). The next question is then: how is point 1 at t_1 selected? The answer is: point 1 at t_1 is the point of highest tensile-stress concentration at the margin of the chamber, namely where the condition of Eq. (10.1) is first satisfied. The third question is then: how is point t_2 selected? The answer is: it is the point (or horizontal line or curve, if the dike forms a volcanic fissure rather than a single crater cone) where the local overpressure or driving pressure (Eq. (10.3)) at the tip of the dike-fracture becomes zero. Presumably, most dikes injected into a homogeneous, isotropic crustal segment will reach the surface – become **feeders**.

Alternatively, the condition for dike arrest at t_2 may be formulated in terms of standard fracture-mechanics parameters. Thus, from Eq. (10.22) and standard formulas for energy release rate for fluid-driven fractures (Gudmundsson, 2011a), the dike-fracture propagation stops when the following conditions apply:

$$G_I = \frac{p_o^2(1 - v^2)\pi a}{E} > G_c = \frac{dW_s}{dA} \tag{10.23}$$

where G_I is the plane-strain energy release rate, p_o is the magmatic overpressure in the dike, a is half the height (dip-dimension) of the dike, E is the Young's modulus of the host rock, G_c is the material toughness of the host rock, and W_s is the surface energy needed to form the dike-fracture as it extends so as to form the new surface area A. So if the energy release rate during step-like dike-tip propagation is less than the material toughness, the dike-fracture propagation stops or becomes arrested.

In detail, it is difficult to forecast exactly where at the boundary of the magma chamber the rupture leading to dike injection will occur. Theoretical models (Chapter 3), analytical and numerical, can be used to infer regions of highest stress concentration at the chamber boundary during unrest periods, based on various properties of the chamber and its host rock. As soon as the chamber ruptures and a dike begins to propagate, however, the associated **earthquake swarm** is a good guide to the location and timing of the rupture, and thus to the location of t_1. Once the location of t_1 is determined, the principles above allow us to **forecast** the likely propagation path, including whether or not the dike will reach the surface to **erupt.**

Heterogeneous, Anisotropic Rock

The above results are useful so far as they go. The assumption of the host rock of the dike being homogeneous and isotropic is very common in geodetic studies in volcanology (Chapter 3). A major theme in this book, however, is that such an assumption is a poor one. By definition, stratovolcanoes are composed of strata, of layers, with widely different mechanical properties. And while the mechanical difference between the layers is much less in basaltic edifices, they are still composed of layers with variations in properties and contacts which may, and often do, affect dike-propagation paths. Similarly, volcanic zones and systems are composed of numerous layers, commonly with contrasting mechanical properties. The rocks of real volcanoes and volcanic zones are not only layered (anisotropic) but also heterogeneous. Here, however, the focus is on the effects of layering on the dike-propagation paths.

There are two main effects that mechanical layering has on dike paths: path arrest and path geometry. The effects of layering on dike arrest have already been discussed in Chapter 7, so here the focus is on the path geometry. The main effect of layering on dike-path geometry relates to the changes in the orientation of the trajectories of σ_1 (and, consequently, the orientations of σ_2 and σ_3 because the three principal stresses are at right angles to each other, Chapter 3). Based on Hamilton's principle, the dike-fracture seeks to be everywhere **parallel** with σ_1 so as to minimise the energy used for the propagation. At the same time, the dike-fracture seeks to minimise the **duration** of the propagation, that is, the time needed to propagate from t_1 to t_2. The propagation velocity of dikes varies somewhat, primarily between about 0.1 m s^{-1} and 1 m s^{-1}. Assuming a typical **velocity** of dike propagation of 0.5 m s^{-1}, it follows that minimising the duration implies **minimising the path length** for the given constraints.

Thus, of all the possible paths following the trajectories of σ_1, the shortest one from t_1 to t_2 would normally be selected. To see how this works out in practice, consider some simple layered numerical models (Fig. 10.12). I have indicated several paths (out of very many possible paths), and also the path most likely to be actually followed by the dike, namely the one in the centre – the shortest path (Fig. 10.12). Here I show feeder-dikes, but if the dikes become arrested, then the location of t_2 will simply be at the point of arrest, commonly at a contact with one of the layers, rather than at the surface.

This numerical model is very simple: it is a two-dimensional model where the stiffnesses of the layers increase gradually with depth (Fig. 10.12). Despite being simple, the model illustrates that the least action or minimum potential energy principles above can be used, in principle, to **forecast likely dike-propagation paths** during unrest periods with dike

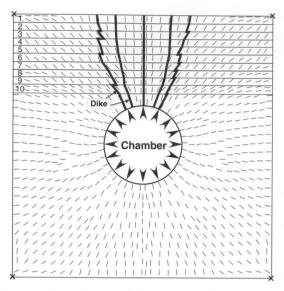

Fig. 10.12　Dike-propagation paths in a layered crustal segment. This is an example of a very simple layering, where there are 10 layers above the layer or unit hosting the chamber. Top layer (layer 1) has a Young's modulus of 10 GPa, and then the stiffness of the layers increases gradually with depth. The increase is 2 GPa for each layer so that layer 10 has a Young's modulus of 28 GPa and the layer hosting the chamber, a Young's modulus of 30 GPa (Gudmundsson and Brenner, 2004). This crustal segment is thus approaching homogenisation as regards mechanical layering, as some segments do when they become older (cf. Fig. 7.16). Thus, the three potential dike paths above the central part of the chamber roof are comparatively smooth, whereas the two outermost two paths show greater variation in geometry and overall length from the source to the surface. Hamilton's principle implies that the paths of least action would be somewhere above the central part of the chamber.

injection. Further illustration of these principles in forecasting potential dike-propagation paths in layered rocks is provided by numerical models in the earlier chapters (Figs. 6.10, 7.6, 7.9, 7.10). Some of these models are more complex than the model in Fig. 10.12 – particularly the models in Figs. 6.10 and 7.10 – but they demonstrate the same energy principles on which the dike forecast rests. Such forecasts, even if crude at this stage, are of great theoretical and applied importance because they allow us to predict the likely paths – say in which direction within a large stratovolcano a dike is likely to propagate, many hours and perhaps days and weeks before the actual propagation is completed. Furthermore, the predicted path can be compared with the **actual path**, as determined by the associated earthquake swarm (Chapter 4), almost in real time.

These forecasts, however, only consider the effects of layering on the dike paths. In addition to layering, all volcanoes and volcanic zones contain numerous fractures of varying sizes. The cooling (columnar) joints are used to form the paths (Gudmundsson, 1986), but these are mostly evenly distributed in the pile and therefore do not encourage the dike to deviate from the path parallel with σ_1. Field observations show, however, that some dikes use faults as parts of their paths, and those parts are normally not parallel with σ_1. Thus, it is time to have a look at the conditions whereby less energy is required (and thus in agreement with Hamilton's principle) for a dike path to follow a fault (and be oblique to σ_1) than to follow the trajectories of σ_1.

10.5.6 Propagation Paths: Effects of Faults

If the tensile strength T_0 (Eq. (10.2)) was uniform, then, as discussed above, the entire path of the dike would be expected to follow the trajectories of σ_1. All rocks, however, contain fractures, most commonly joints – particularly columnar joints. The columnar joints are extension fractures and are roughly uniformly distributed in the lava flows that form the bulk of the host rock for most dikes. Such joints do not normally result in a significant deviation of the dike path from the direction of σ_1. Some dikes (and inclined sheets), however, **use faults** for parts of their paths (Fig. 10.13; Gudmundsson, 1983). Perhaps the best-known examples are **ring-dikes** (e.g. Figs., 5.21, 5.22, 5.24, 8.16, 9.22; Chapters 5, 8, 9). Faults, by definition, are shear fractures and their dip is **oblique** to that of σ_1 (at the time of fault formation or slip) or any of the principal stresses. The **tensile strength** across active or recently active faults is essentially **zero.** Less energy may thus be needed for a dike or a dike segment to use the fault, even though the segment is then not perpendicular to σ_3 but rather to the normal stress on the fault plane σ_n, which by definition is always higher than σ_3 (σ_n is equal in magnitude to σ_3 only for extension fractures, in which case σ_n and σ_3 coincide).

For a dike segment following a path that is perpendicular to σ_n, the effective overpressure available to drive the dike segment open is no longer the one given by Eq. (10.3), because that equation assumes that the dike segment to be perpendicular to σ_3. The propagation criterion therefore needs to be modified to assess under what conditions the energy needed

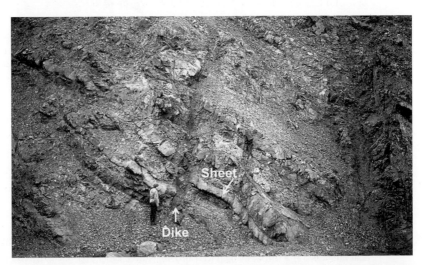

Fig. 10.13   Some dikes follow faults for a while along their paths, particularly steeply dipping normal faults (cf. Fig. 10.14) such as the dike here (indicated). The fault dissects a swarm of gently dipping inclined sheets (one sheet is indicated). The structures seen here are a part of the fossil Reykjadalur Volcano in West Iceland (Gautneb and Gudmundsson, 1992).

to propagate a dike segment along an existing fault is less than that needed to propagate the dike segment along an extension (mode I) fracture formed directly by the magmatic overpressure (but often using existing joints as weaknesses).

The normal stress σ_n on an inclined fracture is given by:

$$\sigma_n = \frac{\sigma_1 + \sigma_3}{2} - \frac{\sigma_1 - \sigma_3}{2}\cos 2\alpha \qquad (10.24)$$

where σ_1 and σ_3 are the maximum and minimum compressive principal stresses, respectively, and α is the angle between the fracture plane and the direction of σ_1 (e.g. Gudmundsson, 2011a). The difference between the normal stress σ_n, given by Eq. (10.24), and the minimum compressive principal stress σ_3 is then:

$$\sigma_n - \sigma_3 = \frac{\sigma_d}{2}(1 - \cos 2\alpha) \qquad (10.25)$$

where $\sigma_d = (\sigma_1 - \sigma_3)$, as defined in Eq. (10.3).

Using Eqs. (10.1) and (10.3), it can be shown under which mechanical condition a propagating dike is likely to use a fault as a part of its path (Fig. 10.14). This condition

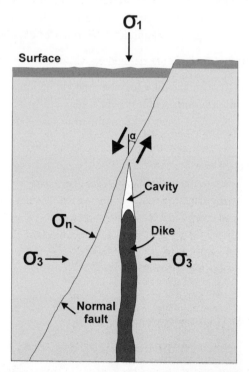

Fig. 10.14   When a propagating dike meets a steeply dipping fault, such as the normal fault seen here, the dike may enter the fault and use it as a part of its path (cf. Fig. 7.14). The conditions for this to happen follow from Eqs. (10.26)–(10.28). Here, σ_1 and σ_3 are the maximum and minimum principal stresses, respectively, σ_n is the normal stress on the fault plane, and a is the acute angle between σ_1 and the fault plane. Dike propagates in steps (Fig. 10.9) and during each step the fracture front propagates much faster than the fluid front, generating a temporary empty (air- or gas-filled) cavity at the dike tip, a cavity which the magma subsequently flows into (cf. Fig. 4.10).

can be stated thus. A dike is likely to **inject an existing fault** and use it as a part of the dike path if the following condition holds:

$$\sigma_n - \sigma_3 = \frac{\sigma_d}{2}(1 - \cos 2\alpha) \leq \Delta T_0 \qquad (10.26)$$

where

$$\Delta T_0 = T_0^{\sigma_3} - T_0^{\sigma_n} \qquad (10.27)$$

Here, ΔT_0 is the difference between the tensile strength along the path that is perpendicular to σ_3 and the path that is perpendicular to σ_n. If the tensile strength along the path perpendicular to σ_n is zero, then:

$$\Delta T_0 = T_0^{\sigma_3} \qquad (10.28)$$

For an active, or recently active, fault, the condition of Eq. (10.28) often applies.

The **overpressure** given by Eq. (10.3) assumes that the dike segment opens up against σ_3. When the opening of the dike segment is against σ_n, the available overpressure is less than that given by Eq. (10.3). Using Eq. (10.3) and Eq. (10.26) the magmatic overpressure in a dike segment with reference to the normal stress σ_n on the segment, denoted by $p_o^{\sigma_n}$, is given by:

$$p_o^{\sigma_n} = p_e + (\rho_r - \rho_m)gh + \frac{\sigma_d}{2}(1 + \cos 2\alpha) \qquad (10.29)$$

where all the symbols are defined above (notice the difference in sign for $\cos 2a$ between Eqs. (10.29) and (10.25)). The difference in overpressure between Eq. (10.3) and Eq. (10.29) is in the last term on the right-hand side. In Eq. (10.3) that term is σ_d whereas in Eq. (10.29) the term is $\frac{1}{2}\sigma_d(1 + \cos 2\alpha)$. So long as α is positive, so that $\sigma_n \neq \sigma_3$, then $P_o^{\sigma_n} > p_o$, that is, the overpressure in Eq. (10.29) is less than that in Eq. (10.3).

Let us now estimate the energy needed to form a dike segment in a direction perpendicular to σ_n. These energies are, first, the surface energy W_s (Eq. (10.16)) for rupturing the rock and, secondly, the energy needed to open up the dike-fracture. The total energies required to rupture the rock and open up a dike segment perpendicular to σ_n and σ_3 can now be stated and compared. Recall that **work** is force × displacement and it is positive if in the direction of the force, but negative if in a direction opposite to the force. For a dike segment with a final **volume** of ΔV_v, the work W needed to form a dike segment of that volume against the normal stress σ_n is:

$$W = \Delta V_v \sigma_n \qquad (10.30)$$

and similarly for the work against σ_3:

$$W = \Delta V_v \sigma_3 \qquad (10.31)$$

The force or overpressure that opens the dike segment varies linearly with the opening displacement so that the corresponding elastic energy must be multiplied by ½. It follows that the **total energy** $U_t^{\sigma_n}$ needed to form the dike-fracture, that is, rupture the rock and then open the rupture/fracture up against σ_n is given by:

$$U_t^{\sigma_n} = W_s + \frac{\sigma_n \Delta V_v}{2} \qquad (10.32)$$

where W_s is the surface energy (Eq. (10.16)) and the other symbols are defined above. A similar equation is obtained for the elastic energy needed to form and open the dike segment against σ_3, namely:

$$U_t^{\sigma_3} = W_s + \frac{\sigma_3 \Delta V_v}{2} \tag{10.33}$$

If the surface energy of rupture W_s is constant for the host rock of a given dike segment then, because $\sigma_3 > \sigma_n$, it follows from comparison between Eq. (10.32) and Eq. (10.33) that less energy is needed to form a segment perpendicular to σ_3 than to σ_n. Thus, as expected, the dike tends to follow the path perpendicular to σ_3, namely the path parallel with σ_1. While the surface energy W_s may be roughly constant at the atomic level, it may not be so at the scale of joints or other discontinuities used by the dike while forming its path. However, such differences in W_s as exist at that level may be regarded as being included in the difference in tensile strength at the same level.

If the tensile strength varies considerably, in particular if the tensile strength across a fault is zero while being several mega-pascals along the path of σ_1 at a given locality along the potential path of the dike, then Eqs. (10.26)–(10.28) suggest that less energy – less action for a dike segment of a given length – is needed for the segment to **follow the existing fault** than to form its own path along σ_1. More specifically, a dike may follow an existing fault along part of its path (Figs. 10.13 and 10.14) if the difference $\sigma_n - \sigma_3$ (Eq. (10.26)) is less than the difference in the tensile strength (Eq. (10.27)) between the rock along σ_1 and along the fault. The stress difference $\sigma_n - \sigma_3$ depends much on the angle α between the fault plane and σ_1 (Eq. (10.25)); the smaller the angle α, the more likely it is that the fault plane will be entered by a dike that meets it. In particular, for normal faults, as are most common in volcanic zones with dike injection, a small angle α means steeply dipping faults, which are the most likely to be used as parts of dike paths (Examples 10.11–10.13).

10.6 Dike-Induced Deformation

As discussed earlier (Chapters 1 and 3), dike-induced surface deformation is routinely recorded through GPS, InSAR, and other geodetic techniques. The results are used to assess volcanic hazards and estimate the geometries of dikes (including their volumes) injected from magma chambers/reservoirs during unrest periods. More specifically, geodetic studies together with seismic studies are the principal methods for estimating the volumes of dikes injected during unrest periods, and thereby the volume of magma supposed to flow out of the chamber/reservoir following its rupture. In addition, deformation studies help determine the dike paths and assess the likelihood of dike-fed eruptions.

Following earlier general **analytical studies** of the deformation and stresses at the surface of an elastic half-space above an elliptical crack (Isida, 1955; Tsuchida and Nakahara, 1970) there have been many **numerical** studies focusing on dikes in an elastic half-space. These include studies by Pollard et al. (1983), Davis (1983), Rubin and Pollard (1988), and Cayol and Cornet (1998). A very different approach is to model dikes as elastic

dislocations, which has also been applied when using (inverting) surface geodetic data to infer the opening/thickness, strike, dip, and depth of dikes, inclined sheets, and sills (Chapter 3). The application of dislocation theory to volcano deformation is reviewed in detail by Okada (1985, 1992), Dzurisin (2006), and Segall (2010).

The numerical and dislocation models in cited publications assume that the volcanic zone/volcano hosting the dike is acting as a homogeneous, isotropic, **elastic half-space**. This means that the models do not consider the effects of mechanical layering or contacts between layers on the dike-induced stresses and deformation. The surface deformation in layered elastic host rocks, however, has been considered by Roth (1993) and by Bonafede and Rivalta (1999a, b). Dislocation and numerical models of dike-induced deformation are reviewed by Rivalta et al. (2015) and by Townsend et al. (2017).

With a few exceptions, all these models focus on the **surface displacement** induced by the dike. Only a few studies have combined field observations of arrested dikes in layered rocks with numerical models of the local stresses induced by the dike in the layers **adjacent to the dike** as well as at the surface. These studies include Gudmundsson and Brenner (2001), Gudmundsson (2003), Gudmundsson and Loetveit (2005), and Philipp et al. (2013). In contrast to the present study, however, none of these studies focuses on the changes in the dike-induced stresses and displacements/deformation inside the volcano/volcanic zone as a function of the variation in the elastic properties of the mechanical layering.

In order to understand dike-propagation paths and, in particular, the conditions for dike arrest, which are of fundamental importance in hazard assessments during unrest periods, we must be able to calculate the stresses and displacements induced by a dike **inside** the crustal segment and volcano during dike emplacement. Here, I provide numerical-model results not only on dike-induced stresses and displacements (or deformation) at the surface but also on the associated **internal stresses and displacements.** The dike-induced stresses and displacements are analysed as functions of variation in the mechanical properties, that is, in the Young's modulus of the layers/units that constitute the host rock (Bazargan and Gudmundsson, 2019). In the models, the dike tip is arrested at a contact between layers at 0.5 km depth below the surface of the volcano/volcanic zone. Above the layer or unit hosting the dike there are four layers of equal thickness, each 100 m. In the models, the Young's modulus of the fourth layer below the surface (the one adjacent to the layer or unit hosting the dike) is varied from 10 GPa to 0.01 GPa, while all the other layers/units have constant Young's moduli in the model runs (Fig. 10.15). The Poisson's ratio of all the layers is the same, 0.25, and the only loading is a magmatic overpressure in the dike of 5 MPa.

The **surface displacements** (upwards and downwards) are important for understanding better the location, geometry, and propagation paths of associated dikes. For understanding and forecasting dike-induced surface fracturing, however, the surface stresses are of main concern. This follows because tension fractures and faults form at the surface – or elsewhere – only if the local stresses reach certain magnitudes. These magnitudes are well known. For tension fractures, the local absolute tensile stress must normally reach at least 2–4 MPa, the most common tensile strength of rocks. Similarly, for faults – here in particular normal faults and grabens – the shear strength is normally about double the tensile strength, so commonly 4–8 MPa, which is, indeed, similar to common stress drops in earthquakes (Kanamori and Anderson, 1975; Scholz, 1990).

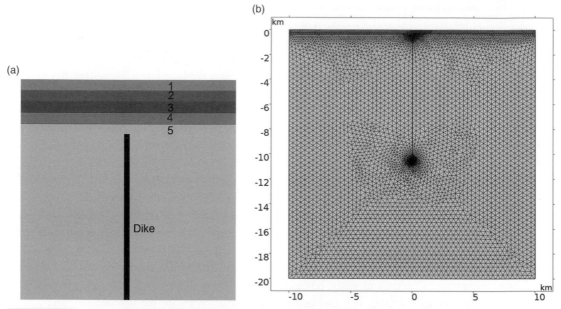

Fig. 10.15 (a) Schematic illustration of the set-up of the numerical model of the dike with the induced stresses and displacements presented in Figs. 10.16–10.20. The dike is vertical with a tip or top arrested at 0.5 km below the free surface of the volcanic zone/volcano within which the dike is emplaced. The model itself is 20 km × 20 km in size (the whole model is shown in (b)). Each of the four layers above the dike tip (layers 1–4) is 100 m thick, whereas layer 5, the unit hosting the dike, extends to the lower tip of the dike (arbitrarily the dike dip-dimension or height is set at 10 km). The indicated dike thickness is not to scale (it is far too thick in comparison with the thickness of the crustal layers). The actual dike thickness depends on the overpressure used, the dike dimensions, and the Young's modulus of the host rock (there is greater dike thickness for given overpressure and dimensions in more compliant layers or units). In the model the entire dike is located in the comparatively stiff unit/layer 5 (with a Young's modulus of 40 GPa). Simple fracture-mechanics models (Gudmundsson, 2011a; Becerril et al., 2013) indicate a model dike thickness of about 2.3 m for an overpressure of 5 MPa and a thickness of 6.9 m for an overpressure of 15 MPa, thicknesses that are similar to common dikes observed in the field (Chapter 2). (b) Set-up of the numerical model, shown here as an example of a typical two-dimensional finite-element model. The 10-km-tall dike is the black, vertical line in the central upper part of the model (with dimensions of 20 km × 20 km, as indicated above). The total number of elements in the model is 12 981 (cf. Bazargan and Gudmundsson, 2019).

The models (Figs. 10.15–10.19) show that as the fourth layer becomes softer or more compliant (0.1 GPa) dike-induced stresses and displacements (lateral and vertical) in the layers/units above the fourth layer, including those at the surface, become **suppressed**. At the same time, the stresses and displacements of the layer/unit hosting the dike (layer 5; cf. Fig. 10.15a) increase and their peaks do not coincide in location or magnitude with those of the other layers. Thus, the dike-induced **internal deformation** of the volcano/volcanic zone **increases** as the fourth layer becomes softer.

The model results also show that the tensile- and shear-stress peaks at the surface occur at locations **widely different** from those of maximum surface uplift. More specifically, for

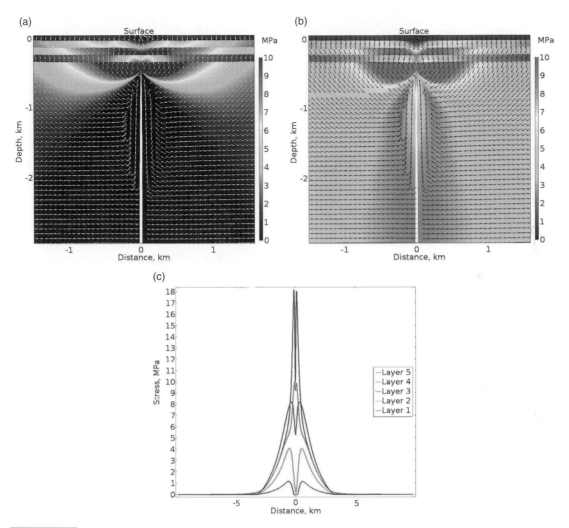

Fig. 10.16 Numerical model results with the following Young's moduli: 3 GPa (layer 1), 20 GPa (layer 2), 30 GPa (layer 3), 10 GPa (layer 4), and 40 GPa (layer 5), all the layers having a Poisson's ratio of 0.25. Overpressure of dike: 5 MPa. (a) Contours of the maximum tensile principal stress σ_3 in mega-pascals (red highest stress, blue lowest), with white arrows (ticks) indicating the direction or trajectories of the maximum compressive principal stress σ_1. (b) Contours of the von Mises shear stress τ in mega-pascals. (c) Plots of the variation in the magnitude of the maximum tensile principal stress σ_3 (in mega-pascals) at the contacts between the layers. Layer 5 denotes the contact between layer 5 and 4; layer 4, that between layer 4 and 3; layer 3, that between layer 3 and 2; layer 2, that between layer 3 and 2; and layer 1, the contact between layer 1 and the atmosphere, that is, the free surface of the volcanic zone/volcano (Bazargan and Gudmundsson, 2019). A black and white version of this figure will appear in some formats. For the colour version, please refer to the plate section.

a comparatively stiff fourth layer (1–10 GPa), the surface tensile and shear stresses peak at lateral distances of 0.5–0.7 km from the projection of the dike to the surface. (Essentially, no tensile/shear stresses reach the surface when the fourth layer is as soft as 0.1–0.01 GPa,

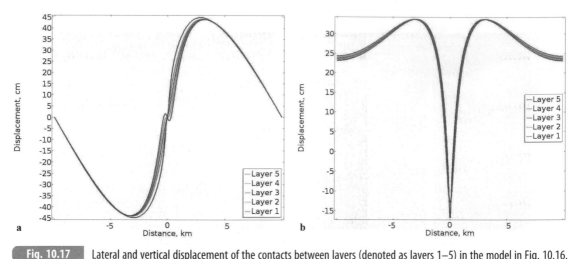

Fig. 10.17 Lateral and vertical displacement of the contacts between layers (denoted as layers 1–5) in the model in Fig. 10.16. (a) Here the lateral displacements at all the contacts, including the surface (layer 1), are similar and reach a maximum of 44 cm at a horizontal distance of about 3 km from the dike (the dike projection to the surface is located at 0 km). Displacements to the right of the dike are regarded arbitrarily as positive whereas those to the left of the dike as negative. (b) All the vertical displacements at the contacts are here also similar. The maximum uplift or vertical displacement of the contacts is about 34 cm and occurs at a horizontal distance of about 3 km on either side of the dike. Right above the tip of the dike there is a general subsidence of 12 cm for all the contacts except for the contact between layers 4 and 5 (denoted as layer 5), where the subsidence reaches 16 cm. A black and white version of this figure will appear in some formats. For the colour version, please refer to the plate section.

so that there are no stress peaks. This is seen in Fig. 10.18c where the fourth layer has a stiffness of 0.1 GPa.) By contrast, the maximum surface displacements (uplift) peak at lateral distances of 2.8–3.3 km from the dike projection to the surface (Bazargan and Gudmundsson, 2019). Thus, the lateral **distance** between the two peaks of the surface tensile (and shear) stresses is much less than that between the two peaks of the surface uplift. The models clearly show that the maximum surface uplifts, the **uplift peaks**, do not show high tensile or shear stresses. This is understandable because the strain associated with uplift of a fraction of a metre over distances of kilometres to tens of kilometres is of the order of 10^{-4} to 10^{-5} and thus too small to result in high stress for the given Young's modulus of the surface layer of 3 GPa. If tension fractures or faults – in particular the boundary faults of a **graben** – are induced by the dike, they should therefore form at the tensile/shear **stress peaks** and not, as is commonly suggested, at the location of the surface displacement peaks

That the maximum tensile and shear stress occur at a lateral distance from the projection of the dike, which is similar to the depth to the tip of the dike, is in accordance with the well-known '**graben rule**' (Fig. 5.9) The disturbance or the source of a graben is commonly seen in experiments to be at a depth similar to that of the half-width of the graben which, in the present case, is the lateral distance between the projection of the dike to the surface and the peak stress. These results are in agreement with other recent results (Chapter 3) that any

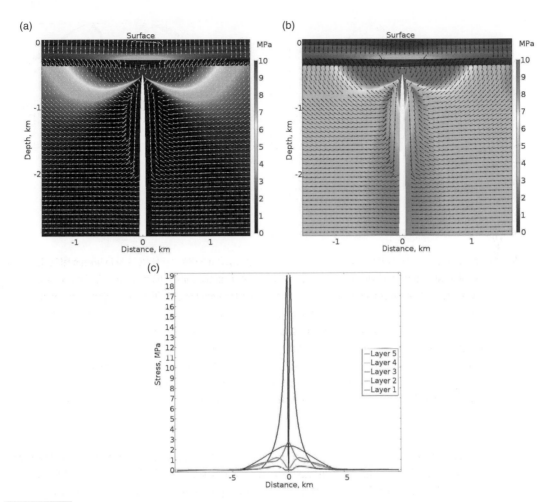

Fig. 10.18 The same model as in Fig. 10.16 except that layer 4 has here a Young's modulus of 0.1 GPa. (a) Contours of the maximum tensile principal stress σ_3 (red, highest stress; blue, lowest), with white arrows (ticks) indicating the direction or trajectories of the maximum compressive principal stress σ_1. (b) Contours of the von Mises shear stress τ. (c) Plots of the variation in the magnitude of the maximum tensile principal stress σ_3 at the contacts between the layers. Layer 5 denotes the contact between layer 5 and 4; layer 4, that between layer 4 and 3; layer 3, that between layer 3 and 2; layer 2, that between layer 3 and 2; and layer 1, the contact between layer 1 and the atmosphere, that is, the free surface of the volcanic zone/volcano. A black and white version of this figure will appear in some formats. For the colour version, please refer to the plate section.

tension fractures or normal faults induced by arrested dikes will tend to form within a zone of a width similar to that of the depth to the tip of the dike (Al Shehri and Gudmundsson, 2018). In particular, any dike-induced graben will likely be of a **width about twice the depth** to the tip of the arrested dike. These findings thus do not lend any support to the common practice of interpreting the boundary faults of grabens as coinciding with the maximum uplift (the upward displacement peaks) at the surface above an arrested dike.

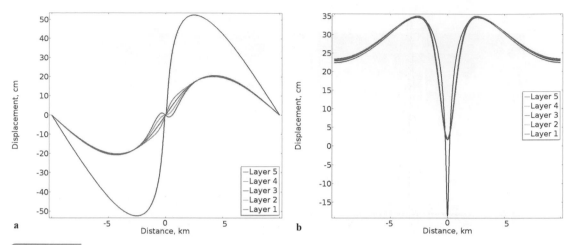

Fig. 10.19 Lateral and vertical displacement of the contacts between layers (denoted as layers 1–5) in the model in Fig. 10.18. Here, the lateral and vertical displacements of the layers/contacts are increasingly different close to the dike. (a) The maximum lateral displacements are still similar for layers/contacts 1–4, about 20 cm at a distance of about 4.3 km from the dike. The maximum lateral displacement of layer 5 (contact 5/4), however, reaches a maximum of about 52 cm at about 3 km from the dike. Displacements to the right of the dike are regarded arbitrarily as positive whereas those to the left of the dike as negative. (b) The maximum upward displacements of all the layers roughly coincide and reach about 34 cm at a distance of about 2.8 km from the dike. Here, however, layers/contacts 1–4 do not show any absolute subsidence above the dike tip, while layer 5 (contact 4/5) shows an absolute maximum vertical subsidence of about 18 cm (Bazargan and Gudmundsson, 2019). A black and white version of this figure will appear in some formats. For the colour version, please refer to the plate section.

The **only exception** to the 'graben rule' and its relation to the depth of the arrested dike tip is if the contact arresting the dike also opens up through Cook–Gordon delamination or debonding (Gudmundsson, 2011a,b). Contact delamination may happen at very shallow crustal depths, and has been modelled for arrested dikes (Fig. 7.22; Gudmundsson, 2003, 2011b). The main result of the modelling is that for delamination the stresses at the surface normally peak above the lateral ends of the opened-up contact. It follows that the double-stress peak does not have any correlation with the depth to arrested tip of the dike (Fig. 7.22). In the models presented here, however, the contacts are mechanically too strong to open up.

In the models above, the magmatic overpressure or driving pressure p_o is 5 MPa and the surface layer has a Young's modulus of 3 GPa. To test the **maximum** possible dike-induced surface effects, a surface layer with the very high stiffness of 20 GPa and an overpressure of 15 MPa was used. The models were run for the two cases of greatest importance, namely layer 4 with a stiffness of 0.1 GPa and stiffness of 0.01 GPa. The main aim was to see how much the surface stresses would increase for high overpressure and very high surface stiffness in a volcanic zone/volcano with one or more soft layers (layer 4).

For layer 4 with stiffness of 0.1 GPa, and overpressure of 15 MPa, and a surface stiffness of 20 GPa, the surface tensile and shear stresses reach about 5 MPa. (No stress plots for this model are provided here, but such plots are in Fig. 15 in Bazargan and Gudmundsson, 2019).Thus tension-fracture and normal-fault formation (including graben formation) would be possible

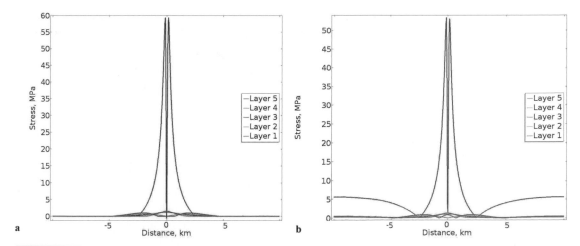

Fig. 10.20 Same basic model as in Fig. 10.16, with the following differences: layer 4 has a Young's modulus of 0.01 GPa (very soft/compliant), the surface layer (layer 1) has a very high stiffness of 20 GPa, and the magmatic overpressure in the dike is 15 MPa (rather than 5 MPa). (a) Plots of the variation in the magnitude of the maximum tensile principal stress σ_3 at the contacts between the layers. (b) Plots of the variation in the magnitude of the von Mises shear stress τ at the contacts between the layers. Layer 5 denotes the contact between layer 5 and 4; layer 4, that between layer 4 and 3; layer 3, that between layer 3 and 2; layer 2, that between layer 3 and 2; and layer 1, the contact between layer 1 and the atmosphere, that is, the free surface of the volcanic zone/volcano (Bazargan and Gudmundsson, 2019). A black and white version of this figure will appear in some formats. For the colour version, please refer to the plate section.

under these conditions. For this very great surface stiffness, the lateral distance between the stress peaks is about 2.2 km, rather than about 1.4 km so that the graben, if formed, would be somewhat wider than in the models with 3 GPa surface-layer stiffness. The results mean that for a dike with an arrested tip at 0.5 km below a very stiff surface of 20 GPa and with the high overpressure of 15 MPa, normal faulting and tension fracturing is just possible.

For layer 4 with stiffness of 0.01 GPa, an overpressure of 15 MPa, and a surface stiffness of 20 GPa, the results show that the surface tensile and shear stresses reach only a fraction of a mega-pascal (Fig. 10.20). Thus, even an unusually high surface-layer stiffness and overpressure for an arrested dike with a tip at the shallow depth of 0.5 km, the stresses at the surface are still far too small to generate either tension fractures or normal faults. The results thus underline the great effects of soft layers in reducing dike-inducing stresses and in suppressing surface stresses and fracture formation.

Generally, the models indicate that for 5 MPa overpressure little or no dike-induced surface deformation would be expected unless the dike tip propagates to depths below the surface of less than a kilometre, and commonly to depths of only **several hundred metres**. The model results thus suggest that when significant dike-induced deformation is seen at the surface, the dike is normally very close to the surface – within several hundred metres of the surface – indicating a high likelihood of the dike **reaching the surface** to erupt.

Elastic half-space models normally **overestimate** the potential dike-induced surface stresses and, consequently, the depth of the tip of the associated dike below the surface. Half-space models also commonly underestimate the dimensions – particularly the thickness – and therefore the volumes of the dike inducing a particular surface deformation during a volcanotectonic episode (e.g. Al Shehri and Gudmundsson, 2018). When the dike volumes are underestimated, the flow of magma out of the source chamber during the episode is also underestimated. It follows that, for an eruption, the combined intrusive and extrusive (erupted) volume of magma injected from the chamber is then underestimated. Since this combined volume is one of the factors used to calculate the volume of the source magma chamber (Chapter 6; Gudmundsson, 2016), that volume, and hence the likely dimensions of the chamber, will also be in error.

In summary, the model results show that soft or compliant layers/units, such as are very common in volcanic edifices and zones, make dike-induced surface fracturing unlikely until the dike is at a **very shallow depth**, with a very high overpressure, or both. This is in agreement with the field observations of arrested dike tips, showing that many dikes arrested at shallow depths did not generate tension fractures or normal faults above their tips (Chapter 7). More specifically, the results suggest that the common use of homogeneous, elastic half-space dislocation models to infer dike geometries and dimensions through the inversion of surface-deformation data is likely to lead to **erroneous results**. Furthermore, the present models indicate that the common practice of associating the calculated maximum theoretical surface displacements with graben formation is not justified.

10.7   Forecasting Large Eruptions

Earth is the home to humankind. If humankind survives as a civilised high-technology society for hundreds of years, space settlements are likely to happen. These can take various forms, including human-made space colonies, settlements on the Moon (and possibly some satellites of other planets in our solar system) and Mars. In the far future there may be settlements on suitable planets in other solar systems. For the coming centuries, however, Earth is likely to be humankind's main, and perhaps only, home.

Risks to human life on Earth are many. The most severe ones are named catastrophic risks, of which the completely devastating ones are existential risks (Bostrom and Cirkovic, 2008). The difference is that a **catastrophic risk** is an event or scenario that could result in global damage to humankind with the potential of destroying modern civilisation, whereas an **existential risk** is an event or scenario that could result in the extinction of humankind. There are many potential human-related (anthropogenic) risks, such as human-made climate change, nuclear wars, and technological accidents. Natural (non-anthropogenic) risks include natural climate change (e.g. ice ages), astronomical threats (gamma rays from nearby supernovas, changes in the orbits of nearby planets), geomagnetic reversals (when Earth may be less protected from solar radiation), asteroid, comet, and meteoritic impacts, and very large volcanic eruptions. The focus here is on the last one, namely very large eruptions: what effects they can have and, in particular, if and how we could prevent them from happening.

10.7.1 Quantifying Eruptions

The eruptions that pose catastrophic – even existential – risk are of two types: explosive eruptions and effusive eruptions. All these eruptions have volumes of the order of hundreds of cubic kilometres or more. Sometimes a distinction is made in that volcanoes capable of producing eruptive materials in excess of 1000 km^3 (solid-rock volume) in a single eruption are referred to as **supervolcanoes.** There is, however, no basic difference between the mechanics of eruptions that produce tens or hundreds of cubic kilometres of eruptive materials and those that produce thousands of cubic kilometres. Most or all large eruptions – those in excess of 10 km^3 (Chapter 8) – require a squeezing out of the magma in the chamber/reservoir. The squeezing is presumably mostly the result of subsidence of the roof of the chamber/reservoir, either through caldera collapse or rapid graben subsidence (Chapter 8). Thus, supervolcanoes are in this sense simply volcanoes with unusually large chambers/reservoirs.

Very large **explosive eruptions** have received more attention in connection with catastrophic risks than effusive eruptions (e.g. Self, 2006; Bostrom and Cirkovic, 2008; Crosweller et al., 2012; Rougier et al., 2018). This is partly because these eruptions are easier to quantify in terms of the commonly used **Volcanic Explosivity Index** (VEI), originally proposed by Newhall and Self (1982). This index combines the eruptive volume, the height of the eruption column, and various qualitative assessments to rank eruptions. It works best for historical eruptions where both the volume produced and the height of the eruption column are reasonably well known, but it has been used for prehistoric eruptions – some happening millions of years ago – where the only reasonably well-known factor is the eruptive volume. The VEI is, as the name implies, primarily a measure of the **explosivity** of an eruption, and has only a marginal relation to other quantitative measures of eruption size. Large effusive eruptions, for instance, do not normally score as high on the VEI as they would if the eruption volume was the main criteria for size (Pyle, 2000). Several other eruption magnitude scales have been developed, such as the **eruption intensity**, given as the mass of eruptive material per unit time, kg s^{-1} (Fedotov, 1985; Pyle, 2000), and the **thermal energy** released or transformed during the eruption (Yokoyama, 1957; Pyle, 2000) but these can only be used for historical (and preferably, well-documented) eruptions, and thus they are of limited use for very large eruptions, all of which are prehistoric.

A different approach to quantifying eruptions focuses on the **elastic energy** released or transformed during the eruption. Such a measure can be compared directly with the quantification of earthquakes through their moments (cf. Section 4.4). In fact, using this approach it has been shown that the energy released or transformed in the largest eruptions (explosive and effusive) is of the **same order of magnitude** as that released in the largest earthquakes, namely 10^{19} J (Gudmundsson, 2014, 2016).

In this approach, use is made of the fact that the volume change or shrinkage of the magma chamber during an eruption (and associated dike intrusion) is a measure of the elastic energy transformed. From Eq. (8.25) it follows that the volume of magma flowing out of the chamber/reservoir during an eruption and the associated intrusion is a measure of the energy transformation. Therefore, the **elastic-energy magnitude (EEM)** scale uses

eruptive volumes as a measure of eruption magnitude. This is fortunate because for many prehistoric large eruptions the eruptive volume is reasonably well known. There are of course uncertainties or errors in the estimated volumes of all eruptions, and particularly those that occurred millions of years ago. For example, part of the eruptive material may have been eroded away and the volume of the intrusive material (the dike volume) may be poorly known, if at all.

The uncertainty or error in the estimated eruptive volume is, however, normally within a factor of two to three. Thus, even if the dike volume is unknown, that would be included in, and thus covered by, the factor of two to three. This follows because for a very large eruption, of the order of 10^2–10^3 km^3, the dike volume is normally less than 10% of the total volume of magma that leaves the chamber/reservoir. For these very large eruptions, the estimated eruptive volumes may thus be regarded as being of the correct order of magnitude – whether or not the intrusive (dike) volume is included. Eruptive volumes that are accurate within an order of magnitude yield the correct order of elastic energy transformed during the eruption. For very large eruptions, the **correct order of magnitude** for the eruptive volume and released energy is all that is needed to forecast their catastrophic risk.

The EEM scale has the general form (Gudmundsson, 2016):

$$M_e = A\log U_{er} - B \tag{10.34}$$

where M_e is the elastic-energy magnitude, U_{er} is the elastic potential energy of the eruption, log is the common logarithm (to the base of 10), and A and B are constants that must be determined from empirical data such as in Fig. 8.1. When the constants have been determined, the magnitude scale (Eq. (10.34)) can be applied to any eruption with an eruptive volume that is known to within an order of magnitude. Thus, this magnitude scale can be applied to pre-instrumental eruptions – roughly eruptions that happened more than a century ago – for which commonly the only quantitative information available is the eruptive volume. For some eruptions associated with caldera or graben subsidences, the subsidence volume can also be used as an additional constraint on the eruption volume (Chapter 8; Gudmundsson, 2016).

10.7.2 Forecasting Large Eruptions

Based on the theory presented in Chapter 8, large and very large eruptions happen only if special mechanical conditions are satisfied. The main conditions are subsidence of the crustal block that forms the roof of the chamber/reservoir so as to maintain the magmatic excess pressure during most of the eruption. As the crustal block subsides (Fig. 8.19), the volume of the chamber/reservoir decreases (shrinks) so that the excess pressure is maintained. Using this theory as a basis for forecasting large eruptions, the main points to consider are the following:

1. An unrest period is not the same as an eruption. More specifically, most unrest periods do not result in eruptions. General unrest periods have many causes, some of which are magma being added to a chamber/reservoir, whereas others are due to changes in the geothermal fields associated with the unrest volcano.

2. Magma-chamber/reservoir rupture is also not the same as an eruption. Rupture normally results in a dike (or sheet) injection. However, as we have learned in earlier chapters, the majority of dikes become arrested and never reach the surface to erupt.

3. Based on the present theory (Chapter 8; Gudmundsson, 2016), a large eruption normally requires the subsidence of the roof block into the chamber/reservoir so as to squeeze the magma out of the chamber. This means that the **conditions for subsidence** must be satisfied in order for the eruption to have the chance of becoming large. Since the subsidence occurs along **faults** – ring-faults for calderas and normal faults for grabens – it follows that the conditions imply **concentration of shear stress** at the appropriate locations.

4. Measurements of the concentration of shear stress can only be made directly through **stress monitoring**. Such monitoring is best made in drill holes with hydraulic fracturing, but is very expensive and applies primarily to the state of stress in the layer being fractured. In the future, many high-risk volcanoes will presumably have stress monitoring, some in real time, but this is not the situation as yet. However, there are currently many methods for indirect stress measurements (cf. Chapters 3 and 4).

5. Perhaps the best indication of shear-stress concentration, in the absence of direct *in situ* measurements, is earthquakes. If **earthquakes begin to concentrate** on existing boundary-faults of large calderas or grabens during volcanic unrest periods, that should be taken as an indication of a potential large eruption (Fig. 10.21).

6. Because stress concentration results in strain concentration and displacement fields, the location of concentrated surface displacement may also be used as a criterion for a potential large eruption. It should be noted, however, that the maximum surface displacement does not necessarily coincide with the maximum shear-stress concentration (Section 10.6).

7. Large-scale uplift or **doming** of the surface of a volcano/volcanic field is a measure of the **strain energy** stored in the associated crustal segment. This strain energy is mostly available to help squeeze out the magma during an eruption.

8. For deep-seated reservoirs, doming is presumably a good indication of the strain energy stored and largely available during dike propagation and eruption. For shallow magma chambers, however, the measured doming may represent only a part of the stored strain energy. For many shallow chambers, part of the expansion when new magma is added to the chamber – as is common before rupture and dike injection/eruption – is accommodated by the floor of the chamber. Thus, only part of the space needed for the chamber expansion is through upward displacement of the roof, and that part is reflected in doming of the surface, but the space generated through the downward displacement of the floor is normally not seen in any surface data.

9. The potential of a large eruption depends, of course, on the **size of the magma chamber/reservoir**. Few chambers, and presumably no reservoirs, are of volumes less than several tens of cubic kilometres. It follows that if available data during unrest indicate a stress concentration at caldera/graben boundary faults (or potential boundary faults), implying that subsidence of the roof is likely into a large chamber/reservoir during an eventual eruption, then a large eruption is probable.

Fig. 10.21　When, during an unrest period, earthquakes and deformation begin to concentrate close to or at a ring-fault (here within the zone between the two crude circles) or the boundary-faults of a graben, there are reasons to expect a major slip on the caldera/graben faults and possibly large eruptions. Here, the earthquakes within the zone in which the ring-fault is located are indicated schematically by crosses (cf. Fig. 8.19).

10. In summary, based on current monitoring technology, the best indication of a likely large eruption during a volcanic unrest period is that earthquakes become concentrated (Fig. 10.21) along actual or potential ring-faults (for calderas) and boundary faults (for grabens). The ring-fault is normally associated with a shallow magma chamber, whereas the graben would more commonly be associated with a deep-seated reservoir. Earthquake concentration at these faults suggests that large-scale subsidence is likely to happen during an eventual eruption. Such a subsidence helps squeeze out magma from the chamber/reservoir, thereby, commonly, generating a large or, depending on the size of the chamber/reservoir, a **very large eruption.**

Now that the monitoring aspects have been summarised, so that we may be able to assess the likelihood of – **forecast** – a large or a potentially very large eruption, the question arises: is there anything we can do to prevent such an eruption from happening? Do we have any methods of stopping a potentially large, and particularly very large, eruption?

10.8 Preventing Large Eruptions

A common view is that there is nothing which we can do to prevent large eruptions – or volcanic eruptions in general. Instead, the focus has been on how to mitigate their effects on human lives and properties. A primary aim of reliable forecasting of eruptions is thus to have time to **evacuate** people and animals from the relevant areas in the vicinity of a volcano – or areas that may possibly be subject to tsunamis – the unrest of which is interpreted as being very likely to result in an eruption. A secondary aim is to provide information to aviation authorities so as to change routes for airlines to avoid ash clouds and related aspects. For comparatively small lava flows, there have also been several attempts to divert the flow from towns and cities.

All these methods to mitigate the effects of eruptions are worthwhile, but unlikely to be of much use in very large eruptions. Eruptions with volumes of the order of hundreds or thousands of cubic kilometres potentially could affect such large areas – some the entire Earth – that mitigation methods of this kind are not likely to be of great value. The question then arises: are there any ways to prevent or stop large eruptions? Here, let us consider this question, previously hardly ever taken seriously, and discuss the possibilities.

There are potentially **several ways** of dealing with, and trying to prevent, large eruptions with currently available techniques. Here, two main methods are discussed. One is through cooling of the magma, either in the magma chamber/reservoir or in the propagating dike injected from the source during an unrest period. The second method aims at gradually decreasing or transforming the accumulated elastic energy in the volcano through slip on existing fractures/faults. This second method has a long history in **enhanced geothermal systems**, formerly referred to as hot-dry-rock systems, and it is commonly referred to as **hydro-shearing**. In the present context, the aim of the second method is to form a stress barrier or a seal in the roof close to the magma, which prevents magma-chamber rupture and dike injection; or, alternatively, if the hydro-shearing is carried out in a shallow layer or unit, to prevent an injected dike from reaching the surface. Both methods are discussed, cooling and hydro-shearing, but the focus is on the latter because it appears to be much more promising. I begin, however, by discussing the potential of cooling the magma by water.

10.8.1 Cooling of the Magma

The basic idea behind artificial cooling of magma in the chamber is that the main reason an eruption becomes large is that much elastic energy has been stored in the associated crustal segment. It is this energy that is primarily responsible for the high proportion of the magma being squeezed out of the magma chamber, and thus for as large an eruption as the chamber can possibly produce. If the elastic energy can be reduced, which means being transformed into other forms of energy, well before a large eruption is imminent, the resulting eruption, if any, would be significantly smaller. (The idea of cooling the entire magma chamber so as to reduce the magma available to erupt is not feasible because it would take such a long time, as discussed in connection with the elastic-energy reduction process below.)

The stored elastic energy at any time in the crustal segment hosting the source chamber/reservoir is primarily a function of the excess **magmatic pressure** in the source. This excess pressure, hence the elastic energy (the strain energy), can be reduced through artificial cooling of the magma.

Cooling of the Magma in the Source

When the magma cools, it shrinks – reduces in volume – and therefore, normally, its excess pressure decreases. When this decreases, the elastic energy in the hosting crustal segment also decreases, thereby lessening the energy available to squeeze out a large volume of magma (Chapter 8; Gudmundsson, 2016).

The magma in all source chambers/reservoirs acts as a heat source for the host rock and is thus gradually cooling all the time. It is only when new magma is received that the average temperature of the magma source may, temporarily, increase. Human-made or **artificial cooling** of the magma is thus just speeding up of an otherwise natural process. Exploitation of natural geothermal (hydrothermal) systems increases the geothermal gradient between the magma source and the surrounding rocks and thus speeds up the cooling of the magma. But the magma can also be cooled by **pumping water** directly into the source (the chamber/ reservoir). For large magma chambers, of the order of hundreds of cubic kilometres, cooling through any such process, natural or human-controlled, would normally take thousands of years. Based on the subsidence of collapse calderas during many caldera eruptions, and other considerations (Chapter 6), the magma sources may be up to tens of kilometres in lateral dimensions and several kilometres in thickness. From Eqs. (7.26) and (7.27) solidification of sill-like basaltic magma chambers with a thickness of many hundreds of metres, or a few kilometres, would take from thousands to **tens of thousands** of years. It is then assumed that the solidification range is from an original temperature of the magma of 1200 °C to a solidified magma at a temperature of 1000 °C.

These figures may vary. Both the original temperature and the solidification temperature may be higher or lower, depending on the magma type. Also, these are the times for solidification of the entire magma chamber, whereas contraction of the chamber, and therefore reduction in the stored strain energy in the surrounding rocks, begins as soon as cooling starts. However, much contraction occurs during the development of cooling or columnar joints, and these form at lower temperatures than 1000 °C, commonly at 700–800 °C for basaltic magma (Chapter 7). Also, in order that contraction due to cooling of the magma is effective in reducing the strain energy in the host rock, the magma source should not receive any new magma. Most active magma chambers/reservoir will receive additional magma over periods of thousands or tens of thousands of years, which would then counteract the reduction in strain energy due to cooling of the existing magma. Cooling by natural or human-made processes is such a slow process that it is **unlikely to be effective** in reducing the stored strain energy in the host rock so as to lower the risk of a large eruption from the magma source.

Cooling of Magma in a Propagating Dike

Once a magma chamber/reservoir ruptures and injects a dike, there are several possibilities for stopping the dike propagation to the surface through human activities. The main method, discussed below, is through hydro-shearing of layers/units on the potential path of the dike to the surface. Other methods that come to mind include cooling of the propagating dike. **Rapid drilling into the dike**, somewhere along its path, and pumping water into the segment so as to solidify the magma may be feasible in some instances. The segment selected for pumping into need not be close to or at the present top of the dike. This follows because solidifying the magma in any dike segment through which magma flows to the dike top (tip) will, theoretically, cut off the flow to the top and stop its propagation. The aim would then be that the rapid cooling would increase the magma viscosity of the cooled segment and eventually stop the flow of magma through that segment and thus the dike propagation.

While this is possible in theory, it is not likely to be of much use in practice. If the flowing magma in a propagating dike meets a barrier of any kind – including a partly or totally solidified part of the dike itself – the magma will simply find another path so long as the dike magma has any significant overpressure. Thus, the dike would simply become somewhat offset across the cooled part, and continue its propagation (Fig. 10.22). The same applies when trying to **solidify its actual tip**. Rapid (natural) solidification of a (commonly thin) dike tip has been suggested as a means of arresting the dike. The tensile strength of a solidified dike tip, however, is similar to that of the solidified host rock. Thus, if the overpressure in the dike is high enough to propagate it through the host rock in the first place, it would normally be high enough to propagate it through a solidified tip. Alternatively, if the solidified tip offers greater resistance to dike propagation than the surrounding host rock, then the dike simply leaves the tip as a horn, becomes somewhat offset, and continues its propagation along a nearby path – a commonly observed process during dike emplacement (Fig. 10.22; cf. Figs. 2.26 and 2.27).

Artificial rapid solidification of the tip segment or other segments of a propagating dike is thus unlikely to arrest the propagation path so long as the magma has sufficient overpressure to rupture the host rock. It is generally only when the overpressure has become effectively zero with respect to the local stress field at the tip that the tip becomes arrested, and, most commonly, in vertical sections, at contacts between mechanically dissimilar layers (Chapters 2 and 7).

10.8.2 Energy Release and Stress-Barrier Formation through Hydro-Shearing

Hydraulic Fracturing

When the stored strain energy reaches a critical level, it generates fractures. Many of the fractures, in turn, cause earthquakes. These aspects have been discussed in detail in Chapter 4. One way by which stored strain energy due to magma chamber/reservoir inflation can be reduced is through fracturing of the rock. Much of the fracturing results in earthquakes, so that **earthquakes** are an indication of strain-energy reduction or release.

Earthquakes occur in all crustal segments hosting active magma chambers/reservoirs. In fact, earthquakes are one of the main indicators of volcanic unrest, which normally implies magma-source expansion and associated **inflation**. Thus, natural fracturing and associated earthquakes are a means of reducing stored strain energy.

Human-made (artificial) fracturing can provide an additional reduction in stored strain energy. There are two basic techniques used to generate fractures and increase the permeability in rock bodies. One is hydraulic fracturing, the other is hydro-shearing. **Hydraulic fracturing**, as discussed above, has been used in the hydrocarbon industry for over 70 years

Fig. 10.22 If the upper part of the dike seen here had been cooled through injection of water (from a drill hole, indicated schematically) while the dike was propagating, the magma would simply find a slightly different path, such as along one or the other of the horns (indicated). This follows because so long as the magma below the cooled/solidified part has overpressure that is high enough to break the rock, the propagation should continue. This basaltic dike is in Tenerife (Canary Islands) and located in pyroclastic host rock (cf. Fig. 2.27).

to increase the permeability of certain target layers. The classic method is to inject a hydraulic fracture under fluid (water) pressure horizontally (from a vertical well) into the target layer (the layer containing the hydrocarbons), to keep the fracture open by (primarily) sand grains ('proppants'), so as to allow the oil to flow along the fracture. In the past few decades, the hydraulic-fracturing technique has been developed so that multiple, vertical hydraulic fractures can now be injected from a single horizontal well. This latter technique is primarily aimed at fracturing gas shales, and its success has been one of the main reasons for the recent shale-gas 'boom' (Davis et al., 2012; Wu, 2017). Hydraulic fracturing generates **earthquakes** in the host rock in a similar way that dikes generate earthquakes, namely through slip on (mostly) existing fractures due to stresses induced by the fluid overpressure (Fisher and Warpinski, 2011; Fisher, 2014; Shapiro, 2018). The strain-energy release of hydraulic fracturing, however, is very local and thus it is unlikely to be of use for general strain-energy release in the host rocks of magma sources.

Hydro-Shearing

A related method, namely **hydro-shearing**, is more likely to be useful for reducing stored strain energy in the host rocks of magma sources. In hydro-shearing, existing fractures in the rock are induced to slip, mostly producing earthquakes, through increased pore-fluid pressure that is generated by pumping water into the rock (from a vertical well). Thus, in contrast to hydraulic fracturing, which implies the formation of a new, single, extension fracture (mode I crack), hydro-shearing implies triggering the slip (in mode II or mode III, or mixed-mode) on numerous existing fractures (Fig. 10.23). The existing fractures, in a pile of lava flows, pyroclastic layers, or both, are mostly cooling joints. These have various orientations, and when the pile has become tilted they are no longer acting as extension fractures, but rather as shear fractures (Gudmundsson, 2011a).

Hydro-shearing has been used for decades to stimulate or, in fact, generate human-made geothermal reservoirs. This technique was initially applied to so-called hot-dry-rock systems, which are currently mostly referred to as **enhanced geothermal systems** (EGS). The technique is straightforward and has been well tested over decades. Water under pressure is pumped into the target layer or unit at depth. The increase in the **pore-fluid pressure** results in fault slips (shearing) on favourably oriented fractures (Gudmundsson, 2011a), thereby (commonly) giving rise to earthquakes. For EGS, the main aim is to increase the permeability of the fractured rocks so as to allow the cold water to migrate through the rock body and take up heat – changing cold water into

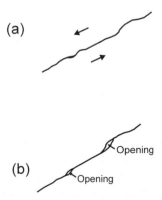

Fig. 10.23 Principles of hydro-shearing illustrated. (a) All natural fractures are to some degree irregular in shape rather than perfectly planar (or straight lines in cross-sections such as here). When the pore-fluid pressure increases so as to result in shear movement along a fracture, that is, fault slip (indicated by the arrows), the irregular shape of the original fracture normally results in some parts of the fracture being open. (b) These openings generate permeability that allows the water to migrate through the rock and take up heat so as to become hot water.

geothermal water. The increase in permeability is primarily due to a mismatch of the fracture walls once a fault slip (shearing) has occurred. More specifically, because of their irregular shapes, the fracture walls do not fit perfectly together (as they may do for an original extension fracture, such as a cooling joint) following slip or shear movement along the fracture. Thus, there will be openings along the fracture following the slip, and such openings on numerous interconnected fractures increase the permeability of the rock body (Fig. 10.23).

Hydro-shearing that aims at reducing the strain-energy concentration in a crustal segment would also increase the rock permeability, thereby increasing the rate of cooling of the magma. The focus of hydro-shearing in a volcano, however, is to trigger numerous smaller fault slips rather than a single large one. More specifically, pumping water into regions of strain concentration, as indicated by earthquakes and crustal doming, would encourage numerous fault slips, thereby reducing the strain energy that would be available to form a ring-fault and squeeze out magma in an eventual eruption.

Hydro-shearing for reducing stored strain energy in volcanoes and associated crustal segments can be done in several ways. Normally, the focus would be on those areas where, during unrest periods, there is a concentration of deformation and earthquakes. This is because the latter occur primarily where strains and stresses concentrate. The deformation is measured through geodetic studies (Section 10.2) and the earthquakes directly through seismometers. Current geodetic studies are confined to surface deformation, and thus they are less helpful in understanding processes taking place at depth in the volcano – say close to the source magma chamber – than the earthquakes. However, both types of data, geodetic and seismic, are an indication as to where strain energy is being stored, and thus they are of great value when deciding where to use hydro-shearing.

Before discussing the details of the hydro-shearing procedure, it is worth emphasising that any change in the strain energy in one large part of the volcano affects the strain energy in the remaining parts. It is perhaps easier to visualise this in terms of stresses. So, rephrasing the previous statement, we can say that any change in the state of stress in a reasonably large part of the volcano (through hydro-shearing) affects the state of stress in the rest of the volcano. The effects on the stress/strain fields are greatest close to the source of the stress change, but they are noticeable out to a distance similar to the diameter of that source (Gudmundsson, 2011a). For example, if hydro-shearing is carried out in the centre of the inflation dome, the relaxation of stress and strain in that area will affect the stress and strain concentration elsewhere – such as at any existing ring-faults associated with the volcano. Thus, in this sense, hydro-shearing that results in a significant reduction in strain energy and stress in a reasonably large body of the volcano and the hosting crustal segment will generally alter the stresses and strains in the rest of the volcano, normally with the result of reducing the likelihood of a large eruption.

Hydro-Shearing Procedure

Consider first hydro-shearing of layers close to a shallow magma chamber. During inflation, the roof of the reservoir concentrates tensile stresses and stores strain energy. Furthermore, it is the state of stress in the roof closest to the upper boundary of the chamber that

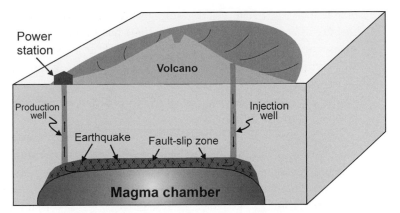

Fig. 10.24   Schematic illustration of hydro-shearing around a magma chamber so as to try to reduce the stress difference in the roof of the chamber (or in a different layer at a shallower depth) and thus to prevent dike propagation and a large eruption. Cold water is pumped into the roof of the chamber, above one of the edges of the chamber, through an injection well. Through increased pore-fluid pressure and slip on existing fractures (Figs. 10.23 and 10.25), the permeability of rock layers that constitute the roof increases and allows the water to circulate through the layers. As it does so, the water increases the pore-fluid pressure, thereby generating a fault slip and earthquakes in gradually the larger part of the roof and, eventually, the entire roof . The fault slip reduces the stress difference, as presented for example by Mohr circles (Fig. 10.25), and brings the state of stress closer to a lithostatic state, which is unfavourable (acts as a barrier) to dike propagation. Eventually, above the other edge of the chamber, a production well may be drilled for pumping up the hot water and using it in a power plant (power station) for space heating and the production of electricity. A black and white version of this figure will appear in some formats. For the colour version, please refer to the plate section.

determines whether or not rupture occurs and a dike or an inclined sheet becomes injected during the unrest period. It is thus logical to begin the hydro-shearing and strain and stress reduction there. The procedure would then normally be as follows:

1. One or more vertical holes are drilled to the depth of the target layer, say the layer very close to the magma chamber (Fig. 10.24). How close to the chamber one drills depends on various factors, such as the geothermal gradient. If, as is common, there is a large geothermal field associated with the chamber, the geothermal gradient is normally not very steep except in a very thin layer very close to the chamber. There exist equipment that can drill at high temperatures (at least up to about 500 °C), and the boiling point of water increases to over 370 °C at high pressures close to the chamber – and then becomes supercritical, at which point distinct liquid and gas phases no longer exist. In fact, some drill holes have penetrated into actual magma, both in Iceland and in Hawaii, so that theoretically there is no limit to how close to the magma in the chamber one can drill.

2. Once the drill hole or well has been made, water under pressure is pumped into the target layer or unit. The pore-fluid pressure moves the region closer to shear failure, according the Navier–Coulomb theory of shear failure, which can also be presented by the Modified Griffith Criterion (Fig. 10.25; Chapter 5). As the water migrates through the

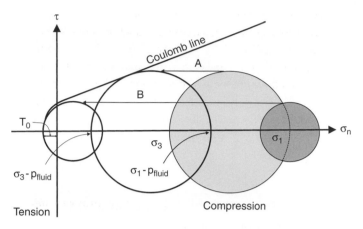

Fig. 10.25 Increasing the pore-fluid pressure in a crustal layer or segment shifts the Mohr circles to the left, towards the Coulomb failure line or curve, thereby triggering a fault slip. If the original principal stress difference ($\sigma_1 - \sigma_3$) is large, then the shifted Mohr circle will touch the Coulomb line and cause a fault slip while well within the compressive stress regime (shift A). By contrast, if the original stress difference is small, then the increased pore pressure may shift the circle all the way into the tensile regime, generating either a pure extension fracture (a hydrofracture) or a hydbrid (mixed-mode) fracture, that is, one formed partly in extension and partly in shear (shift B).

      rock, a gradually larger body of the rock becomes subject to a large number of fault slips, many of which generate earthquakes.

3. The migration of earthquakes throughout the roof of the chamber can be used to monitor exactly where (within the limits of accurate location of the earthquake foci) strain energy is being used to generate a fault slip. Since the volume subject to strain-energy reduction is monitored accurately, we can at any time decide to stop the pumping of water into the layers, if it has been decided that a large-enough volume has been treated in this way.

4. Fluids tend to flow towards sites of minimum potential energy. Areas of high strain/stress concentration are also commonly areas of dilatancy (increased pore space) and of low potential energy. Also, critically stressed faults tend to drag in or collect fluids (Barton et al., 2010). Thus, water tends to migrate to the areas of greatest strain, which are also commonly the areas where earthquakes are already occurring during the inflation period and before any hydro-shearing has begun.

5. From the location of the earthquake foci, we can map the volume of rocks in the roof of the magma chamber where strain energy has been reduced. And from the cumulative or total seismic moment (cf. Section 4.4) we can estimate crudely how much strain-energy reduction has occurred (Chapter 4).

6. In addition, once the hydro-shearing has taken place, it is a logical step to carry out a stress measurement at one or more locations along the well. These are currently mainly done through hydraulic fracturing, but in the future there will, no doubt, be other methods. Also, the focal mechanisms of the last earthquakes recorded during the hydro-shearing experiment are an indication of the state of stress at the end of the experiment.

7. From the results in point 5 we can estimate how much reduction in strain energy has occurred. Of course, part of the reduction is likely to be aseismic, but at least a minimum will be indicated by the cumulative seismic moment. Furthermore, the hydraulic-fracturing stress measurements and the focal mechanisms of the last earthquakes recorded during the hydro-shearing will give a further indication of the state of stress. From these data and interpretations we can estimate (a) the likelihood of magma-chamber rupture and dike injection after the hydro-shearing and (b) the magnitude of the strain energy in the most highly strained part of the roof that is available to squeeze out magma during an eventual eruption.

Hydro-Shearing of Faults

If there is a clear strain and stress concentration at an existing, or a potential, **ring-fault** hydro-shearing of the fault and its surroundings would be an option. It is well known from geothermal studies that when water (mostly wastewater from power plants) is pumped into an existing fault, it generates earthquakes. This is a direct consequence of increasing the pore-fluid pressure in the fault (Fig. 10.25). Thus, the possibility exists of drilling into the potential or actual ring-fault and pumping water into the fault zone in a manner indicated for the general hydro-shearing method discussed above. Alternatively, and much more cheaply, water can simply be pumped into the fault at the surface and from there it will migrate into the fault zone. The aim of both methods is to use increased fluid pressure to bring many of the fractures in the damage zone of the fault zone, as well as certain segments of the fault itself, to failure. Through the failure of numerous small fractures in the damage zone – and possibly in the core – as well as certain segments of the fault itself, the strain energy is reduced.

The aim is thus to **reduce the strain energy** available to drive a large ring-fault displacement, primarily through reduction of the energy available in the damage zone and the core. Thus, by generating numerous small fault slips, and (commonly) associated earthquakes, the goal is to reduce the available energy that could be used to produce a large displacement (hundreds of metres, or a couple of kilometres) on the ring-fault, with a possibility of a ring-dike eruption.

There is certainly more **risk** involved in the hydro-shearing of a ring-fault (and graben faults, in the case of deeper reservoirs) than in hydro-shearing of the general roof layers of the chamber. For example, it might be argued that there is a great chance that pumping water into the actual (or potential) ring-fault zone could trigger the large ring-fault displacement that we are trying to avoid. This is possible, but unlikely, so long as the seismicity associated with the pumping of water into the fault zone is carefully monitored and controlled.

Fault slips normally occur only in zones that have favourable stress fields (Gudmundsson and Homberg, 1999). More specifically, the entire zone within which the fault slip is to occur must be **stress-homogenised**, that is, it must everywhere have basically the same state of stress, and that stress field must be favourable to the type of slipping fault. The carefully monitored migration of microearthquakes within the fault zone indicates crudely the size of the area within the fault zone where the stress field has been modified so as to trigger earthquakes. So long as this area is of a limited size in comparison with the entire size (dimensions) of the fault zone, the probability of a slip of the entire fault remains very low. There are many examples of water flowing into active faults during hydraulic fracturing of gas shales (Davis et al., 2012). And all these earthquakes have remained comparatively small, presumably because the fracturing was stopped immediately, once the earthquake occurred along the fault. The main point is that when the stress concentration and strain energy nearby and within the potential or actual ring-fault are reduced, the chances of a large slip and a **large eruption are diminished.**

Stress Effects of Hydro-Shearing

Basically, what hydro-shearing does is to **minimise the stress difference** in any layer or unit where it is applied. A measure of the effectiveness of hydro-shearing is the **stress drop** during the induced earthquakes. This stress drop is approximately the driving shear stress for the earthquake and indicates the reduction in (relaxation of) the shear stress across the fault following the earthquake slip. The stress drop in earthquakes varies from about 0.3 MPa to about 50 MPa (Scholz, 1990; Allmann and Shearer, 2009), but is commonly 1–12 MPa (Chapter 4; Gudmundsson, 2011a). Relaxing the shear stress means diminishing the difference between the principal stresses, primarily between σ_1 and σ_3. Thus, during a normal-fault slip, σ_3 increases so as to be closer in magnitude to σ_1, whereas in a reverse or thrust-fault slip, σ_1 decreases and becomes closer to σ_3.

It follows that hydro-shearing tends to bring the state of stress in the layer/unit close to **lithostatic.** Lithostatic is the equilibrium state of stress (Chapter 3), where all the principal stresses are the same and equal to the overburden pressure or vertical stress σ_v, and also equal to the magmatic pressure in the chamber p_l, namely:

$$\sigma_1 = \sigma_2 = \sigma_3 = \sigma_v = p_l \qquad (10.35)$$

When the state of stress is lithostatic there is no tendency towards any type of brittle deformation, either through faulting or dike/sheet injection. Thus, hydro-shearing transforms the roof of the chamber into a **stress barrier** or seal that does not allow rupture and dike/sheet injection, thereby effectively preventing any kind of eruption.

The question then arises: how long can the stress barrier or seal be **maintained**? If no new magma is added to the chamber, then the seal can be maintained as long as it is needed. But active chambers normally receive new magma from time to time; receiving new magma is their principal way of maintaining their activity and the main reason for major volcanic unrest periods. So if new magma is intermittently received by the chamber, for how long can hydro-shearing ensure that the chamber roof does not rupture and allow a dike or a sheet to be injected – with the potential for eruption?

The answer is that, theoretically, the stress barrier or seal can be maintained **as long as we wish**. The replenishment or refilling of a magma chamber is normally a slow process, resulting in a low strain rate. Unrest periods with magma-chamber expansion (inflation) commonly last many years or decades. Hydro-shearing over decades, or centuries, during and following unrest periods can thus be used to maintain the stress barrier around the chamber. A given EGS reservoir has a lifetime of several decades and it is thus similar to that of many unrest periods. But in order to maintain a stress barrier condition in the roof of a large chamber many EGS reservoirs are normally needed, implying many injection and production wells. The reservoirs can supply hot fluids to associated geothermal power plants. For large chambers with many EGS reservoirs, these power plants could be operative for many decades or possibly several centuries (with suitable updates in the infrastructure and monitoring technology).

10.8.3 Advantages and Disadvantages of Hydro-Shearing

In addition to diminishing the chances of a larger eruption, hydro-shearing of volcanoes has several other advantages, including the following:

- It produces **geothermal fluids** that can be used for power plants. Large-scale hydro-shearing in the roofs of magma chambers or potential/actual ring-faults (or graben faults) is done through cold water being pumped down into the target layers/zones. This water takes up heat from the surrounding rock and therefore becomes geothermal water or steam. Extensive hydro-shearing around a large magma chamber may be continued for decades, even centuries (at different sites in and around the volcano) and **paid for** by the income from the energy provided by the power plant, both electricity and hot water for space heating. Thus, the use of hydro-sharing to reduce the probability of a large eruption is likely to be not only risk-reducing but also **economically profitable.**
- Because a considerable part of the water that is pumped into the volcano during hydro-shearing through the injection wells is recovered by the production wells and used for the power plants, some of the water can be recycled. It follows that much **less water** is needed for hydro-shearing than that needed, for example, for attempting to cool down (solidify) the magma in the chamber itself through pumping water into the chamber.
- **Aseismic creeping** may be encouraged (through hydro-shearing) on some of the larger faults, provided their core material is known to be suitable. Thus, clay, cataclasite, and other particulate materials encourage creeping, thereby minimising the likelihood of large earthquakes on the larger faults.
- Hydro-sharing basically allows a magma chamber to **continue growing without erupting**. If made in the roof of the chamber, close to the contact with the magma, hydro-shearing effectively prevents magma-chamber rupture and dike/sheet injection. Theoretically, there is **no obvious limit** as to how long the stress barrier/seal can be maintained so as to prevent dike/sheet injection and eruptions. We may choose, however, to let the magma chamber rupture from time to time – with some of the resulting dikes eventually erupting – when the stored strain energy is comparatively small. There will presumably be real-time stress and strain monitoring in dangerous volcanoes in the future, so that dike emplacements and comparatively small eruptions are likely to be processes that can then be **largely controlled**.

There are, however, certain disadvantages with hydro-shearing of volcanoes, including the following:

- The method requires **much water**. While a significant amount of the water can be reused, the method is still water-demanding. Where groundwater is scarce, such as in desert areas, the use of water for hydro-shearing would be in competition with other uses. Fortunately, many of the world's dangerous volcanoes are located at plate boundaries that receive moderate to high precipitation, such as along much of the 'Ring of Fire' (the plate boundaries around the Pacific and the nearby areas) and thus there are sufficient resources of groundwater for this method to be feasible.

- Hydro-sharing produces **earthquakes**, which give rise to hazards and risks. Hydro-shearing, however, is controlled and can be effectively stopped at will. Given the sizes of the fault slips that are most common during hydro-shearing, and based on decades of experience with the method for EGS-reservoirs in many countries, the earthquakes produced are mostly very small. The earthquakes rarely exceed M3, even when the hydro-shearing is made in active fault zones (Bachmann et al., 2011). Given the choice between being subject to thousands of small earthquakes, on the one hand, and a large eruption, on the other hand, presumably most people and decision-makers would choose the earthquakes.

- For some volcanoes, particularly those nearby or at major urban areas, there are various **social and legal issues** that need to be addressed before hydro-shearing can be under-taken. These include decisions by the appropriate authorities as to: (1) if, when, and where to carry out hydro-shearing; (2) how best to educate the general public about the associated earthquakes; and (3) legal aspects of potential damage to buildings and infrastructure from the earthquakes. Groundwater resources also need to be protected from pollution during site operations and also from pollution from geothermal water that is produced during and following the hydro-shearing. With proper care during hydro-shearing and hydraulic fracturing, groundwater pollution can normally be prevented from happening.

10.9 Summary

- The main aim of volcanology, and volcanotectonics in particular, is to provide an under-standing and reliable interpretation of the various geological, geophysical, and geochem-ical signals during an unrest period in a volcano in terms of plausible physical and chemical processes operating inside the volcano. Volcanic unrest is then defined as a significant increase in these signals, indicating changes in the modes or rates of the processes.

- A volcanic hazard is the probability that an eruption will occur in a given volcano or a volcanic system/field within a specific time window. For a realistic assessment of volcanic hazards we must understand the processes responsible for the unrest signals. In particular, there are four hazard-related questions that we should be able to answer when a volcano enters an unrest period. (1) Is the chamber going to rupture and inject a dike (or sheet)? (2) What is the likely propagation path of the dike/sheet? (3) Is the dike/sheet likely to reach the surface to erupt; if yes, then when and where? (4) In the case of an eruption, how large is it likely to be?

- Magma chambers/reservoirs are one key concept in volcanology. Shallow chambers have roofs at 5 km or less, whereas reservoirs are at greater depths and commonly supply magma to shallow chambers. Many chambers are sill-like, either partially or totally molten, and with volumes that reach up to at least 500 km^3, although some are likely to be larger. They rupture when the excess magmatic pressure (p_e) reaches the tensile strength (generally 0.5–9 MPa, and most commonly 2–4 MPa) of the roof or the walls to form dikes/sheets or sills, all of which are extension fractures and modelled as mode I cracks.

- Magmatic overpressure (p_o) makes it possible for dikes to rupture the host rock and propagate. Overpressure is composed of two main terms: excess pressure in the source chamber/reservoir and buoyancy. Buoyancy is due to the density difference between the magma and the host rock and can be positive, zero (neutral), or negative, depending on this difference. The buoyancy term, however, is always close to zero so long as the height of the dike is small, that is, while the vertical dike-propagation distance from the source chamber is short. For typical basaltic magma in a dike injected from a shallow magma chamber (roof depth at 5 km or less), the buoyancy makes little or no significant positive contribution to the magmatic overpressure. Layers with the same density as that of the magma (zero or neutral buoyancy) normally do not stop the vertical propagation of dikes.

- It is proposed that, of the theoretically infinite number of possible paths that a given dike/sheet may follow, it selects the path of least (minimum) action as determined by Hamilton's principle. The path chosen is the one along which the variation of the action is zero. Action has the dimensions of energy × time and the unit of J s (joule-second). For an elastic solid, such as a crustal segment, action (S) is the kinetic energy (T) minus the potential energy due to generalised external forces (V) and minus the strain energy due to internal forces (U), or $S = T - V - U$, where the right-hand side is also referred to as the Lagrangian (the difference between the kinetic and the total potential energy). Thus, in the theory proposed here, which applies to conservative systems, the path taken by a dike/sheet is normally the one along which the energy transformed (released) multiplied by the time taken for the propagation is the least (is a minimum).

- For a (normally) slowly propagating dike, the kinetic energy may be assumed to be close to zero, in which case Hamilton's principle of least action reduces to the principle of minimum potential energy. This latter principle states that of all the possible displacement fields (configurations) of an elastic body that satisfy the constraint conditions and the external and internal forces, the actual (true) displacements are those that make the total potential energy of the body a minimum. This principle applies only to (linear and non-linear) elastic bodies.

- Hamilton's least-action principle and the principle of minimum potential energy can both be derived from the principle of virtual work. This principle states that, for a continuous (e.g. elastic) system in static equilibrium, the virtual work of all external and internal forces associated with the system through reversible virtual displacements is zero. Virtual displacements are imagined (virtual) tiny instantaneous changes in the configuration of the system and are consistent with any constraints acting on the system. The work of real forces acting on a system subject to virtual displacement is named virtual work. The configuration for a continuous system means the position, at a given moment, of all its material points. The principle of virtual work is valid for (1) any constitutive law, (2) conservative and non-conservative systems, and (3) large displacements/strains.

- Dikes/sheets advance the tips/fronts in steps, with a time lag between the fracture front and the magma front. Each step may be similar in length to the thickness of lava flows or other layers through which the dike is propagating. In the theory proposed here, each propagation step is controlled by Hamilton's principle or, if the kinetic energy is omitted (is sufficiently small), by the principle of minimum potential energy. The energy needed to advance the dike-fracture is the surface energy, whereas the energy transformed or released during the dike-fracture propagation is part of the total potential energy (potential energy due to external forces and strain energy due to internal forces).

- For a homogeneous, isotropic, and non-fractured crustal segment, dike/sheet propagation paths, according to the present theory, follow the trajectories of the maximum compressive principal stress σ_1 and are thus everywhere perpendicular to the trajectories of the minimum compressive (maximum tensile) principal stress σ_3. When the dike/sheet propagates, the excess pressure in the chamber/reservoir decreases, and so do the stress magnitudes (the stress-concentration zones around the chamber/reservoir shrink). However, the geometry of the stress orientations (the stress trajectories) remains similar during the excess-pressure decrease, thereby controlling the dike-propagation paths.

- For a fractured and/or a layered (anisotropic) crustal segment, the dike/sheet paths can locally be oblique to the trajectories of the maximum compressive principal stress σ_1 – and therefore also oblique (rather than perpendicular) to σ_3. This means that part of the dike/sheet path is along a shear fracture (a fault). Whether the dike/sheet uses a fault as a part of its path depends primarily on (1) the dip of the fault (steep normal faults are the most likely to be used), and (2) the tensile strength across the fault as compared with the tensile strength of the host rock along a path following the direction of σ_1. The results suggest that dikes/sheets use faults as parts of their paths mainly if the fault is comparatively steeply dipping, preferably a normal fault or currently in an extensional regime, and with close to zero tensile strength.

- Numerical models for a layered (anisotropic) crustal segment support the 'graben rule', which means that the distance between any dike-induced boundary faults of a graben at the surface is similar to half the depth to the tip (top) of the dike below the surface. This follows because the tensile/shear stress peaks at the surface (where the boundary faults are most likely to form) occur at a lateral distance from the projection of the dike to the surface that is similar to the depth to the dike tip. More specifically, for a dike with an arrested tip at 0.5 km below the surface, the surface stresses peak (depending on the exact layering) at a lateral distance of 0.5–0.7 km from the projection of the dike to the surface. A dike-induced graben, if it forms, would thus be 1–1.4 km wide at the surface.

- The maximum surface uplift peaks occur at widely different locations from those of the stress peaks. For a dike with a tip arrested at a depth of 0.5 km below the surface then (depending on the exact layering) the uplift peaks occur at lateral distances of 2.8–3.3 km from the dike projection to the surface. The tensile/shear stresses associated with the uplift peaks are far too small to generate the boundary faults of a graben. The model results indicate that the common practice of associating the calculated maximum theoretical surface displacement (uplift) with graben formation is unjustified; any grabens induced by arrested dikes are associated with the (much more narrowly spaced) stress peaks, not with the (more widely spaced) uplift peaks.

- Elastic half-space (non-layered, isotropic) models commonly overestimate the potential dike-induced surface stresses and, consequently, the depth of the tip of the associated (arrested) dike below the surface. It follows that, for a given surface deformation, the half-space models also underestimate the dimensions of the dike, particularly its thickness. One conclusion of the models presented here is that dikes must normally be at very shallow depths (hundreds of metres or less) and of large dimensions (with thicknesses of many metres) to induce large deformations, including grabens, at the surface.

- Very large volcanic eruptions (of the order of 1000 km^3) pose a catastrophic risk, and some even an existential risk, to humankind. A new scale is proposed for quantifying eruptions, namely the elastic-energy magnitude (EEM) scale. This scale is based on the potential energy of the eruption as determined from the eruptive volume (including the intrusive volume, which may be ignored for very large eruptions). The EEM scale allows for direct comparison with the moment-magnitude scale for earthquakes. The results show that the energy transformed and released in the largest eruptions and in the largest earthquakes are of the same order of magnitude, 10^{19} J.

- Based on current monitoring methods, as well as the theories presented in the book, the best indication of a likely large eruption during volcanic unrest is the concentration of earthquakes along actual or potential ring-faults (for calderas) and boundary faults (for grabens). The earthquake concentration at these faults indicates that a large-scale subsidence is likely to happen during an eventual eruption. The subsidence helps to squeeze out magma from the chamber/reservoir, thereby providing a potential for a large or, depending on the chamber/reservoir size, very large eruption.

- Preventing very large eruptions from happening is of vital importance for humankind. The main method proposed here is hydro-shearing of the host rock, particularly the roof, of the source magma chamber so as to reduce the likelihood of magma-chamber rupture (with dike injection) as well as reducing the strain energy available to drive the eruption – if it happens. Hydro-shearing induces numerous small fault slips (and earthquakes), thereby minimising the stress difference in the chamber roof – or any layer where the method is applied. Hydro-shearing thus brings the state of stress in the roof/layer closer to lithostatic, whereby all the principal stresses are the same and equal to the overburden pressure. For a lithostatic state of stress, there is no tendency to brittle deformation, either through faulting or dike/sheet injection. Thus, hydro-shearing transforms the roof/layer into a stress barrier – a seal – that prevents dike/sheet injections and, thereby, eruptions.

10.10 Main Symbols Used

A area

A constant

B constant

δC constraint

E Young's modulus (modulus of elasticity)

G_c critical strain energy release rate for a crack, material toughness

G_I strain energy release rate during the extension of a mode I crack

g acceleration due to gravity

h height or dip-dimension of a dike (measured up from its magma source)

L Lagrangian (Lagrangian function)

$\log$ common logarithm (to the base of 10)

M_e elastic-energy magnitude

p_e magmatic (fluid) excess pressure

p_l lithostatic pressure (or stress)

p_o magmatic (fluid) overpressure

$p_o^{\sigma_n}$ magmatic overpressure with reference to σ_n

$p_o^{\sigma_3}$ magmatic overpressure with reference to σ_3

p_t total magmatic (fluid) pressure

Q_i generalised forces

q_i generalised coordinates

S action

T kinetic energy

t time

T_0 tensile strength

$T_0^{\sigma_3}$ tensile strength along the path that is perpendicular to σ_3 (parallel to σ_1)

$T_0^{\sigma_n}$ tensile strength along the path that is perpendicular to σ_n

ΔT_0 difference between the tensile strength across a fault and along the path of σ_1

U strain energy

U_0 strain energy per unit volume

U_{er} elastic potential energy of an eruption

U_t total energy

$U_t^{\sigma_n}$ total energy needed to form a dike-fracture and open it up against σ_n

$U_t^{\sigma_3}$ total energy needed to form a dike-fracture and open it up against σ_3

V potential energy

V_v volume

ΔV_v final volume of an emplaced dike

W work needed to form a dike segment of a given volume ΔV_v

W_s surface energy (of a fracture)

δW total virtual work (of the generalised forces Q_i)

δW_I virtual work due to internal forces

δW_E virtual work due to external forces

α the angle between the fracture (fault) plane and the direction of σ_1

δ the variational symbol

ε normal strain

ε_{ij} normal strain (i and j indicate the strain components)

Π total potential energy

ρ_m magma density

ρ_r rock or crustal density

σ normal stress

σ_d differential stress ($\sigma_1 - \sigma_3$)

σ_{ij} normal stress (i and j indicate the stress components)

σ_1 maximum compressive principal stress

σ_2 intermediate compressive principal stress

σ_3 minimum compressive (maximum tensile) principal stress

10.11 Worked Examples

Example 10.1

Problem

Define the term volcanic unrest and list the main indicators or signals that characterise unrest.

Solution

Volcanic unrest denotes a significant change, normally an increase, in geological, geophysical, and geochemical signals from a volcano, indicating changes in the mode or rates of the associated physical and chemical processes inside the volcano. Unrest therefore means that the volcano shows behaviour that deviates from its normal or baseline (background) behaviour.

The signals that characterise volcanic unrest are caused by physical and chemical changes in the volcano. The main signals are changes, normally an increase, in the following:

- Frequency of earthquakes within the volcano.
- Ground elevation and tilting.
- Emission of volcanic gases (fumarole activity).
- Melting of snow and/or ice.
- Volumetric flow rate and chemistry of groundwater and geothermal water.

Example 10.2

Problem

Indicate the four main questions that volcanologists should, and in the future must, be able to answer when a volcano enters an unrest period.

Solution

1. Is the magma chamber going to rupture and inject a dike (or sheet)?
2. What is the likely propagation path of the dike/sheet?
3. Is the dike/sheet likely to reach the surface to erupt; if yes, then at which location?
4. If an eruption occurs, how large is it likely to be?

Example 10.3

Problem

(a) What are the main geophysical methods for detecting active shallow magma chambers?
(b) In what depth range are active shallow magma chambers?
(c) What are the common shapes of shallow magma chambers?
(d) Are shallow chambers partially or totally molten?

Solution

(a) The main methods are seismic and geodetic. S-wave shadows are often an indication of shallow magma chambers. GPS, InSAR, and other geodetic methods provide data on inflation/deflation during volcanic unrest. These data can be used, commonly using the Mogi model but also through numerical modelling, to provide a crude estimate of the depth to (normally the top of the) inflating/deflating magma chamber.
(b) Shallow magma chambers, detected worldwide, have depths mostly less than 5 km below the surface. This figure refers to the depth to the roofs, that is, the tops of the chambers (cf. Section 6.8.2).
(c) Sill-like chambers, idealised as oblate ellipsoids, are probably the most common shapes.
(d) Some shallow chambers are totally molten, others are partially molten. Many chambers are presumably totally molten during some periods of their lifetimes. This is indicated, for example, by kilometre-scale subsidences during some caldera collapses.

Example 10.4

Problem

(a) State the conditions for magma-chamber rupture.
(b) Provide the appropriate equation and explain all the symbols and the appropriate units.
(c) What volcanotectonic structure does magma-chamber rupture normally give rise to?
(d) Explain why there is not a one-to-one correspondence between a magma-chamber rupture and a volcanic eruption.

Solution

(a) The conditions for magma-chamber rupture are given by Eqs. (10.1) and (10.2) below. The equations state that when the total fluid pressure in the magma chamber becomes equal to the combined value of the minimum compressive (maximum tensile) principal stress and the *in situ* tensile strength, the chamber ruptures and a magma-filled fracture is injected.

(b) The conditions for magma-chamber rupture are given by Eqs. (10.1) and (10.2) as follows:

$$p_l + p_e = \sigma_3 + T_0$$

$$p_t = \sigma_3 + T_0$$

Here, p_l denotes the lithostatic stress at the rupture site at the boundary of the magma chamber, p_e is the magmatic pressure in the chamber in excess of σ_3, the minimum compressive (maximum tensile) principal stress, T_0 is the local *in situ* tensile strength at the rupture site, and p_t is the total magmatic pressure in the chamber. All the symbols have the same unit, pascal, that is $N\,m^{-2}$.

(c) The most common structures are dikes and inclined sheets.

(d) Magma-chamber ruptures are much more common than eruptions. This follows because many dikes and inclined sheets become arrested on their paths to the surface and therefore do not erupt.

Example 10.5

Problem

A basaltic feeder-dike in a collapse caldera is injected from a shallow magma chamber at the depth of 5 km. The average crustal density down to that depth is 2700 kg m^{-3} and the average magma density is 2650 kg m^{-3}. The tensile stress measured at the surface of the caldera before the dike erupts is 1 MPa. Estimate the likely magmatic overpressure or driving pressure in the feeder-dike at the surface.

Solution

Magmatic overpressure in a dike p_o can be estimated from Eq. (10.3), namely:

$$p_o = p_e + (\rho_r - \rho_m)gh + \sigma_d$$

In the present example we do not know the excess pressure p_e at the time of rupture. However, we know that the excess pressure is similar to the *in situ* tensile strength of the roof of the chamber, so it is most likely in the range of 2–4 MPa. Here, we use the value 3 MPa for the excess pressure. All the other values in Eq. (10.3) are given, so the overpressure becomes:

$$p_o = p_e + (\rho_r - \rho_m)gh + \sigma_d = 3 \times 10^6\,\text{Pa} + (2700\,\text{kg m}^{-3} - 2650\,\text{kg m}^{-3})$$
$$\times\, 9.81\,\text{m s}^{-2} \times 5000\,\text{m} + 1 \times 10^6\,\text{Pa} = 6.4 \times 10^6\,\text{Pa} = 6.4\text{MPa}$$

This is a very reasonable overpressure, and is similar to what one would commonly expect for feeder-dikes issuing evolved basalt. In the calculations we use an average density for the basaltic magma. As discussed earlier, gas expansion occurs in basaltic magma primarily in the uppermost few hundred metres of the dike path below the surface. Gas expansion reduces the density of the magma, thereby potentially increasing the density difference between the rock and the magma, and thereby the buoyancy. However, we also use an average density for the entire host rock of the feeder-dike, of 2700 kg m^{-3}, while it is known that the density in the uppermost several hundred metres of the crust in volcanic areas is much less than this value, commonly around 2300 kg m^{-3}. Thus, the magma buoyancy may increase in the uppermost few hundred metres, but perhaps less than might be expected. The average density difference used here may therefore be regarded as reasonable for evolved basaltic magma.

Example 10.6

Problem

Derive the buoyancy term (Eq. (10.4)), namely: $buoyancy = (\rho_r - \rho_m)gh$

Solution

Let p_l be the total magma pressure/overburden pressure at the rupture site of the magma chamber/reservoir, as in Eq. (10.1). When the magma-filled fracture – the dike – begins to propagate upwards from the rupture site, the total magma pressure in the dike decreases with height according to the formula:

$$p_l - \rho_m gh$$

where ρ_m is the magma density, g is the acceleration due to gravity, and h is the height above the rupture site at the boundary of the magma chamber/reservoir. Similarly, the overburden pressure in the rock adjacent to the dike decreases with height according to the formula:

$$p_l - \rho_r gh$$

where ρ_r is the rock density. It then follows that the buoyancy term or contribution to the magmatic overpressure is given by:

$$Buoyancy = (p_l - \rho_m gh) - (p_l - \rho_r gh) \tag{10.36}$$

which then reduces to Eq. (10.4), namely:

$$Buoyancy = (\rho_r - \rho_m)gh$$

Example 10.7

Problem

(a) Provide a simple statement of Hamilton's principle of least action as applied to dike paths. What are the dimensions and the unit of the term action?
(b) What is a conservative mechanical system? What is the main disadvantage of using such systems in modelling rock-fracture propagation?

(c) Explain the concepts of virtual displacement and virtual work as well as the principle of virtual work.

(d) Define and explain external and internal forces.

Solution

(a) When applied to dike paths, the principle states that the dike selects that path along which the time integral of the difference between the kinetic and potential energies is stationary (is an extremum) relative to all other possible paths with the same points of initiation and termination. Alternatively, the dike path chosen is such that the variation of the action, from the point of initiation to the point of termination, is zero, that is, $\delta S = 0$, where δ is the variation and S is the action. In practice, the action is normally minimised along the actual path. Action has the dimensions of energy $\times$ time and the unit of joule-second, J s (cf. Example 10.8)

(b) In a conservative system the work done by a force is independent of the path (and thus reversible). More specifically, the work is equal to the difference between the final and the initial values of the potential energy, and the associated force field can be expressed as the gradient of the potential energy. When all the forces acting on a system are conservative, the system itself is conservative. The main disadvantage in using such a system in modelling rock-fracture propagation is that a conservative system cannot include friction (which is a non-conservative force). All faulting (shear fracturing) involves friction.

(c) Virtual displacement is imagined, infinitesimal, instantaneous (occur in no time at all) changes in the coordinates that are consistent with the constraints of the system. Alternatively, the system is imagined to reach a somewhat different position or configuration while time is held constant. By contrast, all real displacements take time. When a system is subject to virtual displacement the work done on it by real forces is called virtual work. The principle of virtual work (or virtual displacement) states that for a system in static equilibrium the virtual work done on the system, and compatible with the constraints, is zero.

(d) External forces are outside the system, say a crustal segment, and are generally attributable to sources outside the system. By contrast, internal forces are inside the system (the crustal segment) and are attributable to sources or processes inside the system. The internal force is a measure of the resistance of one part of the body to being separated from other parts and, per unit area, is referred to as stress, which, in turn, generates strain. External forces can move the system as whole, but internal forces cannot.

Example 10.8

Problem

(a) Provide an equation for Hamilton's principle of least action for an elastic solid (e.g. a crustal segment). Explain all the symbols and their units.

(b) Provide an equation for the principle of minimum potential energy for an elastic solid (e.g. a crustal segment). Explain all the symbols and their units and how this principle relates to Hamilton's principle. Provide a simple statement as to what the principle of minimum potential energy means.

Solution

(a) Combining Eq. (10.11) and Eq. (10.14) gives Hamilton's principle for an elastic solid as:

$$\delta S = \delta \int_{t_1}^{t_2} (T - V - U)dt = \delta \int_{t_1}^{t_2} (T - \Pi)dt = 0 \tag{10.37}$$

where δ is the variational symbol, S is the action, and t_1 and t_2 are two specified and arbitrarily chosen times in the evolution of the system (here corresponding to the time of the magma-chamber rupture and dike injection (t_1) and the dike-propagation termination (t_2), that is, the time it takes the dike to reach the surface in the case of a feeder-dike). Also, T is the kinetic energy, V is the potential energy, U is the strain energy, and Π is the total potential energy (here of the crustal segment holding the magma chamber and associated volcano). The action S has the unit of J s, the times have the unit of s, and all the energies, T, V, U, and Π have the unit of J.

(b) Eq. (10.15) gives the principle of minimum potential energy thus:

$$\delta(V + U) = \delta\Pi = 0$$

where all the symbols have been defined above. The principle of minimum potential energy applies only to elastic systems (and therefore to crustal segments) and follows from Hamilton's principle (Eq. (10.37)) when all the forces are conservative and the kinetic energy is zero, that is, $T = 0$. The principle can be stated as follows. Of all the possible displacement fields or deformations (configurations) of an elastic body that satisfy the constraint conditions and the external and internal loads, the actual (true) displacements or deformations are those that make the total potential energy of the body a minimum.

Example 10.9

Problem

(a) In the simplified case of a homogeneous, isotropic and unfractured crustal segment hosting a shallow magma chamber, based on the Hamilton least action/minimum potential energy principles, what would be the likely dike path?

(b) How would the dike path change in the more realistic case of a layered (anisotropic) crustal segment?

Solution

(a) The dike path would be everywhere parallel with the maximum compressive principal stress σ_1, and thus perpendicular to the minimum compressive principal stress σ_3. Out of the theoretically infinite possible paths between the dike-injection time t_1 and dike-termination time t_2 the chosen path would be straight, namely the shortest distance between t_1 and t_2.

(b) In the absence of fractures, particularly faults (see Example 10.10), from Hamilton's principle it follows that the dike path seeks to be everywhere parallel to σ_1 so as to minimise the energy needed for the propagation. At the same time, the dike-fracture seeks to minimise the time needed to propagate from t_1 to t_2, that is, the duration of the propagation. The propagation velocity of dikes is mostly between 0.1 m s^{-1} and 1 m s^{-1}. If we use an average velocity of dike propagation as being 0.5 m s^{-1}, minimising the duration of the dike-path formation implies minimising the path length, for the given constraints. Thus, of all the possible paths following σ_1, the shortest one from t_1 to t_2 would normally be selected, but it would normally be an irregularly shaped path (Figs. 7.10 and 10.22).

Example 10.10

Problem

Explain under what conditions the dike-propagation path in a fractured crustal segment may deviate from being parallel to the maximum compressive principal stress σ_1, and thus perpendicular to the minimum compressive principal stress σ_3.

Solution

Some dike segments use faults for parts of their paths. Faults, by definition, are shear fractures and their dip is oblique to that of σ_1. The tensile strength across an active or recently active fault may be essentially zero. It follows that less energy may be needed for the dike segment to use the fault than to rupture the rock to form a path parallel to σ_1. This is so even though the dike segment in the fault would not be perpendicular to σ_3 but rather to the normal stress on the fault plane σ_n, which, by definition is always higher than σ_3 (σ_n is equal in magnitude to σ_3 only for an extension fracture). Most dikes observed in faults in the field follow the faults for a short part of the dike path. This is partly because the tensile strength of the fault and the path parallel with σ_1 vary, and partly because the dip of the fault also changes with depth. Thus, while it may require less energy for the magma to open up a fault than form its own fracture parallel with σ_1 at one location along the dike path, more energy may be needed to follow the fault rather than σ_1 at other locations (cf. Examples 10.11–10.13).

Example 10.11

Problem

A propagating dike meets a normal fault with a dip of $70°$ at a depth 800 m below the surface of a rift zone. At that depth the vertical stress σ_1 is 20 MPa, and the minimum horizontal stress σ_3 perpendicular to the dike is 10 MPa. Under what mechanical conditions would the dike enter the fault to use it as a part of its path?

Solution

From Eq. (10.26) we know that the dike is likely to inject the fault and use it for part of its path if the following condition holds:

$$\sigma_n - \sigma_3 = \frac{\sigma_d}{2}(1 - \cos 2\alpha)$$

where (Eq. (10.27))

$$\Delta T_0 = T_0^{\sigma_3} - T_0^{\sigma_n}$$

ΔT_0 is the difference in tensile strength along the path that is perpendicular to σ_3 (and thus parallel with σ_1) and the path across the fault and it is thus perpendicular to the normal stress on the fault plane σ_n. Since the dip is 70°, then the angle between the fault plane and σ_1, α, is $90° - 70° = 20°$ and $2\alpha = 40°$ (cf. Eq. (10.24)). From Eq. (10.26) above we then get the difference as:

$$\sigma_n - \sigma_3 = \frac{\sigma_d}{2}(1 - \cos 2\alpha) = \frac{20 \text{ MPa} - 10 \text{ MPa}}{2}(1 - \cos 40°) = 1.2 \text{ MPa}$$

The total range of measured *in situ* tensile strengths outside faults (that is, $T_0^{\sigma_3}$) is 0.5–9 MPa, with 2–4 MPa being the most common values. From Eq. (10.26) the stress difference has to be less than or equal to ΔT_0 for the dike segment to use the normal fault rather than form its path perpendicular to σ_3. Given the typical range of *in situ* tensile strengths, the dike would therefore become deflected into the fault plane only if the tensile strength across the fault segment was very low, or close to zero. For example, if $T_0^{\sigma_3}$ was 2 MPa, then $T_0^{\sigma_n}$ would have to be 0.8 MPa or less for the conditions of Eq. (10.26) to hold. If the normal fault was recently active, or perhaps reactivated/opened up by the approaching dike, the tensile strength across the fault, $T_0^{\sigma_n}$ could indeed be very low, in which case the dike would be deflected into the fault.

Example 10.12

Problem

A vertically propagating dike in a rift zone meets a normal fault at 2 km depth below the surface. The fault dips 70°, and the vertical stress, σ_1, at 2 km depth is 50 MPa. Calculate the stress difference $\sigma_n - \sigma_3$ if the horizontal minimum stress σ_3 is (a) 40 MPa and (b) 30 MPa.

Solution

(a) From Eq. (10.26) the stress difference $\sigma_n - \sigma_3$ when σ_3 is 40 MPa is:

$$\sigma_n - \sigma_3 = \frac{\sigma_d}{2}(1 - \cos 2\alpha) = \frac{50 \text{ MPa} - 40 \text{ MPa}}{2}(1 - \cos 40°) = 1.2 \text{ MPa}$$

This is the same value as in Example 10.12 above and shows that $\sigma_n - \sigma_3$ remains the same, irrespective of the crustal depth where the dike meets the normal fault, so long as the fault dip and the difference between the principal stresses σ_d remain the same.

(b) Using Eq. (10.26) again, the stress difference $\sigma_n - \sigma_3$ when σ_3 is 30 MPa becomes:

$$\sigma_n - \sigma_3 = \frac{\sigma_d}{2}(1 - \cos 2\alpha) = \frac{50 \text{ MPa} - 30 \text{ MPa}}{2}(1 - \cos 40°) = 2.3 \text{ MPa}$$

This value for the stress difference $\sigma_n - \sigma_3$ is somewhat higher than in (a), simply because of greater difference between the principal stresses, that is, larger value of σ_d. However, to satisfy the conditions of Eq. (10.26), namely that $\sigma_n - \sigma_3 \leq \Delta T_0$, given the typical *in situ* tensile strengths outside faults, the tensile strength across the faults in both cases, (a) and (b), would have to be very low. A recently active fault is the main candidate for a very low tensile strength.

Example 10.13

Problem

Many normal faults in rift zones have dips widely different from 70°. For example, in the Tertiary and Pleistocene palaeo-rift zones of Iceland the dip of normal faults ranges from 49° to 89°, while the mean dip is 73° (Gudmundsson, 2011a). Assess the likelihood that a dike meeting a normal fault at 2 km depth would use the fault as a path if the fault dip is 60° and the principal stresses $\sigma_1 = 75$ MPa and $\sigma_3 = 45$ MPa.

Solution

Since the dip is 60°, then the angle between the fault plane and σ_1, α, is 90° – 60° = 30° and $2\alpha = 60°$. Then from Eq. (10.26):

$$\sigma_n - \sigma_3 = \frac{\sigma_d}{2}(1 - \cos 2\alpha) = \frac{75 \text{ MPa} - 45 \text{ MPa}}{2}(1 - \cos 60°) = 7.5 \text{ MPa}$$

For the typical tensile strengths of 2–4 MPa, the dike would have no chance of entering the fault to use it as a path for the simple reason that 7.5 MPa is much larger than 4 MPa. Only for exceptionally strong rocks, with tensile strength of 9 MPa, would there be a chance of the dike using the fault as a path, and then only if the tensile strength across the fault were less than 1.5 MPa (since 9 MPa – 7.5 MPa = 1.5 MPa). In a pile of lava flows and pyroclastic rocks, the *in situ* tensile strength would hardly ever exceed 6 MPa, in which case the dike would not use the fault as a path.

Example 10.14

Problem

What is the 'graben rule' and how does it relate to dike-induced surface stresses and deformation?

Solution

The 'graben rule' states that the source or disturbance triggering a graben formation is commonly at a crustal depth similar to that of the half-width of the graben at the surface. The graben rule was initially based on analogue experiments. However, recent results, presented above and by Al Shehri and Gudmundsson (2018), show that the lateral distance between the two surface stress peaks induced by dikes is generally in agreement with the graben rule. By contrast, the surface uplift peaks occur at widely different locations and are normally associated with small (tensile and shear) stresses. Since the boundary faults of a graben would be expected to form where the tensile and shear stresses peak, the graben rule is clearly supported by these new numerical results.

Example 10.15

Problem

Summarise briefly the main indications that a volcano may be preparing for a large eruption.

Solution

The first two things to remember are that (a) most unrest periods do not result in dike injection, and (b) most injected dikes (and thus magma-chamber ruptures and associated earthquake swarms) do not reach the surface to erupt but rather become arrested at depth in the volcano. When these points have been taken into account, the main indications of an approaching large eruption in a volcano may be summarised as follows:

1. According to the theories presented in the book, large eruptions are generally associated with block subsidence into the magma source, either a caldera block into a shallow magma chamber or a graben block into a deep-seated reservoir. It follows that a shear-stress concentration at actual or potential boundary faults of a caldera/graben is one of the major indications of a potential larger eruption.
2. In the absence of direct stress measurements (in drill holes), the best indication of a stress concentration at boundary faults is earthquakes. When earthquakes begin to concentrate at the location of actual or potential boundary faults of calderas (ring-faults) or grabens (normal boundary faults), subsidence of the crustal block bounded by the faults is likely to occur.
3. Large-scale doming of the volcano/volcanic field/zone is an indication of an accumulation of strain energy. Most of this energy is available to help squeeze magma out of the chamber/reservoir. Thus, the greater the stored strain energy before eruption, the larger the proportion of magma that flows out of the chamber/reservoir during the eruption.
4. The eventual size of the eruption depends also on the volume of the magma source. The greater the volume of the chamber/reservoir, the greater is its potential for feeding a large eruption when the stress/energy conditions satisfy some or all of the points 1–3.

Example 10.16

Problem

What is hydro-shearing and how can it be used to prevent large volcanic eruptions?

Solution

In hydro-shearing, water is pumped (through drill holes) into rock layers/units so as to increase the pore-fluid pressure and trigger slips on existing faults. Slips on numerous (mostly small) faults reduce the stress difference in the layer and bring its state of stress close to lithostatic. No magma-chamber rupture is possible when the state of stress in its roof (or walls) is lithostatic, so that this state of stress prevents chamber rupture and dike injection if it is reached at or close to the contact between the magma and the host rock (the ceiling). Alternatively, hydro-shearing can be undertaken in a shallower layer/unit so as to arrest propagating dikes/sheets that have already been injected from the chamber during the unrest period. Hydro-shearing has been used for decades on a much smaller scale to generate geothermal reservoirs – increase the permeability of rock bodies – for enhanced geothermal systems. Hydro-shearing on a large scale in the host rock at the margin of large magma chambers can maintain close to lithostatic stress conditions for centuries or millennia so as to prevent large eruptions. Also, hydro-shearing of shallower layers/units above the chamber can be used to arrest dikes or deflect them into sills, thereby preventing them from bringing magma to the surface.

10.12 Exercises

10.1 In a given volcano, several earthquakes are detected over a period of a few days. What necessary information is needed to determine if the volcano has entered an unrest period?

10.2 How are most earthquakes associated with dike propagation in a volcano thought to be generated?

10.3 Explain the main physical role of a magma chamber in (a) the formation of a polygenetic volcano, and (b) the subsequent collapse of that volcano to form a caldera.

10.4 What is a multiple intrusion? Under what conditions can a sill formed through multiple intrusions function as a magma chamber?

10.5 Provide strong evidence that at least during some parts of its lifetime the entire shallow magma chamber is likely to be in a molten stage.

10.6 What is the main reason that a magma-chamber rupture with a dike injection commonly does not result in an eruption?

10.7 What mechanical types of fractures are most dikes? What is the main evidence for this conclusion?

10.8 What is the difference between an excess magmatic pressure and a total magmatic pressure in a chamber? What is the common value of the excess pressure at magma-chamber rupture?

10.9 What are the two main geophysical signals that indicate that a magma chamber has just ruptured and injected a propagating dike?

10.10 Why must the magmatic overpressure in a dike always be referred to a particular crustal level or layer (including the Earth's surface)?

10.11 Why is strike-slip faulting particularly common during dike propagation?

10.12 What is the range of *in situ* tensile strengths of crustal rocks? How is the tensile strength most accurately measured?

10.13 Why is large-scale plastic or ductile behaviour unlikely to be common in the roofs of shallow magma chambers?

10.14 It has been suggested that the *in situ* tensile strengths of host rocks of magma chambers may be as high as 100 MPa. Why is this unlikely to be the case?

10.15 Why does buoyancy make little or no positive contribution to the overpressure of most basaltic dikes injected from shallow magma chambers?

10.16 If buoyancy makes little positive contribution to the overpressure or driving pressure of basaltic dikes injected from shallow chambers, what pressure is then driving them and how is that pressure generated?

10.17 What is the contribution of gas expansion to the overpressure of basaltic dikes?

10.18 Why is the level of neutral buoyancy unlikely to trap vertically propagating dikes?

10.19 Why does the average density of crustal layers/units generally increase with increasing depth?

10.20 What are human-made hydraulic fractures and how do they compare with dikes and other sheet-like intrusions?

10.21 What is the action integral? What does an extremum (or extreme value) of the action integral mean?

10.22 Show that giving the dimensions of action as energy × time is equivalent to giving the dimensions as momentum × distance.

10.23 Define the Lagrangian (the Lagrangian function).

10.24 Discuss friction on a fault in the context of the term constraint of a mechanical system.

10.25 Define body and surface forces of a mechanical system. Are they external or internal forces?

10.26 List and briefly discuss the main characteristics of the principle of virtual work.

10.27 What is the meaning of the term degrees of freedom in mechanics, and how does the term differ when applied to discrete and continuous systems?

10.28 In words, what is the main difference as regards Hamilton's principle when applied to a continuous system rather than a discrete system?

10.29 What is meant by time lag in the context of the propagation of dikes (and hydrofractures in general)?

10.30 Out of a theoretically infinite number of possible paths for a dike, what main factors control the actual path taken by the dike?

10.31 What is the total energy for a dike to be initiated and to propagate?

10.32 Provide an equation for the total strain energy generated during magma-chamber inflation (expansion), which is partly available for driving dike propagation. Define all the terms and give their units.

10.33 Show that the decrease in total potential energy during dike propagation equals the increase in dike-fracture surface energy. Provide the name, the equation, and the unit for the rate of transformation of potential energy into surface energy.

10.34 Based on least-action/minimum potential energy principles, what geometries would dike paths normally have in homogeneous, isotropic crustal segments and why?

10.35 Based on least-action/minimum potential energy principles, what geometries would dike paths normally have in heterogeneous, anisotropic (mechanically layered) crustal segments and why?

10.36 Under what mechanical conditions is a dike likely to use a fault as a part of its path?

10.37 Why is the use of dike-induced surface uplift peaks a poor indication of the likely width of a dike-induced graben? What would be a better indication of the likely width of an induced graben?

10.38 What is the exception to the graben rule for the ratio between the width of a dike-induced graben and the depth to the dike tip?

10.39 What is an existential risk and what types of volcanic eruptions might possible generate such a risk?

10.40 List and discuss very briefly the main suggested methods for preventing large eruptions.

References and Suggested Reading

Agustsdottir, T., Woods, J., Greenfield, T., 2016. Strike-slip faulting during the 2014 Bardarbunga–Holuhraun dike intrusion, central Iceland. *Geophysical Research Letters*, **43**, 1495–1503, doi:10.1002/2015GL067423.

Allmann, B. P., Shearer, P. M., 2009. Global variations of stress drop for moderate to large earthquakes. *Journal of Geophysical Research*, **114**, doi:10.1029/2008JB005821.

Al Shehri, A., Gudmundsson, A., 2018. Modelling of surface stresses and fracturing during dyke emplacement: application to the 2009 episode at Harrat Lunayyir, Saudi Arabia. *Journal of Volcanology and Geothermal Research*, **356**, 278–303.

Anderson, T. L., 2005. *Fracture Mechanics: Fundamentals and Applications*, 3rd edn. London: Taylor & Francis.

Arioldi, G., Muirhead, J. D., Zanella, E., White, J. D. L., 2012. Emplacement process of Ferrar Dolerite sheets at Allan Hills (South Victoria Land, Antarctica) inferred from magnetic fabric. *Geophysical Journal International*, **188**, 1046–1060.

Bachmann, C. E., Wiemer, S., Woessner, J., Hainzl, S., 2011. Statistical analysis of the induced Basel 2006 earthquake sequence: introducing a probability-based monitoring approach for enhanced geothermal systems. *Geophysical Journal International*, **186**, 793–807.

Barton, C. A., Zoback, M. D., Moos, D., 2010. Fluid flow along potentially active faults in crystalline rock. *Geology*, **23**, 683–686.

Bazargan, M., Gudmundsson, A., 2019. Dike-induced stresses and displacements in layered volcanic zones. *Journal of Volcanology and Geothermal Research*, **384**, 189–205.

Becerril, L., Galindo, I., Gudmundsson, A., Morales, J. M., 2013. Depth of origin of magma in eruptions. *Scientific Reports*, **3**, 2762, doi:10.1038/srep02762.

Bedford, A., 1985. *Hamilton's Principle in Continuum Mechanics*. London: Pitman Publishing.

Bergerat, F., Angelier, J., 1998. Fault systems and paleostresses in the Vestfirdir Peninsula. Relationship with the Tertiary paleo-rifts of Skagi and Snaefells (northwest Iceland). *Geodinamica Acta*, **11**, 105–118.

Bonaccorso, A., Aoki, Y., Rivalta, E., 2017. Dike propagation energy balance from deformation modeling and seismic release. *Geophysical Research Letters*, **44**, 5486–5494.

Bonafede, M., Rivalta, E., 1999a. The tensile dislocation problem in a layered elastic medium. *Geophysical Journal International*, **136**, 341–356.

Bonafede, M., Rivalta, E., 1999b. On tensile cracks close to and across the interface between two welded elastic half-spaces. *Geophysical Journal International*, **138**, 410–434.

Bostrom, N., Cirkovic, M. M., 2008. *Global Catastrophic Risk*. Oxford: Oxford University Press.

Boudreau, A., Simon, A., 2007. Crystallization and degassing in the basement sill, McMurdo DryValleys, Antarctica. *Journal of Petrology*, **48**, 1369–1386.

Bradley, J., 1965. Intrusion of major dolerite sills. *Transactions of the Royal Society of New Zealand*, **3**, 27–55.

Cashman, K. V., Sparks, R. S. J., 2013. How volcanoes work: a 25 year perspective. *Geological Society of America Bulletin*, **125**, 664–690.

Cashman, K. V., Sparks, R. S. J., Blundy, J. D., 2017. Vertically extensive and unstable magmatic systems: a unified view of igneous processes. *Science*, **355**, 6331, doi:10.1126/science.aag3055.

Cayol, V., Cornet, F. H., 1998. Three-dimensional modelling of the 1983–1984 eruption of Piton de la Fournaise volcano, Reunion Island. *Journal of Geophysical Research*, **103**, 18025–18037.

Chaussard, E., Amelung, F., 2014. Regional controls on magma ascent and storage in volcanic arcs. *Geochemistry, Geophysics, Geosystems*, **15**, doi:10.1002/2013GC005216.

Chesner, C. A., Rose, W. I., Deino, A., Drake, R., Westgate, J. A., 1991. Eruptive history of Earth's largest Quaternary caldera (Toba, Indonesia) clarified. *Geology*, **19**, 200–203.

Chevallier, L., Woodford, A., 1999. Morpho-tectonics and mechanism of emplacement of the dolerite rings and sills of the western Karoo, South Africa. *South African Journal of Geology*, **102**, 43–54.

Crosweller, H. S., Arora, B., Brown, S. K, Cottrell, et al., 2012. Global database on large magnitude explosive volcanic eruptions (LaMEVE). *Journal of Applied Volcanology*, **1**, doi:10.1186/2191–5040-1–4.

Davis, P. M., 1983. Surface deformation associated with a dipping hydrofracture. *Journal of Geophysical Research*, **88**, 5826–5834.

Davis, R. J., Mathias, S. A., Moss, J., Hustoft, S., Newport, L., 2012. Hydraulic fractures: how far can they go? *Marine and Petroleum Geology*, **37**, 1–6.

Davis, R. J., Foulger, G. R., Mathias, S., 2013. Reply: Davis et al. (2012). Hydraulic fractures: how far will they go? *Marine and Petroleum Geology*, **43**, 519–521.

Dym, C. L., Shames, I. H., 2013. *Solid Mechanics: A Variational Approach*. Berlin: Springer Verlag.

Dzurisin, D., 2006. *Volcano Deformation: New Geodetic Monitoring Techniques*. Berlin: Springer Verlag.

Fedotov, S. A., 1985. Estimates of heat and pyroclast discharge by volcanic eruptions based upon the eruption cloud and steady plume observations. *Journal of Geodynamics*, **3**, 275–302.

Fedotov, S. A., Chirkov, A. M., Guscv, N. A., Kovalev, G. N., Slezin, Yu. B., 1980. The large fissure eruption in the region of Plosky Tolbachik Volcano in Kamchatka, 1975–1976. *Bulletin of Volcanology*, **43**, 47–60.

Fialko, Y., Khazan, Y., Simons, M., 2001. Deformation due to a pressurized horizontal circular crack in an elastic half-space, with applications to volcano geodesy. *Geophysical Journal International*, **146**, 181–190.

Fisher, K., 2014. Hydraulic fracture growth: real data. Presentation given at GTW-AAPG/STGS Eagle Ford plus Adjacent Plays and Extensions Workshop, San Antonio, Texas, February 24–26.

Fisher, K., Warpinski, N., 2011. Hydraulic fracture-height growth: real data. Society of Petroleum Engineers Annual Technical Conference and Exhibition, SPE 145949.

Fleming, T. H., Heimann, A., Foland, K. A., Elliot, D. H., 1997. ^{40}Ar/^{39}Ar geochronology of Ferrar Dolerite sills from the Transantarctic Mountains, Antarctica: implications for the age and origin of the Ferrar magmatic province. *Geological Society of America Bulletin*, **109**, 533–546.

Flewelling, S. A., Tymchak, M. P., Warpinski, N., 2013. Hydraulic fracture height limits and fault interactions in tight oil and gas formations. *Geophysical Research Letters*, **40**, 3602–3606.

Folch, A., Marti, J., 2004. Geometrical and mechanical constraints on the formation of ring-fault calderas. *Earth and Planetary Science Letters*, **221**, 215–255.

Francis, E. H., 1982. Magma and sediment – I: emplacement mechanism of late Carboniferous tholeiite sills in northern Britain. *Journal of the Geological Society*, **139**, 1–20.

Fung, Y. C., Tong, P., 2001. *Classical and Computational Solid Mechanics*. Singapore: World Scientific Publishing.

Galindo, I., Gudmundsson, A., 2012. Basaltic feeder dykes in rift zones: geometry, emplacement, and effusion rates. *Natural Hazards and Earth System Sciences*, **12**, 3683–3700.

Galland, O., Scheibert, J., 2013. Analytical model of surface uplift above axisymmetric flat-lying magma intrusions: implications for sill emplacement and geodesy. *Journal of Volcanology and Geothermal Research*, **253**, 114–130.

Gautneb, H., Gudmundsson, A., 1992. Effect of local and regional stress fields on sheet emplacement in West Iceland. *Journal of Volcanology and Geothermal Research*, **51**, 339–356.

Gelman, S. E., Gutierrez, F. J., Bachmann, O., 2013. On the longevity of large upper crustal silicic magma reservoirs, *Geology*, **41**, 759–762, doi:10.1130/g34241.1.

Geshi, N., Neri, M., 2014. Dynamic feeder dyke systems in basaltic volcanoes: the exceptional example of the 1809 Etna eruption (Italy). *Frontiers in Earth Science*, **2**, doi:10.3389/feart.2014.00013.

Geshi, N., Shimano, T., Chiba, T., Nakada S., 2002. Caldera collapse during the 2000 eruption of Miyakejima volcano, Japan. *Bulletin of Volcanology*, **64**, 55–68.

Geshi, N., Kusumoto, S., Gudmundsson, A., 2010. Geometric difference between non-feeder and feeder dikes. *Geology*, **38**, 195–198.

Geshi, N., Kusumoto, S., Gudmundsson, A., 2012. Effects of mechanical layering of host rocks on dike growth and arrest. *Journal of Volcanology and Geothermal Research*, **223–224**, 74–82.

Geyer, A., Marti, J., 2008. The new worldwide collapse caldera database (CCDB): a tool for studying and understanding caldera processes. *Journal of Volcanology and Geothermal Research*, **175**, 334–354.

Geyer, A., Marti, J., 2014. A short review of our current understanding of the development of ring faults during collapse caldera formation. *Frontiers in Earth Science*, **2**, doi:10.3389/feart.2014.00022.

Goldstein, H., Poole, C. P., Safko, J. L., 2013. *Classical Mechanics*. New York, NY: Pearson.

Gonnermann, H. M., Manga, M., 2013. Dynamics of magma ascent in the volcanic conduit. In Fagents, S. A., Gregg, T. K. P., Lopes, R. M. C. (eds.), *Modeling Volcanic Processes*. Cambridge: Cambridge University Press, pp. 55–84.

Greenland, L. P., Rose, W. I., Stokes, J. B., 1985. An estimate of gas emissions and magmatic gas content from Kilauea volcano. *Geochimica et Cosmochimica Acta*, **49**, 125–129.

Greenland, L. P., Okamura, A. T., Stokes, J. B., 1988. Constraints on the mechanics of the eruption. In Wolfe, E. W (ed.), *The Puu Oo Eruption of Kilauea Volcano, Hawaii: Episodes through 20, January 3, 1983 through June 8, 1984. US Geological Survey Professional Paper, 1463*. Denver, CO: US Geological Survey, pp. 155–164.

Gretener, P. E., 1969. On the mechanics of the intrusion of sills. *Canadian Journal of Earth Sciences*, **6**, 1415–1419.

Gudmundsson, A., 1983. Form and dimensions of dykes in eastern Iceland. *Tectonophysics*, **95**, 295–307.

Gudmundsson, A., 1986. Formation of dykes, feeder-dykes and the intrusion of dykes from magma chambers. *Bulletin of Volcanology*, **47**, 537–550.

Gudmundsson, A., 1988. Effect of tensile-stress concentration around magma chambers on intrusion and extrusion frequencies. *Journal of Volcanology and Geothermal Research*, **35**, 179–194.

Gudmundsson, A., 2002. Emplacement and arrest of sheets and dykes in central volcanoes. *Journal of Volcanology and Geothermal Research*, **116**, 279–298.

Gudmundsson, A., 2003. Surface stresses associated with arrested dykes in rift zones. *Bulletin of Volcanology*, **65**, 606–619.

Gudmundsson, A., 2006. How local stresses control magma-chamber ruptures, dyke injections, and eruptions in composite volcanoes. *Earth-Science Reviews*, **79**, 1–31.

Gudmundsson, A., 2009. Toughness and failure of volcanic edifices. *Tectonophysics*, **471**, 27–35.

Gudmundsson, A., 2011a. *Rock Fractures in Geological Processes*. Cambridge: Cambridge University Press.

Gudmundsson, A., 2011b. Deflection of dykes into sills at discontinuities and magma-chamber formation. *Tectonophysics*, **500**, 50–64.

Gudmundsson, A., 2012. Strengths and strain energies of volcanic edifices: implications for eruptions, collapse calderas, and landslides. *Natural Hazards and Earth System Sciences*, **12**, 2241–2258.

Gudmundsson, A., 2014. Energy release in great earthquakes and eruptions. *Frontiers in Earth Science*, **2**, doi:10.3389/feart.2014.00010.

Gudmundsson, A., 2016. The mechanics of large volcanic eruptions. *Earth-Science Reviews*, **163**, 72–93.

Gudmundsson, A., 2017. *The Glorious Geology of Iceland's Golden Circle*. Berlin: Springer Verlag.

Gudmundsson, A., Brenner, S. L., 2001. How hydrofractures become arrested. *Terra Nova*, **13**, 456–462.

Gudmundsson, A., Brenner, S. L., 2004. How mechanical layering affects local stresses, unrests, and eruptions of volcanoes. *Geophysical Research Letters*, **31**, doi:10.1029/2004GL020083.

Gudmundsson, A., Homberg, C., 1999. Evolution of stress fields and faulting in seismic zones. *Pure and Applied Geophysics*, **154**, 257–280.

Gudmundsson, A., Loetveit, I. F., 2005. Dyke emplacement in layered and faulted rift zone. *Journal of Volcanology and Geothermal Research*, **144**, 311–327.

Gudmundsson, A., Philipp, S. L., 2006. How local stress fields prevent volcanic eruptions. *Journal of Volcanology and Geothermal Research*, **158**, 257–268.

Gudmundsson, A., Bergerat, F., Angelier, J., Villemin, T. 1992. Extensional tectonics of Southwest Iceland. *Bulletin of the Geological Society of France*, **163**, 561–570.

Gudmundsson, A., Friese, N., Galindo, I., Philipp, S. L., 2008. Dike-induced reverse faulting in a graben. *Geology*, **36**, 123–126.

Gupta, A. B., 2015. *Classical Mechanics and Properties of Matter*. Kolkata: Books & Allied.

Hamill, P., 2014. *A Student's Guide to Lagrangians and Hamiltonians*. Cambridge: Cambridge University Press.

Holmes, A., 1965. *Principles of Physical Geology*. London: Thomas Nelson.

Isida, M., 1955. On the tension of a semi-infinite plate with an elliptic hole. *Scientific Papers of the Faculty of Engineering, Tokushima University*, **5**, 75–95.

Jaeger, J. C., 1961. The cooling of irregularly shaped igneous bodies. *American Journal of Science*, **259**, 721–734.

Janssen, V., 2008. *GPS-Based Volcano Deformation*. Saarbrücken: VDM Verlag.

Kanamori, H., Anderson, D.L., 1975. Theoretical basis of some empirical relations in seismology. *Bulletin of the Seismological Society of America*, **65**, 1074–1095.

Karson, J. A., 2017. The Iceland plate boundary zone: propagating rifts, migrating transforms, and rift-parallel strike-slip faults. *Geochemistry, Geophysics, Geosystems*, **18**, 4043–4054.

Kavanagh, J. L., Sparks, R. S. J., 2011. Insights of dyke emplacement mechanics from detailed 3D dyke thickness datasets. *Journal of the Geological Society of London*, **168**, 965–978.

Kavanagh, J., Menand, T., Sparks, R. S. J., 2006. An experimental investigation of sill formation and propagation in layered elastic media. *Earth and Planetary Science Letters*, **245**, 799–813.

Kavanagh, J., Boutelier, D., Cruden, A. R., 2015. The mechanics of sill inception, propagation and growth: experimental evidence for rapid reduction in magmatic overpressure. *Earth and Planetary Science Letters*, **421**, 117–128.

Kumagai, H., Ohminato, T., Nakano, M., et al., 2001. Very-long-period seismic signals and caldera formation at Miyake Island, Japan. *Science*, **293**, 687–690.

Kusumoto, S., Gudmundsson, A., 2014. Displacement and stress fields around rock fractures opened by irregular overpressure variations. *Frontiers in Earth Science*, **2**, doi:10.3389/feart.2014.00007.

Kusumoto, S., Geshi, N., Gudmundsson, A., 2013. Inverse modeling for estimating fluid-overpressure distributions and stress intensity factors from arbitrary open-fracture geometry. *Journal of Structural Geology*, **46**, 92–98.

Lacazette, A., Geiser, P., 2013. Comment on Davis et al., 2012 – Hydraulic fractures: how far will they go? *Marine and Petroleum Geology*, **43**, 517–519.

Lauthold, J., Muntener, O., Baumgartener, L. P., et al., 2014. A detailed geochemical study of a shallow arc-related laccolith: the Torres del Paine Mafic Complex (Patagonia). *Journal of Petrology*, **54**, 273–303.

Lu, Z., Dzurisin, D., 2014. *InSAR Imaging of Aleutian Volcanoes: Monitoring a Volcanic Arc from Space*. Berlin: Springer Verlag.

Manconi, A., Walter, T R., Amelung, F., 2007. Effects of mechanical layering on volcano deformation. *Geophysical Journal International*, **170**, 952–958.

Marinoni, L. B., Gudmundsson, A., 1999. Geometry, emplacement, and arrest of dykes. *Annales Tectonicæ*, **13**, 71–92.

Marti, J., Geyer, A., Folch, A., Gottsmann, J., 2008. A review on collapse caldera modelling. In Gottsmann, J., Marti, J. (eds.), *Caldera Volcanism: Analysis, Modelling and Response*. Amsterdam: Elsevier, pp. 233–283.

Marti, J., Villasenor, A., Geyer, A., Lopez, C., Tryggvason, A., 2017. Stress barriers controlling lateral migration of magma revealed by seismic tomography. *Scientific Reports*, **7**, doi:10.1038/srep40757.

Mason, B. G., Pyle, D. M., Oppenheimer, C., 2004. The size and frequency of the largest explosive eruptions on Earth. *Bulletin of Volcanology*, **66**, 735–748, doi:10.1007/s00445-004-0355-9.

Masterlark, T., 2007. Magma intrusion and deformation predictions: sensitivities to the Mogi assumptions. *Journal of Geophysical Research*, **112**, doi:10.1029/2006JB004860.

Meirovitch, L., 2003. *Methods of Analytical Dynamics*. New York, NY: Dover.

Menand, T., Daniels, K. A., Benghiat, P., 2010. Dyke propagation and sill formation in a compressive tectonic environment. *Journal of Geophysical Research*, **115**, doi:10.1029/2009JB006791.

Michel, J., Baumgartner, L., Putlitz, B., Schaltegger, U., Ovtcharova, M., 2008. Incremental growth of the Patagonian Torres del Paine laccolith over 90 k.y. *Geology*, **36**, 459–462, doi:10.1130/G24546A.1.

Mindlin, R. D., 1936. Force at a point in the interior of a semi-infinite solid. *Physics*, **7**, 195–202.

Mogi, K., 1958. Relations between eruptions of various volcanoes and the deformations of the ground surfaces around them. *Bulletin of the Earthquake Research Institute University of Tokyo*, **36**, 99–134.

Murase, T., McBirney, A. R., 1973. Properties of some common igneous rocks and their melts at high temperatures. *Geological Society of America Bulletin*, **84**, 3563–3592.

Newhall, C. G., Dzurisin, D., 1988. *Historical Unrest of Large Calderas of the World*. Reston, VA: US Geological Survey.

Newhall, C. G., Self, S., 1982. The volcanic explosivity index (VEI): an estimate of explosive magnitude for historical volcanism. *Journal of Geophysical Research*, **87**, 1231–1238.

Okada, Y., 1985. Surface deformation due to shear and tensile faults in a half-space. *Bulletin of the Seismological Society of America*, **75**, 1135–1154.

Okada, Y., 1992. Internal deformation due to shear and tensile faults in half space. *Bulletin of the Seismological Society of America*, **82**, 1018–1040.

Passarelli, L., Rivalta, E., Cesca, S., Aoki, Y., 2015. Stress changes, focal mechanisms, and earthquake scaling laws for the 2000 dike at Miyakejima (Japan). *Journal of Geophysical Research*, **120**, 4130–4145.

Paterson, M. S., Wong, T. W., 2005. *Experimental Rock Deformation: The Brittle Field*, 2nd edn. Berlin: Springer Verlag.

Philipp, S., Philipp, S. L., Afsar, F., Gudmundsson, A., 2013. Effects of mechanical layering on the emplacement of hydrofractures and fluid transport in reservoirs. *Frontiers of Earth Science*, **1**, doi:10.3389/feart.2013.00004.

Pollard, D. D., Delaney, P. T., Duffield, W. A., Endo, E. T., Okamura, A. T., 1983. Surface deformation in volcanic rift zones. *Tectonophysics*, **94**, 541–584.

Pyle, D. M., 2000. Sizes of volcanic eruptions. In Sigurdsson, H. (ed.), *Encyclopedia of Volcanoes*. New York, NY: Academic Press, pp. 263–269.

Reddy, J. N., 2002. *Energy Principles and Variational Methods in Applied Mechanics*, 2nd edn. Hoboken, New Jersey: Wiley.

Richards, T. H., 1977. *Energy Methods in Stress Analysis*. Chichester: Ellis Horwood.

Rivalta, E., Taisne, B., Bunger, A. P., Katz, R. F., 2015. A review of mechanical models of dike propagation: schools of thought, results and future directions. *Tectonophysics*, **638**, 1–42.

Roth, F., 1993. Deformations in a layered crust due to a system of cracks: modeling the effect of dike injections or dilatancy. *Journal of Geophysical Research*, **98**, 4543–4551.

Rougier, J., Sparks, S., Cashman, K., Brown, S., 2018. The global magnitude–frequency relationship for large explosive eruptions. *Earth and Planetary Science Letters*, **482**, 621–629.

Rubin, A. M., 1995. Propagation of magma-filled cracks. *Annual Reviews of Earth and Planetary Sciences*, **23**, 287–336.

Rubin, A. M., Pollard, D. D., 1988. Dike-induced faulting in rift zones of Iceland and Afar. *Geology*, **16**, 413–417.

Ryan, M. P., 1993. Neutral buoyancy and the structure of mid-ocean ridge magma reservoirs. *Journal of Geophysical Research*, **98**, 22 321–22 338.

Sanford, R. J., 2003. *Principles of Fracture Mechanics*. Upper Saddle River, NJ: Prentice-Hall.

Scholz, C. H., 1990. *The Mechanics of Earthquakes and Faulting*. Cambridge: Cambridge University Press.

Segall, P., 2010. *Earthquake and Volcano Deformation*. Princeton, NJ: Princeton University Press.

Segall, P., Llenos, A. L., Yun, S. H., Bradley, A. M., Syracuse, E. M., 2013. Time-dependent dike propagation from joint inversion of seismicity and deformation data. *Journal of Geophysical Research*, **118**, doi:10.1002/2013JB010251.

Self, S., 2006. The effects and consequences of very large explosive volcanic eruptions. *Philosophical Transactions of the Royal Society A*, **364**, 2073–2097.

Shapiro, S. A., 2018. *Fluid-Induced Seismicity*. Cambridge: Cambridge University Press.

Sigmundsson, F., Hreinsdottir, S., Hooper, A., et al., 2010. Intrusion triggering of the 2010 Eyjafjallajökull explosive eruption. *Nature*, **468**, 426–430.

Sobradelo, R., Bartolini, S., Martí, J., 2013. HASSET: a probability event tree tool to evaluate future volcanic scenarios using Bayesian inference. *Bulletin of Volcanology*, **76**, 1–15.

Sobradelo, R., Martí, J., Kilburn, C., López, C., 2015. Probabilistic approach to decision-making under uncertainty during volcanic crises: retrospective application to the El Hierro (Spain) 2011 volcanic crisis. *Natural Hazards*, **76**, 979–998.

Sparks, R. S. J., Aspinall, W. P., Crosweller, H. S., Hincks, T. K., 2013. Risk and uncertainty assessment of volcanic hazards. In Rougier, J., Sparks, R. S. J., Hill, L (eds.), *Risk and Uncertainty Assessment for Natural Hazards*. Cambridge: Cambridge University Press, pp. 365–397.

Spera, F. J., 2000. Physical properties of magmas. In Sigurdsson, H. (ed.), *Encyclopedia of Volcanoes*. New York, NY: Academic Press, pp. 171–190.

Steketee, J. A., 1958. On Volterra's dislocations in a semi-infinite elastic medium. *Canadian Journal of Physics*, **36**, 192–205.

Sun, R. J. 1969. Theoretical size of hydraulically induced horizontal fractures and corresponding surface uplift in an idealized medium. *Journal of Geophysical Research*, **74**, 5995–6011.

Tauchert, T. R., 1981. *Energy Principles in Structural Mechanics*. Malabar, FL: Krieger.

Tibaldi, A., 2015. Structure of volcano plumbing systems: a review of multi-parametric effects. *Journal of Volcanology and Geothermal Research*, **298**, 85–135.

Townsend, M., Pollard, D. D., Smith, R., 2017. Mechanical models for dikes: a third school of thought. *Tectonophysics*, **703–704**, 98–118.

Tsuchida, E., Nakahara, I., 1970. Three-dimensional stress concentration around a spherical cavity in a semi-infinite elastic body. *Japan Society of Mechanical Engineers Bulletin*, **13**, 499–508.

Valko, P., Economides, M. J., 1995. *Hydraulic Fracture Mechanics*. New York, NY: Wiley.

Villemin, T., Bergerat, F., Angelier, J., Lacasse, C., 1994. Brittle deformation and fracture patterns on oceanic rift shoulders: the Esja peninsula, SW Iceland. *Journal of Structural Geology*, **16**, 1641–1654.

Volterra, V., 1907. On the equilibrium of multiply-connected elastic bodies. *Annales scientifiques de l'École Normale Supérieure*, **24**, 401–517 (in French; English translation).

Wadge, G., 1981. The variation of magma discharge during basaltic eruptions. *Journal of Volcanology and Geothermal Research*, **11**, 139–168.

Wallerstein, D. V., 2002. *A Variational Approach to Structural Analysis*. New York, NY: Wiley.

Washizu, K., 1975. *Variational Methods in Elasticity and Plasticity*. Amsterdam: Elsevier.

Woods, A. W., Huppert, H. E., 2003. On magma chamber evolution during slow effusive eruptions. *Journal of Geophysical Research*, **108**, 2403, doi:10.1029/2002JB002019.

Wu, Y. S. (ed.), 2017. *Hydraulic Fracture Modeling*. Houston, TX: Gulf Publishing.

Yew, C. H., Weng, X., 2014. *Mechanics of Hydraulic Fracturing*, 2nd edn. Houston, TX: Gulf Publishing.

Yokoyama, I., 1957. Energies in active volcanoes. *Bulletin of the Earthquake Research Institute Tokyo*, **35**, 75–97.

Zang, A., Stephansson, O., 2010. *Stress Field of the Earth's Crust*. Berlin: Springer Verlag.

Zoback, M. D., Harjes, H. P., 1997. Injection-induced earthquakes and crustal stress at 9 km depth at the KTB deep drilling site, Germany. *Journal of Geophysical Research*, **102**, 18 477–18 491.

Zobin, V. M., 2003. *Introduction to Volcanic Seismology*. Amsterdam: Elsevier.

Appendix A Units, Dimensions, and Prefixes

Scientists and engineers use the International System of Units (in French: Système Internationale d'Unités – hence the acronym SI units), which has **seven base units**, relying on simple physical effects, and many **derived units**, only some of which (such as force and stress) are listed below. Also provided below are the dimensions of some of the quantities, where L is length, M is mass, T is time, and F is force. More details are given by Huntley (1967), Gottfried (1979), Emiliani (1995), Benenson et al. (2002), Woan (2003), and Deeson (2007). The reference list for all the appendices is at the end of Appendix F.

A.1 SI Base Units

Physical quantity	Unit name	Symbol	Dimensions
Length	metre	m	L
Mass	kilogram	kg	M
Time (interval)	second	s	T
Electric current	ampere	A	I
Thermodynamic temperature	kelvin	K	Θ
Amount of substance	mole	mol	N
Luminous intensity	candela	cd	J

A.2 Derived SI Units of Some Quantities

Physical quantity	Unit name	Symbol	Dimensions	SI units
Area		A	L^2	m^2
Energy	joule	J	$L^2\,M\,T^{-2}$	N m
Force	newton	N	$L\,M\,T^{-2}$	$kg\,m\,s^{-2}$
Heat	joule	J	$L^2\,M\,T^{-2}$	N m
Power	watt	W	$L^2\,M\,T^{-3}$	$J\,s^{-1}$
Pressure	pascal	Pa	$L^{-1}\,M\,T^{-2}$	$N\,m^{-2}$
Shear modulus	pascal	G,μ	$L^{-1}\,M\,T^{-2}$	$N\,m^{-2}$
Speed		v	$L\,T^{-1}$	$m\,s^{-1}$

Physical quantity	Unit name	Symbol	Dimensions	SI units
Stress	pascal	Pa	$L^{-1} M T^{-2}$	$N\,m^{-2}$ $kg\,m^{-1}\,s^{-2}$
Velocity		$\bar{v}$	$L\,T^{-1}$	$m\,s^{-1}$
Viscosity (dynamic)		μ, η	$L^{-1} M T^{-1}$	$Pa\,s$ $kg\,m^{-1}\,s^{-1}$
Viscosity (kinematic)		v	$L^2 T^{-1}$	$m^2\,s^{-1}$
Volume		V	L^3	m^3
Weight	newton	W, w	$L\,M\,T^{-2}$	N
Work	joule	J	$L^2 M T^{-2}$	$N\,m$
Young's modulus	pascal	Pa	$L^{-1} M T^{-2}$	$N\,m^{-2}$

A.3 SI Prefixes

Here, the EU name means the name of the factor as commonly used in continental European countries and many others. UK and US names are specified only when they differ from the EU names. Note that SI multiples and submultiples go up and down in steps of thousand (10^3). It follows that hecto, deca, deci, and centi are not strictly SI prefixes. Also, widely used abbreviations for centimetres and millimetres are cm and mm, respectively.

Value/factor	Prefix	Symbol	EU name	UK name	US name
10^{24}	yotta	Y	quadrillion		septillion
10^{21}	zetta	Z	trilliard	thousand trillion	septillion
10^{18}	exa	E	trillion		quintillion
10^{15}	peta	P	billiard	thousand billion	quadrillion
10^{12}	tera	T	billion		trillion
10^{9}	giga	G	milliard	thousand million	billion
10^{6}	mega	M	million		
10^{3}	kilo	k	thousand		
10^{2}	hecto	h	hundred		
10^{1}	deca/deka	da	ten		
10^{0}	none	none	none		
10^{-1}	deci	d	tenth		
10^{-2}	centi	c	hundredth		
10^{-3}	milli	m	thousandth		
10^{-6}	micro	μ	millionth		
10^{-9}	nano	n	milliardth	thousand millionth	billionth
10^{-12}	pico	p	billionth		trillionth

Value/factor	Prefix	Symbol	EU name	UK name	US name
10^{-15}	femto	f	billardth	thousand billionth	quadrillionth
10^{-18}	atto	a	trillionth		quintillionth
10^{-21}	zepto	z	trilliardth	thousand trillionth	sextillionth
10^{-24}	yocto	y	quadrillionth		septillionth

Appendix B The Greek Alphabet

Lower case	Upper case	Name
α	A	alpha
β	B	beta
γ	Γ	gamma
δ	Δ	delta
ε	E	epsilon
ζ	Z	zeta
η	H	eta
θ	Θ	theta
ι	I	iota
κ	K	kappa
λ	Λ	lambda
μ	M	mu
ν	N	nu
ξ	Ξ	xi
o	O	omicron
π/ϖ	Π	pi
ρ	P	rho
σ/ς	Σ	sigma
τ	T	tau
υ	Y	upsilon
ϕ/φ	Φ	phi
χ	X	chi
ψ	Ψ	psi
ω	Ω	omega

Appendix C Some Mathematical and Physical Constants

Quantity	Symbol	Value
Ratio of the circumference to the diameter of a circle	π	3.1516
Exponential constant, base of natural logarithms	e	2.7183
Typical acceleration due to gravity at the Earth's surface	g	9.81 m s^{-2}
Acceleration due to gravity at the surface at the equator	g_e	9.78 m s^{-2}
Acceleration due to gravity at the surface at the poles	g_p	9.83 m s^{-2}
Gravitational constant	G	$6.673 \times 10^{-11} \text{ N m}^2 \text{ kg}^{-2}$
One calendar year, 365 days	a/yr	$3.1536 \times 10^7 \text{ s}$
Earth's equatorial radius	R_e	$6.3781 \times 10^6 \text{ m}$
Earth's polar radius	R_p	$6.3567 \times 10^6 \text{ m}$
Earth's mean radius	R	$6.37 \times 10^6 \text{ m}$
Earth's volume	V	$1.083 \times 10^{21} \text{ m}^3$
Earth's mass	M	$5.974 \times 10^{24} \text{ kg}$
Earth's mean density	ρ	$5.515 \times 10^3 \text{ kg m}^{-3}$
Earth's surface area	A	$5.10 \times 10^{14} \text{ m}^2$

Appendix D Elastic Constants

Below are values of Young's moduli and Poisson's ratios of some common rocks (Appendix D.1). All the other elastic moduli or constants can be derived from these two using the relations in Appendix D.2. These values are derived from (mostly static) laboratory measurements, and these are normally different from the *in situ* values. All rocks show a great variety in elastic properties, in particular when measured *in situ*. The 'typical values' shown below are rounded and approximate and are meant as a rough guide. Where *in situ* values are available, such as for tensile strengths, these are also given. These data on rock properties as measured in the laboratory are derived from tables and information in many books including Jaeger and Cook (1979), Jumikis (1979), Carmichael (1989), Jeremic (1994, for rock salt), Waltham (1994), Hansen (1998), Nilsen and Palmström (2000), Myrvang (2001), Schön (2004), Paterson and Wong (2005), Fjaer et al. (2008), and Mavko et al. (2009). When data are not known, or not measureable (such as the tensile strength of unconsolidated sand or gravel), there is no number in the appropriate box in the tables in Appendices D and E.

D.1 Typical Young's Moduli and Poisson's Ratios

Rock type	Young's modulus: normal range, GPa	Young's modulus: typical values, GPa	Poisson's ratio: normal range	Poisson's ratio: typical values
Sedimentary rocks				
Limestone	7.8–150.0	20–70	0.10–0.44	0.22–0.34
Dolomite	19.6–93.0	30–80	0.08–0.37	0.10–0.30
Sandstone	0.4–84.3	20–50	0.05–0.45	0.10–0.30
Shale	0.14–44.0	1–20	0.04–0.30	0.10–0.25
Marl/mudstone	0.002–0.25	0.025–0.10		
Chalk	0.5–30.0	1–15	0.05–0.37	0.15–0.30
Anhydrate	48.8–86.4	70–80	0.20–0.34	0.25–0.30
Gypsum	0.74–36.0	15–35	0.14–0.47	0.18–0.30
Rock salt	5.0–42.3	15–30	0.09–0.50	0.29–0.38
Conglomerate	24.0–51.0	30–40	0.22–0.23	0.22–0.23
Unconsolidated sand	0.01–0.1	0.05	0.45	0.45
Clay	0.003–0.5	0.05–0.1	0.40	0.40

Rock type	Young's modulus: normal range, GPa	Young's modulus: typical values, GPa	Poisson's ratio: normal range	Poisson's ratio: typical values
Igneous rocks				
Basalt	20.0–128.0	50–90	0.14–0.30	0.23–0.27
Diabase	29.4–114.0	55–100	0.10–0.33	0.24–0.28
Gabbro	58.4–108.0	60–80	0.12–0.48	0.15–0.25
Granite	17.2–70.5	20–50	0.12–0.34	0.15–0.24
Diorite	43.0–92.5	50–80	0.13–0.26	0.18–0.22
Syenite	21.3–86.3	30–50	0.15–0.34	0.20–0.30
Tuff	0.05–5	0.5–1.0	0.21–0.33	0.25–0.30
Rhyolite	15.5–37.6	20–30	0.10–0.28	0.14–0.19
Metamorphic rocks				
Gneiss	9.5–147.0	20–60	0.03–0.346	0.1–0.2
Marble	28.0–100.0	40–90	0.11–0.38	0.24–0.30
Schist	13.7–98.1	20–50	0.01–0.31	0.1–0.2
Quartzite	25.5–97.5	40–60	0.09–0.23	0.15–0.20
Water ice	8.6–12.0	9–10	0.33–0.35	0.33

D.2 Relations among the Elastic Constants for Isotropic Rock

Each elastic constant is here presented in terms of two other constants. For example, Young's modulus E is presented in terms of Poisson's ratio v and shear modulus G. The other two constants used here are Lamé's constant λ and the bulk modulus K. Some of the more complex expressions are omitted (cf. Slaughter, 2002).

	E	v	G	K	λ
λ, G	$\frac{G(3\lambda+2G)}{\lambda+G}$	$\frac{\lambda}{2(\lambda+G)}$	G	$\lambda+\frac{2}{3}G$	λ
λ, E	E				λ
λ, v	$\frac{\lambda(1+v)(1-2v)}{v}$	v	$\frac{\lambda(1-2v)}{2v}$	$\frac{\lambda(1+v)}{3v}$	λ
λ, K	$\frac{9K(K-\lambda)}{3K-\lambda}$	$\frac{\lambda}{2(\lambda+G)}$	$\frac{3(K-\lambda)}{2}$	K	λ
G, E	E	$\frac{E-2G}{2G}$	G	$\frac{GE}{3(3G-E)}$	$\frac{G(E-2G)}{3G-E}$
G, v	$2G(1+v)$	v	G	$\frac{2G(1+v)}{3(1-2v)}$	$\frac{2Gv}{1-2v}$
G, K	$\frac{9KG}{3K+G}$	$\frac{3K-2G}{6K+2G}$	G	K	$K-\frac{2}{3}G$

	E	v	G	K	λ
E, v	E	v	$\frac{E}{2(1+v)}$	$\frac{E}{3(1-2v)}$	$\frac{Ev}{(1+v)(1-2v)}$
E, K	E	$\frac{3K-E}{6K}$	$\frac{3KE}{9K-E}$	K	$\frac{3K(3K-E)}{9K-E}$
K, v	$3K(1-2v)$	v	$\frac{3K(1-2v)}{2(1+v)}$	K	$\frac{3Kv}{1+v}$

Appendix E Properties of Some Crustal Materials

E.1 Rock Densities, Strengths, and Internal Friction

All the values given below refer to dry conditions except that the upper values for gravel–sand and for clay refer to saturated specimens. The sources of the data are mostly the same as in Appendix D.

Rock type	Dry density ρ_r, kg m^{-3}	Compressive strength C_0, MPa	Tensile strength T_0, MPa	Coefficient of internal friction μ
Sedimentary rocks				
Limestone	2300–2900	4–250	1–25	0.70–1.20
Dolomite	2400–2900	80–250	3–25	0.40
Sandstone	2000–2800	6–170	0.4–25	0.50–0.70
Shale	2300–2800	10–160	2–10	0.27–0.58
Marl/mudstone	2300–2700	26–70	1	0.78–1.07
Chalk	1400–2300	5–30	0.3	0.47
Anhydrate	2700–3000	70–120	5–12	
Gypsum	2100–2300	4–40	0.8–4	0.58
Rock salt	1900–2200	9–23	0.2–3	
Sand, gravel	1400–2300		0	0.58–1.20
Clay	1300–2300	0.2–0.5	0.2	0.35–0.52
Igneous rocks				
Basalt	2600–3000	80–410	6–29	1.11–1.19
Diabase	2600–2900	120–250	5–13	1.19–1.43
Gabbro	2850–3100	150–290	5–29	0.18–0.66
Granite	2500–2700	120–290	4–25	0.60–1.73
Diorite	2650–2900	70–180	10–14	
Andesite	2500–2800	70–200		
Syenite	2550–2700	100–340	10–13	
Tuff	1600	0.4–44	0.1–0.8	0.21–0.36
Rhyolite	2150–2500	180–260	16–21	

Rock type	Dry density ρ_r, kg m^{-3}	Compressive strength C_0, MPa	Tensile strength T_0, MPa	Coefficient of internal friction μ
Metamorphic rocks				
Gneiss	2500–2900	80–250	4–20	0.60–0.71
Marble	2600–2800	50–200	5–20	0.62–1.20
Schist	2500–2900	20–100	2	1.90
Quartzite	2600–2700	90–300	3–5	0.48–1.73
Water ice, 0 °C	917 (–934)			

E.2 General Rock and Fluid Properties

The main data presented in the first part of this appendix are from Byerlee (1978), Atkinson (1987), Schultz (1997), Amadei and Stephansson (1997), Bell (2000), Schön (2004), Paterson and Wong (2005), and Gudmundsson (2009). Other references are cited where appropriate.

1. *In situ* **tensile strengths** of most rocks are between 0.5 and 6 MPa (Haimson and Rummel, 1982; Schultz, 1997; Amadei and Stephansson, 1997). The highest value, 9 MPa, was obtained at about 9 km depth in the KTB drill hole in Germany (Amadei and Stephansson, 1997).

2. Experimental **rock friction relationships** for different normal stresses σ_n were obtained by Byerlee (1978). These relationships are sometimes referred to as **Byerlee's law** and, using Eq. (5.14), may be stated as follows:

For normal stress in the range 10 MPa $\leq \sigma_n \leq$ 200 MPa, Eq. (5.14) becomes

$$\tau = 0.85\sigma_n \qquad (E.1)$$

For normal stress in the range 200 MPa $\leq \sigma_n \leq$ 1500 MPa, Eq. (5.14) becomes

$$\tau = 50 + 0.6\sigma_n \qquad (E.2)$$

In Eq. (E.1) the inherent shear strength, τ_0, in Eq. (5.14) would be zero and the coefficient of internal friction μ, 0.85. In Eq. (E.2) the inherent shear strength τ_0, in Eq. (5.14) would be 50 MPa and the coefficient of internal friction μ, 0.6. These laws hold for normal stresses that correspond to crustal depths of some 50–60 km and are largely independent of roughness on the fault surface and the type of rock being faulted (that is, they apply to a variety of rock types). The coefficients of internal friction, 0.6–0.85, are similar to those of many rocks in Appendix E.1. However, the fluid pressure (effective stress) does not enter these laws (Chapters 7, 8) and they are, as yet, purely empirical since they have no clear theoretical explanation.

3. **Laboratory fracture and material toughnesses**. Compilations of laboratory data on fracture and material toughnesses are provided by Fourney (1983), Atkinson and Meredith (1987), Paterson and Wong (2005), Nasseri et al. (2006), and Nasseri and Mohanty (2008).

Typical values for **fracture toughness** of rock K_c for mode I (extensions) fractures are 0.5–3 MPa m$^{1/2}$. Few rock specimens have fracture toughesses as high as 5–20 MPa m$^{1/2}$

(Fourney, 1983). These values do not change much with increasing temperature and pressure. For example, basalt specimens from Iceland and Italy, subject to temperatures of 20–600 °C and pressures of as much as 30 MPa, yield fracture toughnesses of 1.4–3.8 MPa m$^{1/2}$ (Balme et al., 2004). At pressures of 60–100 MPa, however, some limestone and sandstone specimens yield fracture toughnesses of 5 MPa m$^{1/2}$ or higher.

Typical values for **material toughness** of rock, G_c, for mode I fractures are 20–400 J m^{-2}. The highest and lowest specimen values given by Atkinson and Meredith (1987) are 1580 J m^{-2} (for sandstone normal to bedding) and 15 J m^{-2} (for limestone), respectively. The highest and lowest values given by Fourney (1983) are 2298 J m^{-2} (for basalt) and 8 J m^{-2} (for limestone). Material toughness values for shear fractures are normally much higher. For example, for mode II fractures in various rock specimens, most material toughness values are of the order of 0.01 MJ m^{-2} (Li, 1987).

4. ***In situ* fracture and material toughnesses**. Many *in situ* fracture and material toughnesses for various crustal segments and fracture types have been provided, as summarised by Gudmundsson (2009). Consider first estimates for faults (mode II and III). Estimates of materal touchnesses for strike-slip faults range from 260 J m^{-2} to 100 MJ m^{-2}, but the most common values are in the range of 0.15–17 MJ m^{-2} (Li, 1987; Rice, 2006; Rice and Cocco, 2007). For dip-slip faults, the material toughness is estimated at 2.3 MJ m^{-2} and the corresponding fracture toughness (stress intensity) at 150 MPa m$^{1/2}$ (Gudmundsson, 2009).

Many estimates have been made of the *in situ* toughnesses of mode I fractures, primarily using dikes. Most fracture-toughnesses (stress-intensity) values are in the range of 30–150 MPa m$^{1/2}$ (Delaney and Pollard, 1981; Rubin and Pollard, 1987; Parfitt, 1991; Rivalta and Dahm, 2006), although much higher values have been suggested (Jin and Johnson, 2008). For typical dikes, Gudmundsson (2009) obtained material toughnesses of 1.3–47 MJ m^{-2} and corresponding fracture toughnesses (stress intensities) of 114–690 MPa m$^{1/2}$.

5. **Yield strengths and brittle–ductile transition stresses of some rocks**. Bell (2000) and Paterson and Wong (2005) provide the yield strengths and stresses at brittle–ductile transition for some rocks. For gypsum, the yield strengths, that is, the stresses at which plastic yield (or flow) starts, are generally between 7 MPa and 29 MPa (Bell, 2000). Peterson and Wong (2005) give the following brittle–ductile transition stresses:

Rock	Stress at brittle–ductile transition, MPa
Limestone/marble	10–100
Chalk	<10
Dolomite	100–200
Gypsum	40
Anhydrate	100
Rock salt	<20
Quartzite	600
Sandstone	200–300
Shale/siltstone	<100
Basalt	300
Porous lava flows	30–100

Non-porous igneous and metamorphic rocks at **room temperature** (25 °C) are normally brittle up to confining pressures (σ_3) of at least 1000 MPa, corresponding to roughly 30 km depth. And some granites behave as brittle up to confining pressures of 3000 MPa. Increasing the temperature generally increases the ductility, that is, lowers the depth to the brittle–ductile transition. However, the effects of temperature have much more limited effects in increasing ductility if the confining pressure does not also increase. Generally, for non-porous igneous and metamorphic rocks, temperatures of at least 300–500 °C (600–800 K) are needed for them to behave as ductile. For example, limited ductility occurred in specimens of basalt and granite at temperatures of 600–900 °C under a compressive loading of 500 MPa. This pressure corresponds to crustal depths of roughly 18 km. Strain rates, however, also affect the depth to the brittle–ductile transition; lower strain rates decrease this depth.

6. **Densities and viscosities of some crustal fluids**. The data are from Middleton and Wilcock (2001) and Smits (2000).

Fluid	Density, kg m^{-3}	Dynamic viscosity, Pa s
Water at 0 °C	999.9	1.787×10^{-3}
Water at 5 °C	1000	1.519×10^{-3}
Water at 40 °C	992.2	0.653×10^{-3}
Water at 80 °C	971.8	0.355×10^{-3}
Water at 100 °C	968.4	0.282×10^{-3}
Seawater at 20 °C	1025	1.070×10^{-3}
Crude oils	850–950	0.1–0.01
Glacier ice	920	1×10^{13}
Rock salt	1900–2200	3×10^{15}

Appendix F Physical Properties of Lavas and Magmas

The values for the (flowing) lavas refer to their values at the eruption sites (at the crater cones/ fissure openings). The values refer to atmospheric pressure. Please note that the viscosities of the magmas vary over several orders of magnitude depending on their compositions, temperatures, volatile content, and other factors; the values given here are typical ranges. Data are mostly from Kilburn (2000), Murase and McBirney (1973), and Spera (2000).

F.1 Lavas

Composition	Density, kg m^{-3}	Dynamic viscosity, Pa s	Temperature, °C
Basalt	2600–2800	10^{2-3}	1050–1200
Andesite	2450	10^{4-7}	950–1170
Rhyolite	2200	10^{9-13}	700–900
Komatite	2800	10^0	1600

F.2 Magmas

Composition	Density, kg m^{-3}	Dynamic viscosity, Pa s	Temperature, °C	SiO$_2$, weight %
Picritic	2800–2900	10^{1-2}	1400–1500	<45
Basaltic	2650–2800	10^{1-2}	1050–1300	45–52
Andesitic	2450–2500	10^{3-4}	950–1170	52–66
Rhyolitic	2180–2250	10^{7-10}	750–1000	>66

References

Amadei, B., Stephansson, O., 1997. *Rock Stress and Its Measurement*. London: Chapman & Hall.

Atkinson, B. K. (ed.), 1987. *Fracture Mechanics of Rock*. London: Academic Press.

Atkinson, B. K., Meredith, P. G., 1987. Experimental fracture mechanics data for rocks and minerals. In Atkinson, B. K. (ed.), *Fracture Mechanics of Rock*. London: Academic Press, pp. 477–525.

Balme, M. R., Rocchi, V., Jones, C., et al., 2004. Fracture toughness measurements on igneous rocks using a high-pressure, high-temperature rock fracture mechanics cell. *Journal of Volcanology and Geothermal Research*, **132**, 159–172.

Bell, F. G., 2000. *Engineering Properties of Rocks*, 4th edn. Oxford: Blackwell.

Benenson, W., Harris, J. W., Stocker, H., Lutz, H. (eds.), 2002. *Handbook of Physics*. Berlin: Springer Verlag.

Byerlee, J. D., 1978. Friction of rocks. *Pure and Applied Geophysics*, **116**, 615–626.

Carmichael, R. S., 1989. *Practical Handbook of Physical Properties of Rocks and Minerals*. Boca Raton, FL: CRC Press.

Deeson, E., 2007. *Internet-Linked Dictionary of Physics*. London: Collins.

Delaney, P., Pollard, D. D., 1981. Deformation of host rocks and flow of magma during growth of minette dikes and breccia-bearing intrusions near Ship Rock, New Mexico. *US Geolological Survey Professional Paper*, **1202**, 1–61.

Emiliani, C., 1995. *The Scientific Companion*, 2nd edn. New York, NY: Wiley.

Fjaer, E., Holt, R. M., Horsrud, P., Raaen, A. M., Risnes, R., 2008. *Petroleum Related Rock Mechanics*, 2nd edn. Amsterdam:Elsevier.

Fourney, W. L., 1983. Fracture control blasting. In Rossmanith, H. P. (ed.), *Rock Fracture Mechanics*. New York, NY: Springer, pp. 301–319.

Gottfried, B. S., 1979. *Introduction to Engineering Calculations*. New York, NY: McGraw-Hill.

Gudmundsson, A., 2009. Toughness and failure of volcanic edifices. *Tectonophysics*, **471**, 27–35.

Haimson, B. C., Rummel, F., 1982. Hydrofracturing stress measurements in the Iceland research drilling project drill hole at Reydarfjordur, Iceland. *Journal of Geophysical Research*, **87**, 6631–6649.

Hansen, S. E., 1998. Mechanical properties of rocks. Report STF22 A98034, Sintef, Trondheim (in Norwegian).

Huntley, H. E., 1967. *Dimensional Analysis*. New York, NY: Dover.

Jaeger, J. C., Cook, N. G. W., 1979. *Fundamentals of Rock Mechanics*, 3rd edn. London: Chapman and Hall.

Jeremic, M. L., 1994. *Rock Mechanics in Salt Mining*. Rotterdam:Balkema.

Jin, Z. H., Johnson, S. E., 2008. Magma-driven multiple dike propagation and fracture toughness of crustal rocks. *Journal of Geophysical Research*, **113**, B03206.

Jumikis, A. R., 1979. *Rock Mechanics*. Clausthal:Trans Tech Publications.

Kilburn, C. J., 2000. Lava flows and flow fields. In Sigurdsson, H. (ed.), *Encyclopedia of Volcanoes*. Academic Press, New York, NY, pp. 291–305.

Li, V. C., 1987. Mechanics of shear rupture applied to earthquake zones. In Atkinson, B. K. (ed.), *Fracture Mechanics of Rock*. London: Academic Press, pp. 351–428.

Mavko, G., Mukerji, T., Dvorkin, J., 2009. *The Rock Physics Handbook*, 2nd edn. Cambridge: Cambridge University Press.

Middleton, G. V., Wilcock, P. R., 2001. *Mechanics in the Earth and Environmental Sciences*. Cambridge: Cambridge University Press.

Murase, T., McBirney, A. R., 1973. Properties of some common igneous rocks and their melts at high temperatures. *Geological Society of America Bulletin*, **84**, 3563–3592.

Myrvang, A., 2001. *Rock Mechanics*. Trondheim: Norway University of Technology (NTNU) (in Norwegian).

Nasseri, M. H. B., Mohanty, B., 2008. Fracture toughness anisotropy in granitic rocks. *International Journal of Rock Mechanics and Mining Sciences*, **45**, 167–193.

Nasseri, M. H. B., Mohanty, B., Young, R. P., 2006. Fracture toughness measurements and acoustic activity in brittle rocks. *Pure and Applied Geophysics*, **163**, 917–945.

Nilsen, B., Palmström, A., 2000. *Engineering Geology and Rock Engineering*. Oslo:Norwegian Soil and Rock Engineering Association (NJFF).

Parfitt, E. A., 1991. The role of rift zone storage in controlling the site and timing of eruptions and intrusions at Kilauea Volcano, Hawaii. *Journal of Geophysical Research*, **96**, 19 101–10 112.

Paterson, M. S., Wong, T. W., 2005. *Experimental Rock Deformation: The Brittle Field*, 2nd edn. Berlin: Springer.

Rice, J. R., 2006. Heating and weakening of faults during earthquake slip. *Journal of Geophysical Research*, **111**, B05311.

Rice, J. R., Cocco, M., 2007. Seismic fault rheology and earthquake dynamics. In Handy, M. R., Hirth, G., Horius, N. (eds.), *Tectonic Faults: Agents of Chance on a Dynamic Earth*. Cambridge, MA: MIT Press, pp. 99–137.

Rivalta E., Dahm, T., 2006. Acceleration of buoyancy-driven fractures and magmatic dikes beneath the free surface. *Geophysical Journal International*, **166**, 1424–1439.

Rubin, A. M., Pollard, D. D., 1987. Origins of blade-like dikes in volcanic rift zones. *US Geological Survey Professional Paper*, **1350**,1449–1470.

Schön, J. H., 2004. *Physical Properties of Rocks: Fundamentals and Principles of Petrophysics*. Oxford: Elsevier.

Schultz, R. A., 1997. Displacement-length scaling for terrestrial and Martian faults: Implications for Valles Marineris and shallow planetary grabens. *Journal of Geophysical Research*, **102**, 12 009–12 015.

Slaughter, W. S., 2002. *The Linearized Theory of Elasticity*. Berlin: Birkhauser.

Smits, A. J., 2000. *A Physical Introduction to Fluid Mechanics*. New York, NY: Wiley.

Spera, F. J., 2000. Physical properties of magmas. In Sigurdsson, H. (ed.), *Encyclopedia of Volcanoes*. New York, NY: Academic Press, pp. 171–190.

Waltham, A. C., 1994. *Foundations of Engineering Geology*. London: Spon.

Woan, G., 2003. *The Cambridge Handbook of Physics Formulas*. Cambridge: Cambridge University Press.

Index